MASTERING BREWING SCIENCE

MASTERING BREWING SCIENCE
Quality and Production

MATTHEW FARBER
Department of Biology
University of the Sciences
Philadelphia, PA 19063, USA

ROGER BARTH
Department of Chemistry
West Chester University
West Chester, PA 19383, USA

Registered Office
John Wiley & Sons, Inc., 111 River Street, Hoboken, NJ 07030, USA

Editorial Office
111 River Street, Hoboken, NJ 07030, USA

For details of our global editorial offices, customer services, and more information about Wiley products visit us at www.wiley.com.

Wiley also publishes its books in a variety of electronic formats and by print-on-demand. Some content that appears in standard print versions of this book may not be available in other formats.

Library of Congress Cataloging-in-Publication Data

Names: Farber, Matthew, 1984– author. | Barth, Roger, author.
Title: Mastering brewing science : quality and production / Matthew Farber,
    Department of Biology, University of the Sciences, Roger Barth, Department
    of Chemistry, West Chester University.
Description: First edition. | Hoboken, NJ : John Wiley & Sons, Inc., 2019. |
    Includes bibliographical references and index. |
Identifiers: LCCN 2019007045 (print) | LCCN 2019011200 (ebook) | ISBN 9781119456049
    (Adobe PDF) | ISBN 9781119456032 (ePub) | ISBN 9781119456056 (pbk.)
Subjects: LCSH: Brewing. | LCGFT: Cookbooks.
Classification: LCC TP570 (ebook) | LCC TP570 .F33 2019 (print) | DDC 663/.3–dc23
LC record available at https://lccn.loc.gov/2019007045

Cover Design: Marcy Barth
Cover Image: Naomi Hampson and Dorothy Ringler
Illustrations: Roger Barth

Set in 10/12pt TimesTen by SPi Global, Pondicherry, India

Printed in the United States of America

SKY10030798_102521

CONTENTS

ABOUT THE AUTHORS

Matthew Farber, PhD, is the program director of the Brewing Science Certificate program at the University of the Sciences in Philadelphia where he is also an assistant professor of biology. He received a BS in Biology with a minor in writing from Seton Hall University and a doctorate in molecular and cellular biology with a minor in teaching from the University of Pittsburgh. His research focuses on innovative applications of biotechnology for the improvement of beer production, and he is an inventor on two pending patents in brewing. He is an active member of the American Society of Brewing Chemists, the Master Brewers Association of the Americas, and the Brewers Association.

Roger Barth, PhD, has been a faculty member in the chemistry department at West Chester University in Pennsylvania since 1985. He created a course entitled The Chemistry of Beer and is the author of its textbook. Barth's research interests center upon brewhouse processing and beer and malt analytical methods. He has a BA in Chemistry from La Salle College. His doctorate is in physical chemistry from The Johns Hopkins University, and he has done postdoctoral work in the area of catalysis at University of Delaware and Drexel University. He is a member of the American Chemical Society, the American Association for the Advancement of Science, Sigma Xi, the American Society of Brewing Chemists, and the Master Brewers Association of the Americas.

PREFACE

Brewing is a creative art. Underlying that art is a framework of technical knowledge, mostly in chemistry and biology, but with significant contributions from physics and engineering. To make five gallons of drinkable beer requires little scientific background, with the definition of *drinkable* up for debate. To make hundreds of batches of beer, each of which meets the expectations of the customers, each of which is consistent with the last, sound science is required. This is the goal of modern brewers.

This book is written as an instructional resource for teaching or learning brewing practice and theory with a focus on the underlying science. We try to strike a balance between critical scientific concepts, beer production, and day-to-day practical issues in beer quality. By understanding the science of beer production, readers will be better equipped to troubleshoot problems in the brewery, one of the most critical skills for a successful career in beer. We have produced hundreds of illustrations to demonstrate key concepts and to demonstrate the numerous pieces of equipment commonly used in breweries. Unlike drawings provided by equipment suppliers, our drawings do not inform operation or maintenance. Rather, they illustrate essential design elements and concepts as they pertain to the process.

In this book, we first introduce a high-level view of the brewing process. Then we dive into the fundamentals of biology and chemistry with appropriate application to the brewing process. These concepts will be critical to better understanding of subsequent chapters. The remaining material is presented in order, from raw materials, through the brewing process, and on to methods for quality. All employees at a brewery should be trained in basic concepts of quality. Quality is best managed at the source where response time is quick.

As more and more brands line the shelves, the consumer has more and more options. If one batch of beer is flawed in the eyes (and palate) of the consumer, it is easy to move on to another brewery. Brewing with quality requires a high level of awareness of the procedures, the materials, and the equipment used in brewing. Our goal in this book is to address essential concepts in quality and consistency to help the readers become better brewers. At the conclusion of most chapters are review questions to check for understanding, followed by a case study for critical analysis and discussion.

We make extensive use of primary and secondary references, but we deliberately omitted all in-line references and any citations that might be distracting to the student. Useful and critical references are mentioned at the end of each chapter under "Bibliography." Many of the facts that we present were won by the brilliant insights and very hard work of thousands of scientists. We herewith acknowledge their contributions, even if, for the benefit of readers, we did not give them citations. We hope that this approach will be more effective than voluminous citations in putting the work of our colleagues into the hands of students, who will be the next generation of brewers and brewing scientists.

ACKNOWLEDGMENTS

We gratefully acknowledge our wives, Dr. Grace Farber and Marcy Barth, for their love and support. Marcy provided outstanding expertise and artistry in the photography, design, and execution of illustrations. Grace provided guidance and advice in the teaching of essential concepts in biology. We humbly thank our families for their patience, encouragement, and love.

Donna-Marie Zoccoli and Dr. David Barth, our eagle-eyed copy readers, read the manuscript multiple times, corrected errors, and made countless suggestions that make the book easier to read and understand. Dr. Naomi Hampson took some outstanding photographs. Kent Pham, Dave Goldman, Michelle McHugh, and Eric Jorgenson provided additional pictures. Our friends at Deer Creek Malt House, Philadelphia Brewing Company, Sly Fox Brewing Company, Susquehanna Brewing Company, Victory Brewing Company, Yards Brewing Company, and Yuengling Beer Company were generous with their time and gave us insights, explanations, and access for photographs. Our professional societies, The American Society of Brewing Chemists (ASBC) and the Master Brewers Association of the Americas (MBAA) maintain outstanding resources and networking opportunities that were critical to our own development as brewing chemists. Our institutions (University of the Sciences in Philadelphia and West Chester University) and their libraries and librarians provided essential support. Our editors at Wiley, Jonathan Rose, Aruna Pragasam, and Viniprammia Premkumar, have been supportive, helpful, and responsive. Our students motivated the entire project with their enthusiasm and unquenchable desire to learn.

We are forever grateful to the supporting words and actions of our own teachers and mentors in years past. Dr. Farber acknowledges Dr. Peter Berget for his mentorship and Dr. Angela Weisl for the inspiration to pursue writing. Dr. Farber is incredibly thankful for his parents, Dr. Phillip and Larice Farber, for instilling endless curiosity and creativity in him. Dr. Barth acknowledges his physical chemistry teacher and father, the late Dr. Max Barth; his high school chemistry teacher, Mr. Dugan; and his high school English teacher, Mr. Martini, who taught the importance of clarity and precision in writing.

CHAPTER 1

BREWING QUALITY OVERVIEW

We wrote this book to help you to better understand, appreciate, and apply the science behind the materials and processes of making **beer**. The better your grasp of **brewing** science, the more dependably you will be able to make delicious beer, and the more reliably you will be able to devise new beers to meet changing consumer preferences. So what is beer? How does beer differ from its **fermented beverage** brethren? There are legal and marketing definitions, but in a book on brewing science, we will use a scientific definition. Beer is an undistilled **alcoholic beverage** derived from a source of **starch**. "Derived from" covers a complex series of interacting steps, each of which influences the character of the final product and is ultimately the focus of this book. Brewing beer differs from **fermentation** of **wine** in that for brewing, a source of starch must first be converted into **fermentable sugars**. The brewer is responsible for management and control of all steps of the brewing process to produce a beer of reliable and reproducible quality.

There are four main ingredients in beer: water, **malt**, **hops**, and **yeast**. If randomly combined, these four ingredients might turn into an alcoholic beverage of questionable quality, but in this chemical process, the brewer is like an **enzyme**, a substance that guides and speeds up a **reaction**. Mastering

Mastering Brewing Science: Quality and Production, First Edition.
Matthew Farber and Roger Barth.
© 2019 John Wiley & Sons, Inc. Published 2019 by John Wiley & Sons, Inc.

the science of raw materials and the process steps of beer production is essential to making quality beer. Here, we will start with a broad overview of the brewing process followed by a scientific history of beer and the scientific method. In learning how to conduct an experiment, you will begin to understand the process of troubleshooting problems in the **brewery**. And finally, as our major goal is to brew beer of excellent quality and consistency, we will discuss beer quality as defined in several contexts. Each of these topics will be discussed further in depth in the chapters that follow.

1.1 INGREDIENTS

In addition to the main ingredients, beer is often brewed with other ingredients. These can include **adjuncts**, which are sources of starch or **sugar** other than malt, and processing aids, which are materials used to help give the beer desirable characteristics. Some common processing aids are **finings**, which help to clarify the beer; carbon dioxide, which **carbonates** the beer; **foam** enhancers, which provide desirable foam properties; and colored materials, which are used to adjust the color of the beer. In this introduction we will touch upon the main four ingredients. Adjuncts and processing aids are covered in later chapters.

Water

Beer is usually more than 90% water. It can take as much as 12 volumes of water to make 1 volume of beer. Some breweries have been able to cut this ratio to three or less. Less water means less **energy** use, less waste material to dispose of, and less negative impact on the environment. Water itself is a characterless **compound** of fixed composition. Water supplied to breweries is a **mixture** with many desirable and undesirable **components** present in trace amounts. The nature and amount of these trace components is important to the character and quality of the beer. Water is usually processed to adjust the trace components. Water that is to be made into beer is sometimes called **brewing liquor**. Chapter 4 discusses brewing water in detail.

Malt

Brewing beer requires starch, the source of which is **cereal grain**. At least some of the grain is ordinarily processed to give malt, a process called malting. Malt is seeds of grain that are germinated and then dried. The most common grain for malting is **barley**, but wheat, rye, and oat malt are available. Rice and maize (corn) can be malted, but these malts are strictly specialty items; they are rarely used in beer brewing. Since medieval times, malting has been a separate craft from brewing, and malt is produced in specialized facilities. Brewers need a basic understanding of the malting process to make the most effective use of the available varieties of malt.

Malting begins by **cleaning** live seeds of grain over a series of sieves. The grain is then **steeped** (soaked in water) at a controlled **temperature**, typically in two to three stages of steeping and draining. The grain is then permitted to **germinate**. It must be kept in contact with air to support **respiration** and to carry away **heat** generated by the life processes. The seeds are regularly turned to expose them to oxygen and to maintain a uniform temperature, avoiding hot or cold spots. Regular turning also prevents sprouting roots from becoming tangled. The germination process produces several changes in the seeds, collectively called **modification**. **Enzymes** are produced that assist the modification process. Some of these enzymes are also critical to the brewing process in that they are responsible for converting the starch to sugar during **mashing**. Certain **polymers**, including **proteins** and **beta-glucan**, are **hydrolyzed** into smaller **molecules** under the influence of the enzymes. After modification, the seed loses its pebble-like hardness and becomes **friable** (easily crushed). Some of the starch in the seeds is consumed as fuel to power the life processes of the **embryo**. This is called malting loss. When the **maltster** judges that germination has proceeded far enough, the seeds are transferred to an oven, called a **kiln**, and heated with moving air. Different grades of malt are produced by varying the degree of modification and the temperature and duration of heat treatments. Shorter kiln times at lower temperatures yield malt with more starch-hydrolyzing enzymes and less **flavor**. Longer, higher temperature kiln treatment yields darker, more highly flavored malts, but with a lower enzyme content. Some malt is subjected to additional heating, called roasting, to give dark, highly flavored but nonenzymatic malt. Chapter 4 covers the malting process in detail.

Hops

The **hop** is a climbing plant, ***Humulus lupulus***. The fruits of the hop plant, hops, are boiled with the beer **wort** to provide bitterness and other flavors. Hop compounds also have an antibacterial effect that can help preserve the beer. Sometimes hops are added at other points in the brewing process to provide desired flavor effects. There are many varieties of hops with different flavor profiles. In addition, there are products derived from hops that are often used instead of or in addition to the natural hops. Chapter 4 provides details about hops and their processing.

Yeast

Yeast is the single-**cell fungus** that converts sugar to **ethanol** and carbon dioxide. The action of yeast on sugar is fermentation. Most beer fermentation is carried out by one of two species of yeast, ***Saccharomyces pastorianus***, used for **lager beer**, and ***Saccharomyces cerevisiae***, used for **ale**. Some specialty beer styles are fermented with ***Brettanomyces*** *bruxellensis*, *Brettanomyces lambicus*, or related species. Within a particular yeast species, there are many variations, called **strains**. The species and strain of yeast affects the character of the beer. Yeast is often cultivated at the brewery. Processes and practices involving yeast are covered in detail in Chapter 9.

1.2 BREWING OVERVIEW

A graphical overview of the brewing process is provided in Figure 1.1. In brief:

- Malt and other grains are crushed in the **mill**. Crushed grain is called **grist**.
- The grist is loaded into the grist case until mashing.
- The grist is mixed with hot water in the **premasher** on its way into the **mash tun**.
- In the mash tun, enzymes from the malt cause the starch in the grist to be converted to soluble **extract**, which contains sugars that the yeast can ferment.
- The **solution** of extract, called wort, is separated from the remaining grist particles in the **lauter tun**. Extract that sticks to the particles is washed out with hot water in a process called **sparging**.
- The clear wort is boiled in the **kettle**. Hops are added.
- The remains of the hops and solids that form during boiling (**hot break** or **trub**) are removed in the **whirlpool**.
- The clear, boiling hot wort is cooled in a **heat exchanger** called the **chiller**.
- The cool wort is **pumped** into a fermenter. Yeast is added (**pitched**).
- After several days of fermentation and **conditioning**, the yeast is removed from the beer, and the beer is pumped into the **bright beer tank**. Carbon dioxide is added under **pressure**.
- The beer is served or packaged.

A summary of the duration and temperature ranges for each step in the brewing process is provided in Table 1.1. This table represents a general summary and overview; different breweries using different equipment and brewing different styles of beer may have quite different programs.

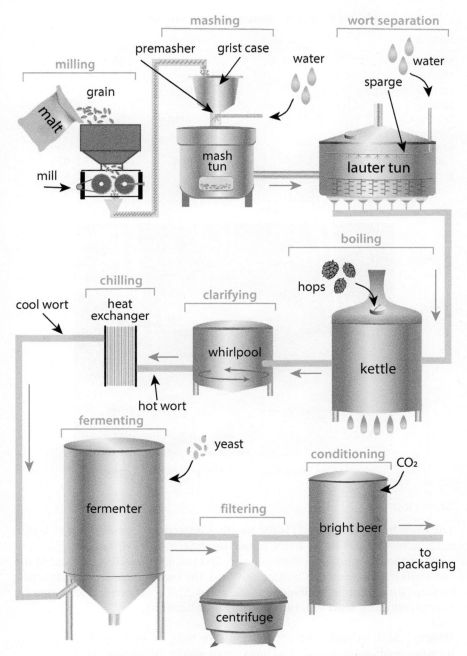

Figure 1.1 Overview of the brewing process for a four-vessel brew house.

TABLE 1.1   **Brewing Steps, Durations, and Temperatures**

Process Step	Duration	Temperature
Milling	1–2 hours	Ambient
Mashing	1–2 hours	45–67 °C
Lautering/sparging	1–2 hours	75–78 °C
Boiling	1–2 hours	105 °C
Whirlpool	15–30 minutes	76–74 °C
Fermentation (ale)	4–10 days	15–25 °C
Conditioning (ale)	2–14 days	–1 to 6 °C
Filtration	2–12 hours	2–6 °C
Packaging	<12 hours	2–6 °C
Duration of typical shelf life	~6 months	2–6 °C

Figure 1.2   Brew house at Victory Brewing Company.

Brew House

The **brew house** (Figure 1.2) is the facility that makes beer wort out of water, malt, adjuncts, and hops. Brew house operations involve hot water or hot wort, so the brew house is sometimes called the *hot side*. Because one of the last steps in this process is boiling the wort, the brew house presents less of a concern for **microbial spoilage** than the **cellar**. The brew house operations are milling, mashing, **wort separation**, boiling, and chilling.

Milling   Malt is delivered to breweries in bulk or in bags. Before use in brewing, malt must be crushed into small pieces to extract the starch. The physical operation involved is milling. Crushed grain is called grist. The device

Figure 1.3   Mill at Susquehanna Brewing Company.

that performs the operation is a mill (Figure 1.3). The primary purpose of milling is to allow starch from the grain, enzymes from malt, and water to come into contact during the mashing step. A seed of grain is protected by a water-resistant **seed coat**, also called the **testa**. Milling breaks open the seed coat and crushes the interior of the seed, producing additional **surfaces** at which water can react with starch. Milling details have a significant effect on the character of the beer and the efficiency of the process. It is essential that the malt **hulls** be split but not pulverized. They will aid in a later step, wort separation.

Mashing   During the mashing step, starch is converted to smaller sugars that brewing yeast can ferment. Yeast cannot ferment starch, so this step is essential. During mashing, hot water, sometimes called brewing liquor, is mixed with the grist to give a temperature in the range of 60–70 °C (140–158 °F). Sometimes mashing starts at a lower temperature, and the temperature is raised continuously or in steps to influence the protein or **carbohydrate** profile. Mashing is conducted in a **mash conversion vessel** (**MCV**), also called a mash tun (Figure 1.4). The mash tun may contain an agitation paddle for gentle mixing. The details of the time–temperature profile, the activities of

Figure 1.4   Inside the mash tun at Victory Brewing Company.

enzymes derived from malt, and the **pH** of mashing have a decisive effect on the character of the beer.

Three processes must occur for effective mashing. The first is **gelatinization**, in which starch granules absorb water, swell, and burst, giving the starch molecules access to water. Some grains, including barley and wheat, gelatinize readily in the normal mashing temperature range. Others, like maize (corn) and rice, must be cooked in a separate vessel before addition to the mash. The second process is **liquefaction**, in which starch molecules are hydrolyzed in the interior of the molecular chain to give soluble, but still too large for fermentation, fragments. The third process is **saccharification**, in which starch chains and fragments are further broken down at the ends of the chains to yield the fermentable sugars: **glucose**, a **monosaccharide**; **maltose**, a **disaccharide**; and **maltotriose**, a **trisaccharide**. Mashing temperature plays a key role in determining the fraction of starch that is liquefied and the fraction of dissolved carbohydrate that is fermentable.

The amounts of unfermentable and fermentable carbohydrates are determined during mashing, influencing the character of the finished beer. The generation of more fermentable sugars results in a thinner, dryer beer with more alcohol. A mash with less fermentable sugars leads to less alcohol but more body and texture.

Wort Separation   After mashing, the wort, the insoluble material, and the broken hulls remain in a **slurry**. Wort separation is required to obtain clear wort. The solids remaining after separation are called **draff** or spent grain. Two methods of wort separation are in common use. The most popular is the lauter process [Ger: clear, pure]. In this process the solids are supported on a perforated **false bottom** above the true bottom of the vessel. Liquid is drawn through the grain and the false bottom via **valves** in the true bottom. The actual **filtration** is accomplished by the grain bed, the split hulls from the malt. The false bottom supports the grain bed and facilitates separation. In the first minutes of wort separation, wort is recirculated to the top of the vessel. Recirculation, called **vorlauf** [Ger: forerun], is maintained until the wort runs clear, indicating the grain bed is set. If the mash and lauter are accomplished in the same vessel, this is called a mash/lauter tun. Often the entire mash, liquids and solids, are pumped into a separate, dedicated vessel called the lauter tun (Figure 1.5). The lauter tun is equipped with knives or rakes that slowly dig into the grain bed to increase the filtration speed. A different lautering device, less common in small breweries, is the **mash filter**. Here the entire mash, including liquids and solids, is pumped into compartments from which the liquid is driven by pressure through filtration material.

During or after lautering, the grain is rinsed with hot water, a process called **sparging**. Sparging recovers sugar that is held up in the grain bed, so more beer can be made from less grain.

Figure 1.5   A peek into the lauter tun at Urban Village Brewing Company. *Source*: Photo: Dave Goldman.

Boiling   The clarified wort is sent to a vessel called a brew **kettle**, also called a **copper**, or a wort boiler (Figure 1.6) and heated to boiling. The wort is usually boiled for 60–90 minutes with evaporation of up to 20% of the wort volume. Boiling consumes the most energy of any step of the brewing process. Hops or hop products are generally added before or during boiling, often in stages so that different portions of the hops are subjected to different boiling durations. Boiling serves several purposes, including the following:

- **Isomerization** of hop compounds for bitterness.
- **Sterilization**.
- Dissipation of **off-flavors**.
- Removal of proteins and **lipids** that affect beer clarity and stability.
- Concentration of wort.

Boiling generates solid material called hot break or trub ("troob"). Sometimes the hot break material is removed before chilling, either by allowing it to settle (**sedimentation**) or in a vessel called a whirlpool, in which

Figure 1.6   Brew kettle at Yuengling Beer Company.

the wort is made to move in a horizontal circular pattern that drives the solids into a compact mound at the bottom center of the vessel. In some breweries the kettle itself also serves as the whirlpool.

Chilling   Before fermentation, the temperature of the wort must be lowered from near boiling (100 °C or 212 °F) to the fermentation temperature (typically 9–20 °C or 48–68 °F), a process called chilling. The standard equipment for chilling is a **countercurrent** plate heat exchanger. The **unit** consists of a series of closely spaced and parallel heat-conducting plates, as shown in Figure 1.7. The hot wort flows through half of the channels between the plates, and a coolant, typically water or an **antifreeze** mixture, flows through the rest. Each plate has wort on one side and coolant on the other. Typically, the outgoing water, which is now hot, is added to the **hot liquor tank**, to be used for later brewing operations.

In some traditional breweries, the hot wort is drawn into a wide, shallow vessel called a **coolship**, where it is slowly cooled by convection. A few breweries use this method to capture wild **bacteria** and yeast, but most brewers prefer a closed chiller to avoid the risk of contamination.

At the beginning of fermentation, the yeast needs dissolved oxygen (as a nutrient, not for respiration) to help prepare cell **membranes**. Because the

Figure 1.7   Chiller at Philadelphia Brewing Company.

boiling process strips the wort of all dissolved **gases** and because gases are more soluble at cooler temperatures, oxygen is injected into the wort as it exits the chiller. The oxygen requirement depends on the solids content of the wort and on the strain of yeast.

Cellar

Before the days of mechanical refrigeration, fermentation and conditioning were often carried out in an underground room, or even a cave, called the cellar. Figure 1.8 depicts the underground caves at the Yuengling brewery in Pottsville, PA, where beer was formerly **lagered** and conditioned. Today, fermentation and conditioning temperatures are usually controlled artificially in the tanks themselves, and the "cellar" can be at any level of the brewery. The cellar is sometimes called the *cold side*.

Fermentation   Cooled, aerated wort from the chiller is transferred to a fermenter, also called a fermentation vessel (**FV**). The most widely used configuration for the fermenter is the **cylindroconical** vessel (CCV), shown in Figure 1.9. A selected strain of yeast is added or pitched into the wort. Fermentation converts certain sugars to ethanol and carbon dioxide. The reaction is carried out by yeast, a single-celled fungus, as a means for the yeast to make cellular energy in the absence of oxygen. Fermentation occurs in 12 distinct steps. The overall reaction is $C_6H_{12}O_6 \rightarrow 2C_2H_5OH + 2CO_2$. In addition

Figure 1.8   Caves at Yuengling Beer Company.

Figure 1.9   Fermentation vessels in the cellar at Susquehanna Brewing Company.

to the main reaction, fermentation is accompanied by a variety of side reactions whose products can affect the flavor profile of the beer. The types and amounts of flavor-active side products depend strongly on the fermentation temperature, the species and strain of yeast, and the presence of bacteria. The fermentation reaction generates heat, so fermenters usually have provision for cooling.

Beer that is fermented at a temperature higher than 15 °C (59 °F) is classified as ale. Ale is usually fermented with a species of yeast called *S. cerevisiae*, also called top fermenting yeast. Beer fermented at lower temperature is lager beer, usually made with a species called *S. pastorianus*, also called bottom fermenting yeast.

Conditioning   After fermentation, the new beer, called **ruh beer**, or **green beer**, is held in contact with the yeast for a period that can be as short as a few days for a low-strength ale to several months for some types of lager beer. This is the first part of the conditioning process, sometimes called **secondary fermentation**. In lager beer, the secondary fermentation is called lagering. During this period the flavor of the beer matures, mainly because the yeast absorbs off-flavor compounds. This part of conditioning can take place in the original fermenter or in a dedicated conditioning vessel. Once flavor maturation is achieved, the beer is cooled, which facilitates separation of yeast and clarification of the beer.

Filtration   Beer is often, but not always, subjected to one or more clarification processes. Materials called finings may be added to beer to bind and remove **haze**-forming compounds. The beer may be kept in a tank to allow solids to sediment. It may be clarified in a **centrifuge**. It may be filtered, often through a bed of **diatomaceous earth** (DE) or **cellulose**-containing membranes. It may be treated for microbial stability before packaging by filtering out microbes or after packaging by a heat process called **pasteurization**.

After filtration, the beer is pumped into the bright beer tank, so named because the beer at this stage is free of yeast and haze, that is, it is bright. Usually carbon dioxide is dosed into the beer, either in the bright beer tank or in the line to the packaging unit, to provide characteristic carbonation. At this point the beer is ready to be served directly from the bright tanks or to be packaged in **kegs**, bottles, or cans.

Unfiltered beer is common in small breweries that lack filtration equipment. Here beer is chilled after fermentation, carbonated, and served directly. Brewers should make sure that fermentation is complete before packaging unfiltered, unpasteurized beer to avoid the risk of excessive pressure from secondary fermentation.

Packaging

The major purpose of beer packaging is to protect the beer until it is served. Beer must be kept under pressure to maintain carbonation. Light and oxygen must be excluded to avoid (or at least defer) the development of off-flavors. **Small pack** refers to packaging that is intended for single servings or direct consumer use. About 2 L (~0.5 US gallon) is considered the upper limit of small pack. Standard small packaging is aluminum cans and **glass** bottles. Plastic PET bottles are also on the rise. In addition to protecting the beer, small pack has the very important function of enhancing sales. Small pack is invariably decorated with branding material. Bottles are festooned with paper or plastic labels. Some have front labels, back labels, neck labels, and cap covers (often made of foil). Cans, if purchased in quantity, can be preprinted directly on the aluminum. Alternatively, breweries may apply a label or plastic shrink wrap to an unlabeled aluminum can. Bottles and cans are packed in branded secondary packaging such as six packs and cases, usually made of cardboard or plastic.

The other type of packaging is kegs and **casks** (Figure 1.10). Kegs typically contain 50 L or 15.5 US gallons (58.7 L), although smaller sizes are available. Casks usually contain 40.9 L or 1 firkin (9 imperial gal). Casks and kegs are used to serve beer in bars or at parties where large volumes of a particular brand of beer will be dispensed. A full US-size keg of beer weighs about 73 kg (160 lb); the empty keg alone weighs 13.5 kg (30 lb) (Table 1.2).

The handling of the beer and the packaging process are designed to minimize oxygen entry. Oxygen causes staleness and off-flavors in beer. Because air is 20% oxygen by volume, the requirement to exclude oxygen is technically demanding. The packages are purged with carbon dioxide before and after filling and are sealed within seconds.

Figure 1.10   Kegs (straight-sided vessels) and casks at Philadelphia Brewing.

TABLE 1.2   **US Beer Packages**

	12 oz bottle	12 oz can	15.5 gal keg	1 firkin cask
Gross weight per liter beer	1.61	1.05	1.24	1.25 kg/L
Admit light	Yes	No	No	No
Admit oxygen	Slow	No	Slow	Yes
Recycle value per liter beer	0.01	$0.044/L	Reuse	Reuse

The packaging process involves unpacking the containers, rinsing and sanitizing them, conveying them to the filling station, purging out air, filling, and then sealing the packages. Often the beer is pasteurized just before or after packaging or subjected to microbial filtration before packaging. Labels and their adhesives are applied. Secondary packaging is unloaded from its packaging and folded into shape. The filled cans or bottles are gently loaded into the cases, which are then sealed with adhesive and stacked on **pallets**. The complexity of the packaging operation and its potential for breakdowns rival all the rest of the brewery combined (see Figure 1.11).

Serving

Beer service can be as simple as handing the customer a bottle or can, but the usual expectation is that the beer will be delivered in a glass. Glasses for beer must be extraordinarily clean. Small traces of **fats** found on nominally clean glassware can interfere with the desirable appearance of the **head** of foam.

Figure 1.11   Packaging line at Victory Brewing Company.

For this reason, special procedures are needed for cleaning beer glasses. The beer must be served at the proper temperature and with the correct presentation of foam. The elaborate rituals in some establishments for wine service are trivial in comparison to the routine requirements for serving beer.

For economics as well as esthetics, beer is often held under pressure in bulk containers, like kegs, transmitted through tubing called a **beer line** to a dispensing valve called a beer faucet. Beer served from casks or kegs is called **draft** or draught beer. Although the standards for beer from casks forbid it, beer from kegs is driven from the keg to the **tap** by gas pressure, usually carbon dioxide. The requirements for keg service include a cold locker for the kegs, pressure tanks and **regulators** for the driving gas, lines that hold pressure and exclude **permeation** by oxygen, and faucets. It is often necessary to provide chilling to the beer lines to maintain the proper service temperature and carbonation. The entire system must be amenable to regular and thorough cleaning to maintain beer quality.

1.3   A SCIENTIFIC HISTORY OF BREWING

The word *beer* is derived from the Latin verb *bibere*, which means "to drink." But what is the true definition of beer? A modern, Western-culture definition of beer might be an alcoholic beverage produced from malted cereal grain,

TABLE 1.3   **Traditional Beers from Around the World**

Beer	Country	Description
Aca	Peru	Maize beer
Bilbil	Ancient Egypt	Sorghum beer, also known as Indian millet
Bi-se-bar	Ancient Sumeria	Light barley beer
Boza	Ancient Babylonia and Egypt	Millet beer
Chi	India	Millet beer
Chibuku	Southeastern Africa	Sorghum beer, also made with maize and millet
Chicha	South and Central America	Maize beer, also made with quinoa, peanut, cassava, palm fruit, or potato
Chiu	China	Wheat beer
Dolo	Burkina Faso, Africa	Millet or sorghum beer, flavored with sisal, castor oil, cassia, pimento, and tobacco
Kaffir	Southern Africa	Sorghum beer
Kava	Polynesia	Dried roots of the *Piper methysticum* tree are chewed, spat, and fermented
Kvas	Russia	Low alcoholic beer made by fermenting rye bread
Maltøl	Norway	Farmhouse beer brewed with Kveik yeast
Okolehao	Hawaii	Fermented ti root
Pachwai	India	Rice beer
Pulque	Mexico; Central America	Fermented juice of maguey agave cactus
Tiswin	North America	Maize beer

flavored with hops, and produced through fermentation. But to consider the history of beer, the use of this modern definition severely limits our scope of understanding. By requiring the use of "malted grains" and "hops," we limit the historical context for which modern beer was derived. These are the major ingredients in modern European-style beer, but not necessarily in all beer. Recall from Section 1.1 that brewing beer requires the conversion of starch into a fermentable sugar. Therefore, a more appropriate, historically accurate definition of beer is "an alcoholic beverage derived from a source of starch." This seemingly simple definition of beer covers the breadth of alcoholic beverages indigenous to regions across the world, local beers that can vary widely from the modern European-style beer with which we are familiar (Table 1.3).

To understand the history of beer, we must rely upon the availability of artifacts and documents. Presumably Paleolithic (Old Stone Age) humans experienced an otherworldly euphoria after accidentally eating fermented fruit or drink, with mind-altering affects both captivating and terrifying. But when did humans learn to harness the power of fermentation? The earliest written documents display familiarity with beer, indicating that beer was being brewed before there was a written language. The origins of beer lie in prehistory.

Origins of Beer

The domestication of plants and animals was first undertaken about 12 500 years ago in what was considered the *Fertile Crescent*. This "cradle of civilization" spanned the region of the Middle East from the Persian Gulf to northern Egypt and through Iraq, Syria, Lebanon, Jordan, and Israel. Agriculture made it possible for the land to support larger communities, then cities.

There are three requirements for large-scale beer-making. First, there must be a means to grow and process fermentable grains in quantity. Second, there must be a controllable source of energy via a fireplace. Finally, there must be appropriate brewing vessels for fermentation, such as pottery. In the Fertile Crescent region, pottery is believed to have been invented around 8500 years ago.

The first chemical evidence of barley beer, consisting of deposits of oxalic acid (**beer stone**) in pottery jars, comes from a 5500-year-old Bronze Age site called Godin Tepe in present-day Iran. Earlier evidence of a mixed fruit–grain–honey beverage was discovered as residue in pottery from Jiahu in north central China, dating to 9000 years ago. Very recently, analysis of several 13 000-year-old microscopic starch granules recovered from stone mortars in a burial cave at Mount Carmel in northwestern Israel suggested that the mortars had been used to make beer. If further study confirms these preliminary findings, the horizon for beer will have receded to the Paleolithic Age (Old Stone Age), showing that beer was brewed (presumably on a small scale) before grains were even cultivated as an agricultural product.

Fermentation and Science

In addition to its important function at gatherings of political revolution, beer and other fermented beverages have played a central role in science and technology. Because alcoholic products were embedded in societies across the world, particularly Europe, economic pressures for consistency and reliability were central to economic success, prompting serious study and innovation. In the 1700s commercial innovations such as cast iron kettles, steam for heating, the **thermometer**, and the **hydrometer** were directly due to the brewing industry. Steam engines were used in brewing before they were used in weaving.

The science behind the process of fermentation was not described until 1789. The process of making alcohol from sugar was simply referred to as the "putrefaction of sugar." In 1697, Georg Ernst Stahl (1659–1734), the founder

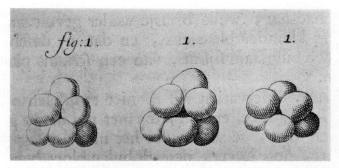

Figure 1.12 Drawing of yeast by Antonie van Leeuwenhoek in 1680 after viewing beer through his primitive microscope. *Source*: Image courtesy of the National Library of Medicine.

of the phlogiston theory of combustion, postulated that the violent activity and heat generation associated with fermentation caused a "loosening" of particles present in the medium; thus the formation of alcohol was simply a process of separation. The microscopic nature of yeast cells was first recorded by Antonie van Leeuwenhoek (1632–1723) in 1680 after examination of fermenting beer, but the role of yeast in fermentation was still unknown (Figure 1.12). Yeast was considered a "carrier of activity," participating in, but not responsible for, the separation of alcohol during putrefaction.

The alchemy theory of fermentation held firm until 1789 when Antoine Lavoisier (1743–1794), the founder of modern chemistry, realized that during fermentation, sugar was directly converted to carbon dioxide and ethanol. He performed an **elemental** analysis of sugar, alcohol, and yeast. Significantly, he noticed 19% nitrogen in dry yeast. Lavoisier considered fermentation to be separation of sugar into two parts, carbon dioxide in an oxidized state and the other, ethanol, in a reduced (deoxygenated) state. Lavoisier, being a member of the French nobility, studied fermentation in wine rather than beer. Also, being a nobleman, he was beheaded during the French Revolution.

Around 1803, British chemist John Dalton (1766–1844) put forward the modern atomic theory that serves as the basis for all chemistry today. In the period from 1803 to 1815, French academic chemists Joseph Gay-Lussac (1778–1850) and Louis Thenard (1777–1857) developed improved analytical methods and used them to refine Lavoisier's analysis of fermentation. Nonetheless, the **chemical equation** for fermentation could not be determined until the molecular **formula** of glucose was published in the 1870s.

In the seventeenth and eighteenth centuries, there was a major controversy in biology, spilling into related chemistry, between the **mechanists** and the **vitalists**. Mechanists held that life processes were governed by the same physical laws as those of inanimate matter. Vitalists held that life possessed a special life force, or *élan vital*, that could not be explained by laws derived from dead matter. Alcoholic fermentation served as a weapon for both sides of the controversy and, ultimately, was its resolution. One argument favoring **vitalism** was that many compounds that occurred in living organisms could not be made in the laboratory from dead matter. In 1828 Friedrich Wöhler (1800–1882) delivered a setback to vitalism by synthesizing urea, an **organic** substance, from ammonium cyanate, considered dead matter. Another piece of evidence for vitalism was that only living organisms were able to make and use *ferments*, which we now call enzymes. In 1833, Anselme Payen (1795–1871) and Jean Peroz (1805–1868) delivered another setback to vitalism by precipitating **amylase** (which they called **diastase**) from barley malt. They demonstrated that isolated diastase could hydrolyze starch. Several other enzymes were found shortly thereafter. The vitalists then made a distinction between *simple ferment* that catalyzes simple hydrolysis reactions and *organized ferment* that is alive and responsible for complex reactions in organisms.

In the 1830s improvements in light **microscopes** paved the way for the study of fermentation as a biological process. From 1836 to 1838, Charles Cagniard de la Tour (1777–1859), Friedrich Traugott Kützing (1807–1893), and Theodor Schwann (1810–1882) used microscopes to demonstrate that yeast is a living organism and is required for fermentation. These contributions are often mistakenly attributed to Louis Pasteur.

The great chemists of the nineteenth century such as Justus von Liebig (1803–1873), Friedrich Wöhler (1800–1882), and Jacob Berzelius (1779–1848) vehemently opposed and ridiculed the idea that fermentation was a life process, even at first rejecting evidence that yeast was a living organism. Their arguments were implausible and their experimental evidence was nonexistent. Nonetheless, the weight of their authority set the field of biochemistry back by two decades. Ironically, 60 years later, fermentation was achieved without intact yeast cells (but with a yeast extract).

Another argument against the mechanist approach was optical rotation of compounds. Certain compounds, all of which originate in living organisms, are *optically active*; they can rotate a plane of polarized light. The same compounds, when prepared synthetically, do not rotate light. The first major insight into this phenomenon came in 1847 from the graduate thesis of Louis Pasteur (1822–1895). Pasteur separated synthetic sodium ammonium tartrate into two fractions, one of which rotated polarized light to the left and the other rotated it to the right. This showed that optically active compounds can exist in two forms, right and left handed. Living systems can selectively make one of these, but synthetic methods invariably make a mixture of both. This difference still has no generally accepted explanation. It may have been this issue that led Pasteur, whose training was in chemistry and physics, to become a founder of the field of microbiology.

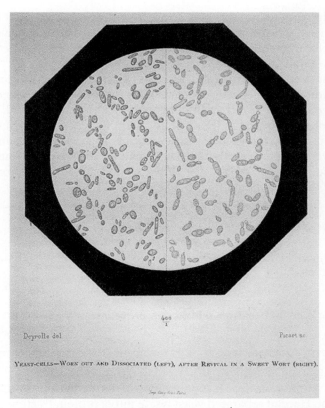

YEAST-CELLS—WORN OUT AND DISSOCIATED (LEFT), AFTER REVIVAL IN A SWEET WORT (RIGHT).

Figure 1.13   An illustration in Louis Pasteur's book, *Études sur la Bière* (*Studies on Beer*), 1876, described the physical appearance of healthy and worn out yeast cells during fermentation. *Source*: Image courtesy of the National Library of Medicine.

Pasteur was an outstanding experimentalist. He established that fermentation carried out by different microorganisms produces different products, usually giving off-flavors to beer and wine. He convinced English brewers to get microscopes to monitor the quality of their yeast (Figure 1.13). Pasteur regarded fermentation as life without oxygen. He discovered the process of pasteurization for the reduction of microbial load in wine and beer. Ultimately, his experiments reduced the fermentation theories of the German chemists to rubble. He further argued that fermentation is not possible without life. But like Liebig, Wöhler, and Berzelius, Pasteur would be proven wrong; alcoholic fermentation can occur without live cells, given the right enzymes.

The final blow to vitalism, cell-free fermentation, was discovered by lucky accident in 1887. Eduard Buchner (1860–1917), while developing methods for an entirely different project, ground several kilograms of brewer's yeast in a large mortar with fine quartz powder as an abrasive. He wrapped the resulting paste in cloth and pressed it in a hydraulic press. Buchner called the resulting cell-free liquid *press juice*. Today we call it *yeast extract* or *lysate*. Sugar was added to the lysate to suppress the growth of bacteria. Afterward, Buchner noticed the formation of bubbles, which he correctly interpreted as fermentation of the sugar by an enzyme in the yeast lysate. He called this enzyme *zymase*. We now know that there are 12 enzymes involved. Evidently, these enzymes produced by yeast performed this reaction independently of the live cells. Fermentation was carried out by isolated enzymes in the same way as Payen and Peroz carried out starch hydrolysis. No mysterious *élan vital* was needed. Importantly, Buchner demonstrated that fermentation could be studied by the ordinary methods of chemistry and biology.

MUSINGS ON THE HISTORY OF BEER...

"It is much more productive to study a process in chemical glassware than in living cells."

—Dr. Barth, chemist

"After drinking beer from said process, my neurons say otherwise."

—Dr. Farber, biologist

Building upon Buchner's discoveries, Arthur Harden (1865–1940) discovered that **phosphate** is a requirement of fermentation. In another key experiment, Harden separated yeast lysate into a protein fraction that would not go through a membrane and nonprotein fraction, which Harden called *coferment*, that did pass through the membrane. Fermentation required both fractions. We now identify *coferment* as **NAD⁺**, a key molecule responsible for transfer of hydrogen in cells. This discovery led to an understanding of the role of **ATP** as energy currency. The discovery of NAD^+ also led to a revolution in the understanding of biological oxidation processes. One of the factors that contributed to this progress was the discovery that the fermentation reactions are identical to those occurring in muscles when oxygen is not available. Muscles turn chemical energy into mechanical work, making them useful for tracking energy. In 1934 the link between glycolysis and ATP production was identified. By 1938, the fermentation reactions, now called the **glycolysis** pathway, had been revealed, although some of the enzymes had not yet been isolated. It was not until 1941 that the role of ATP as energy currency was proposed. Many outstanding scientists made decisive contributions to elucidating glycolysis, the first complete biochemical pathway to be revealed. Glycolysis is often called the Embden–Meyerhof–Parnas pathway after three of its discoverers, Gustav Georg Embden (1874–1933), Otto Fritz Meyerhof

(1884–1951), and Yakov Oskarovich Parnas (1884–1949). Glycolysis in some form occurs in every living cell. It could be said that the discipline of biochemistry was born in a glass of beer.

The Scientific Method

Today, brewing is a scientific process. The term *science* is derived from the Latin word *scientia*, meaning "to know." The scientific method is a systematic way of thinking about and investigating processes to generate new knowledge. In academic science, this might be the discovery of new enzymes and applications, the development of new hop varieties, or the engineering of a new aeration device. In these examples, new information or new knowledge is generated. In brewing, this might be problem-solving or troubleshooting, in which needed information is identified and new information is sought to help correct an issue.

To generate new knowledge, scientists must take a systematic and logical approach. Often the observations made, the questions posed, and experiments designed require creativity from the scientist, an ability to think outside of the box. The process is creative, but the approach must be systematic. The scientific method involves six steps:

1. Make an observation.
2. Pose a question.
3. Generate a **hypothesis** and testable prediction.
4. Design and run an experiment. Record results.
5. Analyze the data. Determine whether it supports or refutes the hypothesis.
6. Repeat the process as needed to further support or refute the hypothesis.

Let us use troubleshooting an issue in the brewery as an example of the scientific method.

1. *Make an observation.* You have noticed that fermentation stalled in the brewery. Most beers finish at 3 °P, but this beer stopped fermenting at 6 °P.
2. *Pose a question.* Why did fermentation stall?
3. *Generate a hypothesis and testable prediction.* A hypothesis is a *testable* statement. It must also be rational and based upon well-established facts. A prediction is a deductive consequence of a hypothesis, typically an "if, then" statement. In this example, the hypothesis is "Fermentation stalled because there were not enough yeast cells." The prediction is "If too few yeast cells were pitched before fermentation, then the

fermentation will stall." It is important to consider that the hypothesis may or may not be correct. And there may be additional hypotheses. With appropriate experimentation and analysis, we can decide if the hypothesis is supported or refuted.

4. *Design and run an experiment; record results.* A hypothesis must be testable. In this example, for a subsequent brew, we triple check that yeast is pitched at the appropriate rate ensuring that there are enough cells for fermentation.

5. *Analyze the data.* If the fermentation stalls again, the hypothesis is incorrect or refuted. Some other process change affected fermentation. If the fermentation is now completed as normal, the hypothesis is supported.

6. *Repeat the process to further support or refute the hypothesis.* The results from a test never *prove* an idea as correct but rather "support the hypothesis." On the other hand, if a test fails and refutes the hypothesis, then a new hypothesis and experiment are proposed. Let us assume that after ensuring a proper yeast count, the fermentation stalls again. In this case, the hypothesis is refuted, and additional hypotheses are generated. What are some other issues that may cause fermentation to stall? The scientific method and the process of troubleshooting are cyclical; it may take several cycles of hypothesis and experiment to move closer to a solution.

For many problems, there are multiple plausible explanations for the issue. After generating additional hypotheses and possibilities, scientists need to prioritize which ideas should be tested. Unfortunately, time and money typically are major limitations to the most thorough testing, especially in a production environment such as a brewery. Therefore, scientists must test the most probable hypotheses while also considering good experimental design.

Good experimental design requires the following:

- A *testable* hypothesis.
- One or more dependent variables.
- Only *one* independent variable or change.
- Experimental controls.
- Statistical significance.

Clearly, if a hypothesis cannot be tested, it cannot be supported by evidence. For an experiment to serve as a test of a hypothesis, an outcome of the experiment is measured. This is called the dependent variable. In the stalled fermentation experiment described above, the dependent variable is the measurement of the wort density during fermentation. Other measurements, or dependent variables, that might be taken during this experiment are yeast cell count in suspension, carbon dioxide production, and pH.

The independent variable describes what is being manipulated or changed in the experiment. In the stalled fermentation experiment described above, the independent variable is the yeast count in the pitch. It is the only parameter being changed or manipulated. A successful scientific experiment must have only one independent variable. This also applies when troubleshooting a problem. Use of a single variable is critical because if several variables are changed, how will we know which was responsible? To troubleshoot a problem, a brewer might try to fix an issue by changing four conditions. While the problem may have been solved, how will he or she learn from the problem and prevent its occurrence in the future? The exact cause of the issue is still unidentified because of a poorly designed experiment. In experimentation or troubleshooting, only change one variable at a time.

A well-designed experiment must be controlled. In typical experiments an *experimental group* would get various levels of a certain treatment, and a *negative control group* would not get the treatment under test, but its treatment would be otherwise identical. The two groups are compared to determine the effect of the treatment. In some experiments it is useful also to include a *positive control*, which is a treatment known to influence the dependent variable. If the positive control fails to yield the expected result, we suspect that there is something wrong with the experiment. For example, we might study the question of whether the addition of zinc chloride increases the fermentation rate. We would set up several flasks with identical wort composition. To the experimental group, we could add various concentrations of zinc chloride solution, but to the negative control group, we could add an equal amount of pure water. The positive control group could be treated with yeast nutrient, known from previous experiments to increase the fermentation rate. We place all the flasks in baths at the same temperature. We add the same amount of the same yeast into each flask. We do all we can to make sure that the experimental group and the two control groups are treated identically except for the independent variable, zinc chloride. If the positive control flasks do not ferment faster than the negative control flasks, we would suspect that there is something wrong with the way the experiment was run. Maybe there is a leak, or the yeast was no good. If the zinc chloride flasks differ from the negative control flasks, it would be evidence that zinc chloride influences the fermentation rate.

In addition to control groups in an experiment, good experimental design includes a tightly controlled environment. Every condition of the experiment, other than the independent variable, should be kept as precisely consistent as possible. In studying the effect of zinc chloride on fermentation, what are some of the environmental controls? This experiment is best controlled by running all experiments at the same time, using the same wort, and at the same temperature. The list continues, but the point is that only the independent variable should be different. Any other difference could influence or change the results. Eliminating uncontrolled variables in laboratory-scale experiments can be difficult, but it is much more difficult in practical settings like a brewery where time, space, and money are critical.

During experimentation scientists gather data, interpret results, and formulate conclusions. What if the experiment was only run once? What is the significance of the experiment? For an experiment to be significant, it must have some type of statistical probability of being correct, and it must be repeatable by others.

Many measurements rely on a representative sample. If, for example, you were checking package oxygen, you could only test a small fraction of the bottles or cans. If you tested them all, you would have no beer to sell. The packages selected for testing are the *sample*. For a sample to be representative, it must be random. If you select 15 bottles in a row as they emerge from the filler, you may miss a problem that emerges later in the run or that is intermittent. If you pick the whole sample from one side of the conveyor, you may introduce a bias into the sample that could affect the conclusion. Here, it is best to select a representative sample of bottles randomly throughout the production run from start to finish.

The other sampling issue is sample size. Larger samples give more accurate results, but they are more expensive in terms of analysis cost and lost beer. To illustrate the effect of sample size, we will use an artificial example involving a large bin of glass marbles, 60% of which are red and the rest blue. Figure 1.14 shows the total percentage of red marbles as we randomly draw marbles one by one. It takes over 100 draws to reach a steady-state value near 60%. Larger sample sizes give more trustworthy results, but even for a

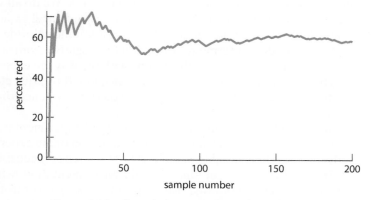

Figure 1.14 Cumulative outcomes of marble draws.

large sample, the outcome is subject to error. In this circumstance, the word "error" refers to the difference between the accepted, "true," or expected value and the measured value. "Error" does not imply that anyone did anything wrong.

Throughout this book, you will have an opportunity to apply the scientific process through troubleshooting potential problems in the brewery. Case studies are provided in select chapters as a way to critically think about problems that may arise in the brewing process and to propose potential solutions. The ability to troubleshoot, to think scientifically about problems, is one of the greatest skills you can bring to any job.

1.4 INTRODUCTION TO BEER QUALITY

It is often said that a glass of beer is best at the source. This is unfortunately true; beer is a perishable product whose quality slowly diminishes with time after production. As brewers our goal is to provide as close to "brewery fresh" beer as possible to all consumers whether on draft at the brewery or at home from a bottle or can. With increasing competition in the brewing industry comes a need for consistency and quality product for the consumer. If we want to provide fresh and high-quality beer to the consumer, we must first define quality.

In 2014, a Brewers Association subcommittee on quality defined quality beer as "…a beer that is responsibly produced using wholesome ingredients, consistent brewing techniques, and good manufacturing practices, which exhibits flavor characteristics that are consistently aligned with both the brewer's and beer drinker's expectations." In this definition, consistency is key as is responsibility and safety. To make quality beer, a brewer must master the brewing process, have a deep familiarity with the raw materials, and have a solid understanding of the underlying science.

Mary Pellettieri (cited in the Bibliography) lays out three fundamental aspects of beer quality that take the definition of quality one step further. First, beer should be free from **defect**. Second, beer should be well defined as fit for use. And finally, beer quality should match the brand values of the company; in other words, the artistic side of brewing also represents quality.

Quality as Freedom from Defect

Beer quality as freedom from defect is defined by laws and government regulations. The best-known historical food safety and quality government regulation was the Bavarian *Reinheitsgebot* or German Purity Law, decreed in 1516. Although not the first such regulation, it was the first to cover more than a single city. Still in place today, this law placed strict specifications on the production of beer, allowing only three ingredients – barley, hops, and water. The fourth ingredient, yeast, is now included in the law, but was absent in the original, because yeast was not understood as the causative agent of fermentation until the nineteenth century. It is thought that the purity law was put into place because some German brewers used alternatives to hops, such as gruit, and alternative cereal grains such as wheat and rye. At the time, brewers were taxed on malt and not on beer produced, so crafty brewers could skirt taxation by brewing with alternative sources of starch. Furthermore, the use of wheat and rye depleted the supply for the baking industry; thus the Reinheitsgebot protected the German economy. The far-reaching impact was to define a specific flavor profile for German beer with a standard of quality still held in high esteem and tradition in Germany today.

Beer quality as freedom from defect has evolved in the United States over the years. Today beer is defined as a food, and breweries are required by law to produce products that are free from defects. These laws are designed to protect consumer safety. US laws require four specific parameters to define beer as free from defect:

1. The Tax and Trade Bureau (TTB) requires accurate reporting of fill levels in bottles and cans.
2. The TTB also requires accurate reporting of **alcohol by volume** (ABV) to within 0.3% if printed on labels.
3. The US Food and Drug Administration (FDA) requires sulfite concentrations to be less than 10 ppm, unless specifically reported on the package.
4. American breweries are now required by the Food Safety Modernization Act, administered by the FDA, to follow good manufacturing practices (GMPs) with strong recommendations for a hazard analysis and critical control points (**HACCP**) plan. GMPs provide safety standards for the facility, employees, and visitors to help ensure that products are free from defect. HACCP is a system of risk assessment that evaluates risk potential across all production processes with a focus on chemical, physical, and biological risks.

Quality as Fitness for Use

"Fitness for use" describes quality from the perspective of the consumer. These are the traits that drive a particular beer or brand. Establishing a definition of quality as fitness for use emphasizes consumer preferences. Often the consumers' preferences are less stringent than the brewer's. For example, variation in slight to moderate haze might be perceptible, but otherwise disregarded by the general consumer population. By contrast, if a slight haze is accompanied by large particles of protein that sink to the bottom of the glass, consumer opinion of quality may plummet. If consumer perception of quality dips into unfavorable territory, the brewery will lose customers. If you consistently produce beers with high quality in fitness for use, consumers will know what to expect in your product, potentially building brand loyalty. Those traits commonly defined under fitness for use include flavor, shelf stability, and aspects of perception like color, clarity, and foam. Understanding these traits, how they are influenced by the brewing process, and how they can change over time in finished beer is essential for the delivery of a high-quality product.

For each beer, a brewery should describe these traits as being "true to type." Being true to type means that for a given brand and style of beer, each production lot of that beer is as similar as possible. Key aspects of the beer such as perception and flavor should be considered before a beer is released to market. Any defects or deviations should be subject to troubleshooting and root-cause analysis.

Perception   Consumers first drink with their eyes. Perception, or the physical quality, of the beer makes the first impression and can underlie the overall reception of the beer. Beer should always be poured into an appropriate and clean glass to fully appreciate this quality. Drinking directly from a bottle or can obscures judgment (which in some cases might be intentional!). Specifically, a number of metrics play a key role in the overall physical quality and perception of the beer, metrics that influence the overall **organoleptic** quality of beer, specifically color, clarity, foam, aroma, and texture.

Color – Beer color can range from light yellow to black with varying shades of red. Generally, beer color should be appropriate to the style. A study by Carvalho et al. demonstrated that when two identical beers are artificially colored so that one resembles a pale beer and the other a dark beer, significant differences were found in consumers' expectations of flavor and cost.

Clarity – Bright or clear beer is a beautiful thing. Significant effort during production ensures a consistently clear product. Nonetheless, some styles like wheat beers, hefeweizens, and New England IPAs are designed to be hazy. In addition to physical appearance, haze, also called **turbidity**, can alter the flavor profile of the beer. Beer haze generally results from aggregates of protein from barley and **polyphenols** from hops. Anecdotally, hazy IPAs are said to be more flavorful than their clarified counterparts. A recent hypothesis with supporting data from Dr. John Paul Maye has demonstrated that certain hop oils with low solubility are stabilized within haze particles, keeping them in solution and thus changing the profile of the product. This observation also explains why flavor profiles of beer can be altered following filtration. The consistent haze found in certain styles is

also called permanent haze. **Chill haze** is a quality issue where haze particles are formed at cold temperatures but disperse as the beer warms. Chill
haze that becomes worse over time is referred to as age-dependent haze.

Foam – Beer is expected to have an attractive layer of foam, or head, that
persists during drinking. As the beer is consumed, **lace** or cling should gently coat the sides of the glass. Foam enhances aroma as **volatile** compounds
are released as effervescence when the foam bubbles pop. Foam also
enhances the texture of the beer and has been shown to dampen the waves
in the beer, so it is less likely to spill.

Aroma – The aroma of beer is heavily influenced by raw material selection,
process, and handling of the beer. In addition, service temperature and carbonation level influence aroma and consumer perception.

Texture – Beer provides different tactile sensations, typically sensed by the
trigeminal nerve. The trigeminal nerve is the main sensory nerve of the head,
innervating the face, mouth, and nasal cavity. It is responsible for sensations
of heat from peppers, the burn from alcohol, and the fizz from carbonation.
Beer is described as dry (thin) or full-bodied, traits related to the residual
carbohydrate content of the beer remaining after fermentation.

Flavor   Flavor is the greatest factor in consumer expectations. It might also
be considered an element of perception, and its impact on overall organoleptic
quality places it in high importance. A beer might look appetizing, but if it
tastes bad, the brewery's reputation is at risk. Beer flavor is a complex synergy
of more than a thousand compounds that derive from raw materials and the
overall process. Some flavors are dominant and distinguishable in isolation,
but others fall below the limit of perception. Some flavors arise only as
combinations of multiple lesser flavors that would be difficult to distinguish
alone. To further complicate the issue, perception of flavor varies greatly
between individuals, as different people can discern different concentrations
of key flavor molecules.

In the brewery, flavors should always be matched as true to type. Flavors
not characteristic of a style are considered "off-flavors." Off-flavors arise
from variations in raw materials, process, or microbial contamination. They
also develop over time as a beer ages, a process known as **staling**. A beer's
exposure to oxygen, high temperatures, and mechanical agitation all affect
the flavor stability of a product.

Stability   Beer flavor will start to decline the moment it leaves the brewery.
Beer served at a brewery taproom can be easily controlled, but once it leaves
the brewery in a package, its quality is much more difficult to manage. In
addition, the chemical processes that create staling off-flavors develop more
quickly with heat. For this reason, beer has a typical shelf life of six to eight
months when kept refrigerated and shorter if kept at room temperature or
heated during summer months. Beer stability will depend on process and also

on style. Stale beer does not have a single characteristic flavor but rather a suite of off-flavors depending on the style, age, and handling.

As a beer ages, hop bitterness and flavor decline steadily. Key **esters** may be reduced. The breakdown of **free amino nitrogen** (FAN) via Strecker degradation can create off-flavors such as sweetness with unpleasant notes of floral, toffee, meat, or bourbon. Oxidation leads to sherry-like aromas and ultimately cardboard flavors. As hop compounds oxidize, they yield **ribes** aromas, often described as catty, tomato leaves, or blackcurrant. The oxidation of lipids leads to the formation of cardboard or paper flavors. Once oxidation and flavor staling occur, there is no recovery. To forestall staling, best brewing practices minimize oxygen uptake throughout the brewing process, including packaging, to prolong beer stability. And of course, beer must be free from microbial contamination of wild yeast and bacteria. Microbial contamination in a packaged product can lead to off-flavors or overcarbonation.

BEER QUALITY AND SCHLITZ BEER

At the start of the twentieth century, the Joseph Schlitz Brewing Company (established in Milwaukee, WI, in 1849) was the largest producer of beer in the United States. Their famous tagline was "The beer that made Milwaukee famous." But the company suffered great decline in the 1970s. A few years earlier, at the edge of innovation, Schlitz championed a new production process called accelerated batch fermentation (high temperature fermentation) to increase yield and improve efficiency. To lower the cost of raw materials, Schlitz substituted corn syrup for some of the malted barley and hop pellets instead of whole hops (a practice that is now widely accepted). Clarity issues were resolved by the addition of a clarifier, possibly **papain**. The resulting foam issue was resolved by adding **propylene glycol alginate** (PGA). Looking back on it today, it seems as though a business decision was made to cut costs and allow the quality to decline in small increments that the consumers would not notice. This approach is derisively called "salami slicing." An unexpected interaction between the clarifier and the PGA sometimes resulted in formation of clouds of snowy particles after a few months in the packages. Schlitz management was slow to deal with the issue. Customers found the bits to be unacceptable. Ultimately Schlitz returned to its original recipe, but it was too late. Sales never recovered.

The downfall of Schlitz is an unfortunate example of how quality as fitness for use was not given adequate consideration in process changes and innovations to improve efficiency. Furthermore, problem-solving strategies and attention to quality management were seemingly inadequate with the result that poor quality product entered the market.

Quality as Art

While adherence to beer quality as freedom from defect is required in all breweries by law, there is more flexibility in quality as fitness for use. But what about the art of brewing? There are certainly small breweries who relish the benefits of being small scale such as faster beer turnover, taproom-only beers, and the ease of innovation. Here, brewers may highly value the opportunity to use new ingredients, to form exciting collaborations, and to push the boundaries of traditional brewing practices. Is there quality in creativity? Absolutely. Here quality as art is defined by how each beer matches the *brand values* of the business. In this case, the leadership team at the brewery should take time and care to define the overall brand values of the business, clearly documenting them and sharing them with all employees. In addition to beers being "true to type," they should also be "true to brand." Then as each new product or innovation is planned, the team can assess its compatibility with the overall brand of the company. In this sense, the quality of art and creativity help define the quality of the beer.

CHECK FOR UNDERSTANDING

1. What is the definition of beer and how does it differ from other alcoholic beverages?

2. Arrange the following units in order of use from start to finish, identify the brewing step in which it is used, and describe the key purpose(s) of the step in beer production.

Lauter tun	Fermenter	Centrifuge
Plate chiller	Mash tun	Mill
Whirlpool	Boil kettle	Bright tank

3. What are the four major ingredients in beer? At which stage(s) of the brewing process is each added? Discuss all areas for variation.

4. Describe the major process inputs during each step of beer production (i.e. time, temperature, etc.)

5. What are the most important quality goals for packaging operations?

6. Where and when were the origins of beer?

7. What were the three major technological advances that made routine beer brewing possible?

8. What were some of Louis Pasteur's major contributions to brewing science?

9. What is a *controlled experiment*? Discuss the difference between an independent and a dependent variable.

10. What are some similarities and differences between the scientific method and troubleshooting?

11. Why is it critical to change only one variable at a time during an experiment or while troubleshooting?

12. What government agencies regulate the beer industry in the United States? By law, what metrics must brewers report?

13. Define the Reinheitsgebot and describe its influence on beer quality.

14. Quality is in the eye of the beer holder. What is meant by this statement?

15. What is meant by the phrase "true to type," and how would you incorporate it into a brewery?

16. You overheard someone at a bar say, "Hazy beer is poor quality beer." Agree or disagree and explain your position.

17. Describe some of the key changes in flavor during beer staling and comment on their causes.

CASE STUDY

A brewery has made the same hefeweizen for years. But in the last several production batches, flavor differences have been noted by a series of trained panelists, particularly an increase in isoamyl **acetate** (banana) and an increase in higher alcohols (unpleasant heat from alcohol). In thinking carefully about what changes may have occurred that could have caused the flavor change, the Director of Quality realized a new sound system was recently installed in the brewery. Several speakers and a subwoofer were placed about a meter from the fermentation tanks, and since installation, the staff enjoyed listening to music throughout the day. In reviewing production records and sensory notes, the Quality Director realized that the changes in flavor corresponded to the date of the speaker installation. The Director then tried to explain to the Operations Manager that the music could be affecting fermentation, but the Manager argued that the music was good for employee productivity. The Director then decided to conduct an experiment with four hefeweizen fermentations in the laboratory. Two were subjected to electro-swing music via a waterproof speaker and two were kept in a quiet corner of the laboratory. Fermentation rates were tracked each day by measuring the beer density. When fermentation was complete, the same sensory panel evaluated the flavors in the finished beers. Analysis of the data revealed that the two samples subjected to music fermented faster. They reached terminal **gravity** a day sooner, and the sensory panel noted an increase in banana flavor and alcohol burn as compared with the quiet fermentation. These results convinced to Operations Manager to remove the speakers from the fermentation cellar.

CASE STUDY QUESTIONS

1. What was the observation that prompted this scientific experiment? What was the hypothesis? What was the prediction statement?
2. What were the independent and dependent variables?
3. Describe how this experiment was controlled. Are there any other controls you might include in the experiment?
4. How confident are you in the results of this experiment?
5. How could the experiment be improved?

BIBLIOGRAPHY

Bamforth C. 2003. *Beer Tap into the Art and Science of Brewing, 2nd ed*. Oxford University Press. ISBN 978-0-19-515479-5.

Barnett JA. 1998. A history of research on yeasts 1: work by chemists and biologists 1789–1850. *Yeast* 14:1439–1451.

Barnett JA. 2003. A history of research on yeasts 5: the fermentation pathway. *Yeast* 20:509–543.

Barth R. 2013 *The Chemistry of Beer: The Science in the Suds*. Wiley. ISBN 978-1-11867497-0.

Barth R. 2015. The role of alcoholic fermentation in the rise of biochemistry. In Barth R, Benvenuto M. (editors). *Ethanol and Education: Alcohol as a Theme for Teaching Chemistry*. American Chemical Society. ISBN 978-0-8412-3059-0. Chap. 3. p. 25–46.

Briggs DE, Boulton CA, Brookes PA, Stevens R. 2004. *Brewing Science and Practice*. CRC. ISBN 0-8493-2547-1.

Carvalho FR, Moors P, Wagemans J, Spence C. 2017. The influence of color on the consumer's experience of beer. *Front. Psychol.* 8:2205–2214. doi:10.3389/fpsyg.2017.02205.

Dalgliesh CE. 1977. Flavour Stability. *Proc. Eur. Brew. Conv. Congr. Amsterdam*. DSW. p. 623–659.

Eßlinger H, Editor. 2009. *Handbook of Brewing Processes, Technology, Markets*. Wiley-VCH. ISBN 978-3-527-31674-8.

Hornsey IS. 2012. *Alcohol and its Role in the Evolution of Human Society*. RSC. ISBN 978-1-84973-161-4. p. 89–111.

Liu L, Wang J, Rosenberg D, Zhao H, Lengyel G, Nadel D. 2018. Fermented beverage and food storage in 13,000 y-old stone mortars at Raqefet Cave, Israel: investigating Natufian ritual feasting. *J. Archaeol. Sci. Rep.* 21:783–793.

McGovern PE. 2009. *Uncorking the Past*. University of California Press. ISBN 978-0-520-25379-7.

Pasteur L. 1879. *Studies on Fermentation*. Translation Faulkner F, Robe DC. Reprinted BeerBooks.com. ISBN 0-9662084-2-0. p. 148.

Pellettieri M. 2015. *Quality Management Essential Planning for Breweries*. Brewers Publications. ISBN 978-1-938469-15-2.

Sauret A, Boulogne F, Cappello J, Dressaire E, Stone HA. 2015. Damping of liquid sloshing by foams. *Phys. Fluids.* 27:022103.

Spedding G. *Best Practices Guide to Quality Craft Beer*. Brewers Association. *Educational Publications*. https://www.brewersassociation.org/educational-publications/best-practices-guide/

CHAPTER 2

CHEMISTRY FOR BREWING

Chemistry is the science of matter. During beer production, we are most interested in the substances in beer, its raw materials, and its processing aids and in the equipment used to make it. In many instances, substances interact to form new substances, processes termed chemical **reactions**. In most chemical reactions, **energy** is exchanged. **Hydrolysis** of **starch** to **sugar** during **mashing**, **isomerization** of **hop alpha acids** during boiling, production of **ethanol** and carbon dioxide during **fermentation**, and even the formation of **haze** particles in packaged beer are examples of chemical reactions. Understanding the principles of chemistry helps the brewer control beer production and effectively troubleshoot any issues that may arise.

2.1 ATOMS

Chemistry is largely governed by the interactions of particles with electrical charges, positive and negative. A pair of particles that are both positively charged or negatively charged **repel** one another, that is, they experience a force that drives them apart. A pair of particles with opposite charges, one positive and one negative, experience an **attractive** force that pulls them together.

Mastering Brewing Science: Quality and Production, First Edition.
Matthew Farber and Roger Barth.
© 2019 John Wiley & Sons, Inc. Published 2019 by John Wiley & Sons, Inc.

Atoms, Elements, and Atomic Structure

Matter is organized into **atoms**. An atom is the smallest, most fundamental **unit** of bulk matter. There are about 100 different types of atoms, each representing an **element**. Of these elements, 13 are of primary importance in **brewing**, 6 of which account for nearly every atom in every compound. These are hydrogen, carbon, oxygen, nitrogen, sulfur, and phosphorus. The elements and, by extension, their atoms are represented by symbols made up of letters of the Latin alphabet, that is, the same alphabet you are reading now. The symbols are not merely abbreviations; they are used to make pictorial representations of the underlying atomic structures of substances. The symbol for an element consists of a capital letter, sometimes followed by a lowercase letter. Not coincidentally, the six elements mentioned above have one-letter symbols. Table 2.1 shows the names, symbols, and other information that we will discuss below for the 13 elements of primary importance.

Atoms are not featureless balls; they have an internal structure. The two main parts of an atom are the **nucleus** and the **electrons**. The nucleus has a positive electrical charge and accounts for more than 99.9% of the mass of an atom. The size of the nucleus is so small that we can think of it as virtually a geometric point marking the center of the atom. The electrons, which have a negative charge, surround the nucleus. The opposite charges of the nucleus and the electrons give rise to a force holding the electrons near the nucleus. The electrons orbit about the nucleus, behaving like a cloud of negative charge; the extent of their travels, the size of the cloud, defines the size of the atom, as shown in Figure 2.1.

TABLE 2.1   **Elements**

Name	Symbol	Atomic Number	Valence electron	Molar Mass (g/mol)	Valency
Hydrogen	H	1	1	1.008	1
Carbon	C	6	4	12.011	4
Nitrogen	N	7	5	14.007	3
Oxygen	O	8	6	15.999	2
Sodium	Na	11	1	22.990	1
Magnesium	Mg	12	2	24.305	2
Aluminum	Al	13	3	26.982	3
Phosphorus	P	15	5	30.974	3, 5
Sulfur	S	16	6	32.066	2, 6
Chlorine	Cl	17	7	35.453	1
Calcium	Ca	20	2	40.078	2
Iron	Fe	26	8	55.845	2, 3, 6
Copper	Cu	29	11	63.546	1, 2, 4

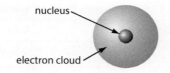

Figure 2.1   Atom.

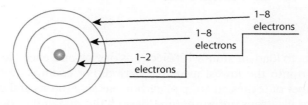

Figure 2.2   Electron periods.

The nucleus also has distinct properties; it contains positive particles called **protons** and uncharged (**neutral**) particles called **neutrons**. Protons and neutrons have roughly the same mass, which is about 1800 times that of an electron. The number of protons in an atom is called its **atomic number**. The atoms of a particular element have a particular atomic number, shown in Table 2.1. The positive charge on a proton exactly **balances** the negative charge on an electron. An atom that has the same number of protons and electrons is electrically neutral. It is possible for atoms of the same element to have different numbers of neutrons. Atoms that contain an equal number of protons but different numbers of neutrons are called **isotopes**.

Electrons exhibit unfamiliar behavior when **bound** to atoms. They can only take certain values of energy. These permitted values are called **energy levels**. When an electron has an energy defined by a particular energy level, we say that the electron is in that energy level. The electrons in low energy levels are close to the nucleus and require a large amount of energy to pull them out of the cloud and away from the atom. Electrons in higher energy levels are farther from the nucleus and easier to remove from the atom. An unexpected property of energy levels is that they can accommodate only a limited number of electrons. For example, the lowest energy level in an atom holds no more than two electrons. Any additional electrons go into the next higher energy level, which can hold no more than eight. The concept of electron energy levels is depicted in Figure 2.2. The energy levels have sublevels. Sometimes the sublevels of one level overlap those of another. We will group the accessible levels/sublevel into groupings termed **periods**. The maximum numbers of electrons that the first four periods can accommodate are given in Table 2.2.

TABLE 2.2   **Electron Period Capacities**

Period	Capacity
1	2 electrons
2	8
3	8
4	18

If we were to load an atom with electrons one by one, each electron would ordinarily go into the lowest available period. For example, carbon has an atomic number of 6; thus an atom of carbon has six protons and six electrons. The first two electrons are in the first period; the remaining four are in the second period. The highest occupied period is called the **valence** period. The electrons in it are valence electrons. The rest of the atom, including the lower-period electrons and the nucleus, is called the **core**. Carbon has two core electrons and four valence electrons.

Periodic Table

It takes a very large amount of energy to remove a core electron, so core electrons are not usually involved in chemical reactions. The number of valence electrons in a neutral atom (one with the same number of electrons as protons) is a key factor that determines the chemical behavior of the atom. This number is so important that the **periodic table**, which is the standard tabular arrangement of the elements, is organized in columns called **groups**, which can be used to infer the number of valence electrons.

The periodic table, shown in Figure 2.3, is a listing of all known elements organized primarily by atomic number (number of protons). The upper part of the table has seven rows, representing the periods, arranged in 18 columns called groups. There are two additional rows shown at the bottom. These rows actually belong in the gaps between elements 56 and 71 in the sixth period and between elements 88 and 103 in the seventh period. There are some other complications, but fortunately our elements of interest, with the exceptions of iron (Fe: 26) and copper (Cu: 29), are confined to elements 1–20, whose behavior with respect to the periodic table is relatively straightforward. These most important atoms fall into 1 of 8 columns (1A through 8A), which are considered the **main groups**. In the main groups, the number of electrons in the valence **period** is the same as the group number. For example, oxygen is in group 6A and contains 6 valence electrons. Carbon is in group 4A and has 4 valence electrons. Like oxygen, sulfur is also in group 6A; thus it also contains 6 valence electrons.

Periodic table of the elements, arranged by group:

Group 1A

Z	Sym	Mass	Name	EN
1	H	1.008	Hydrogen	2.2
3	Li	6.941	Lithium	1.0
11	Na	22.99	Sodium	0.9
19	K	39.10	Potassium	0.8
37	Rb	85.47	Rubidium	0.8
55	Cs	132.91	Cesium	0.8
87	Fr	223	Francium	0.8

Group 2A

Z	Sym	Mass	Name	EN
4	Be	9.012	Beryllium	1.6
12	Mg	24.31	Magnesium	1.3
20	Ca	40.08	Calcium	1.0
38	Sr	87.62	Strontium	1.0
56	Ba	137.33	Barium	0.9
88	Ra	226	Radium	0.9

Transition metals (Groups 3–12)

Z	Sym	Mass	Name	EN
21	Sc	44.96	Scandium	1.4
22	Ti	47.87	Titanium	1.5
23	V	50.942	Vanadium	1.6
24	Cr	51.996	Chromium	1.6
25	Mn	54.938	Manganese	1.6
26	Fe	55.845	Iron	1.8
27	Co	58.933	Cobalt	1.9
28	Ni	58.693	Nickel	1.9
29	Cu	63.55	Copper	1.9
30	Zn	65.39	Zinc	1.6
39	Y	88.91	Yttrium	1.2
40	Zr	91.22	Zirconium	1.3
41	Nb	92.91	Niobium	1.6
42	Mo	95.94	Molybdenum	1.3
43	Tc	98	Technetium	1.4
44	Ru	101.07	Ruthenium	2.2
45	Rh	102.91	Rhodium	2.3
46	Pd	106.42	Palladium	2.2
47	Ag	107.87	Silver	1.9
48	Cd	112.41	Cadmium	1.7
71	Lu	174.97	Lutetium	1.3
72	Hf	178.49	Hafnium	1.3
73	Ta	180.95	Tantalum	1.5
74	W	183.84	Tungsten	2.4
75	Re	186.21	Rhenium	1.9
76	Os	190.23	Osmium	2.2
77	Ir	192.22	Iridium	2.2
78	Pt	195.08	Platinum	2.3
79	Au	196.97	Gold	2.5
80	Hg	200.59	Mercury	2.0
103	Lr	262	Lawrencium	
104	Rf	267	Rutherfordium	
105	Db	268	Dubnium	
106	Sg	269	Seaborgium	
107	Bh	270	Bohrium	
108	Hs	269	Hassium	
109	Mt	278	Meitnerium	
110	Ds	281	Darmstadtium	
111	Rg	281	Roentgenium	
112	Cn	285	Copernicium	

Groups 3A–8A

Z	Sym	Mass	Name	EN
5	B	10.811	Boron	2.0
6	C	12.011	Carbon	2.6
7	N	14.007	Nitrogen	3.0
8	O	15.999	Oxygen	3.4
9	F	18.999	Fluorine	4.0
2	He	4.003	Helium	
10	Ne	20.180	Neon	
13	Al	26.98	Aluminum	1.6
14	Si	28.09	Silicon	1.9
15	P	30.97	Phosphorus	2.2
16	S	32.066	Sulfur	2.6
17	Cl	35.453	Chlorine	3.2
18	Ar	39.948	Argon	
31	Ga	69.723	Gallium	1.8
32	Ge	72.61	Germanium	2.0
33	As	74.922	Arsenic	2.1
34	Se	78.96	Selenium	2.6
35	Br	79.904	Bromine	3.0
36	Kr	83.80	Krypton	3.0
49	In	114.82	Indium	1.8
50	Sn	118.71	Tin	2.0
51	Sb	121.76	Antimony	2.1
52	Te	127.60	Tellurium	2.1
53	I	126.90	Iodine	2.7
54	Xe	131.29	Xenon	2.6
81	Tl	204.38	Thallium	2.0
82	Pb	207.2	Lead	1.9
83	Bi	208.98	Bismuth	2.0
84	Po	209	Polonium	2.0
85	At	210	Astatine	2.2
86	Rn	222	Radon	2.2
113	Nh	286	Nihonium	
114	Fl	289	Flerovium	
115	Mc	289	Moscovium	
116	Lv	293	Livermorium	
117	Ts	294	Tennessine	
118	Og	294	Oganesson	

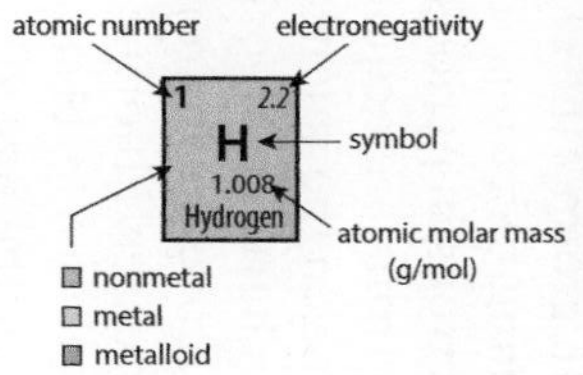

Lanthanides and Actinides

Z	Sym	Mass	Name	EN
57	La	138.90	Lanthanum	1.1
58	Ce	140.12	Cerium	1.1
59	Pr	140.91	Praseodymium	1.1
60	Nd	144.24	Neodymium	1.1
61	Pm	145	Promethium	
62	Sm	150.36	Samarium	1.2
63	Eu	151.96	Europium	
64	Gd	157.25	Gadolinium	1.2
65	Tb	158.92	Terbium	
66	Dy	162.50	Dysprosium	1.2
67	Ho	164.93	Holmium	1.2
68	Er	167.26	Erbium	1.2
69	Tm	168.93	Thulium	1.2
70	Yb	173.04	Ytterbium	
89	Ac	227	Actinium	1.1
90	Th	232.04	Thorium	1.3
91	Pa	231.04	Protactinium	1.5
92	U	238.03	Uranium	1.7
93	Np	237	Neptunium	1.3
94	Pu	244	Plutonium	1.3
95	Am	243	Americium	
96	Cm	247	Curium	
97	Bk	247	Berkelium	
98	Cf	247	Californium	
99	Es	252	Einsteinium	
100	Fm	257	Fermium	
101	Md	258	Mendelevium	
102	No	259	Nobelium	

Legend:
atomic number — electronegativity — symbol — atomic molar mass (g/mol)

Example: 1 H 2.2 1.008 Hydrogen

- ▢ nonmetal
- ▢ metal
- ▢ metalloid

Figure 2.3 Periodic table of the elements.

Most periodic tables include useful descriptive information about each element. The elemental symbol and atomic number are invariably provided. The periodic table in Figure 2.3 also includes the full elemental name, atomic **molar mass** (g/mol), **electronegativity**, and color coding showing metallic character, as shown in the key at the lower left. The arrangement of elements by atomic number within the periodic table relates to their physical and chemical properties. One major division is the **metals**, shown in Figure 2.3 with blue backgrounds, and the **nonmetals**, shown with orange backgrounds. The **metalloids**, with green backgrounds, are placed in between, possessing some properties of metals and some of nonmetals. Hydrogen, at the top left, seems misplaced as a nonmetal; its placement in this sense is governed by the need for one electron to fill its valence period, like the group 7A elements.

The valence electrons in metals are loosely bound; metal chemistry is dominated by a tendency to lose electrons. The loosely bound electrons in metals can move about, carrying electric charge and energy. This accounts for the familiar metallic properties like conduction of electricity and **heat**. Even **metallic luster**, chemistry jargon for shininess, and ductility (ability to deform without shattering) results from the mobility of the valence electrons. Nonmetals hold their electrons tightly and lose them only with difficulty. Many nonmetals tend to gain electrons.

If a single group (vertical column) of elements is selected, as we move down the group an additional layer of core electrons in each succeeding period (row) is added. The valence electrons are farther from the nucleus. For example, starting with hydrogen in group 1 and moving down to francium, the valence electron of lithium is farther from the nucleus than that of hydrogen; that of sodium is farther still, as so it goes with each succeeding element as we move down. The more distant electrons are more loosely bound, hence easier to remove. In the case of group 1 elements, whose chemistry is dominated by removing electrons, reactivity increases with placement lower on the table. For example, hydrogen does not react with water at all; lithium reacts visibly with warm water; sodium reacts violently even with cold water; and potassium reacts explosively with water at any temperature.

In group 7, the elements are nonmetals whose reactivity is characterized by gaining electrons. The elements at the top of the chart accommodate extra electrons in periods that are close to the nucleus. Further down the table, the electrons are acquired further away from the nucleus. Because the distance between the valence level and the nucleus is smaller in elements at the top of the chart (i.e. fluorine), electrons are quickly acquired with high energy. As we move down from fluorine to chlorine and then to bromine, the reactivity decreases. Fluorine has such an extreme tendency to grab an electron to fill its valence level that it is dangerous to handle.

For the first 20 elements, the main group number (i.e. 7A) is equal to the number of valence electrons in the neutral atom. Elements in group 8A have a filled valence level. This group, called the **noble gases**, shows little tendency to engage in chemical reactions, even at high temperature. Helium and argon are often used as blanketing gases to protect metal objects from air during high temperature welding.

If a main group atom is made to fill (or empty) its valence period, it shows extra stability, similar to the noble gases. For example, sodium (Na) is element 11. Its configuration is 2|8|1, with one electron in the valence period. If we can induce the sodium atom to give up its single valence electron forming a positive ion symbolized as Na^+, it would have a 2|8 configuration identical to that of neon (Ne). If an ion is formed, atoms in the main groups generally take the most energetically favorable approach to gaining or losing electrons. If the valence shell has fewer than four electrons, it is more energetically favorable to lose the electrons. If the valence period has more than four electrons, it is more energetically favorable to gain the electrons. The expected behavior of main group elements is to react to lose, gain, or share electrons to give a configuration like that of a nearby group 8 element. This is called the *octet rule*. Sharing electrons results in **covalent bonds**, discussed in Section "Covalent Bonds."

An atom that has gained or lost one or more electrons is called an **ion** (Figure 2.4). If an atom gains one or more electrons, the atom becomes negatively charged. If an atom loses one or more electrons, the atom becomes positively charged. The charge on an ion is displayed as a superscript. If multiple electrons are gained or lost, the number is included. For example, the ion of O is O^{2-}. The ion of Na is Na^+.

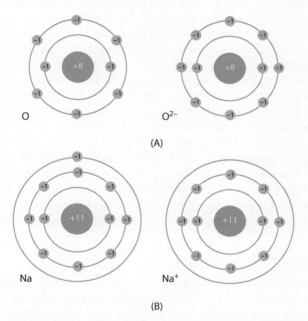

Figure 2.4   (A) Oxygen atom and oxide ion. (B) Sodium atom and sodium ion.

2.2 BONDING AND COMPOUNDS

Atoms stick to one another by forces called **chemical bonds**. The bonded atoms have a lower potential energy than the separated atoms; otherwise there would be no bond. Conversely, breaking a bond requires energy. This amount of energy is called the **bond energy**. When atoms of different elements form bonds, new substances called **compounds** are formed. A compound is a substance made up of more than one element. The relative amounts of each element in a compound are fixed. For example, the elemental **composition** of water is always 88.8% oxygen and 11.2% hydrogen by mass, whether the water comes from the **cytoplasm** of a yeast cell or from the outer moons of Saturn. This is because the atoms combine in fixed amounts, in the case of water, one atom of oxygen to two atoms of hydrogen, yield the familiar **formula** H_2O.

A **mixture**, by contrast, contains more than one substance, each called a component. The substances can be elements or compounds, but they are not chemically bound to one another, so the mixture has no fixed composition. Oxygen and hydrogen are colorless, odorless gases that can be mixed in any proportion, from 100% oxygen to 100% hydrogen (WARNING: mixtures between 4 and 95% hydrogen in oxygen by volume are explosive). The behavior and properties of the mixture are intermediate between those of the components. For example, if 10 g each of hydrogen (density = 0.08 g/L at 1 atm pressure and 25 °C) and oxygen (density = 1.38 g/L) are mixed, the mixture would be a colorless odorless gas with a density of 0.15 g/L. If the mixture is placed in a very strong vessel and ignited, after a dangerously violent reaction, all of the oxygen is consumed, 11.3 g of water, a liquid whose density is 998 g/L, is produced, and 8.7 g of hydrogen remains. In this scenario, the number of hydrogen and oxygen atoms is the same before and after the reaction. Some became bound together in water, while some remained as elemental hydrogen. The mass of the system before and after the reaction (20 g) is unchanged; this is called the *law of conservation of matter*. The properties of the **product**, water, are completely different from those of hydrogen, oxygen, and the mixture. Several key points are illustrated in this scenario. No matter how much of each **reactant** is initially available, the elemental composition of the product is not affected. The properties of a compound are not related in any simple way to those of the elements that make it up. In a chemical reaction some or all of the starting materials are used to produce new compounds. The number and types of atoms are not changed in a chemical reaction.

Ionic Bonds

The simplest type of chemical bond is the attraction between ions (charged particles) of opposite charge. Sodium ions, Na^+, are attracted to chloride ions, Cl^-, because of their opposite charges. This force of attraction is called an **ionic bond**. To maintain overall charge balance, there must be the same

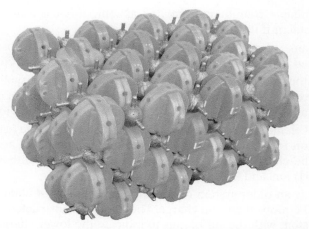

Figure 2.5 Sodium chloride model. Green: chloride ion. Gray: sodium ion.

amount of positive charge as negative charge – hence the same number of sodium ions as chloride ions. The sodium and chloride ions form a new substance called sodium chloride (NaCl) that does not resemble the original elements, sodium and chlorine, at all.

The strength of an ionic bond increases with the amount of charge and decreases with the center-to-center distance between the ions. Small ions that can get very close to one another form strong bonds. Big **polyatomic ions** form weak ionic bonds. Compounds whose atoms are held together by ionic bonds are called ionic compounds. Most of the substances that lend character to brewing water are ionic compounds.

The ions in ionic compounds take up positions that balance the attractions of the oppositely charged ions with the repulsions of the like charged ions. The result is a specific three-dimensional pattern of ions to form a **crystal**. Figure 2.5 shows the **crystal structure** of sodium chloride. Ionic crystals are usually **brittle**; they cannot change **shape** without moving the ions out of their optimal positions, so they usually break instead. When an ionic compound melts or is dissolved in water, the crystal structure is disrupted. Most ionic compounds are high-melting solids, but it is possible to make very large polyatomic ions yielding ionic compounds that are liquid at or near room temperature.

The principle of charge balance governs the formulas of ionic compounds. There must be the same amount of positive charge as negative charge. For example, aluminum oxide, a compound of aluminum ions (Al^{3+}) and oxide ions (O^{2-}), has the formula Al_2O_3, giving a +6 charge from the two aluminum ions and a –6 charge from the three oxide ions. The formulas Al_4O_6 or Al_6O_9 fit the principle of charge balance, but it is conventional practice to simplify the formulas of ionic compounds to the lowest whole number subscripts.

INORGANIC NOMENCLATURE: COMPOUNDS NAMED AS IONIC

1. In naming, a compound is treated as ionic if it has a metal and a nonmetal or if it includes a polyatomic ion.
2. The name of an ionic compound is the name of the positive ion followed by that of the negative ion, leaving off the word "ion."
3. One-atom positive ions of the main group elements usually take a charge equal to the group number. In this case, the name of the ion is just that of the element: Na^+ sodium ion and Ca^{2+} calcium ion.
4. Ions of elements that have more than one positive charge (a common situation for transition elements like iron and copper) are named with the charge shown as a Roman numeral in parentheses: Fe^{3+}, iron(III) ion, and Cu^+, copper(I) ion.
5. There is an older system of naming positive ions of elements that can take more than one charge. The name is constructed from the Latin stem with the suffix -ous to indicate the lower charge and -ic to indicate the higher charge, Cu^+ = cuprous ion. Cu^{2+} = cupric ion and $CuCl_2$ = cupric chloride. This system is no longer approved by chemists, but it is still sometimes used in trade.
6. Single-atom negative ions take a suffix -ide, often with the end part of the element name dropped (-ine, -ogen, -orous, etc.): O^{2-} = oxide ion and Cl^- = chloride ion.
7. Some ions, called **polyatomic ions**, have more than one atom bound together by covalent bonding as discussed in the next subsection. These have an irregular pattern of names shown in Table 2.3.
8. Numerical prefixes (di-, tri-, etc.) are not used unless they are part of the name of one of the ions.

TABLE 2.3 Polyatomic Ions

Formula	Name
$CH_3CO_2^-$	Acetate ion
NH_4^+	Ammonium ion
CO_3^{2-}	Carbonate ion
HCO_3^-	Bicarbonate ion
OH^-	Hydroxide ion
NO_3^-	Nitrate ion
NO_2^-	Nitrite ion
$C_2O_4^{2-}$	Oxalate ion
O_2^{2-}	Peroxide ion
PO_4^{3-}	Phosphate ion
HPO_4^{2-}	Hydrogen phosphate ion
$H_2PO_4^-$	Dihydrogen phosphate ion
SO_4^{2-}	Sulfate ion
HSO_4^-	Bisulfate ion
SO_3^{2-}	Sulfite ion
HSO_3^-	Bisulfite ion
O_2^-	Superoxide ion

Covalent Bonds

An atom that needs one or more electrons to complete its valence period can share electrons with another atom. The electrons are shared between atoms. Hydrogen has one valence electron; it needs one more to complete its first period. Chlorine has seven valence electrons; it also needs one electron to complete its valence period. Each atom can provide an electron to a **shared pair**. The shared pair is localized to some extent between the atoms, and it binds them together in a covalent bond, as shown in Figure 2.6. The resulting particle containing one chlorine atom and one hydrogen atom is a **molecule** of hydrogen chloride. The forces holding the atoms together in a molecule are much stronger than any forces between different molecules. Covalent bonding is prevalent in nonmetal elements, but metals can also form covalent bonds.

To help keep track of the valence electrons in covalent (and ionic) bonds, we can use Lewis dot diagrams, named after G.N. Lewis, a founder of the **shared pair** concept. In a Lewis dot diagram, we show the core of each atom with its element symbol surrounded by dots representing the valence electrons. Figure 2.7 shows Lewis dot diagrams for several elements. The number of electrons that an element shares is called the **valency** of that element. For many nonmetals, the usual valency is equal to 8 minus the group number. Some elements in the third or higher periods have several common valencies, usually including the group number itself. Common valencies are given in Table 2.1. If a single electron pair is shared, this is a single bond. If two or three pairs of electrons are shared, these are **double** or **triple bonds**. Only atoms of carbon (C), oxygen (O), nitrogen (N), sulfur (S), and phosphorus (P) are routinely involved in the formation of multiple bonds.

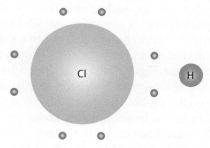

Figure 2.6   Covalent bond.

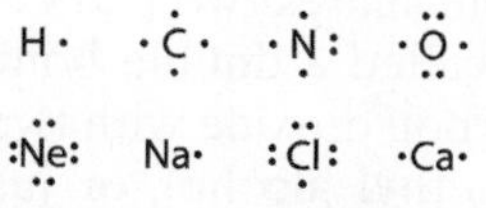

H·   ·C̤·   ·N̈:   ·Ö·

:N̈e: Na·   :C̈l: ·Ca·

Figure 2.7   Lewis dot diagrams.

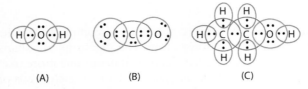

Figure 2.8   Covalent bonding in compounds: (A) water, (B) carbon dioxide, and (C) ethanol.

The concepts of bonding and of compounds are closely related but not identical. It is common for atoms of the same element to form bonds and to exist in the form of molecules. Oxygen, nitrogen, and chlorine form two-atom molecules, sulfur forms eight-atom ring-shaped molecules, and one form of phosphorus exists as four-atom molecules. Although there are bonds that do not form compounds, for a compound to exist, atoms of one element must be chemically bonded to those of another element.

Hydrogen has a valency of 1; oxygen has a valency of 2. The most common compound of these elements is water: H—O—H. Here a dash represents a shared pair of electrons, that is, a single covalent bond. Hydrogen and oxygen can also form hydrogen **peroxide**, H—O—O—H, which also satisfies the valencies of hydrogen and oxygen. Hydrogen peroxide is less stable than water; it decomposes slowly under normal conditions. What about H—O—O—O—H? This compound, called trioxidane, is very unstable.

The Lewis dot diagrams in Figure 2.8 demonstrate electron sharing in water, ethanol, and carbon dioxide, all of which are significant components in beer. This method of representing compounds more clearly demonstrates the filling of valence periods for each atom. The overlapping circles show the allocation of electrons; shared electrons are allocated to two atoms. In Figure 2.8C, the hydrogen atoms (red circles) get two valence electrons, and the carbon atoms (blue circles) and oxygen atom (green circle) get eight.

The dot diagram of water, Figure 2.8A, shows that two pairs of electrons on the oxygen atom are shared with hydrogen atoms and two pairs belong exclusively to the oxygen atom. The pairs of electrons that are shared between two atoms are called **bonding pairs**; those that are not shared are called **unshared pairs** (sometimes called **lone pairs**). Although the unshared pairs do not participate in bonding, they can have important effects on the **geometry** of the molecule. The diagram of carbon dioxide (Figure 2.8B) shows that each oxygen atom shares two pairs of electrons with the carbon atom. This arrangement is called a double bond. Carbon has a valency of four, which is satisfied in carbon dioxide with two double bonds. Figure 2.8C shows ethanol, also called ethyl alcohol, or just "alcohol" in which each carbon atom has four single bonds.

INORGANIC NOMENCLATURE: BINARY COMPOUNDS OF NONMETALS

Binary compounds are those with two elements. The binary compounds of nonmetals have their own naming system, because they do not have ions that are easily described by charge balance. Many of these compounds have well-established non-systematic names. Compounds with C—C or C—H bonds are called **organic** compounds; they follow yet another naming system.

1. Each element is named with a prefix giving the number of atoms of that element in a molecule, as shown in Table 2.4.
2. The element with the lower electronegativity is mentioned first. The element with the higher electronegativity (Figure 2.3) gets an -ide suffix.
3. The mono- prefix is only used when some other compound with the same elements is better known: CO = carbon monoxide. PCl_3 = phosphorus trichloride (not monophosphorus trichloride).
4. As with negative ion names, the stem of the element that gets the -ide is often shortened: CO_2 = carbon dioxide (not carbon dioxygenide).
5. Before a vowel, if the number prefix ends in "a," it is dropped: P_2O_5 = diphosphorus pentoxide (not pentaoxide).
6. Do not use number prefixes with compounds named as ionic: Na_2O = sodium oxide (not disodium oxide).
7. Many binary compounds, especially those with hydrogen, have well-established non-systematic names: CH_4 = methane, NH_3 = ammonia, H_2O = water, and SiH_4 = silane.

TABLE 2.4 **Number Prefixes**

Number	Prefix	Number	Prefix
1	Mono-	7	Hepta-
2	Di-	8	Hexa-
3	Tri-	9	Nona-
4	Tetra-	10	Deca-
5	Penta-	11	Undeca-
6	Hexa-	12	Dodeca-

Figure 2.9   Polar bond.

The electrons in a covalent bond are not always equally shared. Often the electrons are attracted to one of the bound atoms more than to the other, represented in Figure 2.9. In this case the bond is said to be **polar**. The tendency of an atom to draw shared electrons toward itself is called electronegativity. Electronegativity generally decreases in successive periods (rows) of the periodic table, hence F > Cl > Br > I, and it increases in a period going to higher atomic number (to the right) but not including group 8A, hence B < C < N < O < F. Electronegativities are quantified on a scale originated by Linus Pauling and shown on the periodic table in Figure 2.3. The more electronegative atoms pull shared electrons close. If the electronegativities differ by more than 0.3, the bond is regarded as polar. If the electronegativity difference exceeds 1.7, the bond is predominantly ionic. This shows us that the distinction between ionic and covalent bonding is not perfectly sharp; all bonds have some ionic and some covalent character.

2.3   MOLECULES

Most compounds important in brewing consist of molecules. A molecule is a discrete particle made of specific atoms covalently bound together in a specific way. The forces holding the atoms together in a molecule are much stronger than the forces between different molecules. The properties of molecules and the interactions between them are a central concern in beer chemistry.

MOLES AND MASS

Atoms and molecules are too small to weigh and too numerous to count. The atoms of the elements have different masses, so a gram of one element does not have the same number of atoms as a gram of another. In the case of oxygen and hydrogen, the elements in water, an oxygen atom weighs 16 times as much as a hydrogen atom. Looking at it another way, 1 g of hydrogen has 16 times as many atoms as 1 g of oxygen. One gram of hydrogen, the lightest element, has about 6.022×10^{23} atoms, each weighing 1.661×10^{-24} g. Chemists deal with particles like atoms and molecules in groups that are large enough to weigh. The amount of a substance that has as many atoms, molecules, or ions as there are atoms in 1 g of hydrogen (officially, exactly 12 g of the isotope of carbon that has 6 neutrons) is called a **mole** of that substance.

The mass of a mole of anything is its molar mass. It is important to clearly specify what particle is counted for the mole. Elemental hydrogen exists in two-atom molecules. A mole of hydrogen molecules is twice as much material as a mole of hydrogen atoms. The molar masses of atoms of each element at their usual isotopic distribution have been measured carefully and are reported to three decimal places in Table 2.1 and to two places in Figure 2.3. The molar mass of any combination of atoms is the sum of the molar masses of the atoms. The molecular formula of ethanol is C_2H_6O. The molar mass of ethanol is 2×12.011 g/mol + 6×1.008 g/mol + 15.999 g/mol = 46.069 g/mol. The number of moles, n, of a substance can be calculated from the mass and molar mass:

$$n = \frac{\text{mass}}{\text{molar mass}}$$

Ten grams of ethanol comes to

$$\frac{10\,\text{g}}{46.07\,\dfrac{\text{g}}{\text{mol}}} = 0.217 \text{ mol}$$

The number of ethanol molecules in a certain number of moles of ethanol is identical to the number of water molecules, or carbon dioxide molecules, or anything else with that same number of moles. Moles refer to the number of particles.

Composition

The composition of a compound or mixture is the relative amount of the elements or components, expressed in terms of mass, moles, volume, or any other relevant measurement. The mass fraction of an element in a compound or of a component in a mixture is given by

$$\text{mass fraction} = \frac{\text{mass of stuff}}{\text{total mass}}.$$

The mass fraction can be multiplied by 100 to give the mass percent or by one million to give parts per million by mass (ppm). The carbon in a mole of ethanol weighs 2×12.011 g = 24.022 g. The ethanol in a mole of ethanol weighs 46.069 g (the molar mass of ethanol). The mass fraction of carbon in ethanol comes to $\dfrac{24.022 \text{ g}}{46.069 \text{ g}} = 0.5214$, which

(continued)

corresponds to 52.14% carbon by mass. Because mass is conserved, the sum of the mass fractions of each element or component must be exactly one. Suppose we need to know the mass of ethanol that contains 10 g of carbon, we solve the mass fraction for total mass:

$$\text{Total mass} = \frac{\text{mass stuff}}{\text{mass fraction}} = \frac{10\ \text{g}}{0.5214} = 19.18\ \text{g}$$

Volume fraction, often used to express ethanol content of beer, is given by

$$\text{Volume fraction} = \frac{\text{volume of stuff}}{\text{total volume}}$$

In contrast to mass, volume is not conserved; the sum of the volume fractions does not come to exactly one. What volume of ethanol is in one pint of 4.5% ABV beer? We solve the volume fraction equation for *volume of stuff*:

$$\text{Volume of stuff} = \text{volume fraction} \times \text{total volume} =$$
$$0.045 \times 1\ \text{pt} = 0.045\ \text{pt}$$

In fluid ounces, this comes to 16 oz/pint × 0.045 pint = 0.72 oz.

A **concentration** is a measure of composition involving amount of stuff per unit of volume. Trace materials in water and **flavor** compounds in beer are often reported in milligrams per **liter**, also called ppm w/v (**weight**/volume). The water treatment industry often uses grains per gallon. Many chemical calculations use **molar concentration**, also known as **molarity**, which is the number of moles per liter of **solution**. The defining equation is $c = n/V$, where c is the molar concentration, n is the number of moles, and V is the volume in liters. The units mol/L are often symbolized M, but we will not follow this practice. If 330 mL of beer contains 15 g of ethanol, what is the molar concentration of ethanol?

Moles of ethanol: 15 g/46 g/mol = 0.326 mol
Volume in liters: 330 mL/1000 mL/L = 0.330 L
Molar concentration = 0.326 mol/0.330 L = 0.988 mol/L = 0.988 M

Molecular Geometry and Shape

The geometry of a molecule is governed by the repulsive forces between valence electrons. In this context, we consider the valence electrons to be in groups that travel together. One **electron group** can comprise a single electron, an unshared pair, a shared pair (single bond), two pairs shared between

TABLE 2.5   **Molecular Geometry**

Number Electron Groups	Geometry	Bond Angle (°)	Figure
2	Linear	180	Figure 2.10A
3	Trigonal planar	120	Figure 2.10B
4	Tetrahedral	~109.5	Figure 2.10C

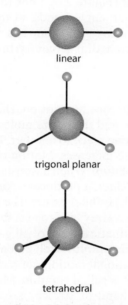

linear

trigonal planar

tetrahedral

Figure 2.10   Molecular geometry.

the same atoms (double bond), or three pairs shared between the same atoms (triple bond). The repulsion of like charges drives the electron groups of an atom away from one another, giving the molecule its **underlying geometry**. To determine the geometry about a central atom (one that is bound to two or more other atoms), we count the electron groups and apply the geometry given in Table 2.5. The geometries are shown in Figure 2.10.

The central atom in water (Figure 2.8A) is oxygen. The oxygen atom has two bonds and two unshared pairs for a total of four electron groups. The geometry is **tetrahedral**; to a first approximation the expected H—O—H bond angle is 109.5°. The actual bond angle is 105° because the unshared pairs repel the bonding pairs more than the bonding pairs repel one another. If we consider only the atoms, and not the unshared pairs, water has a bent **shape**. The bent shape gives water an oxygen end, which is negatively charged, and a hydrogen end, which is positively charged. This polarity of the water molecule accounts for many of water's important properties.

Carbon is the central atom in carbon dioxide (Figure 2.8B). The carbon atom has two double bonds, each of which is an electron group. Two electron groups give a **linear** geometry, so the O—C—O bond angle is 180°. Because there are no unshared pairs, the shape is the same as the geometry. Although the O—C bonds are polar, with the oxygen atoms more negative and the carbon atom more positive, the bond polarities are in opposite directions, so they cancel one another. Because of its symmetric geometry, carbon dioxide is a nonpolar molecule. Ethanol (Figure 2.8C) has three central atoms. The two carbon atoms each have four single bonds, so they are tetrahedral both in geometry and in shape. The oxygen atom has two single bonds and two unshared pairs, adopting tetrahedral geometry, but its shape is bent.

Molecular Motion

Every part of a molecule is in constant motion. The atoms in a molecule move with respect to one another, and the entire molecule rotates about axes and travels through space from one place to another. The intensity of the motions increases as the temperature increases. Motion of the entire molecule from place to place is called **translation**. When a translating molecule strikes something, like a wall of the container, it produces a force. The force on a unit area (square meter or square inch) is the pressure. The kinetic energy of a translating molecule is given by $KE = \frac{1}{2}mv^2$ where m is the mass and v is the speed of the molecule. The average kinetic energy of all gas molecules is directly proportional to the absolute temperature (Kelvin temperature = °C + 273.15). In the case of a gas, the translational motion defines the temperature. Spinning of a molecule about an axis is called **rotation**. Motion of the atoms within a molecule, in which bonds bend or stretch, is called **vibration**. Translation, rotation, and vibration of the molecules can cause molecules to collide, react, and assume new forms characteristic of new substances. Because these motions become more intense as the temperature increases, temperature is a critical issue in controlling the reactions behind the entire brewing process.

2.4   INTERMOLECULAR FORCES

The physical properties of beer-related substances play a large role in their performance. If something has a taste, like hop bitter compounds, it must be soluble in water. If it has an aroma, like malt, it must be able to get into the gas phase. These properties result from the forces that molecules and ions exert on one another, the **intermolecular forces**, also called van der Waals forces. In principle, these forces can be attractive or repulsive, but attraction is always dominant because the particles are pushed or pulled into positions of net attraction. Intermolecular forces are like the behavior of a pile of small

Figure 2.11   Ion–dipole force.

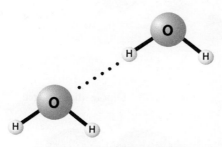

Figure 2.12   Hydrogen bond.

magnets. When you pick up one, it attracts others, even though individually the magnets can both attract and repel one another.

The strongest intermolecular force is that between ions. In a pure ionic substance, the force is strong enough to be a chemical bond; hence it does not count as an intermolecular force. In solutions of an ionic substance, the force is weakened because the ions are not as close together. The ion–dipole force is a net attractive force between ions and polar molecules. The molecules tend to pivot to bring the positive end close to negative ions and the negative end close to positive ions, as shown (with some exaggeration) in Figure 2.11.

The **ion–dipole force** is strong enough to allow many ionic compounds to dissolve in water. It is these dissolved ions that give brewing water its character. Perfectly pure water does not make good beer.

Hydrogen bonding, shown in Figure 2.12, is a potentially strong force of attraction involving —NH, —OH, or F—H groups. The small, strongly electronegative atom pulls the single electron from the hydrogen atom leaving the nucleus virtually bare. This resulting concentrated center of positive charge is attracted to any atom that has an unshared electron pair. Unlike other intermolecular forces, hydrogen bonds take specific directions in space, leading to highly organized structures. The forces behind **base pairing** in **DNA** and **RNA** are hydrogen bonds. If it were not for hydrogen bonding, water and ethanol would be gases near room temperature.

The **dipole–dipole force** (Figure 2.13) arises from the interaction of polar molecules with one another. The attraction of unlike charges pulls the positive

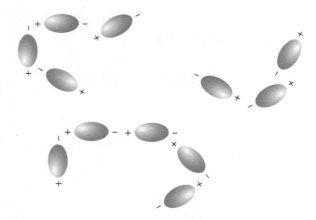

Figure 2.13 Dipole–dipole force.

Figure 2.14 Dispersion force.

ends of molecules to the negative ends of neighboring molecules giving a net force of attraction. This force, as well as hydrogen bonding, plays a significant role in solubility and **surface** interactions.

All molecules are affected by the **dispersion force**, also called the London force or the induced dipole force. Dispersion forces result from temporary polarity in ordinarily nonpolar regions of a molecule. The temporary polarity arises from fluctuations in the positions of electrons, moving the center of negative charge away from the center of positive charge. Alternatively, dispersion forces may result from temporary changes in the relative positions of the atoms caused by molecular vibration. This temporary polarity causes the electrons in nearby molecules to shift toward the positive end of the induced dipole, as shown in Figure 2.14, causing an attraction. The strength of the dispersion force depends on how easily the electrons in a molecule can be displaced by a nearby electric field and by the extent of molecular contact with one another. Electrons in high energy levels, which are distant from the nucleus, are most susceptible to displacement. Molecules containing atoms with high atomic numbers (above 20) have electrons that are easily displaced. Even without distant electrons, long molecules or those with flat rings can approach one another over a broad area, maximizing the potential for dispersion forces. Compounds with rings containing double bonds will stack like dishes to yield a dispersion force called the **stacking force**. Certain finings make use of the stacking force to bind and remove haze-forming compounds in beer.

When a substance dissolves, its molecules push aside the molecules of material in which it dissolves (called the **solvent**). If the solvent molecules are polar and held together by strong dipole–dipole interactions, pushing them aside requires a good deal of energy. If the interactions among the solvent molecules are replaced by new intermolecular forces of about equal strength, the energy cost is offset. In general, strongly polar and hydrogen-bonded substances dissolve in one another. In addition, weakly polar and nonpolar substances dissolve in one another. In contrast, strongly polar and weakly polar substances are typically insoluble in one another. This generality is captured by the phrase "like dissolves like." The same arguments apply to parts of large molecules. Beer and beer wort are mostly water, which is highly polar and hydrogen bonding. **Proteins**, which come mostly from the **malt**, are large molecules that include polar and nonpolar regions. In water, protein molecules coil and fold in ways that allow the polar regions to be in contact with the water and the nonpolar regions to stick to one another. During boiling, the proteins unfold, allowing the nonpolar regions on different molecules to find one another. The protein molecules stick together and drop out of solution as **hot break**. Polar or hydrogen bonding molecules or parts of molecules that have strong intermolecular forces with water are said to be **hydrophilic** (Greek: water loving). Molecules or regions that interact weakly with water are said to be **hydrophobic** (Greek: water fearing).

2.5   STRUCTURE OF MOLECULES

In many cases, the same set of atoms can be bound together in different ways to make different compounds. Compounds with the same atoms in their molecules but that are arranged differently are called **isomers**. The molecule C_4H_8O can form over 30 different isomers! The system of notation that distinguishes isomers is the **structural formula** or "structure." There are several types of structural formulas, each with advantages and disadvantages. Figure 2.15 shows different types of structural formulas for (E)-but-2-ene-1-ol, one of the isomers whose molecular formula is C_4H_8O.

Lewis Structure

A **Lewis structure** shows every atom and bond and often unshared pairs. Sometimes an effort is made to realistically show the geometry in two dimensions. Figure 2.15A shows a Lewis structure of (E)-but-2-ene-1-ol, also called trans-but-2-ene-1-ol. The advantage of the Lewis structure is that everything is shown without simplification or abbreviation. Some disadvantages are that they can be cluttered and hard to use and that Lewis structures must be drawn; they cannot be typed.

Condensed Structure

The condensed molecular structure fits on a line of type. The main chain of atoms, not including hydrogen, is shown with bonds between them. Atoms that are attached to the main chain atoms are shown after the main chain atoms. The condensed structure for (E)-but-2-ene-1-ol is $CH_3-CH=CH-CH_2-OH$. One major disadvantage is that the condensed structure does not always distinguish isomers. In this case, there are two isomers that have the same condensed structure. Condensed structures are difficult to use when the molecule has branches or rings.

Skeletal Structure

The skeletal structure greatly simplifies the formula by using the fact that carbon atoms always share four valence electrons to make four bonds. The structural formula is reduced to the bonds themselves. If the end of a bond is not labeled as another element, it is assumed to be a carbon atom. Hydrogen atoms that are bound to carbon are not shown. Any carbon atom with fewer than four bonds is understood to have hydrogen atoms to make up the missing bonds. The skeletal structure for (E)-but-2-ene-1-ol is shown in Figure 2.15B. The advantages of skeletal structures are that they can nearly always represent each isomer uniquely, especially when some conventions representing 3D structure are added; the omission of hydrogen and carbon makes them less cluttered; and the format emphasizes the **heteroatoms** (atoms other than C and H), which often account for the main chemical properties of the substance. The disadvantage is that preparing and interpreting skeletal structures requires practice because of the implied atoms and bonds.

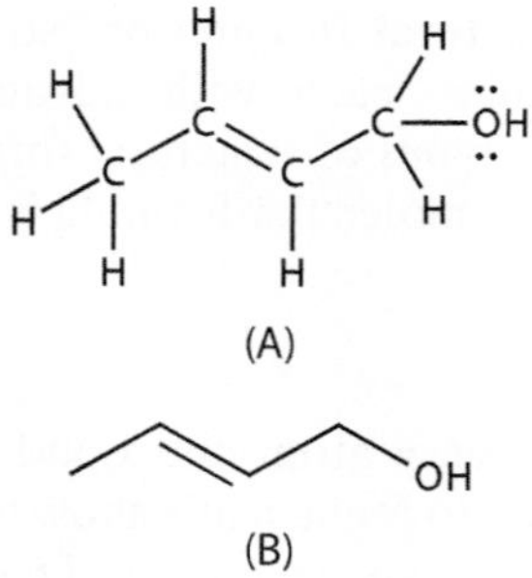

(A)

(B)

Figure 2.15 (E)-But-2-ene-1-ol. (A) Lewis structure. (B) Skeletal structure.

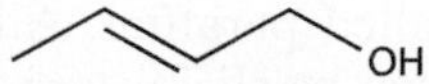

Figure 2.16 Isomers. (A) 2-Methylprop-1-ene-1-ol. (B) 3-Methyl oxetane. (C) (Z)-But-2-ene-1-ol. (D) (E)-But-2-ene-1-ol.

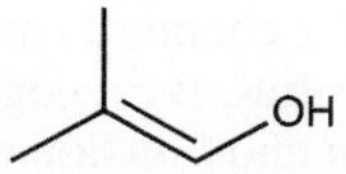

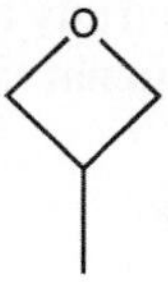

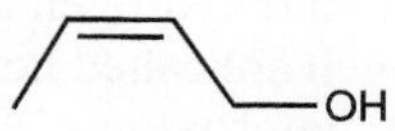

Structural and Stereoisomers

Structural isomers are compounds with the same molecular formula but whose atoms are connected in a different way. **Stereoisomers** are a specific type of isomer in which the atoms are connected in a similar way but their bonds are differently orientated in space, leading to a different 3D form. Figure 2.16A–C shows structural isomers, all with the formula C_4H_8O. The molecules in Figure 2.16A and C both have an OH group and a double bond, but they are connected in different ways. The molecule in Figure 2.16B is completely different; it has a ring, no double bond, and no OH group. The molecule in Figure 2.16C is connected in the same way as that in Figure 2.16D, but the bonds are differently directed in space. In Figure 2.16C the carbon atoms on the left and right side of the double bond are both directed down; in Figure 2.16D the carbon on the left is down and the one on the right is up. The compounds in Figure 2.16C and D are stereoisomers. Stereoisomers typically have similar chemical properties. Structural isomers can belong to totally different classes of compounds with widely different properties.

2.6 ORGANIC CHEMISTRY AND FUNCTIONAL GROUPS

Organic chemistry is the chemistry of compounds that have carbon–carbon bonds (single, double, or triple) or carbon–hydrogen bonds (always single). Carbon has the virtually unique ability to form chains, rings, and branches of any size and shape. There are millions of organic compounds, hundreds of which are important in beer quality in a positive or negative sense. To make the study of these compounds possible, we need to classify them by parts of their structure that give the molecule particular chemical characteristics. These structural fragments are called **functional groups**. By recognizing the functional groups, you will better understand the form and function of essential chemicals in the brewing process such as ethyl hexanoate (a fruity flavor from yeast), hydrochloric acid (a **cleaning** chemical), and **trans-2-nonenal** (a staling flavor from oxidation).

Hydrocarbons

Compounds with only carbon and hydrogen are called **hydrocarbons**. Atoms other than carbon and hydrogen are called heteroatoms, so hydrocarbons are organic compounds without heteroatoms. Hydrocarbons with only single bonds are **alkanes**, also called paraffins. Alkanes are unreactive, although they burn well. They are seen by chemists as the framework upon which the interesting parts, called functional groups, are attached. A hydrocarbon with a double bond is an **alkene**, sometimes called an olefin. The double bond serves as a center of reactivity; it is a functional group. Myrcene, shown in Figure 2.17, is an important flavor alkene from **hops**. This alkene makes up nearly 50% of the essential oil in Cascade hops.

The characteristic reaction of a double bond is **addition**, shown in Figure 2.18. The double bond between E and D is now a single bond, and groups X and Y become part of the original double-bonded molecule. If, for example, X–Y is H–OH (water), the product will have an OH group, making it an **alcohol**. Addition is not restricted to carbon–carbon double bonds; it can occur with any double bond, especially the C=O double bond. Hydrocarbons with a carbon–carbon triple bond are called alkynes. Alkynes are also susceptible to the addition reactions, but they are not very important in beer brewing.

Aromatic Compounds

Aromatic compounds have rings with alternating single and double bonds. Systems of single and double bonds are said to be conjugated. The rings are flat, and the electrons in the double bonds are delocalized, that is, they are spread out over the entire ring. Delocalization makes these bonds much less

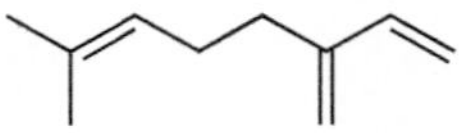

Figure 2.17   Myrcene.

Figure 2.18   Addition to a double bond.

Figure 2.19   Benzene.

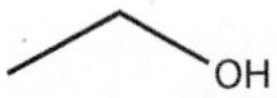

Figure 2.20   Ethanol.

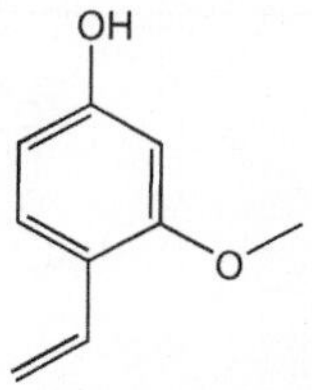

Figure 2.21   4-Vinylguaiacol.

susceptible to addition reactions than isolated double bonds. Figure 2.19 shows benzene, the simplest aromatic compound. Many compounds important in all living systems, like the DNA **nucleobases**, are aromatic compounds.

Alcohols

Alcohols are organic compounds with an —OH group bound to a carbon atom that is not bound to any other heteroatom. The OH group can form hydrogen bonds. The two-carbon alcohol, ethanol, shown in Figure 2.20, is the **psychoactive** component in beer and other **fermented beverages**. Alcohol with one carbon is called methanol, which is naturally produced in small quantities by yeast. Methanol is extremely harmful if concentrated through distillation, which is one reason why home distilling is more tightly regulated than home brewing. Alcohols with three or more carbon atoms are called **fusel alcohols**. They can be precursors to important flavor compounds in beer. Alcohols in which the —OH group is bound to an aromatic ring with alternating double and single bonds are called **phenols**. Figure 2.21 shows the structure of 4-vinylguaiacol, a phenol that gives a characteristic flavor to weizenbier, a German-style wheat ale.

Carbonyl Compounds

The C=O double bond is called a **carbonyl** [CAR-ben-ill] group. Molecules that contain a carbonyl group are called carbonyl compounds. These compounds fall into several families, and many of them are of great importance in brewing. Figure 2.22 shows the general formula for a carbonyl compound.

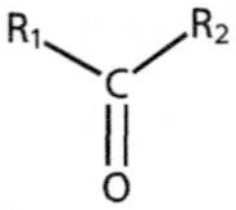

Figure 2.22   Carbonyl compound.

TABLE 2.6   **Carbonyl Compounds**

R_1	R_2	Functional Group	Example	Name (Official Name)
$-H$	$-H$	Aldehyde	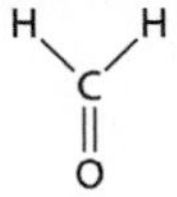	Formaldehyde (methanal)
$-C_nH_m$	$-H$	Aldehyde	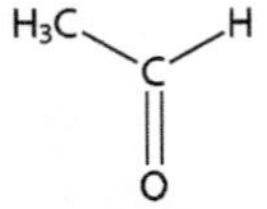	Acetaldehyde (ethanal)
$-C_nH_m$	$-C_nH_m$	Ketone	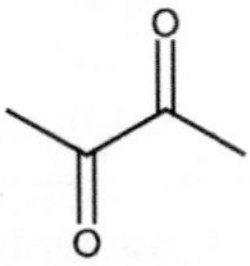	Diacetyl (butanedione)
$-C_nH_m$	$-OH$	Carboxylic acid	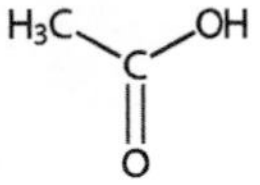	Acetic acid (ethanoic acid)
$-C_nH_m$	$-OC_nH_m$	Ester	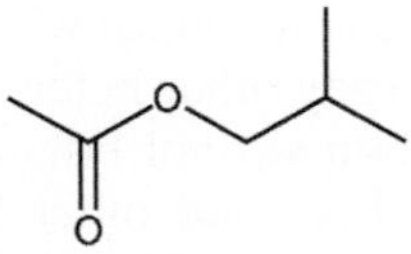	Isobutyl acetate (2-methylpropyl ethanoate)
$-C_nH_m$	$-NH_2$	Amide	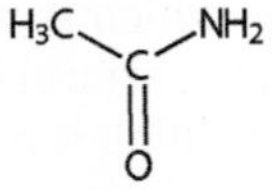	Acetamide (ethanamide)

The specific functional group is determined by the nature of R_1 and R_2. Some important cases with examples are given in Table 2.6.

2.7   CHEMICAL REACTIONS

Chemical Equations

A chemical reaction is a process in which one or more substances, the reactants, are converted to one or more different substances, the products. The reaction occurs by exchange and rearrangement of the atoms; no atoms are created or destroyed. Chemical reactions are represented by **chemical equations** in which the formulas of the reactants and products are shown with an arrow pointing to the products, as shown in the equation for alcoholic fermentation of glucose ($C_6H_{12}O_6$) to give ethanol (C_2H_6O) and carbon dioxide: $C_6H_{12}O_6 \rightarrow 2C_2H_6O + 2CO_2$. This equation shows that one molecule of glucose reacts to give two molecules of ethanol and two of carbon dioxide. It can also be interpreted as one mole (180 g) of glucose reacts to give two moles of each product. The numbers before the formulas are called **coefficients**. Equations are normally **balanced**, where the same number of each atom is on both sides of the equation. Suppose 1000 g of glucose is fermented, which is 1000 g/180.2 g/mol = 5.55 mol. Based on the fermentation equation, 11.1 mol each of ethanol and carbon dioxide are produced. That comes to 11.1 mol × 46.07 g/mol = 511 g of ethanol and 11.1 mol × 44.01 g/mol = 489 g of CO_2. Any calculation with a chemical equation must be done in moles.

Equilibrium

Suppose we have a sealed jar partly filled with ethanol and the rest with vacuum. Ethanol is **volatile**, so the liquid begins to evaporate. This gives an increasing concentration of ethanol molecules in the gas phase. Some of these strike the surface of the liquid and **condense**. The tendency for ethanol to evaporate depends on the amount of energy needed for evaporation and on the temperature. The tendency for ethanol to condense depends on the concentration in the gas phase; more molecules flying around means more will condense every second. Eventually, the concentration of ethanol in the gas phase will be high enough so that number of molecules that condense is equal to the number that evaporate. At this point the **vapor** concentration of ethanol, and hence the pressure, is constant. The system is at **equilibrium**. The pressure or **partial pressure** of a liquid at equilibrium with its vapor is called the **vapor pressure** of the liquid. Even though the ethanol is actively evaporating and condensing, the processes balance one another, so there is no net change.

The principal of equilibrium is important in many chemical reactions. High concentrations of reactants give the reaction a tendency to proceed forward, toward products. High concentrations of products give the reaction a tendency to proceed backward, to reactants. Somewhere in between is a position of equilibrium. The forward and reverse rates of the reaction are equal. The position of a reacting system with respect to equilibrium is evaluated with an expression called the **reaction quotient**. For the generalized reaction $aA + bB \rightarrow cC + dD$, the reaction quotient, Q, is given by the equation below:

$$Q = \frac{[C]^c [D]^d}{[A]^a [B]^b}$$

where $[X]$ represents the concentration of reactant or product X and $a, b, c,$ and d are the coefficients of the reactants and products in the balanced equation. Only concentrations of substances in solution or in the gas phase enter the expression. Pure (or nearly pure) solids or liquids, including the solvent, do not enter the expression. As the reaction proceeds from reactants to products, Q increases. The equilibrium value of the reaction quotient is called the **equilibrium constant**, designated K, often with a subscript indicating the type of reaction. For example, acid dissociation equilibrium constants are designated K_a. For technical reasons, neither Q nor K are considered to have units. When Q is lower than K, the reaction is spontaneous in the forward direction. When Q is higher than K, the reaction is spontaneous in the reverse direction.

Example: For the reaction $2H_2O \rightarrow H_3O^+ + OH^-$, a water sample at equilibrium is found to be $55\,mol/L$ in H_2O, $1.0 \times 10^{-8}\,mol/L$ in H_3O^+, and $1.0 \times 10^{-6}\,mol/L$ in OH^-. Calculate K.

Answer: $K = [H_3O^+][OH^-] = [1.0 \times 10^{-8}][1.0 \times 10^{-6}] = 1.0 \times 10^{-14}$ (no units). Water, being the virtually pure solvent, does not come into the equilibrium expression. This equilibrium constant is designated K_w.

Le Châtelier's principle states that if a system at equilibrium is subjected to a change, the system will respond in a way that, to some extent, undoes the change. Consider the reaction CO_2 (dissolved) $\rightarrow CO_2$ (gas). If we increase the pressure, the system will shift to the low-volume side, so more CO_2 will dissolve. Heat is absorbed when CO_2 comes out of solution, so increasing the temperature will cause CO_2 to come out of solution, and removing some heat will cause more CO_2 to dissolve. This is exactly how carbonation behaves, which is important in setting the pressure and temperature in **bright beer tanks** and in beer dispensing lines.

Reaction Rate

The speed at which a reaction takes place is called the *rate*. Brewers often measure the rate of fermentation by the change in sugar concentration, which could be reported in **degrees Plato** per hour. Reaction rates usually depend strongly on temperature; in many cases a few degrees of temperature change can cause the reaction rate to increase by a significant factor. Reaction rate is a different issue from reaction equilibrium. A stainless steel vessel has a spontaneous tendency to react with oxygen to yield metal oxides. If you wait long enough, all metal **brewery** fittings will corrode. Fortunately, the rate of **corrosion** of stainless steel is low enough to give the equipment a useful lifetime. Nonetheless, if the brewery allows conditions to develop that enhance the corrosion rate, expensive damage will result.

In addition to temperature, the concentrations of reactants, physical contact among the reactants, pressure, mixing, and **catalysts** will all influence the reaction rate. A catalyst is a material that causes a reaction rate to increase,

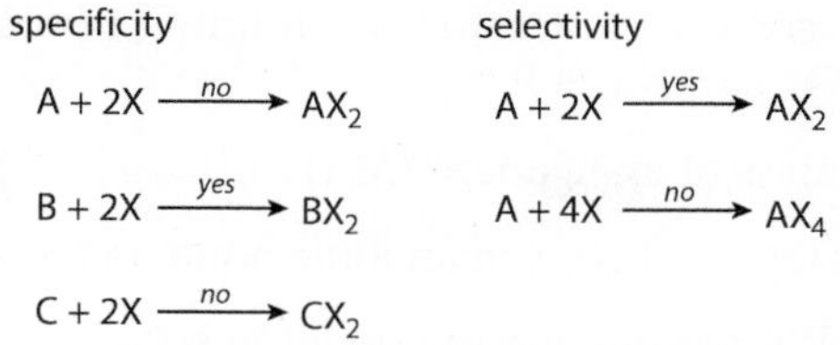

specificity

$$A + 2X \xrightarrow{no} AX_2$$

$$B + 2X \xrightarrow{yes} BX_2$$

$$C + 2X \xrightarrow{no} CX_2$$

selectivity

$$A + 2X \xrightarrow{yes} AX_2$$

$$A + 4X \xrightarrow{no} AX_4$$

Figure 2.23   Specificity and selectivity.

although the catalyst is not a reactant or product. Catalysts work by providing lower-energy pathways for the reaction or by helping to line up the reactants into a favorable arrangement allowing them to react. Two key characteristics that many catalysts possess to some degree are **specificity** and **selectivity**. Specificity is the ability of a catalyst to enhance the rate of reaction of one or a limited set of potential reactants to a much greater extent than other potential reactants. Selectivity is the ability of a catalyst to enhance the rate of one reaction over several to which the reactants could be susceptible. Figure 2.23 shows this distinction graphically. Under specificity, A, B, and C are reactants that can react with X. The catalyst only enhances the B reaction. Under selectivity, reactant A could react with X to give either AX_2 or AX_4. The catalyst enhances only the reaction to AX_2. The major brewing processes of mashing and fermentation depend on protein-based catalysts called **enzymes**. A key feature of many enzymes is outstanding specificity and selectivity, which allow reactions in living cells to be controlled. All catalysts, including enzymes, speed up the rates of reactions, but they have no effect on the position of equilibrium. The catalyst will help Q (reaction quotient) reach K (equilibrium constant) faster, but K is not changed. There is no such thing as a one-way catalyst; if a catalyst can speed up the forward reaction, it can also speed up the reverse reaction.

CHECK FOR UNDERSTANDING

1. Identify the element in group 4A in the fifth period. Give its atomic number, molar mass, and electronegativity.

2. How many protons, valence electrons, and core electrons are present in each of the following elements or ions: potassium, sulfur, Na^+, O^{2-}, I^-, and Sr^{2+}?

3. Rank the following metals in order of increasing reactivity: barium, beryllium, calcium, magnesium, and strontium.

4. Write formulas for calcium chloride, sodium carbonate, ammonium sulfate, barium hydroxide, and aluminum oxide. What are their formula weights (g/mol)?

5. What are the correct names for the following compounds: $NaHSO_4$, N_2O_4, KO_2, ClO_2, and Na_3PO_4?

6. Calculate the mass of manganese (Mn) in 10.0 g of $KMnO_4$.

7. What is the difference between an ionic bond and a covalent bond?

8. Draw the Lewis structure and the skeleton structure for dimethyl ether, CH_3-O-CH_3.

9. Draw the following compounds as Lewis structures:

10. Draw the following compounds as skeleton structures:

11. Identify (name) the principal functional group for the following molecules:

12. Ammonia (NH_3) is made from the elements H_2 and N_2. Balance the reaction by inserting the coefficients __N_2 + __ H_2 → __NH_3.

13. For the reaction $3BaCl_2 + 2H_3PO_4 \rightarrow Ba_3(PO_4)_2 + 6HCl$, calculate the mass of H_3PO_4 required to react with 100.0 g of $BaCl_2$.

14. Name and explain the three types of molecular motion.

15. What is the function of a catalyst? What is the difference between specificity and selectivity?

BIBLIOGRAPHY

Barth R. 2013. *The Chemistry of Beer: The Science in the Suds*. Wiley. ISBN 978-1-118-67497-0.

Snyder CH. 2003. *The Extraordinary Chemistry of Ordinary Things, 4th ed*. Wiley. ISBN 0-471-41575-8.

CHAPTER 3

BIOLOGY FOR BREWING

Biology is the science of life; its role in **brewing** is unmistakable. Brewers rely upon the biology of **barley** to produce seeds with traits that are optimized for brewer's malt. They rely upon **hop** plants to produce lupulin-laden hops for **flavor** and bitterness. They rely upon equipment **cleanliness** and control of contaminating **bacteria**. And most importantly, without **yeast**, there is no **beer**. They rely upon yeast and the biological process of **fermentation** to produce **ethanol**, carbon dioxide, and a variety of **flavors** in beer.

3.1 MACROMOLECULES

Much of biology is rooted in chemistry. **Organic** compounds, along with some inorganic compounds, are produced in a variety of structural configurations depending on the **atoms** and arrangements of their **bonds**. The resulting three-dimensional structures lead to functionality that influences downstream chemical change through chemical **reactions**. Many organic compounds found in living systems have large molecules, termed **macromolecules**, that consist of thousands of atoms. Most macromolecular substances are **polymers**. Polymers are composed of long chains of molecularly linked repeating

Mastering Brewing Science: Quality and Production, First Edition.
Matthew Farber and Roger Barth.
© 2019 John Wiley & Sons, Inc. Published 2019 by John Wiley & Sons, Inc.

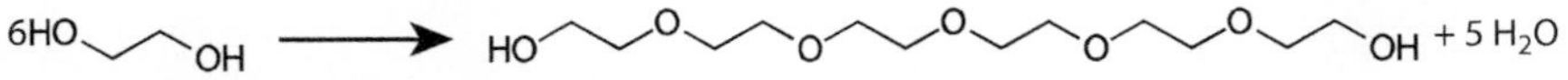

Figure 3.1   Synthesis of polyethylene glycol (antifoam) from ethylene glycol.

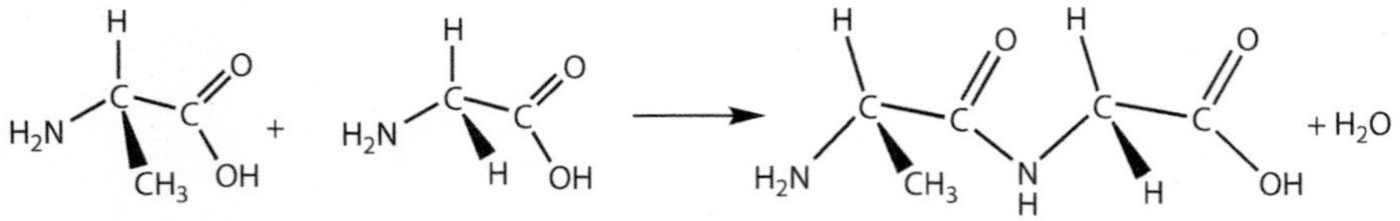

Alanine   Glycine   Alanylglycine

Figure 3.2   Condensation of two amino acids.

subunits termed **monomers**. This concept is illustrated in Figure 3.1 with six ethylene **glycol** monomers. Polymers synthesized by cells are called **biopolymers**. Most biopolymers can be broken down into monomers by the **hydrolysis** reaction. This reaction, like most biological reactions, is enhanced and controlled by large **protein** molecules called **enzymes**, which serve as **catalysts**. A specific enzyme recognizes the polymer and uses water to break the bonds. The hydrogen from water is attached to one monomer, and the remaining **hydroxyl group** is attached to the neighboring monomer. This hydrolysis reaction breaks the **covalent** bond between the subunits. Details of the enzymatic hydrolysis of **starch** are discussed in Chapter 6. The breakdown of complex molecules in biology is called **catabolism**.

The covalent bond between subunits of a polymer is formed by a **condensation** reaction. A specific enzyme binds to a growing polymer, recruits a new monomer, and joins them by removing a water molecule. Unlike hydrolysis, condensation reactions usually require **energy**. The synthesis of complex molecules in biology is called **anabolism**. The condensation of the **amino acids** alanine and glycine to give a dipeptide is shown in Figure 3.2.

In the following sections, we will discuss the four major macromolecules in biology: **carbohydrates**, proteins, **lipids**, and nucleic acids (**DNA** and **RNA**). All are formed by anabolic reactions and broken down by catabolic reactions.

Carbohydrates

Carbohydrates are a family of biological compounds containing carbon, hydrogen, and oxygen. They have multiple $-OH$ groups and either **carbonyl** or related functional groups. Many carbohydrates have names with the suffix "-ose." Simple **sugars** have the **formula** $(CH_2O)_n$, where n is 3 or more. The most important sugars for brewing are hexoses: $n = 6$. Of these, D-glucose, shown in Figure 3.3A and B, is the most important. Polymers of sugars include

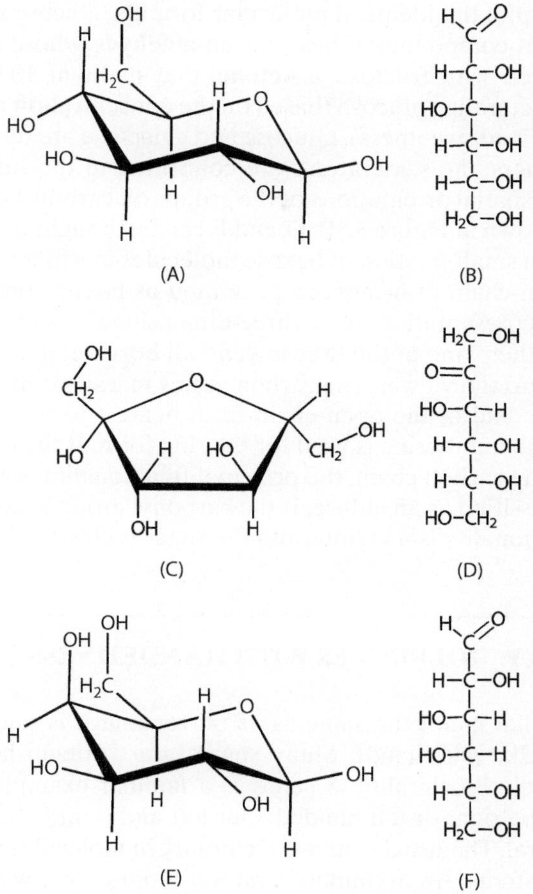

Figure 3.3 Hexose structures. (A) Alpha-glucose, ring form. (B) Glucose, open chain. (C) Alpha-fructose, ring form. (D) Fructose, open chain. (E) Alpha-galactose, ring form. (F) Galactose, open chain.

starches that provide energy reserves for cells and other carbohydrates like **cellulose** and **beta-glucan** (β-glucan) that provide structural support in the **cell walls** of plants like barley. Ultimately the catabolism of starch into simple sugars will be the main process goal during the **mashing** step of beer production.

Figure 3.3 shows the most common six-carbon sugars in both **linear** and ring structures. All three sugars have the same formula, $C_6H_{12}O_6$, but differ in their atom arrangement and structure. Glucose and fructose are structural **isomers**; their atoms are connected in different ways to give different structures. Despite the identical molecular formula, glucose and fructose are clearly different compounds. Glucose is an **aldehyde** whose melting point is 141 °C compared with fructose, a **ketone**, that melts at 103 °C. Fructose is noticeably sweeter than glucose (based on the concentration needed to evoke the same perceived sweetness). Glucose and galactose are more similar. They are **diastereomers**; the same atoms are connected in the same way in both sugars, but the spatial orientations of the groups on carbon 4 differ. The open-chain forms, shown in Figure 3.3B, D, and F, can be thought of as parent forms, although only a small fraction of hexose molecules in water **solution** take this form. The open-chain drawings are presented as Fischer projections, a two-dimensional representation of a three-dimensional molecule. All vertical bonds go into the plane of the drawing and all horizontal bonds come out of the plane toward the viewer. The carbon atoms in a sugar molecule are numbered from the end of the open-chain form nearest to the **carbonyl** (C=O) group. The same numbering is used for the ring form. If the carbonyl group is at the end of the carbon chain, the principal functionality is an **aldehyde**, and the sugar is classified as an **aldose**. If the carbonyl group is not at the end, the principal functionality is a **ketone**, and the sugar is classified as a **ketose**.

CHIRALITY: MOLECULES WITH HANDEDNESS

An object that is not the same as its mirror image is said to be **chiral** [KYE r'l: Gk: χείρ, hand]. Many sugars are distinguished from one another solely by chirality. A screw is a familiar example of chirality; screws can be right or left handed. Our left and right hands themselves are also chiral. The usual sources of chirality in molecules are **asymmetric carbon** atoms. An asymmetric carbon atom is one with four single bonds, each to a different group. Figure 3.4 shows an asymmetric carbon atom and its mirror image. Groups B and D occupy corresponding

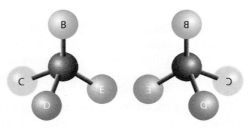

Figure 3.4   Asymmetric carbon.

positions, but C and E are reversed. The molecules cannot be overlaid on one another; they are different compounds. Chirality is not just a matter of the first atom bonded to the asymmetric carbon, but rather the differences in the positioning of the *different* groups bonded to the asymmetric carbon. So if A is —CH₃ and B is —C₂H₅, they are different groups.

Isomers that are mirror images are called **enantiomers**. Enantiomers have identical physical and chemical properties except for their interactions with other chiral phenomena, such as other chiral molecules or polarized light. Usually the enantiomers are given the same name with prefixes denoting the specific enantiomer as "L-" or "D-." In biological systems, most carbohydrates are "D" form, while most amino acids are "L."

Sugar molecules often have several asymmetric carbon atoms. For example, the ring forms of the six-carbon aldose sugars have five asymmetric carbons. In a simpler example for this discussion, the four-carbon aldoses have two asymmetric carbons, C-2 and C-3. We refer to the configurations of the asymmetric carbon atoms as R and S (for an explanation of absolute configuration, see Barth 2013 p 117). There are four possibilities for the two asymmetric carbons: RR, RS, SR, and SS. In a mirror image, each R turns to an S and vice versa. So RR and SS are enantiomers, and RS and SR are enantiomers. RR and RS are not mirror images of one another; they are another sort of isomer called **diastereomers**. They have different chemical and physical properties, including different melting points. The two pairs of enantiomers are called erythrose and threose. They are shown in Figure 3.5.

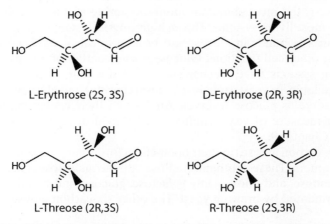

Figure 3.5 Diastereomers: threose and erythrose. Solid wedges are bonds coming toward the viewer; dashed wedges go away from the viewer.

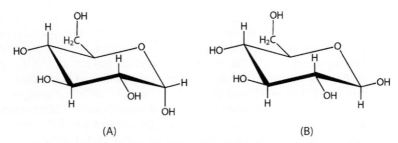

(A)   (B)

Figure 3.6   Glucose anomers. (A) Alpha-D-glucose and (B) beta-D-glucose.

Open-chain sugars with five or more carbon atoms spontaneously react to form a ring. In the case of six-carbon aldoses, like glucose, the most common ring form results from carbon number five (counting from the C=O carbon) connecting to the carbonyl oxygen. This chain-ring interconversion is reversible. When the ring is formed, the C=O carbon becomes asymmetric, so instead of 8 isomers, there are 16. The carbon atom that originated as the C=O is called the **anomeric carbon**; it is **bound** to two oxygen atoms. The configuration about the anomeric carbon changes when the ring cycles between ring- and open-chain forms. Figure 3.6 shows the glucose anomers, forms that vary by the configuration about the anomeric carbon. In the alpha (α) anomer, the —OH group on the anomeric carbon extends in the **axial** direction, that is, in the direction of the "axle" of the ring. In the beta (β) anomer, the —OH group extends in the **equatorial** direction, in the direction of the "spokes" of the ring. About two-thirds of dissolved glucose is in the beta form. The remaining one-third is mostly in alpha form with a small amount in the open-chain form.

Glucose is the most abundant **monosaccharide** and is used as an energy source in most living systems. Through **glycolysis**, fermentation, and **aerobic respiration**, the bonds in glucose are broken down or **oxidized** to provide energy for other cell functions. Glucose is so essential for energy production that living systems have mechanisms to ensure that glucose reserves are always available. In yeast, **trehalose** (a **disaccharide**) and **glycogen** (a **polysaccharide**) serve as glucose reserves. Analysis of their **concentrations** can be used as a measure of yeast **vitality**, as discussed in Section "Viability and Vitality," Chapter 9.

The six-carbon aldoses in their open-chain forms have 16 isomers in 8 pairs of enantiomers. These are named allose, altrose, galactose, glucose, gulose, idose, **mannose**, and talose. Only galactose, glucose, and mannose are commonly fermented by brewing yeast. The other six-carbon aldoses are rare in nature.

More complex carbohydrates are condensation products of simple sugar molecules. These involve condensation of the anomeric —OH group on one sugar molecule with an —OH group on another sugar molecule, releasing

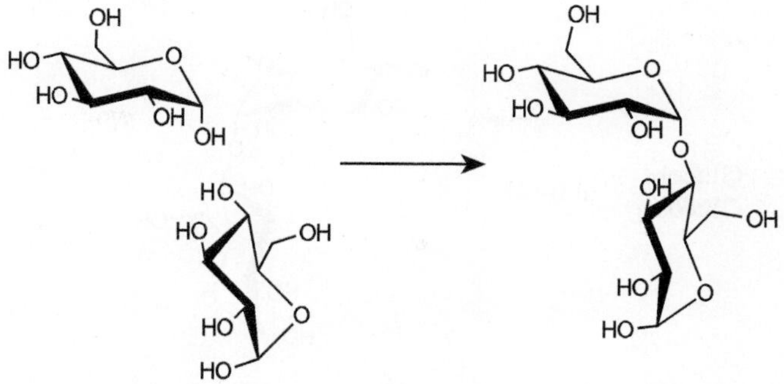

Figure 3.7   Alpha-glucose condensation to maltose.

water. The connection is called a **glycosidic link**. Figure 3.7 shows two glucose molecules condensing to yield a disaccharide called **maltose**, the most abundant sugar in barley **wort**. In maltose, the alpha-configured −OH group on carbon 1 of one glucose molecule has condensed with the −OH group on carbon 4 of another; the link is named alpha(1→4). Because there is no −OH group on the anomeric carbon of the upper glucose **unit**, the upper ring can no longer open, and the unit is locked in the alpha form, leaving the maltose molecule bent. When there is a glycosidic link between the anomeric carbons of two sugar units, the alpha/beta configurations of both are locked. Thus in nomenclature of disaccharides, the configuration of both must be specified, as in sucrose: glucose-alpha(1→2)-alpha-fructose. Therefore, specific disaccharides will be defined by two parameters, the location of the glycosidic link and the identity of the sugars. Table 3.1 shows common disaccharides. The differences among them seem subtle, but they behave very differently. For example, maltose is easily assimilated by yeast, in contrast to lactose, which is not usually assimilated at all.

Any number of sugar molecules can be joined by glycosidic links to give polymers that are important in brewing, such as starches, **gums** like β-glucan, and structural materials like **cellulose**. A polysaccharide consists of repeating units of one or more sugar monomers, typically glucose. Although the precise number of glucose units of individual polysaccharides can vary, they often number in the thousands! Polysaccharides may be branched or unbranched. This structural arrangement, coupled with the glucose isomer involved and the location of the glycosidic bond, defines the polysaccharide and its functionality.

The most important polysaccharide in brewing is starch (α-glucan). Starch is the most abundant carbohydrate used for energy storage in plants such as barley, making up 63% of all barley seed dry matter with granules containing up to 98% starch. Starch is a polymer of glucose subunits connected by

TABLE 3.1 Most Common Disaccharides

Name	Sugars	Bond	Structure
Maltose	Glucose Glucose	$\alpha(1\rightarrow4)$	
Trehalose	Glucose Glucose	$\alpha(1\rightarrow1)\text{-}\alpha$	
Cellobiose	Glucose Glucose	$\beta(1\rightarrow4)$	
Lactose	Galactose Glucose	$\beta(1\rightarrow4)$	
Sucrose	Glucose Fructose	$\alpha(1\rightarrow2)\text{-}\alpha$	

alpha(1→4) bonds. There are two major types of starch in barley, **amylose** and **amylopectin**. Of all starch in barley seeds, amylose makes up only about 20%. It consists of long, unbranched chains of glucose subunits, all in the alpha(1→4) configuration. Because of the angle of the glycosidic bond, amylose forms a helical (spiral) structure (Figure 3.8A). Amylopectin makes up about 80% of the total starch in barley. Unlike amylose, amylopectin contains branched molecules composed of alpha(1→4) glucose polymers with occasional alpha(1→6) branch points (Figure 3.8B). Glycogen, a starch storage molecule in yeast and animals, is similar in structure to amylopectin but is more highly branched.

Barley seed contains another carbohydrate called cellulose. Cellulose is an insoluble polymer of beta(1→4) bonded glucose that serves as a structural carbohydrate. It is an important part of plant cell walls and is found in the biofilms produced by some bacteria. Cellulose makes up about 5–6% of the dry **weight** of a barley seed and is located mostly in the husk. Unlike starch, the orientation of the beta(1→4) bonds yields a straight-chain polymer. As a result, the hydroxyl groups on one chain form **hydrogen bonds** with oxygen atoms on parallel chains, tightly holding adjacent polymers together (Figure 3.9). This intermolecular bonding, coupled with the rodlike structure of the polymer, contribute to the mechanical strength of cellulose. Because the bonds of cellulose are structurally different than starch, the enzymes used to break down starch into simple sugars for fermentation cannot act on cellulose. After mashing, cellulose remains behind in the spent **grain**. Some biotechnology companies are looking at ways to reuse spent grains by adding an enzyme to break down the cellulose into simple sugars for subsequent fermentation.

Hemicellulose resembles cellulose in having a backbone of sugar molecules connected mainly by beta(1→4) linkages, but it contains sugars other than glucose, especially mannose, galactose, and five-carbon sugars including L-**arabinose**

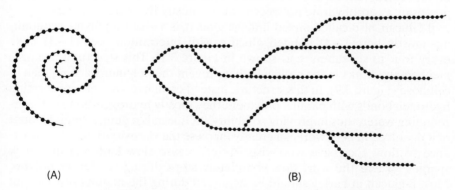

(A)   (B)

Figure 3.8   Starch structure schematic. (A) Amylose and (B) amylopectin.

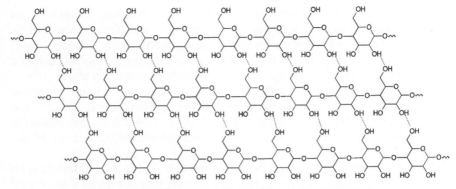

Figure 3.9   Cellulose stacking: hydrogen bonds dotted.

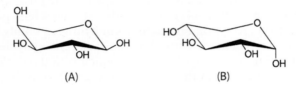

(A)   (B)

Figure 3.10   Pentoses. (A) α-L-Arabinose and (B) α-D-xylose.

and D-**xylose** (Figure 3.10) in addition to some modified sugars. Whereas the sugars in cellulose are connected entirely by beta(1→4) linkages, those in hemicellulose have frequent instances of other connections, like beta(1→3) links, which introduce kinks in the chain. Because of the variety of sugars and linkages in hemicellulose, it is much less **crystalline** than cellulose, and it is readily degraded into soluble fragments. Hemicellulose is a major component of the barley endosperm cell walls, with 80–90% β-glucan and 10–20% pentosan, a carbohydrate polymer made of mostly five-carbon sugars.

β-Glucan, also called mixed-linkage glucan, is a glucose polymer containing mostly beta(1→4) bonded glucose with intermittent beta(1→3) links every four to six subunits, as shown in Figure 3.11. This structural arrangement creates kinks in the molecule that prevent the polymer stacking seen in cellulose (Figure 3.9). In this structure, instead of all hydroxyl groups forming hydrogen bonds with parallel polymers, they mostly hydrogen-bond with surrounding water, thus improving solubility. β-Glucan is a gum; a small amount of it dissolved in beer wort can greatly increase the **viscosity**, defined as resistance to flow. Excessive wort viscosity can cause slow **lautering** or "stuck sparges" where the wort flow completely stops during the **lauter process**. Most β-glucan in barley should be degraded during the malting process, but insufficient malt **modification** or the use of **adjuncts** high in β-glucan such as unmalted barley, wheat, oats, or rye can increase wort viscosity. Raising the

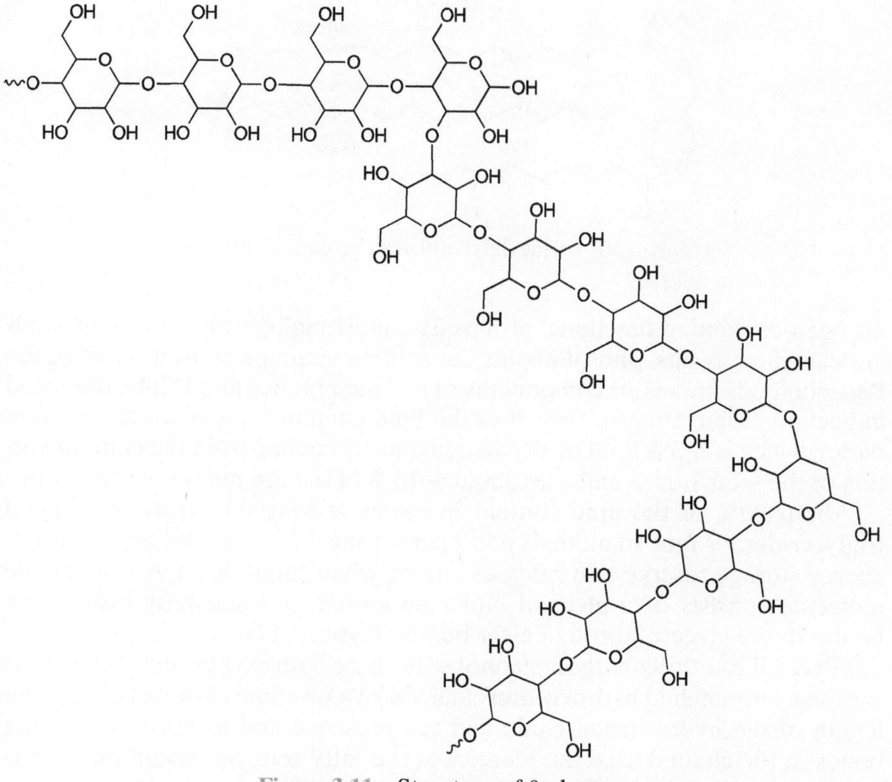

Figure 3.11   Structure of β-glucan.

mash temperature to a "mash-out" temperature of ~76 °C (168 °F) helps to decrease viscosity caused by gums and thus shortens total lauter time.

The pentosans in barley are polymers of pentose saccharides, typically xylose and arabinose. Pentosans are key structural components of the starchy endosperm cell walls; they must be degraded during malting to allow sufficient access to the starch granules. Insufficient malt modification and enzymatic degradation of pentosans can cause low mash extract and filterability issues during the lauter. The degradation of pentosans in a ferulic **acid rest** during mashing promotes the release of ferulic acid. The ferulic acid rest will be further discussed in Chapter 6.

Lipids

Lipids are **hydrophobic** biological compounds defined by their solubility in nonpolar solvents and relative insolubility in water, rather than by any consistent molecular features. Their hydrophobicity is brought about by an abundance of nonpolar groups with relatively few oxygen- or

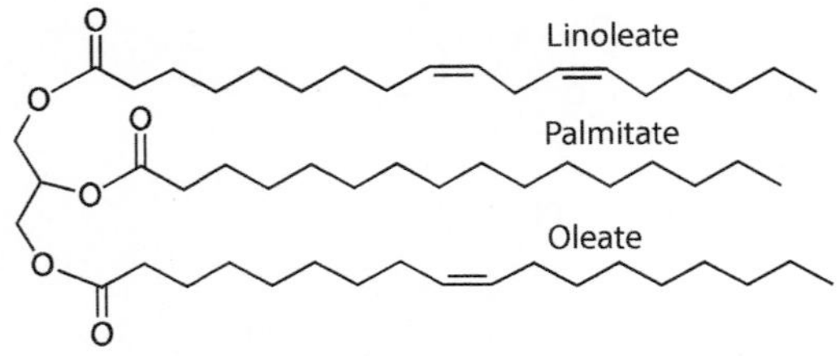

Figure 3.12   Triacylglycerol, acyl groups identified.

nitrogen-containing functional groups. The most biologically important lipids are classified as **fats, phospholipids,** fat-soluble vitamins, **steroids,** and waxes. Phospholipids are major components of cell membranes and will be discussed in Section "Lipid Bilayer." Much of the lipid content of beer is derived from barley, which is 2–3% lipid by dry weight, mostly coming from the **embryo** portion of the seed. Lipids make up about 6–12% of the dry matter in yeast cells.

About 90% of the lipid content in barley is **triacylglycerols,** also called **triglycerides,** or **fats.** In animals and plants, triacylglycerols are an important energy storage reserve that releases energy when catabolized. A triglyceride molecule consists of a **glycerol** molecule joined by three **fatty acids,** each bound to the glycerol through **ester** bonds (Figure 3.12).

Glycerol is a three-carbon alcohol with three hydroxyl groups. Fatty acids are long, unbranched **hydrocarbon** chains with a terminal carboxyl group. The length of the hydrocarbon chain and the presence and locations of double bonds in the chain define the identity of the fatty acid. **Saturated** fatty acids have no double bonds; they are "saturated" with the most hydrogen atoms possible. **Unsaturated** fatty acids include one or more carbon–carbon double bonds; each double bond lowers the number of hydrogen atoms by two.

Fatty acids are sometimes described with a shorthand notation called lipid numbers. The lipid number takes the form $C:D$, where C is the number of carbon atoms in the molecule and D is the number of double bonds. The presence of double bonds affects the chemical and physical properties of fatty acids and their derived compounds. Unsaturated fatty acids, which are those with one or more double bonds, are much more reactive, especially with oxygen, than saturated fatty acids; their physical properties are also greatly affected. Saturated fatty acid molecules are mostly straight, as shown for stearic acid (lipid number 18 : 0) in Figure 3.13A. They fit together well to give well-ordered structures. Double bonds introduce kinks in the molecules of unsaturated fatty acids, as shown for linoleic acid (18 : 2) in Figure 3.13B. They form disordered structures that remain fluid even at fairly low temperatures. The melting point of stearic acid is 69.3 °C, compared with −5 °C for linoleic acid. There are about 30 biologically relevant fatty acids; those most abundant in barley are described in Table 3.2.

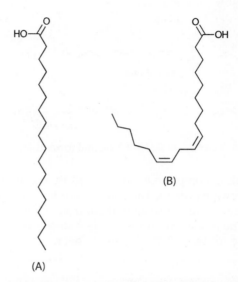

(A)

(B)

Figure 3.13   Fatty acids. (A) Stearic acid, saturated, and (B) linoleic acid, unsaturated.

TABLE 3.2   **Fatty Acids in Barley**

Name	Lipid number	Relative amount (%)
Linoleic acid	18 : 2	58
Palmitic acid	16 : 0	20
Oleic acid	18 : 1	13
Linolenic acid	18 : 3	8
Stearic acid	18 : 0	1

LIPID DEGRADATION AND STALING FLAVORS

The best-known, and perhaps the most important, beer staling compound is (E)-2-nonenal (also called **trans-2-nonenal**), an unsaturated nine-carbon aldehyde that contributes a stale cardboard flavor to beer at concentrations as low as 50 ng/L (parts per trillion). Figure 3.14 shows the overall reaction for the formation of (E)-2-nonenal from linoleic acid, the most abundant fatty acid in barley. This reaction occurs in multiple steps, some of which involve reactive oxygen species, like the hydroperoxyl radical, HO_2.

It is becoming clear that (E)-2-nonenal does not result from reactions involving oxygen in the product packaging. Instead, it seems likely that

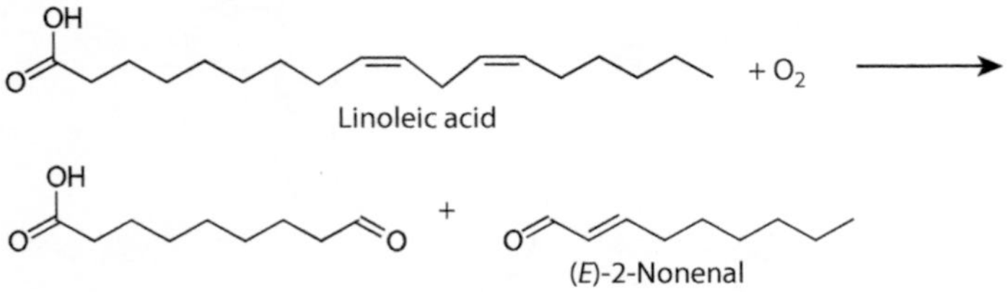

Figure 3.14 (E)-2-Nonenal formation.

it and other staling compounds are formed by reactions with oxygen earlier in the brewing process. The compounds become bound, possibly to sulfites or amino acids, protecting them from being destroyed during fermentation. The free compounds are then slowly released after packaging, giving rise to stale off-flavors in the beer.

Certain lipids are **amphiphilic**, meaning that they have both hydrophobic and **hydrophilic** regions. Amphiphilic lipids can be water soluble, yet they tend to associate with insoluble material. The majority of lipids are removed during the brewing process with up to a 90% reduction in the lauter step. **Wort separation**, specifically the lauter step, is the most important process step for reduction of lipids and promotion of wort clarity. Wort **turbidity** is specifically correlated with lipid content. The higher the concentration of long-chain fatty acids is in wort, the higher is the turbidity. Though greatly reduced during the lauter, additional lipids are concentrated in the **trub** after boiling. Lipids make up more than 50% of trub in the **whirlpool**. Some trub carryover is beneficial to fermentation, because lipids are necessary for yeast health. However, excessive lipids (i.e. turbid wort) will compromise beer flavor and stability. For this reason, consistency in trub carryover from the whirlpool to the fermenter is essential to ensure consistency in fermentation and beer flavor.

Proteins

Proteins are easily the most complex of the biopolymer molecules, although not the largest. Some proteins serve as structural components; some serve as selective **channels** for molecules to pass through **membranes**; some respond to molecules, light, or **voltage** and transmit information; some, called enzymes, serve as catalysts that speed up reactions; and some of the smaller ones serve as signaling molecules carrying messages from one part of an organism to another.

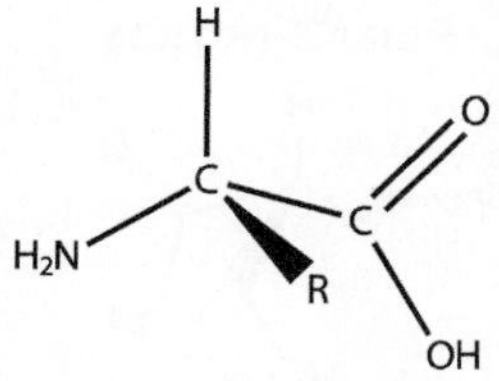

Figure 3.15   Amino acid.

The basic structure of a protein is a linear polymer made from a sequence of 20 different monomers called **amino acids**. The amino acids all have a **carboxylic acid** group, an **amine** group, and a group that is different for each amino acid, known as the **R-group**, because it is represented by the letter "R" in the structure of a generic amino acid (Figure 3.15). The R-group can be classified by a variety of characteristics such as hydrophilic, hydrophobic, positively or negatively charged, hydrogen bonding, bulky, or small. Structures of the 20 amino acids and their chemical properties are shown in Figure 3.16. Each amino acid may be specified by a single- or three-letter code, which is provided in Table 3.3. The amino ($-NH_2$) group of one amino acid can condense with the carboxyl ($-COOH$) group of another, as shown in Figure 3.2. The resulting molecule still has a free amino group and a free carboxyl group, so more amino acids can condense, giving a polymer of any length. Proteins of any length have directionality; there is a single amino end, the N-terminus, and a single carboxyl end, the C-terminus.

A linear, unbranched chain of amino acids is called a **polypeptide**. Short polymers of 20–30 amino acids are called **peptides**. Polypeptides that are long enough to fold into functional three-dimensional structures can range from 100 to over 1000 amino acids and are called proteins.

In living systems, most proteins fold into three-dimensional structures based upon the specific sequence of amino acids and their chemical properties. There are four key components to a protein's structure, which are illustrated in Figure 3.17:

1. The simple sequence of the amino acids in a protein is called the **primary structure**. Figure 3.17A shows the primary structure for the first 20 (of 639) amino acids in amylase from *Aspergillus niger*, a mold.

2. Defined, repeating structures within the whole protein are called the **secondary structure**. Figure 3.17B shows an alpha **helix** (α-helix) structure. Figure 3.17C shows a beta-pleated sheet (β-sheet) structure. The α-helix and β-sheet are common motifs in protein secondary structure.

3. The folded three-dimensional shape of the entire protein is called the **tertiary structure**. Tertiary structure involves a specific arrangement of secondary structures and results from interactions between amino acid

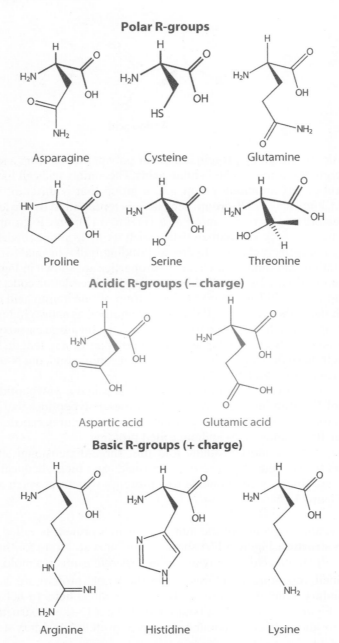

Figure 3.16   The 20 amino acid monomers and their chemical properties.

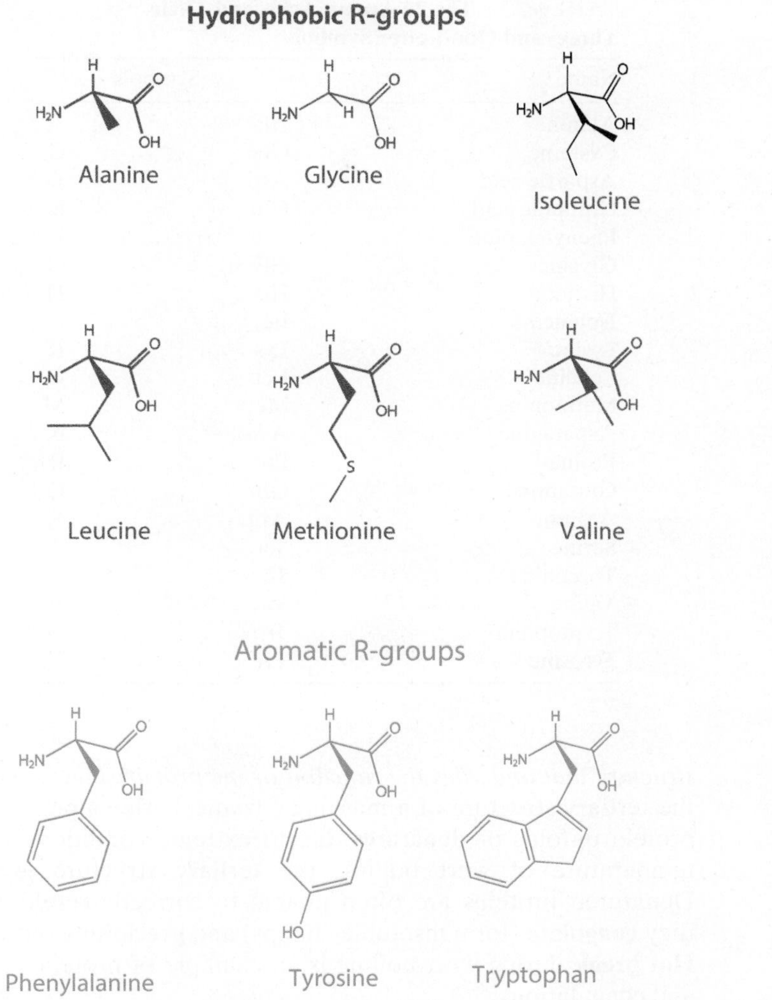

Hydrophobic R-groups

Alanine

Glycine

Isoleucine

Leucine

Methionine

Valine

Aromatic R-groups

Phenylalanine

Tyrosine

Tryptophan

Figure 3.16 (*Continued*)

R-groups that hold parts of the chain together. These interactions may consist of a hydrophobic core, electrostatic interactions, hydrogen bonding, or disulfide bonds. In all cases, these key structural elements are guided by the biochemical properties of the individual amino acids in the sequence. For example, in a hydrophobic core, nonpolar amino acids will cluster together to exclude water, while polar amino acids will orient to the water at the protein periphery. *It is ultimately the tertiary*

TABLE 3.3   **The 20 Amino Acids and Their Three- and One-Letter Symbols**

Name	Symbols	
Alanine	Ala	A
Cysteine	Cys	C
Aspartic acid	Asp	D
Glutamic acid	Glu	E
Phenylalanine	Phe	F
Glycine	Gly	G
Histidine	His	H
Isoleucine	Ile	I
Lysine	Lys	K
Leucine	Leu	L
Methionine	Met	M
Asparagine	Asn	N
Proline	Pro	P
Glutamine	Gln	Q
Arginine	Arg	R
Serine	Ser	S
Threonine	Thr	T
Valine	Val	V
Tryptophan	Trp	W
Tyrosine	Tyr	Y

structure that underlies the function of the protein. Figure 3.17D shows the tertiary structure of a mashing enzyme, barley alpha-amylase. If a protein unfolds or **denatures** due to extreme conditions such as the temperature of wort boiling, the tertiary structure is destroyed. Denatured proteins are often unable to correctly refold; sometimes they **coagulate** (form insoluble clumps) and **precipitate** out of solution. **Hot break** during wort boiling is an example of protein denaturation and coagulation.

4. Some protein complexes require multiple protein subunits to function. This is called **quaternary structure**.

Free amine groups are essential to yeast health during fermentation. They provide nitrogen to be used for synthesis of new organic molecules. Therefore, it is essential to ensure that wort contains sufficient **free amino nitrogen (FAN)**. A large protein only contains one free amino group and thus does not contribute significantly to FAN measurement. FAN primarily consists of amino acids and short peptides. A minimum of 200 mg/L of FAN is needed in wort to support healthy fermentation. FAN concentration greater than

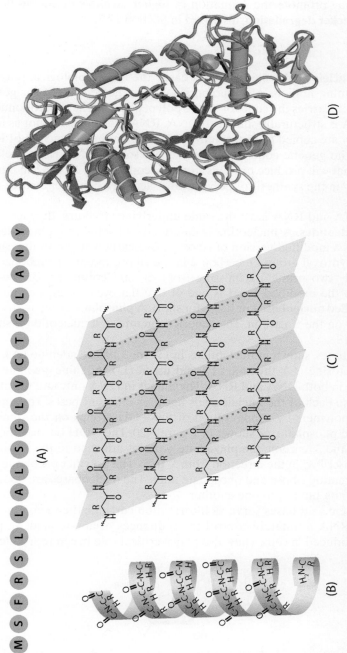

Figure 3.17 (A) Primary structure – first 20 amino acids in *Apergillus niger* amylase. (B) Secondary structure – alpha helix. (C) Secondary structure – beta-pleated sheet. (D) Tertiary structure – barley alpha-amylase. Secondary motifs highlighted: red arrows pleated sheet and green cylinder helix. *Source*: Data for drawing D: Kadziola et al. (1994).

350 mg/L may promote the formation of **higher alcohols** or **staling** flavors through **Strecker degradation**, discussed in Section 12.3.

Nucleic Acids: DNA and RNA

Deoxyribonucleic acid, called DNA, is easily the largest biological polymer, often containing millions of monomers. DNA is the repository for genetic information; it carries the code for all the polypeptides in the cell. Ribonucleic acid or RNA is structurally similar to DNA. RNA molecules are not as large, because they are copied from segments of DNA. A primary function of RNA is to carry the genetic code from the original source DNA to the cellular structures that will produce proteins from the code. Other RNAs can participate directly in the synthesis of polypeptides, act as enzymes, or control **gene** regulation.

Both DNA and RNA have the same underlying structure; they are polymers of **nucleotides**. A nucleotide is described as having three modules. At the center is a modified version of ribose, a five-carbon sugar. In the case of DNA, the hydroxyl group on carbon 2 in the ribose module (symbolized 2′, pronounced "two prime") is missing from the nucleotide, hence "deoxy." Attached to the modified sugar is one of five flat cyclic groups containing nitrogen, called **nucleobases**, or bases for short. A **phosphate** group is attached to the sugar on the 5′ carbon. The generic structure of a nucleotide is shown in Figure 3.18.

The five bases, shown in nucleotides in Figure 3.19, are adenine (A), guanine (G), cytosine (C), thymine (T), and uracil (U). Adenine, guanine, and cytosine are in both DNA and RNA. Thymine is in DNA only and uracil is in RNA only, so each of the nucleic acids has four different bases. The nucleotides form polymers by condensation of the −OH groups on the ribose to those on the phosphate, as shown in Figure 3.20. The −OH on the ribose is attached to the 3′ carbon; the phosphate group is attached to the 5′ carbon. DNA is transcribed in the direction from 5′ to 3′. The polymer backbone consists of alternating ribose and phosphate units. The nucleobases are bound to the ribose units, but not to one another.

The nucleic acid bases serve as information carriers; their sequences in DNA and RNA ultimately control the sequences of amino acids in polypeptides produced in cells. They also play a critical role in cell reproduction.

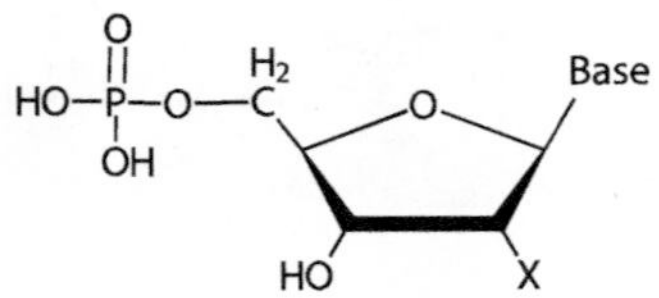

Figure 3.18   Nucleotide. X = OH: ribonucleotide. X = H: deoxyribonucleotide.

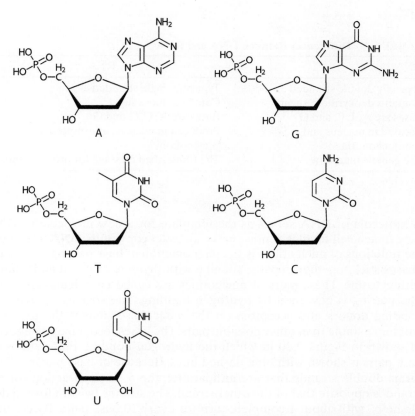

Figure 3.19 Nucleotides. (A) Deoxyadenosine phosphate, (G) deoxyguanosine phosphate, (T) deoxythymidine phosphate, (C) deoxycytosine phosphate, and (U) uridine phosphate.

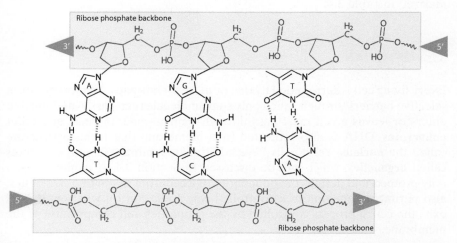

Figure 3.20 DNA section showing three deoxynucleotides – T, C, and A – and their complementary base pairs.

TABLE 3.4 **Differences Between DNA and RNA**

DNA	RNA
Typically double stranded	Typically single stranded
Contains deoxyribose sugar	Contains ribose sugar
Bases are A, T, C, and G	Bases are A, U, C, and G
Located in nucleus and mitochondria	Produced in nucleus; transported into cytoplasm
The genetic blueprint	The transcribed message for protein synthesis

A molecule of DNA serves as the template for a new molecule of DNA. Any time a cell divides, it must make an exact copy of its DNA. Central to the functions of nucleotides is the phenomenon of **base pairing**. The nucleobases stick together in pairs: adenine with thymine or uracil and guanine with cytosine. These pairs of nucleotides are called **complementary** pairs. The pairing is governed by hydrogen bonding. The spacing of hydrogen bonding donors and acceptors on the nucleobases makes the preferred pair more stable than other possible pairs. The bases recognize one another, as shown in Figure 3.20 in which the hydrogen bonding that defines the base pairs is shown with blue dashed lines. Because of base pairing, DNA forms double strands that are **antiparallel**; the 5′ to 3′ direction for one strand is opposite that of the other strand. The strands twist to form a double helix, with about 1 complete turn for every 10 base pairs. By contrast, RNA is stiffer than DNA because of the extra −OH group on each ribose unit, so RNA does not form a double helix, though it can form base-paired secondary structures. The differences between DNA and RNA are summarized in Table 3.4.

3.2 MEMBRANES

Every living cell is surrounded by one or more **membranes**. Membranes form selective barriers through which only some molecules can pass. Anything that enters or leaves a cell must pass through a membrane. In yeast and other **eukaryotes**, DNA is also protected by a membrane that forms a structure called the **nucleus**. Yeast cells have additional membrane-bound structures called **organelles**, which will be discussed in Section 3.4. A membrane not only protects and defines the boundaries of a cell from its environment, but it also permits interaction in a controlled way. Cells control what enters and exits the cell, a process controlled by the properties and composition of the membrane.

Lipid Bilayer

The basic structure of a membrane consists of two layers of phospholipid molecules, called a **lipid bilayer**, assembled as shown in Figure 3.21. The phospholipid molecules, an example of which is shown in Figure 3.22, have a hydrophilic head and two hydrophobic tails. The hydrophilic character is provided by a phosphate ester with a negative charge and, in many cases, by a tertiary amine with a positive charge. The tails are long-chain fatty acids such as oleic (18 : 1) and palmitic (16 : 0) acids. The phospholipids are organized with the heads in the watery medium inside and outside of the cell or organelle, and the tails make up the 7–10 nm thick interior of the membrane.

The fluidity, which is the ease of flow, of the membrane is influenced by the presence of double bonds in the tails that introduce kinks and weaken the interactions among adjacent tails and by the presence of **sterols**. Sterols

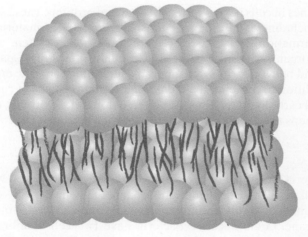

Figure 3.21   Membrane structure. *Source*: From Barth (2013). © 2013 Wiley. Used with permission.

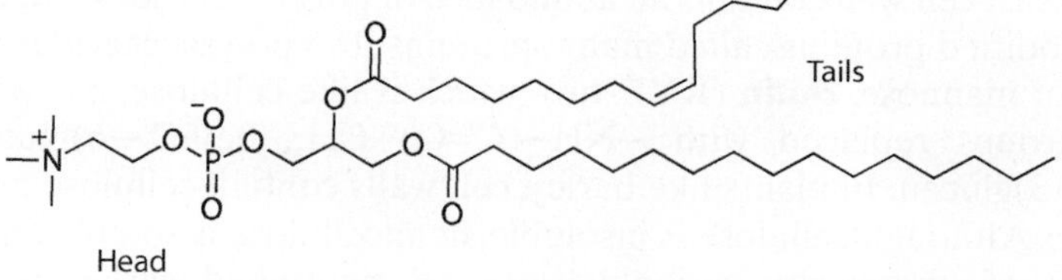

Tails

Head

Figure 3.22   Membrane phospholipid.

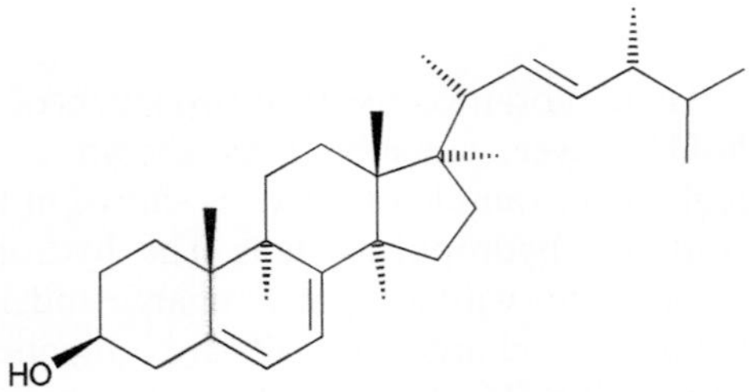

Figure 3.23   Ergosterol.

are flat, four-ring, mostly hydrophobic molecules that congregate among the phospholipid tails. The most important sterol in yeast membranes is ergosterol, shown in Figure 3.23. Oxygen is required during early fermentation to promote phospholipid and sterol synthesis and integrity during cellular division.

Membranes in cells have a variety of attached proteins. These serve several functions including catalyzing reactions, transmission of material, transmission of information, and anchoring the *cytoskeleton*, a system of molecular ropes that organizes the cell contents. Certain proteins may span the entire phospholipid bilayer, forming channels through which specific molecules can pass. In addition to proteins, the membrane contains chains of carbohydrates or modified carbohydrates. Present on the outer **surface** of cells, these carbohydrate groups are attached to proteins, forming **glycoproteins**.

Cell Wall

Plants, bacteria, and fungi have another structure called the cell wall that surrounds each cell. The cell membrane is sometimes mistakenly called a cell wall. In plants like barley and fungi like yeast, the cell wall is outside of the cell membrane. Gram-negative bacteria have two cell membranes, one of which is inside the cell wall and the other is outside. The major function of the cell wall is to provide strength and structure to the cell within. Yeast and bacterial cell walls help the cells resist deformation or bursting under **osmotic pressure**. Cell walls of land plants also provide mechanical strength to allow the plant to stand against gravity and wind.

The yeast cell wall contains an abundance of polysaccharides and carbohydrate-modified proteins called mannoproteins. The polysaccharides typically consist of mannose, **chitin** (KYE-tin), which is like cellulose, but with some —OH groups replaced with $-NH-(C=O)-CH_3$, beta(1→6)glucan, and beta(1→3)glucan. In plants like barley, cell walls contain cellulose and hemicellulose. Although cellulose is insoluble, hemicellulose is soluble and can be a source of undesirable carbohydrates, as mentioned earlier in Section "Carbohydrates."

3.3   CELLULAR STRUCTURES

A cell is the basic biological and structural unit of life. First discovered in 1665 by Robert Hooke, the term was coined after he examined cork under a **microscope**, as the shapes reminded him of the cells in a monastery.

There are two major types of organisms, **prokaryotes** and **eukaryotes**, that are distinguished by their cells. Prokaryotic cells do not contain a membrane-bound nucleus, while eukaryotic cells do. Bacteria like *Lactobacillus* and *Pediococcus* are prokaryotes; animals, plants, yeast, and human brewers are eukaryotes. Another difference between prokaryotes and eukaryotes is the organization of their DNA. Eukaryotes contain multiple linear strands of DNA organized into **chromosomes**; prokaryotes contain one large, circular piece of supercoiled DNA. Finally, eukaryotes contain membrane-bound organelles; prokaryotes do not. An organelle is a specialized membrane-enclosed structure within a cell. Each organelle performs a specific biological function. The membrane around each organelle isolates or protects its functions from other parts of the cell, serving to compartmentalize key enzymes and signaling pathways. For example, the nucleus protects the DNA and serves as the site for **replication** and **transcription**.

In addition to organelles, the cell is filled with liquid cytosol (also called **cytoplasm**). Table 3.5 lists organelles found within eukaryotes like yeast. A model yeast cell is depicted in Figure 3.24.

The nucleus is a round organelle about 1.5 μm in diameter in which the genomic DNA is protected from the rest of the cell. The nucleus is separated from the cytoplasm by its own phospholipid bilayer membrane called the **nuclear envelope**. In this membrane are **nuclear pore complexes**, which are proteins that form channels across the membrane for controlled exchange of materials. From the DNA-based genes, messenger RNA (**mRNA**) is produced

TABLE 3.5   **Organelles in Eukaryotes**

Organelle	Function
Nucleus	DNA storage and protection
Mitochondria	Energy production (aerobic respiration)
Smooth endoplasmic reticulum	Lipid production; detoxification
Rough endoplasmic reticulum	Protein production and processing
Lysosome	Protein breakdown and recycling; not found in yeast
Peroxisome	Lipid breakdown; oxidative
Golgi apparatus	Protein tagging and export
Vacuole	Stores excess water and waste; serves a lysosome-like function in yeast

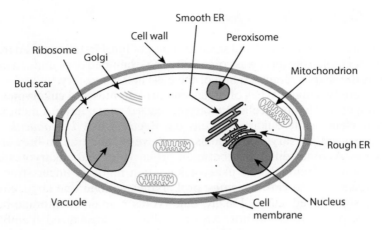

Figure 3.24   Structure of a yeast cell.

by transcription within the nucleus. mRNA is then transported across the nuclear envelope and translated into protein in the cytoplasm. Within the nucleus is the **nucleolus**. This dense region of DNA encodes and transcribes the RNA that will become the **ribosomes**.

Ribosomes are not membrane-bound organelles but rather complex particles consisting of ribosomal RNA (**rRNA**) and proteins. They bind mRNA and are responsible for the process of **translation**, the formation of proteins.

The **endoplasmic reticulum** (ER) is a highly continuous and branched network of tubules attached to the nuclear envelope. All transmembrane proteins and proteins exported outside of the cell are processed within the ER. Ribosomes loosely associated with the cytosolic side of the ER translate proteins into the ER. From there, these proteins are folded and packaged into small membrane-bound **vesicles** that translocate to the **Golgi apparatus**.

The Golgi apparatus contains several stacks of membrane-bound compartments. After receiving newly folded proteins from the ER, the Golgi facilitates **posttranslational modifications** such as glycosylation, lipidation, and phosphorylation. Vesicles containing improperly folded proteins will return to the ER for further folding assistance. This is an important checkpoint to ensure that misfolded proteins are not released.

The **vacuole** is a compartment in yeast in which proteins are recycled. In this organelle, **proteases** break down proteins into amino acids that can be reused by the cell. As the yeast cell ages, the vacuole increases in size. In a healthy yeast cell with sufficient nutrients, the vacuole may take up 20–30% of the cellular space, but in stressed or nutrient-starved yeast, the vacuole can grow to be 70% of the cellular volume.

The **mitochondrion** (pl. mitochondria) is known as the powerhouse of the cell. This organelle is surrounded by an outer membrane and contains an inner membrane that folds upon itself to increase surface area. This surface area is important because the inner membrane is the location of aerobic respiration and the production of **ATP**. Mitochondria also contain DNA that encodes proteins specifically for the mitochondria.

The **peroxisome** is a small organelle that breaks down lipids and facilitates the reduction of reactive oxygen species, like hydrogen **peroxide**, that would otherwise be harmful to the cell.

3.4 THE CENTRAL DOGMA

The central dogma of molecular biology describes the movement of genetic information in biological systems and was first defined by Francis Crick in 1958. As an oversimplification, DNA makes RNA, and RNA makes protein. A portion of a DNA molecule that contains the information for a single protein is called a gene. Using these macromolecules, biological information is passed from the code (DNA) through a messenger (mRNA) and into a functional protein.

All living cells use three primary steps to execute this transfer of information. DNA can make a copy of itself (DNA replication), genes can be written into a messenger RNA called mRNA (transcription), and mRNA can be interpreted into a protein (translation). As we will discuss below, this transfer of information protects and preserves the DNA code in the nucleus. At each stage of information transfer, there are several opportunities to change the activity of genes and the amounts of proteins they produce. Understanding these basic concepts of biology in yeast will help the brewer better understand the importance of yeast health, as changes in yeast lead to changes in beer. Though all mechanisms are generally shared between prokaryotes and eukaryotes, here we primarily focus on eukaryotes to which yeast belong.

DNA Replication

DNA replication is the process by which a single DNA macromolecule is identically copied into two. As a cell divides, it must replicate its DNA; thus the major purpose of DNA replication is genetic inheritance. Because DNA contains genes, accurate and highly precise copying of DNA is critical to the persistence of new cells. DNA replication is *semiconservative*, which means that replicated DNA contains one new strand and one strand that is conserved from the original molecule. This occurs because both strands of the original DNA double helix act as templates for the new strands. Semiconservative replication is modeled in Figure 3.25. DNA replication *only* occurs when a cell is actively dividing.

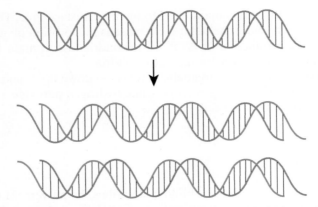

Figure 3.25   Semiconservative DNA replication. Green: original strand. Red: replicated strand.

As DNA makes RNA, which makes protein, the specific sequence of nucleotides in the DNA begets the final sequence of all proteins and thus is critical to the function and identity of the organism. If the DNA sequence changes, the protein sequence changes. Because the amino acid sequence is essential for protein structure and function, changes to the DNA sequence could negatively impact yeast performance. It is therefore essential that the process be precise and perfect, introducing no base pair changes or **mutations**. DNA replication is so precise that experimental methods in *Saccharomyces cerevisiae* measured a genome-wide single-nucleotide mutation rate at $1.67 \pm 0.04 \times 10^{-10}$ per base per generation. The *S. cerevisiae* genome is 12.1 million base pairs in length; thus for one yeast cell division, there is only about 1 chance in 500 of a mutation.

The reason for DNA replication precision is the molecular machinery of the process. A **helicase** separates the double helix, while a **topoisomerase** unwinds the double-stranded template DNA to prevent supercoiling. *Single-stranded binding proteins* keep the individual strands separated. *RNA primase* is an enzyme that adds a short RNA primer that is complementary to the DNA sequence. **DNA polymerase** III then binds to the double-stranded DNA–RNA section and initiates DNA replication by adding nucleotides to the 3′ end of the polymerizing DNA strand. *DNA and RNA are always synthesized in the 5′ to 3′ direction.* The nucleotide added to the new strand will be complementary to the nucleotide of the old strand being copied. The energy for this addition is generated by hydrolysis of triphosphate on the free nucleotide being added. In linear DNA molecules, like the chromosomes in yeast, the process ends when the DNA polymerase falls off the end of the DNA molecule. DNA polymerase I then replaces the RNA primer with DNA. DNA replication is modeled in Figure 3.26.

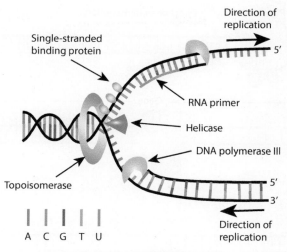

Figure 3.26   DNA replication.

Transcription

Transcription is the process through which a segment of DNA, called a gene, is copied into mRNA. A gene is a specific sequence of DNA that is responsible for the amino acid sequence of a specific polypeptide. In yeast and other eukaryotes, DNA is protected within the cell's nucleus. The machinery responsible for making protein is found in the cytoplasm, which is outside of the nucleus. Transcription synthesizes mRNA that carries the code from the nucleus to the cytoplasm.

Transcription is orchestrated by an enzyme called RNA polymerase. RNA polymerase reads one strand of the DNA helix as a template to create a complementary RNA strand. As with all nucleotides, polymerization occurs in the 5′ to 3′ direction. There are three major stages of transcription.

First, during **initiation**, RNA polymerase binds to a specific region of the gene called the promoter. Binding separates the double-stranded DNA helix, making one strand accessible for transcription. Second, during **elongation**, the RNA polymerase will scan the strand of DNA, making a complementary copy of RNA. RNA uses the same bases as DNA except that uracil (U) is used instead of thymine (T). Finally, during **termination**, a specific DNA sequence causes the RNA polymerase to stop. This is often caused by the formation of an RNA stem loop structure in which internal base pairing of the RNA transcript halts the processing by the RNA polymerase. When the RNA polymerase stops moving, it falls off the DNA template. This process is modeled in Figure 3.27. mRNA is further distinguished by addition of a 7-methylguanosine cap on the 5′ end, called the *5′ cap*, and a long polyadenylated

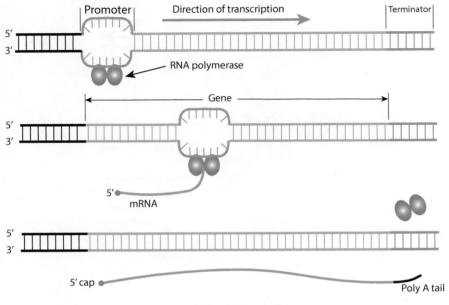

Figure 3.27   Transcription.

3′ end, called the *poly A tail*. The DNA gene will act as a template for many more mRNA copies, leading to an amplification of the signal, to be further amplified during the next step, translation.

Translation

Translation interprets the mRNA sequence and turns it into a polypeptide from within the cytoplasm. This information is encoded in codons, which are sets of three consecutive nucleotides. The genetic code is made up of 64 distinct codons. Of these, 61 encode an amino acid, and three serve as stop signals, shown in red, that terminate translation. Only one codon, AUG (which also encodes methionine), acts as the starting translation signal (shown in green). The genetic code for mRNA is shown in Table 3.6.

Translation involves two main pieces of machinery, both of which contain RNA. Transfer RNAs (**tRNAs**) interpret the mRNA codons into an amino acid sequence. Within the tRNA is a complementary anticodon, which recognizes the mRNA codon through complementary base pairing. Each specific tRNA, defined by its anticodon, is attached to an amino acid, the identity of which is specified in the genetic code. The removal of the amino acid from the tRNA and the subsequent formation of the peptide bond on the amino-terminus of the growing polypeptide are catalyzed by the ribosome.

TABLE 3.6   The mRNA Genetic Code

		Second letter						
		U		C		A		G	
First letter	U	UUU	Phe	UCU	Ser	UAU	Tyr	UGU	Cys
		UUC	Phe	UCC	Ser	UAC	Tyr	UGC	Cys
		UUA	Leu	UCA	Ser	UAA	Stop	UGA	Stop
		UUG	Leu	UCG	Ser	UAG	Stop	UGG	Trp
	C	CUU	Leu	CCU	Pro	CAU	His	CGU	Arg
		CUC	Leu	CCC	Pro	CAC	His	CGC	Arg
		CUA	Leu	CCA	Pro	CAA	Gln	CGA	Arg
		CUG	Leu	CCG	Pro	CAG	Gln	CGG	Arg
	A	AUU	Ile	ACU	Thr	AAU	Asn	AGU	Ser
		AUC	Ile	ACC	Thr	AAC	Asn	AGC	Ser
		AUA	Ile	ACA	Thr	AAA	Lys	AGA	Arg
		AUG	Met	ACG	Thr	AAG	Lys	AGG	Arg
	G	GUU	Val	GCU	Ala	GAU	Asp	GGU	Gly
		GUC	Val	GCC	Ala	GAC	Asp	GGC	Gly
		GUA	Val	GCA	Ala	GAA	Glu	GGA	Gly
		GUG	Val	GCG	Ala	GAG	Glu	GGG	Gly

The ribosome is a multi-complexed entity that contains both rRNA and protein. The ribosome contains a large subunit and a small subunit that sandwiches the mRNA, forming the **active site**.

In the first stage of translation called initiation, ribosomes are loaded onto mRNA on the 5′ end, moving in a 5′ to 3′ direction. The ribosome is activated when the first start codon, AUG, is encountered. During the second stage of translation, elongation, the ribosome continues in a 5′ to 3′ direction along the mRNA, catalyzing the formation of the polypeptide. In the ribosome, there are three active sites, called A (acceptor), P (peptidyl), and E (exit), through which the polypeptide is formed. In the A site, the tRNAs cycle in and out of position until the correct combination of anticodon and codon is found. The ribosome catalyzes the formation of a peptide bond from the amino acid bound to tRNA at the P site (at the near end of the peptide chain) to the new amino acid in the A site, transferring the chain to the A site. The tRNA in the P site becomes empty. Then the ribosome moves in the 3′ direction to bring the A site to a new codon, the P site to the growing chain, and the E site to the empty tRNA, which loses affinity for the ribosome and exits. In this manner, the ribosome will translate the mRNA into a full-length polypeptide by moving 5′ to 3′ from the start codon to the first stop codon. The first stop codon causes ribosome disassembly through recruitment of a termination factor. A model of translation is shown in Figure 3.28.

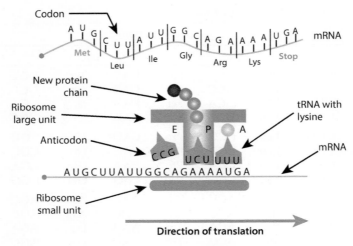

Figure 3.28   Translation.

So now that we have a full-length polypeptide, are we finished? Sometimes. During translation, as a polypeptide exits the ribosome, it can sometimes fold on its own. But other polypeptides need folding assistance by way of **chaperone proteins**. A full-length polypeptide rarely exists without tertiary structure; there is always some degree of protein folding; thus when a protein is **denatured**, by boiling for example, it is nearly impossible for it to refold to its original structure. Hot break results from the inability of proteins to refold.

Posttranslational modifications can further influence protein folding, stability, and functionality. These include attachment of small molecules through condensation with phosphate (phosphorylation), with **acetate** (acetylation), or with glucose or its polymers (glycosylation). Covalent disulfide bonds may form between sulfur-containing amino acids, stitching together distant parts of the molecule. And finally, some proteins are expressed as precursors called pro-proteins, which are not active until parts are chopped out by proteolysis (hydrolysis of proteins) at a specific amino acid sequence. A new polypeptide also needs to find its proper location within or outside the cell. Specific signal sequences on the N- or C-terminus of a polypeptide may be used to direct shuttling of the protein to intracellular locations or to extracellular secretion.

Regulation of Gene Expression

The process of creating an mRNA copy of a gene and translating it into protein is called **gene expression**. Although all cells of the human body have the same DNA, they become differentiated by the types and relative amounts of

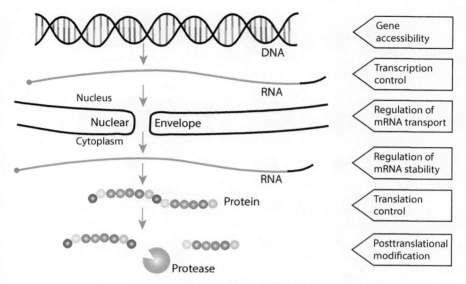

Figure 3.29   Control of gene expression in eukaryotes.

proteins produced. For example, a muscle cell has different proteins than a brain cell. These differences are the result of **gene regulation**. Yeast, being single-cell organisms, do not need to differentiate into tissues; they must nonetheless regulate their gene expression to meet external and internal conditions. For example, yeast cells that are actively dividing on the first day of fermentation express a completely different set of genes than yeast after the end of fermentation. On the first day of fermentation, a yeast cell needs to express, or turn on, genes responsible for cellular division, but after fermentation is complete, the yeast cell must focus on survival by activating stress response genes, promoting survival in an environment with insufficient nutrients. If the yeast were unable to regulate or change gene expression, the cells would continue to divide without sufficient nutrients, leading to cell death. Plus, they would make bad beer! Every step of gene expression, from the DNA sequence to the final protein, is fine-tuned by mechanisms of gene regulation. A summary of different areas in which a gene can be regulated is illustrated in Figure 3.29.

But how does a cell "know" which genes to activate and how to regulate them? Molecular pathways in the cell convey information that influences gene expression. Signaling from inside or outside of the cell can activate these pathways. The inherited DNA of the yeast as well as existing proteins that came from the mother cell will influence inside signaling. But a major part of gene regulation is the ability of a cell to respond to signals from the environment outside of the cell. These signals may be **hormones**, proteins, nutrients,

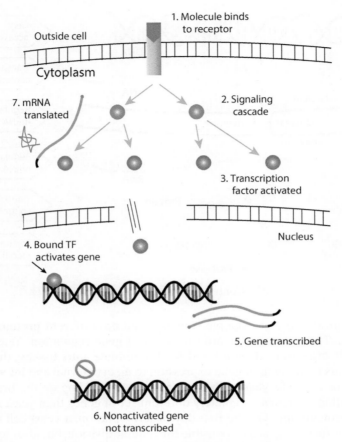

Figure 3.30   Transcription factor signaling.

or other chemicals. Here a specific signal reacts with a protein on the cell surface called a *receptor*, which activates a chemical cascade inside the cell that reaches into the nucleus. The molecular signaling cascade activates **transcription factors** that bind to specific regulatory DNA sequences before a gene. Transcription factor binding recruits and activates the transcriptional machinery, creating new mRNA and leading to new protein expression. This process is generalized in Figure 3.30 and is the primary example of transcriptional control of gene regulation.

While transcription factor signaling is a general concept shared by many genes, the details of each signaling pathway vary for every gene and are still active areas of study. This type of control illustrates how different genes can be regulated at the transcriptional level. Genes may be either "turned on" or "turned off" by environmental stimuli but can also be fine-tuned as to how much mRNA or protein is made. More mRNA often means more protein. mRNAs are also regulated by stability. Some mRNA transcripts are very stable, while others quickly degrade. Stable mRNAs can be translated repeatedly, producing more and more protein. One of the factors that affects mRNA stability is the presence of **small regulatory RNAs** (sRNAs). These noncoding RNAs are genetically encoded in DNA but do not produce a functional protein. Instead, they act through complementary binding to an mRNA, leading to its degradation. **MicroRNAs** (miRNA) were among the first sRNAs to be discovered. These short transcripts form a hairpin loop through complementary base pairs within the molecule. The loop structure is recognized by enzymes that then create a small single-stranded miRNA of about 22 base pairs. The miRNA binds to target mRNAs through complementary base pairing. Perfect matches are quickly degraded. Thus miRNAs provide a means to regulate gene expression through the targeted degradation of specific mRNAs.

And finally, gene expression is regulated at the translational and protein levels. Generally, there are support proteins that bind to specific mRNAs that can either promote or inhibit translation. These proteins typically affect ribosome loading onto the mRNA and subsequent translation initiation. Once translated, proteins will be quickly degraded by the **proteasome** if they are unfolded or tagged with a posttranslational modification called **ubiquitin**. Other posttranslational modifications include proteolysis, phosphorylation, acetylation, and glycosylation.

CHECK FOR UNDERSTANDING

1. What are the four classes of biological macromolecules? For each, describe the nomenclature, structure, and function of the most common monomers and polymers.

2. For each of the macromolecules, describe its role or influence on the brewing process, and give specific examples of each.

3. There are two types of starch in barley malt. Give their names, and explain how they differ in structure, relative abundance, and contribution to beer production.

4. Maltose is fermentable. Lactose is unfermentable. What is the molecular difference between these two disaccharides?

5. Which common barley carbohydrate is most undesirable during the brewing process due to its effect on wort viscosity? How does its structure lead to this behavior?

6. Describe the structure and function of the plasma membrane. What biochemical properties of lipids enable the formation and function of cellular membranes?

7. What is the flavor of the staling molecule trans-2-nonenol? How can it be avoided in beer?

8. For each of the following amino acids, describe their biochemical properties (i.e. polar, nonpolar, acidic, basic, etc.). With which type of intermolecular force are they likely to influence protein folding?

   a. Glycine

   b. Alanine

   c. Glutamic acid

   d. Serine

   e. Cysteine

9. What is the relative composition of hot break material after the boil? Why does it form?

10. What is the purpose of the following organelles in yeast?

    a. Nucleus

    b. Mitochondrion

    c. Lysosome

    d. Golgi body

11. What are the differences between DNA and RNA?

12. Describe the central dogma using essential terminology for each step.

13. Considering the various methods for regulation of gene expression in a cell, describe some differences in the central dogma which could vary by yeast strain and influence fermentation performance or flavor.

BIBLIOGRAPHY

Barth R. *The Chemistry of Beer: The Science in the Suds*. 2013. Wiley. ISBN 978-1-118-67497-0. p. 117.

Feldman H (editor). 2012. *Yeast: Molecular and Cell Biology*, 2nd ed. Wiley-Blackwell. ISBN 978-3-527-33252-6.

Kadziola A, Abe J, Svensson B, Haser R. 1994. Crystal and molecular structure of barley alpha-amylase. *J. Mol. Biol.* *239*(1): 104–121.

Kuhbeck F, Back W, Krottenthaler M. 2006. Influence of lauter turbidity on wort composition, fermentation performance and beer quality – a review. *J. Inst. Brew.* *112*(3):215–221.

Zhu YO, Siegal ML, Hall DW, Petrov DA. 2014. Precise estimates of mutation rate and spectrum in yeast. *Proc. Natl. Acad. Sci. U. S. A.* *111*(22):E2310–E2318.

CHAPTER 4

RAW MATERIALS

There are just four essential ingredients in **beer**: water, **malt**, **hops**, and **yeast**. Yeast is an ingredient, but not really a raw material because most of it is grown in the **brewery**. We discuss its biology and processing in Chapters 3 and 9. Sources of starch or sugar other than malt are **adjuncts**. Other items used for **brewing** such as **enzymes**, **filtration media**, and **finings** (clarifying agents) are processing aids. We discuss adjuncts and processing aids in subsequent chapters.

Quality beer cannot be made without quality ingredients. It is essential to understand the impact that each of the four major ingredients has on the finished product. Raw materials should be regularly inspected for quality. Brewers should take every opportunity to evaluate their raw materials. They should chew on malt, smell hops, taste water, and visually inspect yeast. The impact of quality and more detailed methods of assessment for each of the raw materials are discussed in this chapter.

Mastering Brewing Science: Quality and Production, First Edition.
Matthew Farber and Roger Barth.
© 2019 John Wiley & Sons, Inc. Published 2019 by John Wiley & Sons, Inc.

4.1 WATER

Water is the matrix of all life as well as the major component of beer. Water and ethanol are the only two beer components that exceed 1%. In contrast to **wine**, the water in beer is an ingredient; it is added during brewing. This makes it convenient to ship the dry ingredients to a brewery near the consumers and use locally available water. Wine, on the other hand, is prepared near the source of grapes, and the finished product must be shipped at great expense to the point of use.

Water Chemistry

Water is unique in many ways. It is the only common substance that can exist as a solid, liquid, or **gas** at ordinary **temperatures**. It is one of the few substances that expands on freezing; as a solid, it floats. Water requires an unusually high **energy** input to increase its temperature (specific **heat capacity** = 4.18 J K^{-1} g^{-1}), melt it (**heat of fusion** = 334 J/g), and boil it (**heat of vaporization** = 2230 J/g). These properties and its low price make water an ideal material for heat transfer.

Hydrophilic/Hydrophobic   The water **molecule** in the liquid and gas phases is bent with an H—O—H **bond** angle of 105°; water in ice has a nearly ideal **tetrahedral** angle of 109.5°. This bond angle gives water molecules an oxygen end and a hydrogen end, as shown in Figure 4.1. Because oxygen is much more **electronegative** than hydrogen, the shared electrons are **attracted** to the oxygen end, making water a **polar** molecule. Water molecules can form **hydrogen bonds** with one another and with other molecules, especially those with —OH or —NH bonds. The **dipole–dipole** and hydrogen bonding interactions between water molecules hold them together strongly in the liquid phase. To allow a foreign molecule to take a position among the water molecules, these forces must be overcome, so the water molecules can be pulled away from one another to make room. If the foreign molecule can form strong dipole–dipole or hydrogen bonding interaction with water, the water–water forces are replaced with water–foreign molecule forces, keeping the energy in balance. Substances whose molecules form strong intermolecular

Figure 4.1   Water.

interactions with water are said to be **hydrophilic**. Substances that do not form strong intermolecular forces with water are squeezed out of the water matrix, allowing the water molecules to interact with one another. Such substances are **hydrophobic**. Sometimes parts of a molecule are hydrophilic and other parts are hydrophobic. In water, molecules tend to take positions that put the hydrophilic parts in contact with the water and the hydrophobic parts in contact with one another. This effect is responsible for the organization of **membranes** around cells, the details of **protein** folding, and the stabilization of **foam** on beer.

Acids, Bases, and pH   Acid–base chemistry plays a dominant role in processes in water **solutions**, like beer brewing. An acid–base reaction involves the transfer of a hydrogen ion, H^+. The significance of the hydrogen ion is that a hydrogen atom consists only of one **proton** and one **electron**. When the hydrogen atom loses the electron, only the proton remains. So an acid–base reaction involves the transfer of a fundamental subatomic particle, the proton. An **acid** *provides* a proton (H^+); a **base** *accepts* a proton. A complete acid–base reaction needs both an acid and a base.

The reaction of a generalized acid, HA, with water is

$$HA + H_2O \rightarrow A^- + H_3O^+$$

H_3O^+, the **hydronium ion**, is the signature of an acid in water. If this reaction goes to near completion, so that no HA remains, the acid is considered a strong acid. If the reaction only goes part way, the acid is a weak acid.

The reaction of a generalized base with water is

$$B + H_2O \rightarrow HB^+ + OH^-$$

OH^-, the hydroxide ion, is the signature of a base in water. There is a similar distinction between strong and weak bases. In both cases, the charges are relative. The acid starts out neutral or charged, but after it loses a proton, its charge decreases by one. The charge on the base increases by one when it accepts the proton. In the acid reaction, the proton is accepted by water; water acts as a base. In the base reaction, water provides the proton; water acts as an acid. Water can behave either as an acid or as a base. One water molecule can even accept a proton from another water molecule:

$$H_2O + H_2O \rightleftharpoons H_3O^+ + OH^-$$

The double arrow shows that the reaction goes only part way to products. This reaction proceeds to a tiny but important extent, giving H_3O^+ and OH^- **concentrations** of 1.0×10^{-7} mol/L at room temperature in pure water.

Le Châtelier's principle tells us that increasing the concentration of H_3O^+ will cause a shift to reactants, so the concentration of OH^- decreases. Similarly, a high concentration of OH^- will be accompanied by a low concentration of H_3O^+. Le Châtelier's principle also applies to the acid and base equations above. For the equation $HA + H_2O \rightarrow H_3O^+ + A^-$, increasing H_3O^+ causes the reaction to shift toward reactants, lowering the concentration of A^- and raising that of HA.

An important practical consequence of Le Châtelier's principle is illustrated in Figure 4.2. In Figure 4.2A the concentration of H_3O^+ is high. The **acidic** groups retain their protons, and the **basic** groups accept protons. In Figure 4.3B the concentration of H_3O^+ is low; acidic groups lose protons and basic groups are unchanged. The result is that under acid conditions (high H_3O^+), the molecule becomes more positive and less negative. The charging is reversed in base (low H_3O^+) conditions. If the molecules in this example were enzymes needed to convert starch to sugar, the correct balance between acidity and basicity would have a big effect on enzyme activity.

pH is a measure of acidity in solution. If the concentration (technically an effective concentration called the *activity*) of H_3O^+ is 10^{-4} mol/L, the pH is 4. If the concentration is 10^{-9} mol/L, the pH is 9. A neutral solution at 25 °C has a pH of 7. Acidic solutions have a low pH; basic solutions have a high pH. To extend the calculation to numbers that are not whole number powers of 10, we define pH by an equation:

$$pH = -\log\left[H_3O^+\right]$$

The square brackets denote the concentration of the item. The **pH** concept was put forward in 1909 by Søren Sørensen of the Carlsberg laboratory. The modern pH meter was invented by Arnold Beckman in 1934. Unlike earlier scientific instruments that occupied an entire lab bench or more, Beckman's meter was self-contained, portable, and fit into a small wooden box. It used an electronic amplifier to capture the weak signal from the pH probe

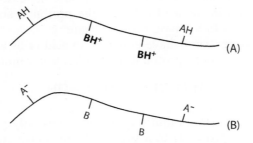

Figure 4.2   Acidity effect. (A) High H_3O^+ and (B) low H_3O^+.

and displayed the result on a meter. Today's pH meters use the same type of sensor, but some of them are small and rugged enough to fit into one's pocket.

The lowest water pH possible, but not actually attainable, is −1.7. This would require every water molecule to bind a hydrogen ion. The highest pH possible, also not attainable, is 15.7. This would require every water molecule to donate a hydrogen ion. Most pH meters measure in the range of pH 1–14, the typical pH range encountered in brewing applications.

ACID–BASE EQUILIBRIUM

The reaction $2H_2O \leftrightarrows H_3O^+ + OH^-$ is governed by the equilibrium expression

$$\left[H_3O^+\right]\left[OH^-\right] = K_w$$

At 25 °C K_w is the **equilibrium constant** for water ionization, 1.0×10^{-14}. This means that if the concentration of hydronium ion is high, the concentration of hydroxide ion must be low. In a neutral water solution, the concentrations are equal and, at 25 °C, are 1.0×10^{-7} mol/L. Using some properties of logarithms and coining the expression $p(\text{anything}) = -\log(\text{anything})$, we can write the above equation as

$$pH + pOH = pK_w$$

At 25 °C, $pK_w = 14$ so pH + pOH = 14. The concentrations of hydronium ion and hydroxide ion are not independent. If we know one, we know the other.

The acid reaction is

$$HA + H_2O \rightarrow A^- + H_3O^+$$

We can write the equilibrium expression

$$\frac{\left[H_3O^+\right]\left[A^-\right]}{[HA]} = K_a$$

keeping in mind that the solvent, H_2O, does not enter the equilibrium expression. K_a is the acid dissociation equilibrium constant. We can solve this equation for $[H_3O^+]$ and perform some logarithmic algebra to get

$$pH = pK_a + \log\left(\frac{\left[A^-\right]}{[HA]}\right)$$

(continued)

This equation, called the Henderson–Hasselbalch equation, tells us that the pH depends only on the ratio of the acid and **anion** concentrations. If these are much larger than the concentrations of hydronium and hydroxide ion, we can use the initial concentrations without having to correct for conversion of acid to base or base to acid. When the A^- and HA concentrations are equal, the argument of the log (the number we take the log of) is 1. The log of one is zero, so $pH = pK_a$.

Example: pK_a for lactic acid, often used to give beer a tart taste, is 3.86. If the concentration of lactic acid is 0.05 mol/L and that of lactate ion is 0.25 mol/L, calculate the pH of the solution.

Answer:

$$pH = 3.86 + \log \frac{0.25}{0.05} = 4.56.$$

Example: A sample of beer has pH = 4.2. The lactate ion concentration is 0.003 mol/L. Calculate the concentration of lactic acid (LA).

Answer:

$$4.2 = 3.86 + \log \frac{0.003}{LA} \cdot 4.2 - 3.86 = \log \frac{0.003}{LA} = 0.34$$

$$10^{0.34} = \frac{0.003}{LA} = 2.19; \; LA = 0.0014 \; mol/L$$

Ions in Water   Many **ionic** compounds are soluble in water, at least to some extent. Although the separation of ions of opposite charge to disperse them in solution requires a large input of energy, much of this energy can be recovered because of **ion–dipole** interactions with water, as shown in a simplified way in Figure 4.3. Stabilization of ions or molecules by water is called **hydration.**

Water used for brewing contains dissolved ions constituting tens to hundreds of milligrams per **liter** (ppm). Some of these ions affect beer **flavor** depending on style, and some create **off-flavors** or processing issues. Some ions are required to keep yeast healthy. Common ions in water, their effects on beer, and recommended concentrations are listed in Table 4.1.

Alkalinity   **Alkalinity** is the measure of the amount of basic substances dissolved in water. It is a different concept from pH. Water containing 0.001 mol/L hydroxide ion has a pH of 11. Water containing the same molar

Figure 4.3   Hydration of ions.

concentration of sodium **bicarbonate** has a pH of 8.7. Both have the same alkalinity. The alkalinity is defined by the amount of hydronium ion needed to neutralize the base. In both examples it would take 1 mmol (0.001 mol) of hydronium ion to neutralize 1 L of water. We will discuss alkalinity measurement and **units** of alkalinity in the next section.

The acceptable range of pH for **mashing** is 5.2–5.6. Malt provides acids, particularly in the form of **phosphate** esters, that neutralize some of the water alkalinity, lowering the pH to a range that provides good activity of the mashing enzymes. If the alkalinity is too high, there may not be enough acid in the malt to neutralize it, and mashing effectiveness will suffer as the pH will be higher than 5.6. Darker malt is more acidic than pale malt, so localities with alkaline water often favor darker styles of beer. There are several ways to lower alkalinity, some of which are discussed in Section "Water Quality and Processing."

Hardness   **Hardness** in water refers to the concentration of metal ions whose charge is +2 or higher, typically calcium and magnesium. Hardness has both desirable and undesirable effects on beer brewing. Hardness causes a modest decrease in the mash pH by reacting with phosphate-containing compounds from the malt according to the simplified reaction $3Ca^{2+} + 2H_2PO_4^- + 4H_2O \rightarrow Ca_3(PO_4)_2 + 4H_3O^+$. The situation is not so simple; phosphate in malt exists mostly in **bound** forms, and there are several calcium phosphate and calcium hydroxyphosphate compounds that can form. The net result is that the acidifying effect of hardness is much less than would be predicted by the simplified equation.

Calcium ions help the mashing enzymes avoid **denaturation**. Hardness can also remove undesirable oxalate ions $(C_2O_4^{2-})$, which are derived from **grain hulls**. When subjected to boiling, oxalate ions can form difficult-to-remove deposits called **beer stone** in the kettle or chiller. Oxalate ions in finished beer promote excessive foaming, called **gushing**. The calcium ion and to some extent the magnesium ion remove the oxalate by forming insoluble solids, like CaC_2O_4, that harmlessly **precipitate** with the grain during mashing. Calcium ions are also important to enhance **flocculation** of yeast during **fermentation**, making it easier to clarify the beer. It is recommended that the concentration of calcium ion in the mash be no less than 50 ppm.

TABLE 4.1   Common Ions Important for Beer Production

Ion	Formula	Brewing Range	Effects and Issues
Bicarbonate	HCO_3^-	0–50 ppm for pale beers 50–100 ppm for amber or malt-forward beers 150–250 ppm for dark or roasted beers	Alkalinity. Raises pH
Calcium	Ca^{2+}	50–100 ppm	Provides hardness. Lowers pH. Precipitates oxalate. Forms deposits in pipes and vessels. Cofactor of many enzymes and is essential for yeast health and flocculation
Chloride	Cl^-	0–250 ppm	Sweetness. Fullness. Balances sulfate. Corrosive to stainless steel
Iron(II)	Fe^{2+}	<0.2 ppm	Metallic. Astringent. Inhibits conversion of starch. Increases color of wort. Increases staling. Forms deposits
Hydronium	H_3O^+		Lowers pH. Enhances bitterness. Suppresses bacteria.
Magnesium	Mg^{2+}	10–30 ppm	Contributes to hardness. Lowers pH. Excessive amounts (>50 ppm) create a sour bitterness. Very excessive amounts (>100 ppm) acts as a laxative
Nitrate	NO_3^-	<10 ppm	Above 10 ppm suggests contamination by agricultural runoff. At high concentrations can be converted to potentially harmful nitrite
Sodium	Na^+	0–150 ppm	At 70–150 ppm will round out flavor and provide sweetness. Salty flavor will develop >200 ppm
Sulfate	SO_4^{2-}	50–150 ppm 150–350 ppm to accentuate bitterness	Enhances bitterness, making the sensation more dry and crisp. At levels >400 ppm can act as a laxative

On the negative side, hardness ions can form insoluble deposits in brewing equipment, called **pipe scale**, especially where water is heated. The layer of deposit narrows the effective diameter of pipes, making flow more difficult. It interferes with heat transfer by forming an insulating layer. Hardness ions can also interfere with the action of **detergents**, leading to ineffective **cleaning** or the need for increased use of chemicals, higher temperatures, and longer treatment times. Hardness is highly undesirable in steam boiler feed water because of the formation of deposits. Boiler water is often treated to remove or sequester (bind in a soluble form) hardness ions.

Water Quality and Processing   At a minimum, water for brewing must meet the standards for drinkable water. This is mostly an issue for breweries that do not use municipal water supplies. Water from any source usually requires some level of processing at the brewery. Disinfectants like chlorine and chloramine must be removed. It may be necessary to remove iron and manganese. Excessive hardness or alkalinity may require treatment. Water for mashing and **sparging** ends up in the beer, so flavor issues are at the forefront. Water used for cleaning should either be free of interfering ions, like calcium and iron, or should be treated with substances that bind these ions and render them harmless to the cleaning process. Water for boilers needs to be treated to minimize **corrosion** and the deposit of **pipe scale** in the boiler. The most demanding specifications are on water that that is used to dilute **high-gravity** beer before packaging. Dilution water must have the same mineral and carbon dioxide content as the beer, and it must be free of dissolved oxygen.

Water coming into a brewery is filtered, that is, it is passed through a medium with holes measuring in the micrometer or tenths of micrometer range. Filtration removes suspended solids, like debris from the plumbing and bits of sand or soil. Filtration protects subsequent purification processes and brewery mechanisms from damage. Disposable cartridges, often of **polymer** yarn wound around a core, are suitable for small breweries. Larger operations benefit from filters with renewable **media**. These often use beds of sand or other inert granular material. The filter is cleaned by flowing water through it in the reverse direction and discarding the wash water.

After filtration to remove suspended solids, all breweries should use **activated carbon** to remove chlorine, chloramine, odors, and colors from the water. Activated carbon is made by partial combustion of a carbon source, like coal, peat, fruit pits, and the like. The resulting char is treated ("activated") at high temperature, usually with steam, to remove some of the carbon and leave a system of pores with a very high **surface** area for binding impurities. Activated carbon is effective for removing neutral molecules like chlorine and chloramine, which are often added to municipal water systems to suppress harmful **microbes**. Very small amounts of chlorine in brewing water will produce undesirable chlorophenols in beer. Chlorophenols have a medicinal, hospital, plastic, or band-aid off-flavor. Chlorine is easy to smell and is easily volatized. It can be removed by boiling your water or simply letting it sit out overnight. Because chlorine escapes from water so easily, many water treatment plants have switched to chloramines. Chloramine is more difficult to detect by the senses and is not as **volatile**. The most effective way to quickly remove chloramine (and chlorine) is with an activated carbon filter. Activated carbon can also remove nonionic off-flavor compounds and colored substances. It does not remove ionic impurities, including heavy metal ions and those that cause hardness, alkalinity, and salinity.

Chlorine may also be removed by **ultraviolet** (UV) light, often installed in-line, upstream of an activated carbon filter. UV light breaks down the chlorine or chloramine molecules into chloride, ammonia, and water, with the added benefits of killing 99.99% of **bacteria** and viruses. Chlorine and chloramine can also be broken down with sodium metabisulfite ($Na_2S_2O_5$) or potassium metabisulfite ($K_2S_2O_5$), sold as Campden tablets. Potassium metabisulfite may also be used for **sanitization** of fruit and as an antioxidant. Metabisulfite releases sulfite ion (SO_3^{2-}), which causes an allergic reaction in some people.

In the past, if hardness and alkalinity were a concern, they were reduced by boiling the water. The chemical reaction is $Ca^{2+} + 2HCO_3^- \rightarrow CaCO_3(s) + CO_2(g) + H_2O$. The solid calcium **carbonate** settles out or is filtered out. Hardness that can be removed by boiling is called **temporary hardness**. The boiling method is seldom used today in commercial breweries because it requires heating, cooling, and sedimentation or filtration, which are expensive in energy and equipment.

Lime softening is another old process that may still be in use at some locations. Calcium hydroxide (slaked lime) is added to the water, converting soluble calcium bicarbonate to insoluble calcium carbonate: $Ca(HCO_3)_2 + Ca(OH)_2 \rightarrow 2CaCO_3(s) + 2H_2O$. The resulting water is treated with carbon dioxide to lower the pH. The process has declined in popularity because it needs a large reactor for sedimentation of the calcium carbonate. Advances in separation technology may make lime softening more attractive.

Ion exchange can be used to alter the ionic composition of water. The exchanger contains one or more types of resins (plastics) with **functional groups** that bind positive or negative ions. In a **water softener**, hardness is removed by a **cation** (positive ion) exchanger, which binds hardness ions like magnesium ion and calcium ion and releases an equivalent (by electrical charge) amount of sodium ion. Negative ions are removed by an anion (negative ion) exchanger and replaced with chloride ion. This process is often used for removal of bicarbonate ion, called dealkalization. The cation and anion exchange materials may be combined in the same vessel. These units are regenerated by treatment with a concentrated solution of sodium chloride, which drives off the captured ions and replaces them with sodium and chloride ions. In another variation, called demineralization, the positive ions are replaced with hydrogen ion and the negative ions with hydroxide ion (see Figure 4.4). The hydrogen and hydroxide ions combine to give water, so complete deionization is possible. In this case, the cation exchanger and the

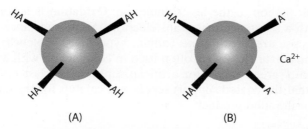

Figure 4.4   (A) Cation exchange resin. (B) Resin exchanged with calcium ion.

anion exchanger must be in separate vessels. The cation exchanger is regenerated with a strong acid like hydrochloric acid, and the anion exchanger is regenerated with sodium hydroxide (**caustic** soda).

Reverse osmosis can, in theory, remove virtually any impurity in water, generating *reverse osmosis water* or *RO water*. It involves driving the water through a membrane with pores on the order of the effective size of a water molecule. In practice, it is usually necessary to pretreat the water to avoid damaging the membrane. **Pressure** is applied to the impure side to drive the water to the pure side, leaving the impurities behind. Without pressure, water would prefer to move toward the impure side. The amount of pressure needed to overcome this force is called the **osmotic pressure**. The osmotic pressure increases with increasing concentration of substances that do not pass through the membrane. The flow rate through the membrane depends on the pressure in excess of the osmotic pressure and the area of the membrane. As water is driven through the membrane, the concentration of impurities on the inlet side increases, so the osmotic pressure increases. To maintain a steady state, up to 20% of the inlet water with concentrated impurities is discarded. Practical membranes are usually made in the form of long, thin tubes to provide adequate mechanical strength and area. Applied pressure is generally in the range of 7–15 bar (100–220 **psi**). It is possible to run a reverse osmosis system using only the pressure of the municipal water system. These systems discard 85–95% of the feed water and are not regarded as suitable for brewery use. Related purification methods include nanofiltration, ultrafiltration, and microfiltration, in order of increasing pore size. There are no sharp boundaries between these methods. Microfiltration is used to remove bacteria from beer. Membranes for nanofiltration have pores small enough to block hardness and alkalinity ions, but not as small as those of osmotic membranes.

Iron and manganese ions are problem **pollutants** in brewing water because they form deposits in pipes and equipment, they give rise to off-flavors, and they promote staling reactions in packaged beer. Both **elements** exist in water as +2 ions. The hydroxide of the +3 ion of iron, $Fe(OH)_3$, is very insoluble; water provides the hydroxide ion. Manganese +4 ions form insoluble MnO_2. The strategies for removal of iron and manganese ions involve **oxidizing** the more soluble +2 ion to the less soluble ion. **Oxidation** is accomplished by reaction with a substance that grabs electrons. Some substances often used for this purpose are air, chlorine, sodium hypochlorite (**bleach**), and potassium permanganate ($KMnO_4$), often used in conjunction with MnO_2 coated on a mineral ("green sand") or artificial support. The last option has the advantage that the coated support serves as a catalyst for the oxidation and as a filter for the solid product.

Water Profiles and Adjustment for Brewing

Describing Hardness and Alkalinity   Systems describing hardness and alkalinity are very different from those describing other potential solutes. One reason for this is that hardness and alkalinity are effects that can be caused by several ions. The contribution to hardness or alkalinity goes by the number of ions, not by their mass; 24.305 g of magnesium ion (1 mol) makes the same contribution to hardness as 40.078 g of calcium ion (1 mol). Many sources are lax about specifying exactly what units they are using. For individual ions, substances, and combinations like total dissolved solids (TDS), the usual measurement is milligrams per liter (mg/L), also called parts per million (ppm). Some water treatment companies use grains per gallon (gr/gal or gpg) = 17.12 mg/L. Table 4.2 provides conversions among units used to express hardness and alkalinity. To convert the unit at the top to that on the left, multiply by the entry in that column and row. For example, to convert 12 German degrees (dGH) to ppm as $CaCO_3$, 12 dGH × 17.848 = 214 ppm as $CaCO_3$.

Water Reports   Annually updated municipal water reports should be available from water suppliers. Unfortunately for brewers, municipal water reports tend to focus more on safety and quality regulations as required by the government, often lacking beer-centric metrics. For some regions, multiple

TABLE 4.2   Hardness and Alkalinity Conversions

	mEq/L	ppm	dfH	dGH	°Clark	gr/gal	Name
mEq/L	1	0.01998	0.1998	0.3567	0.2848	0.3421	Milliequiv per L
ppm	50.043	1	10	17.848	14.254	17.118	ppm as $CaCO_3$
dF	5.0043	0.1	1	1.7848	1.4254	1.7118	French degrees
dGH	2.8039	0.05603	0.5603	1	0.7986	0.9591	German degrees
°Clark	3.5109	0.07016	0.7016	1.252	1	1.2010	Clark degrees
gr/gal	2.9234	0.05842	0.5842	1.0426	0.8327	1	Grains per US gal

water sources are blended in response to demand and availability, resulting in a frequently changing water profile. For these reasons, municipal water reports are a useful approximation but cannot always be depended upon to establish the character of brewing water. There are commercial laboratories who will provide detailed water analysis for a fee. Some laboratories specialize in brewing water chemistry. Do-it-yourself water analysis kits are also available but typically only have sensitivities in the 10–50 ppm range depending on the test. More quantitative tests are described in Section "Analytical and Quality Control Procedures for Water."

A water report for brewing should measure the concentration of the following ions:

- Total hardness (unnecessary if calcium and magnesium are provided)
- Calcium
- Magnesium
- Alkalinity (unnecessary if bicarbonate is provided)
- Sulfate
- Chloride
- Sodium
- Nitrate

Advanced water chemistry, and truly mastering the impact of water on your beer, involves understanding the ionic content of the water. For most beer styles, brewing water should have moderate hardness and low to moderate alkalinity. As we will discuss below, these two water qualities greatly influence mash pH, arguably the most important impact of brewing water. Table 4.3 provides a range of hardness values in water.

Influence of Water Profile on Beer   The grain bill of a mash and the underlying water chemistry play a major role in establishing an acceptable mashing pH of 5.2–5.6. For example, when 100% pilsner malt is mashed with distilled water, the pH is typically 5.7–5.8. The natural acidity of roasted specialty malts can greatly affect the pH. Using a dark crystal or roasted malt

TABLE 4.3   **General Hardness Values in Water**

Concentration as $CaCO_3$	Indication
<60 ppm	Soft water
60–120 ppm	Moderately hard water
120–180 ppm	Hard water
>180 ppm	Very hard water

at 20% of the grist can lower the pH by as much as 0.5. In distilled water, 100% chocolate malt can yield a pH as low as 4.3. The starting pH of water is not very important. If the pH needs to be adjusted, steps should be made to adjust the mash pH rather than the water pH.

The synergy of malt, water ionic content, and pH is best illustrated in areas of the world famous for their particular beer style. The town of Pilsen (Plzeň) in the Czech Republic is the birthplace of the pilsner style of beer. Pilsner is a very pale, clear **lager** with a light, clean, noble hop flavor. The water in Pilsen is very low in bicarbonates, very soft, and free of most minerals. Historically, because the brewers in Pilsen used only pale malts, they needed an **acid rest** to reach an appropriate mash pH. The acid rest is a several-hour-long step that permits the malt enzyme phytase to release phosphoric acid, thus lowering the pH.

In contrast to pilsners from Pilsen is stout from Dublin, Ireland. The water of Dublin is high in bicarbonate (HCO_3^-), but without enough calcium to balance it. This yields hard, alkaline water that overcomes malt acidity. With this water profile, a 100% base malt mash may produce a mash pH higher than 5.8, which can extract harsh **phenolic** and **tannin** compounds from the grain husks. The water profile of Ireland is well suited to stout **ales**, because the roasted black malts add acidity to the mash, enabling an appropriate mash pH of 5.2–5.6. A comparison of the water profiles of Pilsen and Dublin is shown in Table 4.4.

In summary, light lagers cannot be made with the extremely **hard water** of Dublin, and dark beers cannot be made with the very **soft waters** of Pilsen. In brewing beers to style, a brewer might find it appealing to try to match the geographic water profile of an area known for that style. Unfortunately, this is a bit like shooting a target in the dark, as the exact conditions of the brew are usually unknown. It may be hard to exactly mimic the same grain bill in that brew, but more importantly, brewers, especially in areas with very hard or alkaline water, may modify the water profile.

Adjustment of Brewing Water   The addition of food-grade compounds, referred to as "brewer's salts" in the industry, allow a specific water profile to be attained. A specific profile may be needed to hit a target mash pH or accentuate particular beer flavors such as bitterness, maltiness, sweetness, or saltiness. A summary of ionic contributions to beer flavor is provided in

TABLE 4.4 **Comparison of Historical Water Profiles of Pilsen and Dublin (in mg/L)**

	Ca^+	Mg^{2+}	HCO_3^{2-}	Cl^-	Na^+	SO_4^{2-}
Pilsen	10	3	3	4.3	4	—
Dublin	119	4	319	19	12	53

Table 4.1. A summary of brewer's salts used for water adjustment is provided in Table 4.5. Note that these adjustments specifically refer to increasing ionic content. Decreasing ionic content was covered in the previous section.

In addition to the effects of sulfate on bitterness or bicarbonate on perceived maltiness, the brewer should pay attention to the *sulfate-to-chloride ratio*. The combination of high chloride and high sulfate can generate a harsh bitterness, and so one should be kept lower relative to the other, preferably the chloride. Recommended sulfate-to-chloride ratios are 4 : 1 for a hoppier beer and 0.5 : 1 for a fuller, maltier balance.

When adding brewer's salts to water, remember that one salt addition provides two ions, one cation and one anion. This will be an important consideration when calculating the impact of each addition on the overall water profile. As we will see below, two ions are not present in the salt in equal masses. This means that the ionic contribution of each ion in water is different and must be considered. Certain salts have water molecules in their formulas. These salts are referred to as hydrated salts. For example, calcium sulfate

TABLE 4.5 Common Brewer's Salts Used for Water Adjustment

Scientific Name	Common Name	Notes	Composition
Calcium carbonate $(CaCO_3)$	Chalk	Raises pH. Very limited solubility in water; should be added directly to mash. Used for dark beers	40.04% Ca^{2+} 59.96% CO_3^{2-}
Calcium sulfate $(CaSO_4 \cdot 2H_2O)$	Gypsum	Lowers pH. Accentuates crispness and bitterness for hop-forward beers	23.28% Ca^{2+} 55.79% SO_4^{2-}
Calcium chloride $(CaCl_2 \cdot 2H_2O)$	—	Lowers pH. Increases calcium as needed for soft water	27.26% Ca^{2+} 48.23% Cl^-
Magnesium sulfate $(MgSO_4 \cdot 7H_2O)$	Epsom salt	Minimal effect on pH. Accentuates crispness and bitterness for hop-forward beers	9.86% Mg^{2+} 38.97% SO_4^{2-}
Sodium bicarbonate $(NaHCO_3)$	Baking soda	Raises pH through increased alkalinity. Good addition for dark beers	27.37% Na^+ 72.63% HCO_3^-
Sodium chloride $(NaCl)$	Table salt	No effect on pH. Promotes saltiness	39.34% Na^+ 60.66% Cl^-
Magnesium chloride $(MgCl_2 \cdot 6H_2O)$	—	Minimal effect on pH. Used to increase magnesium in deficient profiles without affecting sulfates	11.96% Mg^{2+} 34.88% Cl^-

is commonly sold as gypsum, which is a dihydrated form, $CaSO_4 \cdot 2H_2O$. Magnesium sulfate is commonly sold as Epsom salt, which has seven water molecules, $MgSO_4 \cdot 7H_2O$. A salt with no water molecules is called an *anhydrous* form. Some salts, especially the chlorides, have several **hydrates**, each with a different composition.

Proper safety precautions should always be taken when adding salt additions to hot liquor tanks or the mash. Appropriate gloves and eye protection to protect against chemicals and splashes are required. *Extreme caution is warranted for calcium chloride additions because heat is released that will cause spattering unless the salt is added slowly to plenty of water. Salt should always be added to water, not water to salt.*

Calculating Salt Additions to Reach a Target Water Profile   The impact of salt additions on water chemistry is relatively easy to calculate. We will illustrate this process through an example. You work for a brewery in Philadelphia with the following water profile as identified by the Philadelphia Water Department:

Philadelphia, PA – Queen Lane Reservoir

Ca^{+2}	Mg^{2+}	Na^+	Cl^-	SO_4^{2-}	Alkalinity	pH
43	13	35	92	50	65 (HCO_3^-)	7.1

Ionic content in ppm

You are interested in making an IPA with a very bright, firm bitterness like that of the famous ales made in England at Burton-on-Trent. To approach the **astringent** bitterness of this region, you would like to do a single chemical addition in 100 L of water to increase your sulfates to about 300 ppm. This process, informally called "Burtonizing," requires the addition of gypsum. Gypsum is calcium sulfate dihydrate ($CaSO_4 \cdot 2H_2O$). There are three steps to water addition calculations. (1) Determine the mass percentages of the salt. (2) Determine the mass of salt needed to reach the given level. (3) Determine the impact of the balancing ion on the final water profile.

Step 1: We first calculate the mass percentages of the elements in gypsum. The molar mass of each element is listed on the periodic table of the elements (Figure 2.3). Referring to the periodic table, we know the following:

$$Ca = 40.1 \, g/mol; \, S = 32.1 \, g/mol; \, O = 16.0 \, g/mol; \, H = 1.0 \, g/mol.$$

Given the formula for gypsum ($CaSO_4 \cdot 2H_2O$), we now determine the total molar mass and the molar mass of sulfate.

Item		MM (g/mol)	Mass (g/mol)	Mass Fraction
Ca	1	40.1	40.1	40.1/172.2 = 0.233
SO_4	1	$32.1 + 4 \times 16.0$	96.1	96.1/172.2 = 0.558
H_2O	2	$2 \times 1.0 + 16.0$	36.0	36.0/172.2 = 0.209
$CaSO_4{\cdot}2H_2O$			172.2	

The **mass fraction** is easier to use than the mass percent in mass calculations. Mass fraction = mass percent/100. The mass percent of the ions in various brewer's salts is given in Table 4.5.

Step 2: We now need to calculate the amount (in grams) of gypsum required to raise the sulfate level to 300 ppm in 100 L of water. Starting with a water profile that already contains 50 ppm of sulfate, we need to increase the amount by 250 ppm in 100 L of water. Remember that ppm = mg/L. First, determine the mass of sulfate required to reach 250 ppm in 100 L. Then convert mg to g:

$$250 \text{ mg}/\text{L} \times 100\,\text{L} \times \frac{1\,g}{1000\,mg} = 25\,g$$

We need to calculate the mass of gypsum that provides 25.0 g of sulfate:

$$\frac{\text{Mass sulfate}}{\text{Mass gypsum}} = \text{mass fraction} = 0.558$$

$$\text{Mass gypsum} = \frac{\text{mass sulfate}}{\text{mass fraction}} = \frac{25.0\,g}{0.558} = 44.8\,g$$

Step 3: We now need to account for the increase in calcium upon addition of 44.8 g of gypsum in 100 L of water. What is the final amount of calcium (in ppm) in the 100 L? From step 1, we know the mass fraction of calcium in gypsum is 0.233. To determine the amount of calcium added (in g),

$25.0\,g \times 0.233 = 5.8\,g$.

To convert this to ppm (mg/L), we convert the mass to mg and divide by the liters:

$$\frac{5.8\,g \times 1000 \text{ mg/g}}{100\,L} = 58\,mg/L = 58\,ppm$$

According to the water report, we started with 43 ppm calcium. The gypsum provided another 58 ppm for a total calcium concentration of 101 ppm.

Additional salt additions can be calculated in the same manner. Brewers who do this type of calculation often prepare spreadsheets with the relevant data inserted.

Analytical and Quality Control Procedures for Water

Hardness The standard method for measuring hardness is **titration** with ethylenediaminetetraacetate (EDTA), which binds strongly to hardness ions. The method, in brief, is that a measured volume of water is treated with an ammonia buffer to give a pH of 10. A small amount of an **indicator** dye, typically Eriochrome Black T®, is added. The indicator gives a red color when hardness is present. A solution of disodium EDTA of known concentration is added in a controlled way, a procedure called titration, so that the amount added at any point can be determined. When the hardness ions have all reacted with the EDTA, the dye turns from red to blue. Addition is stopped and the moles of EDTA are equated to the moles of hardness ions. Fast but approximate measurement can be done with test strips impregnated with proprietary **mixtures** including EDTA and indicator dyes that give a color change used to estimate the hardness.

Alkalinity The standard method for alkalinity involves titration with a solution of sulfuric acid while simultaneously measuring the pH. Briefly, a measured volume of water is titrated with a sulfuric acid solution of known concentration. The volume of acid and the pH are monitored. When the pH reaches 4.5 (other values are sometimes used), the volume of acid is recorded. Each mole of sulfuric acid can neutralize two equivalents of alkalinity. The milliequivalents of alkalinity are equal to twice the number of moles of sulfuric acid needed to lower the pH to 4.5. Hydrochloric acid can also be used for the titration; one mole of hydrochloric acid corresponds to one equivalent of alkalinity.

Total Dissolved Solids The defining laboratory method for total dissolved solids (TDS) analysis is to filter the water, evaporate a weighed sample, heat it to 180 °C, then cool, and weigh the residue. Dissolved ions also give rise to electrical conductivity, which is measured by a TDS meter. There are several issues that limit the reliability of TDS meters. Only dissolved ionic compounds conduct. A contaminating molecule like sugar would not register on the meter at all. Volatile ionic substances conduct but are not dissolved solids. A small amount of hydrochloric acid, which is not a dissolved solid, gives a large signal on a TDS meter. Each type of ion makes a different contribution to the conductivity. A 10 ppm solution of potassium sulfate gives a different conductivity reading from a 10 ppm solution of sodium chloride. Because of these issues, the TDS meter gives reliable results only for solutions whose compositions are closely related to the calibrating solutions. TDS meters are useful for tasks like determining changes in brewing water, indicating potential problems with equipment or supply.

Taste/Smell/Color   Water quality can fail. Perhaps a gasket deteriorated into it; or there was an error in the use of treatment or regeneration chemicals; or there was bacterial contamination in the pipes. Water for brewing should be inspected every brew day. Put water from your treatment system into a clean, clear **glass** container – the same one every time. Visually inspect the quality against an evenly lit white surface for cloudiness and any discernable color. Taste and smell the water and have others who have good flavor judgment do the same. If there are any problems, spend the day working on your water treatment system instead of brewing. If you solve the problem, it will be a very profitable day.

4.2 MALT

Barley (***Hordeum vulgare***) malt is barley seed that has been **steeped, germinated**, and dried in a process called malting. The process of malting enables the creation of a very specific set of tools that the brewer will be able to use in the brewery. Malt provides such essential components as:

- Starch (complex carbohydrates).
- Enzymes to generate simple carbohydrates from starch during the mash.
- Proteins – foam and haze.
- Proteases to generate free amino nitrogen (FAN), promoting extract.
- Lipids.
- Polyphenols/tannins.
- Vitamins.
- Filter material for the lauter.
- Color and flavor.

Grain Biology

Barley is a **cereal** grain that is classified as winter or spring and as **two-row** or **six-row** varieties. Winter barley is planted in the late fall and harvested in the spring or early summer. It sprouts in the fall but does not flower until it is exposed to winter temperatures. Spring barely is planted in the early spring and is harvested near the end of the summer. Flowers of barley plants are carried in groups of three spikelets at **nodes** (joints) on a stem, one group per node. The groups alternate from one side of the stem to the other. In six-row barley, each flower yields a seed, so there are three rows of seeds on one side of the stem and three on the other. In two-row barley, only the middle spikelet of the group of three yields a seed; the other two are sterile. This gives one row of seeds on each side of the stem. Surprisingly, this distinction is controlled by only two genes. Both two-row and six-row varieties are used in brewing. Two-row barley is usually preferred for craft and all-malt brewing. Six-row varieties find use in beers made with starchy, nonenzymatic adjuncts because the higher protein content of six-row leads to higher levels of starch **hydrolysis** enzymes in the malt. Table 4.6 provides a general overview of the differences between two-row and six-row barley. This table is a generalization; the development of new varieties has enabled selection of favorable traits for the brewing industry in both barley varieties.

A barley seed is illustrated in Figure 4.5. The seed or kernel has three main parts: the germ or **embryo**, the **endosperm**, and the **seed coat**, which includes the hulls. The germ region, containing the embryo, will develop into the barley plant. During the malting process it produces an **acrospire** that grows up the dorsal side toward the tip and rootlets that emerge from the base. The dorsal and ventral sides of a barley seed are distinguished by the ventral furrow, a large groove that runs the length of the ventral side.

The endosperm consists of dead cells filled with starch granules, each surrounded by a protein-rich matrix. The granules are semicrystalline (partly

TABLE 4.6   Differences Between Two-Row and Six-Row Barley

Two-Row	Six-Row
Mostly used for malting	Mostly used for animal feed
Uniform size	Variable sizes
Thinner hull	Thicker hull
Less polyphenol	More polyphenol
Less total protein	More total protein
Higher extract potential	Lower extract potential
Less beta-glucan	More beta-glucan

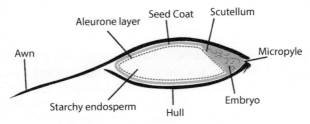

Figure 4.5   Barley corn. *Source*: From Barth (2013) © 2013 Wiley. Used with permission.

crystal, partly **glassy**). They serve as an energy stock for the embryo, providing a source of sugar until the seedling can grow leaves and begin photosynthesis. But during the malting process, care is taken to limit consumption of the starch by the embryo.

Each endosperm cell is surrounded by a cell wall made mostly from complex carbohydrates and protein. The outer layer is mostly **hemicellulose**. The middle **lamella** is a mostly proteinaceous layer between the cells. Figure 4.6 compares cell walls from the shoot (acrospire) with those from the endosperm. The endosperm walls have no detectable pectin (a polymer of mostly galacturonic acid) or xyloglucan (polymer of glucose with **xylose** side chains). Endosperm walls have seven times as much beta-**glucan** and somewhat less, though still significant arabinoxylan, a polymer of xylose with **arabinose** side chains. Some of the OH groups on the xylose and arabinose rings form **esters** with ferulic acid and **acetic acid**. Ferulic acid is potentially important because it can be released by hydrolysis and subsequently decarboxylated to highly flavor-active 4-vinylguaiacol. Endosperm cell walls are easier to hydrolyze during malting than walls of other cells. One of the key objectives in the malting process is to break these cell walls down to gain access to the starch. Timing is everything. There must be enough modification to break down the cell walls but not so much as to consume starch. Surrounding the endosperm is a triple layer of cells derived from the embryo called the **aleurone**. The aleurone will be important for producing starch-degrading enzymes during the malting process.

THE FERULIC ACID REST

In the endosperm cell wall, ferulic acid forms ester bonds with pentosans. During mashing, a ferulic acid rest at 45 °C (113 °F) for 15–20 minutes has been shown to liberate ferulic acid through the activity of beta-glucanase, which degrades both beta-glucan and pentosans. Free ferulic acid is important in wheat beers because it is decarboxylated into 4-vinylguaiacol during fermentation (by POF+ yeast), which contributes to the characteristic phenolic flavor of the style. Furthermore, 4-vinylguaiacol is converted by certain ***Brettanomyces*** yeast species to 4-ethylguaiacol, described as having a clove or smoky flavor.

The seed coat of a barley seed consists of three layers. Moving from inner to outer, the **testa** is flush against the aleurone layer and is permeable only to water. The next layer is the **pericarp**. The outermost layer consists of cellulose-rich **hulls**, the palea (front hull) and the lemma (back hull). The **awn**, also called the *beard*, is at the tip of the lemma; it is removed before malting during

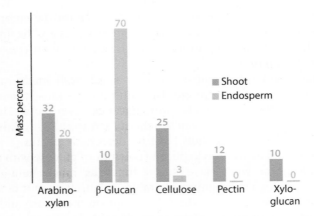

Figure 4.6   Endosperm cell wall carbohydrates.

cleaning. The awn protects barley crops from herbivores like deer. The ventral side of a barley seed has a groove called the ventral furrow (Figure 4.7).

Malting   Malt is grain that has been soaked in water, allowed to sprout, and is then dried. Each of these steps has a key role in the malting process. An overview of the malting process is shown in Figure 4.8. Grain is harvested and dried on the farm to 12% moisture. The barley is cleaned in a series of sieves to remove stones, dust, and undersized seeds. The malting process begins with steeping during which the seed is hydrated and the embryo is activated. Then the hydrated seeds are transferred to a germination chamber. During germination the embryo begins to grow, releasing **hormones** that stimulate the aleurone layer and **scutellum** to produce important enzymes that enable modification. Finally, modification activity is stopped in the drying or **kilning** process, which produces key flavor and color compounds.

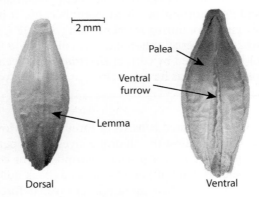

Figure 4.7   External view of a barley kernel.

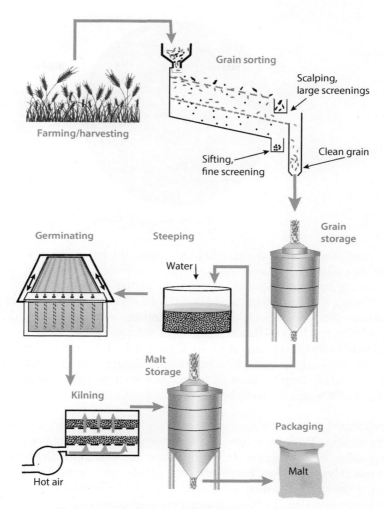

Figure 4.8   Overview of the malting process.

Steeping   Raw barley seed is dried to about 12% moisture before storage. The grain, having been cleaned and freed of debris, like broken kernels, stones, weed seeds, and bits of wire, is added to a tank of water. The grain is soaked for several hours, an operation called steeping. Then the water is drained, and the barley is allowed to rest in air (often in flowing air).

The purpose of steeping is to activate enzymes and begin germination. The embryo is activated at a moisture content of around 30–35%. This step triggers **aerobic** respiration indicated by the following equation: sugar + O_2 → CO_2 + H_2O + energy (**ATP**). As a result, there must be a source of oxygen and a method to dissipate carbon dioxide and heat. This is

Figure 4.9   Barley chit. Photograph: Naomi Hampson.

accomplished through the air rests. Otherwise, the barley seeds would drown. Steeping also helps to wash bacteria and dust from the raw grain.

During steeping, the moisture content of the grain increases to 44–48%. Details like the temperature and duration of each steep and air rest are adjusted to match the requirements of the characteristics of the grain and the type of malt being produced. Usually there are a total of two or three steeps over a period of two days, typically at 13–18 °C (55–64 °F). Each steep is approximately 4 hours, and each air rest is approximately 20 hours, with variations as needed. By the end of steeping, most of the seeds will show a white spot at the base, called a **chit**, where the root sheath breaks through the hull, shown in Figure 4.9.

Germination   After steeping, the malt is allowed to sprout, a process called germination. To grow a new plant, the embryo needs energy provided by respiration. Before it grows large enough to make sugar through photosynthesis, it must rely on its own starch reserves in the endosperm. A critical process during this step is the generation of enzymes called **amylases** that will convert the starch into simple sugars. Another major process goal is the breakdown of the endosperm cell walls to gain access to the starch. Collectively, this process is called **modification**. The germination step must be fine-tuned to maximize enzyme production and cell wall modification but minimize carbohydrate consumption. The more sugar used for growth, the less remains for beer!

The traditional method of germination is to spread the steeped grain in a layer on a floor and to turn the grain with rakes and shovels. Modern methods transfer the steeped grain into a separate container, such as a Saladin box, which provides automated mechanical turning enabling the grain to be piled deeper. During germination, the grain breathes oxygen, releases carbon dioxide, consumes starch reserves, releases heat, and grows rootlets, called **culms** (Figure 4.10). In all germination systems, the malt must be turned regularly to allow oxygen to reach the seeds and to allow carbon dioxide and heat to escape. Regular turning is also needed to prevent the rootlets from becoming entangled, which would make the malt difficult to handle. High air flows provide oxygen and facilitate removal of heat and carbon dioxide. Germination is typically conducted at 18–22 °C (64–72 °F) for a period of four days.

Hydration as low as 30% will foster germination. Typically 40% moisture is targeted. The hydrated embryo produces hormones that migrate to the aleurone layer to regulate enzyme synthesis. During modification, key enzymes are produced such as **beta-** and **alpha-amylase** (collectively referred to as **diastase**), **limit dextrinase**, proteases, glucanases, lipases, lipoxygenases, and phosphatases. The glucanases break down the outer lamella, proteases hydrolyze the middle lamella, and the endosperm becomes more porous, going from **steely** to **mealy**. The amylases will now have access to the starch. Because of the location of the embryo at the base of the seed, modification proceeds from the base to the tip. After modification, the kernels become more **friable**, that is, easy to crush. Culms emerge, as shown in Figure 4.10.

During germination, the shoot of the plant, called the acrospire, travels up the dorsal side of the seed. It may puncture the testa but should not be allowed to puncture the hull. The acrospire can be observed as a swelling under the hull on the dorsal side and is often used as a rudimentary measure of germination progress. For most pilsner and pale malts, the acrospire is permitted to grow no more than two-thirds to three-quarters of the seed length. For darker malts, the acrospire's length may be ¾ to full length. The process of growing the acrospires and rootlets requires the breakdown of starch and consumption of sugar for energy; thus over-modification leads to loss in **extract** or fermentable sugar in finished malt.

Figure 4.10   Culms.

The main plant hormones released by the embryo belong to a family called **gibberellins**. It is possible to supplement the natural gibberellin with artificially produced gibberellic acid (GA), usually sprayed on during germination, if used. Modification time is expedited by using GA for "two-way" modification, whereby the tips of the seed are gently sheered, promoting natural modification from the embryo at the base and from the added GA at the tip. Most maltsters do not use GA. Furthermore, the use of GA would be a violation of the *Reinheitsgebot*.

Kilning   When modification is adequate for the type of malt being made, the germination process is stopped by drying the malt, also called kilning. The details of the drying process are the most important factor affecting the flavor of the malt. Malt before drying is called **green malt**. Drying is accomplished by blowing hot air through a bed of malt in a device called a **kiln**. The details of the drying process vary depending on the type of malt desired. The drying process to make base malt follows three stages. In the free-drying stage, easily accessible water from the surface of the grains is dried in a fast stream of air whose inlet ("air-on") temperature is around 50 °C. The air gives up energy to evaporate the water, cooling down and getting more humid as it passes through the bed of grain. The air emerging from the kiln is **saturated** with water. Respiration continues until the inside temperature of the grains exceeds 50 °C. At the end of free drying, the air temperature rises, because all the easily evaporated water has been removed. This happens first near the air inlet and progresses through the bed. When all, or nearly all, of the grain has lost its unbound water, the moisture content falls to about 10–15%. This stage is diffusion drying; the rate of drying depends on how quickly the water vapor can diffuse from inside to the surface of the grain kernels. Air flow is decreased, and the inlet air temperature is increased to 65–75 °C. The exit air is relatively dry and can be used to dry another batch of malt. The malt enzymes are particularly vulnerable to degradation during this stage because of the presence of moisture and potentially high temperature. At the end of diffusion drying, the moisture content reaches about 5%. The final stage, called **curing**, raises the temperature to around 80–110 °C, depending on the how pale or dark the malt is intended to be. The final moisture content is close to 4%. The drying process is regulated to maintain the activity of the amylase enzymes, which will be used by the brewer in the mashing process. The entire kilning step takes about two days.

After drying, the culms, rendered brittle by the heat, are mechanically removed. The malt is cooled, cleaned by sieving, and stored for three weeks or more before it is used for brewing. The storage period improves the performance of the malt in milling and wort separation. It is believed that storage evens out the distribution of moisture in the batch and promotes flavor stabilization.

Types of Malt

There are three broad classes of malt. **Base malt** retains enough amylolytic (starch-hydrolyzing) enzyme activity to allow it to convert starch to sugar during mashing. The ability of malt enzymes to convert starch is called the **diastatic power** (DP), measured in the United States and the United Kingdom in degrees Lintner (°L). The minimum diastatic power that a malt requires to convert its own starch to sugar is about 40 °L. Variations of base malt are produced by changing the degree of modification and the temperature and duration of curing. Because base malt retains amylolytic activity, it is typically used at 70–100% of the grain bill. Crystal malt, also called **caramel malt**, has been processed to convert most of its starch into sugar in the kiln. Depending on processing conditions, some of the sugar is allowed to react with amino acids to give unfermentable but flavorful compounds. Crystal malts are often used at 2–10% of the total grain bill. **Roasted malt** is subjected to an additional heating step in a specialized roaster to give dark colors and intense flavors. Crystal and roasted malt have little, if any, diastatic power, because their amylolytic enzymes have been inactivated during manufacture. Roasted malts are used at 1–8% of the grain bill. A fourth category of malt that we will call special **processed malt** has characteristics that derive from sources external to the grain.

There are many gradations and variations for malt. A type of malt from one company may be significantly different from one with the same name from another. Let the brewer beware. Many malt names are trademarks, so virtually identical malts may have a different designation. Table 4.7 gives color according to the **Lovibond** scale or the more or less equivalent **American**

TABLE 4.7 Typical Malt Data

Malt Type	Color (SRM or Lovibond)	Diastatic Power (°L)
Distiller's malt	2.4	240
Pilsner/lager	1.2–2.5	140
Pale ale	3.5	85
Vienna	3.5	80
Munich	10	40
Dark Munich	20	20
Wheat	2.2–4	170
Crystal/caramel	5–120	0
Dextrin	1.5	0
Brumalt/melanoidin	25	0
Biscuit/amber	20–36	0
Chocolate	200–500	0
Black	400–600	0
Roasted barley	300–500	0

Society of Brewing Chemists (ASBC) **SRM** scale and diastatic power in degrees Lintner for a variety of malt types.

Base Malts   *Distiller's malt* has the highest diastatic power of any standard malt. As the name implies, it is used to make whisky with a high content of non-malt starch ingredients, like maize (**corn**). Barley for distiller's malt is high in protein to allow for more enzyme production. The barley is allowed to germinate longer than brewer's malt, resulting in an elevated level of amylolytic enzymes. Curing is done very gently to preserve the enzymes. The high level of protein could cause haze and gushing problems in beer. Distillation leaves the protein behind, so these malts are appropriate for whisky.

More common to beer production are pilsner, lager, pale, two-row, and six-row malt (the latter distinction relates to the barley itself). Low temperature curing at 54–70°C (129–158°F) leaves pilsner malt with high diastatic power that allows its enzymes to saccharify (make into sugar) its own starch plus as much as 35% non-diastatic starchy materials like maize and rice. This enables the brewer to produce a product with a very light, crisp flavor and a pale color. Beers of this style are the most popular worldwide.

Pale ale malt is cured at a slightly higher temperature than pilsner malt, 60–80°C (140–176°F). The higher temperature curing drives off dimethyl sulfide (DMS), so it is less susceptible to vegetable off-flavors. Pale ale malt has a subtle malty and biscuit profile especially suitable to British ales and their stylistic offspring.

Vienna malt is similar to pale ale malt in color and diastatic power. The curing process is short but at high temperature around 100–110°C (212–230°F). The flavor it gives to beer has been described as toasty and nutty. Vienna is a major ingredient in the Vienna, Marzen, and Oktoberfest (VMO) beer style.

Munich malt has a unique drying schedule. The green malt is warmed under humid recirculating air to 38°C (100°F) prior to drying. Curing is at 105–120°C (221–248°F), depending on the desired color. Munich is the most highly cured base malt. It retains just enough diastatic power to saccharify its own starch. The key contribution of Munich is a malty flavor. It is the base malt for Bock, Dunkel, and Oktoberfest styles, and it is sometimes used to boost the maltiness in others, including pale ale. Many maltsters offer two types of Munich, regular and dark.

Crystal Malt   Several varieties of malt are subjected to "stewing" in which the green malt is given a high humidity heat treatment before drying. This heat treatment causes the amylase enzymes to convert the starch to sugar within the endosperm. The kilning process then "crystalizes" the sugar. Crystal malts have no diastatic power.

Standard *crystal malt*, also known as *caramel malt*, is briefly dried to remove surface water. The kiln is then sealed and warmed up to 60–72 °C (140–162 °F) for saccharification of the starch. The malt is then cured in dry air in the range of 120 to nearly 180 °C (248–356 °F), depending on the color desired. During curing, the sugars mostly combine to give **unfermentable dextrins**, which add body to the beer. Crystal malt provides a variety of flavors such as malty, caramel, and toast.

Dextrin malt is a variety of crystal malt in which the stewing and curing are carried out at lower than the usual temperatures to give a product with little color or flavor, but which enhances mouth feel and foam.

Brumalt (Brühmalz), also called *melanoidin malt*, is subjected to high temperature of 50 °C (122 °F) at the end of germination. The green malt is stewed at 60 °C (140 °F), dried, and then cured for a long time at a low temperature in the range of 80–90 °C (176–194 °F). It imparts honey-like flavors. Brumalt is used as a substitute for prolonged boiling in the preparation of Bock and related styles of beer.

Roasted Malt   Malt and even unmalted grain can be subjected to heat treatment of varying intensity and duration to give a variety of products. Usually roasting is accomplished in rotating drums.

Biscuit malt, *amber malt*, and *brown malt* provide nutlike, bread, and toffee aromas. They are typically used in British-style ales. Today's brown malt is a completely different product from the historical version formerly used as a base malt for porter ale.

Chocolate malt is roasted to a very dark brown color at 150–200 °C (302–392 °F). It gives color and chocolate and coffee notes to dark beers. *Black malt*, also called black patent malt, is roasted nearly to the point of ignition (315 °C, 599 °F) and then cooled with a spray of water. It contributes acrid flavors and an astringent mouth feel to styles like stout ale. Prior to roasting, *roasted barley* is not malted, but it is steeped before intense roasting at 250 °C (482 °F). It is a defining addition to stouts as compared with porters.

Special Processed Malt   Special processed malts are subjected to processing or treatment that adds flavors or characteristics that are not inherent to the malt itself.

Acidulated malt [Sauermalz] is handled in a way that encourages growth of lactic acid bacteria either during germination or after drying. The resulting product is used in small quantities to lower mash pH and to provide a tart taste to the beer. Acidulated malt, unlike lactic acid itself, is permitted by the Reinheitsgebot.

Various flavors of *smoked malt* can be made by exposing the malt to the smoke of different types of wood. *Peated malt* is dried in a peat fire under conditions that expose the malt to the smoke. It is usually used for Scotch whisky, especially that made on the island of Islay.

Other Grains   Barley is the most common grain used for malt, but it is possible to malt other grains as well. After barley, the most commonly malted grains are wheat, oats, and rye. None of these are as easy to malt as barley, but they are used to make malts for certain styles of beer. Wheat loses its hull on threshing (separating the seed from the plant), which affects malting and mashing. Wheat kernels are larger than those of barely. The protein content is higher. Wheat malt is used in various styles like *weissbier*, *weizen*, and *wit*. Small amounts of wheat malt can be used to enhance foam in barley beer. Rye also loses its hull on threshing. Rye malt is used to give a spicy flavor to the beer. Oat malt is used in some styles, like oatmeal stout. **Gums** and fats in the oats impart a unique silky mouth feel. They also require adjustments to the brewing procedures. Several grains are attracting interest for potential use in gluten-free beer. These include sorghum, millet, and even quinoa, which is not a cereal grain at all.

Malt Flavor and Brewing Issues   Malt flavor is created during the kilning step. Sugars and amino acids, in the presence of heat, yield a Schiff base. The reaction is shown in Figure 4.11. The Schiff base is unstable; it undergoes a variety of reactions, collectivity known as **Maillard reactions**. One pathway gives rise to **melanoidins**, which are complex polymers with intense brown colors. The other pathway gives rise to small molecules, some of which are highly flavored. Just as toast develops more color and flavor with more intense heat treatment, so too does malt color and flavor develop with increasing temperatures and heating times. Two examples of the hundreds of flavored molecules known to form in malt are shown in Figure 4.12. Because

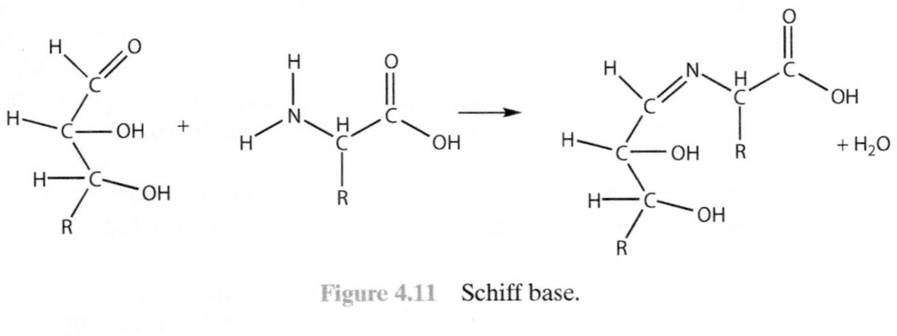

Figure 4.11   Schiff base.

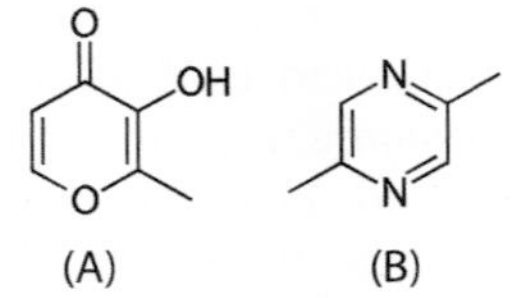

(A)   (B)

Figure 4.12   Flavor molecules. (A) Maltol and (B) 2,5-dimethylpyrazine.

the colored melanoidins and the flavored compounds originate from the same Schiff base reaction, high color and high flavor often go together. The inclusion of such flavorful malts has a decisive effect on the color of the beer and on some aspects of the flavor.

There is another outcome of the Schiff base reaction. Figure 4.11 shows that the amino ($-NH_2$) group on the amino acid is destroyed by the reaction. Because the amino group is basic, the reaction makes the malt more acidic. The use of highly kilned (usually dark) malt lowers mash pH. In the past, brewers whose water was alkaline, like London, got better results with more highly kilned malt. Those who had less alkaline water, like Pilsen in Bohemia, could get better results with paler malt. This is one of many factors, such as taxes, regulations, availability of supplies, transportation issues, and others, that gave rise to traditional regional or local beer styles.

Grain Quality for Malting

Grain, both before and after malting, must be of very high quality. A major part of quality maintenance is careful handling and storage. Issues of fungus, animal pests, skinned and broken kernels, and the like can interfere with malting and brewing processes.

Grain for malting must meet certain standards regarding the grain variety, level of protein, and quality of the seeds. Malting grain requires more effort and expense to grow and entails more risk to the farmer, so it commands a significant premium in price over grain of standard quality. Proper handling of grain begins before it is harvested. Malting barley must be grown with a minimum of fertilizer to keep protein levels low. The use of herbicides ("desiccants") to dry the plants prior to harvest is not permitted. Fungicides may be applied to prevent **fusarium head blight** and other fungus diseases that can make the grain unsuitable. Of all diseases that may affect barley, fusarium head blight, also called head scab, is one of the most detrimental. One species of *Fusarium*, *F. graminearum*, produces a **mycotoxin** called deoxynivalenol (**DON**), also called "vomitoxin" because of its effect on animal digestive systems. US regulations require DON to be less than 1 ppm in food for animal or human consumption. Severe fusarium infections can lead to DON levels exceeding 20 ppm. All barley malt should be tested for DON and reported by the maltster. Although not all *Fusarium* species produce DON, most produce polypeptides called **hydrophobins**. Hydrophobins can coat many materials and form high-potency bubble **nucleation sites**, leading to gushing.

The grain should be harvested as soon as possible at maturity, because rain on mature grain **heads** leads to early sprouting and other problems. The combine harvester should be set to minimize damage to the kernels. Broken and skinned seeds do not malt properly. The grain must be dried at low temperature to maintain the viability of the seeds. Excessive moisture can lead to mold growth. A moisture content of 12% is considered ideal for barley storage. It is important to remember that the seeds are alive; they consume oxygen and they release carbon dioxide and heat. Cool storage, adequate air flow, control of humidity, and freedom from insects and rodents are needed to maintain grain quality.

Raw grain should demonstrate good kernel shape and consistency. Smaller kernels absorb water faster, so variability in size leads to heterogeneity in modification across the batch. Raw barley seed should demonstrate a fresh, straw-like smell without any notes of mold or earth. Raw barley should be light yellow or straw color. A dull gray color can indicate rain damage. A green color indicates early premature harvest. Brown tips indicate water sensitivity, which is the inability of viable seeds to germinate in excess water. Red endosperms are caused by infection from *Fusarium*. Raw barley should also be subject to a float test. An excessive number of floating kernels indicate pest infection.

Analytical and Quality Control Procedures for Malt

There are 15 official Methods of Analysis of the ASBC dealing with malt, plus an additional 12 for raw barley. Similar methods are available from the **European Brewing Convention (EBC)**. As most brewers work with finished malt, we focus here on the most important methods to understand malt quality, discussing how each impacts the brewing process.

Malt specifications are reported on a Certificate of Analysis (COA). COAs should be obtained and analyzed for every lot of malt received in the brewery. Deviations in malt specification can lead to deviations in beer quality. Table 4.8 shows a representative COA. Table 4.9 summarizes the acceptable ranges for each quality trait. Note that these ranges represent superior quality. Beer can be made from malt with qualities that fall outside of this range, but this might impact process or flavor standards.

Most malt COA measurements are conducted on wort produced by the **Congress mashing** procedure. Congress mashing consists of a series of temperature increases and steps, usually with a Congress mashing machine that has capacity for many samples to be mashed at the same time. The ASBC version of the method starts with 50 g of fine-ground malt and 200 mL of distilled water. The malt is mixed with the water, and the mixture is stirred at 45 °C (113 °F) for 30 minutes and heated at 1 °C (1.8 °F)/min to 70 °C (158 °F); then 100 mL of 70 °C water is added. Stirring is maintained at 70 °C for 60 minutes. Additional cold distilled water is added gradually to cool the sample for filtration at room temperature. Enough water is added to bring the total mass to 450 g (net: 50 g malt and 400 g water). Filtration is conducted with a fluted filter paper, including a **vorlauf** step but excluding a sparge step. This method does not mimic a brewer's mash, but rather is used to standardize analysis of malt samples.

With any malt analysis, the brewer is concerned with five major traits:

1. Physical quality
2. Protein modification
3. Carbohydrate modification
4. Enzyme potential
5. Color and flavor

TABLE 4.8 Certificate of Analysis from a Fictitious Company

FarBar Maltings
"We wrote the book"

**Certificate of Analysis: FarBar Pale Malt.
Lot number 19383. Analysis Date: 8/27/2018**

Assortment	
On 7/64″ screen	64.4%
On 6/64″ screen	28.2%
On 5/64′ screen	5.6%
Thru 5/64″ screen	1.8%
Moisture	**4.4%**
Mealy/glassy	
Mealy	98.4%
Half-glassy	1.6%
Glassy	0%
Friability	79.2%
Extract	
Fine grind db	80.2%
Coarse grind db	79.3%
F–C difference	0.9%
Protein	
Total protein	10.42%
Soluble protein	5.19%
S/T	49.8%
Diastasis	
Diastatic power	116 °ASBC
Alpha-amylase	84.1 DU
Conversion time	<5 minute
Wort	
Color	3.5 °SRM
Viscosity	1.2 cP
Beta-glucan	188 ppm
FAN	262 ppm

TABLE 4.9   **Typical Ranges for Base Malt of Superior Brewing Quality**

Moisture	3.5–4.5%
Extract DBFC	80–82%
Extract DBCG	78–80.5%
Alpha-amylase	44–48 DU
Dextrinizing potential	70–100 °Lintner
Conversion	10–15 minute
Protein	9–11%
Mealy	90–97%
Glassy	0–1%
Friability	80–85%
On 7/64	75–85
On 6/64	10–20
On 5/64	0–3
Thru	0–2

Source: Noonan (1996).

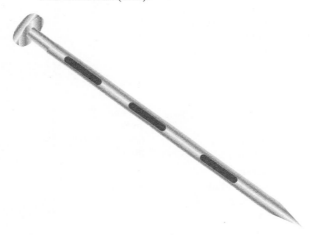

Figure 4.13   Grain trier.

The first task in any malt analysis is to get a representative sample. A shovelful from the top or the first bucket of malt from the silo hopper is not likely to be representative of the rest of the malt. Several samples should be collected from several different locations in the truck, bin, silo, or sack. Samples are taken with a grain **trier**, also called a *grain thief*, shown in Figure 4.13. The trier is stuck into the container, and, once it is in position, its doors are opened, allowing samples to be collected in compartments at different heights in the bed of grain. The doors are closed, and the samples are retained for analysis. The trier should be used at several locations within the malt shipment.

Physical Quality   Rapid physical tests are easily conducted by the brewer to check the quality of the incoming malt shipment. These include examination of foreign material, test **weight** per bushel, **1000 kernel weight** (KW), and the float test. Such tests are not typically reported on malt COAs.

To test for foreign material, weigh out 50 g of sample. Spread out sample on a flat surface and pick out all particles, husks, and contaminants. Classify materials removed, weigh on an analytical scale (±0.01 g), and report percentage by weight.

To test weight per bushel, weigh out 110 g representative sample of malt to the nearest 0.1 g and pour evenly into a funnel placed in a 250 mL graduated cylinder. Pour evenly and do not tap or jar the cylinder. Report the volume to the nearest 2 mL. Calculate the bushel weight: bushel weight (in lb/bu) = 8545.4/volume. Weight per bushel is a crude estimate of the plumpness of the malt. The use of assortment screens as described below is a better method.

To test the *1000 kernel weight*, obtain a representative sample of malt and count out 500 kernels, excluding debris. Weigh to the nearest 0.1 g. Multiply the weight by 2 and report at the 1000-kernel weight (KW), **as-is** basis.

To test for floaters, place an exact number (100–200) of malt kernels in water. Count the number floating and express as a percentage. Malt should sink. Floating malt may be compromised by infestation by insects which bore holes into the kernel.

Assortment refers to the size distribution of the grain or malt kernels, as measured by standard screens (Figure 4.14). The screens are plates with 0.75 in. (19 mm) slots whose width defines the seed size. For barley, the fraction (by weight) that is retained by a 6/64 in. (2.38 mm) screen is designated plump. The fraction that passes through the 6/64 in. screen but is retained by a 5/64 in. (1.98 mm) screen is thin. The fraction that passes through the 5/64 in. screen is thru. Sometimes retention on a 7/64 in. (2.78 mm) screen is also included to give a more complete indication of the size range. High fractions of thin and thru malt are undesirable. Properly sizing malt is critical to ensure that the mill gap settings are appropriate for crushing. This is particularly important for two-roller mills, as variability in malt size results in inadequate crushes (Table 4.10).

20 mm

Figure 4.14   Grain screen.

TABLE 4.10   **Screen Sizes**

ASCB		EBC
7/64 in.	2.8 mm	2.8 mm
6/64	2.4	2.5
5/64	2.0	2.2

Moisture is reported as a percentage of weight. It is measured by weighing a given mass of malt before and after drying in an oven. If moisture is too low (<3%), the malt is brittle and prone to breakage during milling, leading to poor lauter performance. If the moisture is too high (>6%), the malt is **slack**. Slack malt is susceptible to **spoilage** during storage and is difficult to mill, leading to poor extract yield in the mash. Moisture analysis is used in the calculation of dry basis analytical results.

Protein Modification   In measurement of protein modification, we are looking for adequate digestion of the lamella by proteases in the endosperm cell wall. As barley is modified during germination, enzymes hydrolyze some of the protein, cutting the large molecules into smaller, more soluble molecules. Proteolysis of the lamella helps glucanases to break down beta-glucan. Because digestion of the lamella is required to gain access to starch, protein modification is also tied to the maximum extract potential of the malt. Lower protein modification leads to less fermentable carbohydrates. Protein modification also provides FAN, foam-positive proteins, malt flavor, and color (melanoidins). Malt COA specifications involving protein modification include *total protein, soluble protein*, the *S/T* ratio, and *FAN*.

Protein measured in malt is called **total protein**. Protein measured in wort prepared in a laboratory mash is called **soluble protein**. The ratio of soluble to total protein is called the **S/T ratio**, or the **Kolbach index**. Because protein is the principal source of nitrogen, protein is often estimated from the nitrogen content: % protein = % nitrogen × 6.25.

Free amino nitrogen (FAN) is the measurement of substances containing the amino group, $-NH_2$. The method involves a reagent called *ninhydrin*, which interacts with terminal amino groups to give a characteristic purple color which is measured with a spectrophotometer. The absorbance is compared with that from a standard sample of glycine, one of the common amino acids. The results are reported in milligrams per liter (ppm). Adequate FAN is needed for good yeast health. Excessive FAN may result in the formation of higher alcohols (fusel alcohols) or **Strecker** aldehydes, which may impart off-flavors to the beer. Most adjuncts have little or no FAN, so higher levels of FAN from the malt are needed for adjunct brewing.

In general, brewers prefer malt that is low in total protein, yet is well modified (reasonable FAN and S/T). Malt with low protein modification may lead to:

- Poor yeast health and erratic fermentation performance.
- Reduced extract yields.
- Slower runoffs and more turbid wort.
- Increased lauter cuts and more tannin extraction.
- Beer filterability and physical stability problems.
- Bland and drying wort flavor.

Malt with high protein modification may lead to:

- The formation of fusel alcohols.
- Promotion of staling flavors from Strecker degradation products.
- Thinner beer body.
- Poor foam retention.

Carbohydrate Modification   For carbohydrate modification, we are looking for adequate digestion of the barley lamella and endosperm structure. Such modification is required for quality milling operations, high extract potential, and easy lautering. Malt COA specifications involving carbohydrate modification include *extract, fine/coarse extract difference, viscosity, beta-glucan,* and *friability.*

Extract is the amount of soluble material released by the malt in a laboratory test mash. Malt extract values are used along with brewhouse efficiency to estimate the original gravity of wort mashed from a specified grain bill. Extract may be expressed as percent extract fine-ground **dry basis** (fgdb) or as percent extract coarse-ground dry basis (cgdb). To measure percent extract fgdb, a defined amount of malt is milled in a laboratory mill at 0.2 mm to produce a fine flour. It is then mashed under strict Congress mash parameters, and the solids content is determined from the wort specific gravity. To measure percent extract cgdb, a defined amount of malt is milled in a laboratory mill at 0.7 mm to produce a more brewery-relevant crush. It is then mashed under strict Congress mash parameters and extract is determined via density. For extract reported as "dry basis," the mass of the malt is reduced by the moisture content, as water weight provides no extract potential. If moisture is not taken into account, percent extract is reported "**as-is**" (fgai and cgai).

Unlike fgdb, which uses malt ground to flour, the cgdb method is meant to better model brewing conditions. Because the crush is coarser, the extract may be lower. If the malt is well modified, the two measurements should be close. A comparison of these two measurements is reported as the fine grind–coarse grind (FG–CG) extract difference.

To measure extract, the specific gravity of the wort from the Congress mash is measured, and the percent solid is read from a table or calculated from an equation. The equation given in ASBC Method Wort-3 is

$$°\text{Plato} = 135.997 \times SG^3 - 630.272 \times SG^2 + 1111.14 \times SG - 616.868$$

where SG is the specific gravity measured at 20 °C and °Plato is the mass percent solids. The total mass of extracted solid divided by the weight of the grain is the fractional extract, which is usually reported as a percentage. Abbreviated versions of the Congress mash have been proposed. British maltsters sometimes report hot water extract (HWE) in deg·L/kg as defined by the equation

$$HWE = \frac{1000(SG - 1) \times Vol}{mass\ malt}$$

where SG is the specific gravity of a wort prepared under Institute of Brewing and Distilling (IOB) test conditions, Vol is the volume of wort in liters, and mass malt is in kilograms. The factor $1000(SG - 1)$ is the number of points, also called degrees Oechsle (°Oe). One brewer's point corresponds to approximately 0.25 °P. There is no direct conversion between malt extract percent derived from a Congress mash and HWE derived from an IOB mash.

MALT EXTRACT EXAMPLE

A batch of pale ale malt is found to have a moisture content of 4.0% by ASBC Malt-3. A 50.0 g sample of the malt is mashed according to ASBC Malt-4. The specific gravity of the resulting wort is measured to be 1.0352 at 20 °C. Calculate the malt extract, dry basis.

Solution

For convenience, we will do the calculation based on a double recipe of 100.00 g of malt and 800.0 g of water. The dry weight of the malt is 96.0 g. The total water is the 800.0 g used to mash, plus the 4.0 g of water from within the malt, for a total of 804.0 g.

Application of the equation above gives the solids content of the wort as 8.830 °P = 8.830%. The mass of solids divided by the mass of wort (water plus solids) is the mass fraction solids = 0.08830. In equation form,

$$mfs = \frac{mass\ solids}{mass\ solids + mass\ water}$$

where mfs is the mass fraction of solids. The algebraic solution for the mass of solids is

$$Mass\ solids = \frac{mfs \times mass\ water}{1 - mfs}$$

Application of this equation with mfs = 0.08830 and mass water = 804.0 g gives mass solids = 77.87 g. The mass percent extract in the 96.0 g of dry malt is 77.87 g/96.0 g × 100% = 81.1%.

Beta-glucans are gums that increase the viscosity of beer wort. During modification, they should be sufficiently degraded by glucanases. High beta-glucans are indicative of poor modification. They are analyzed by introducing a **stain** called Calcofluor, which binds the beta-glucan to form a fluorescent adduct. Ultraviolet light at 365 nm excites the adduct molecules, which radiate blue light at 420 nm. This test has been subject to scrutiny because it has moderate precision at best, and it is complicated and expensive to run.

Viscosity of wort, by contrast, is quite easy to measure. Viscosity is resistance to flow. The wort sample can be made to flow through a narrow tube or orifice. The time for a certain volume to flow a specified distance is a measure of the viscosity. Another method is to roll a small ball through a tube filled with the wort sample. The time for the ball to roll a certain distance is a measure of the viscosity. Excessive viscosity gives severe flow problems with processes that involve the fluid passing through a porous medium, like filtration or wort separation. The base unit for viscosity is the pascal-second (Pa·s). The centipoise (cP), which is 0.001 Pa·s, is more commonly used. The viscosity of water at 20 °C is 1.002 cP.

Friability is the ease of crumbling a sample. The friability of malt increases with modification. Malt friability can be measured with an instrument called a **friabilimeter** (Figure 4.15). The instrument has a perforated drum that acts as a **sieve**. A rubber roller presses against the inside of the drum under spring tension. A 50.0 g sample of malt is loaded into the drum, which rotates for a period of exactly eight minutes. The friable parts of the malt kernels will break under the pressure of the roller and be driven through the drum. The portion retained in the drum is weighed and accounted as non-friable. It has been argued that if a brewer invests in equipment to check malt quality, the friabilimeter is a great place to start.

FRIABILITY EXAMPLE

A 50 g sample of malt is loaded into the drum of a friabilimeter. After 8.0 minutes, 6.24 g of malt is recovered from the drum. Calculate the percent friability.

Solution

Of 50 g of malt, 6.24 g is non-friable, so 43.76 g is friable. The percent friable is 43.76/50 × 100% = 87.5%.

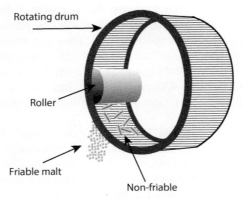

Figure 4.15   Friabilimeter.

Malt must exhibit high carbohydrate modification for optimal brewhouse performance. Malt with low carbohydrate modification may lead to:

- Milling problems.
- Poor extract yields.
- Elevated **final gravity**; reduced **attenuation** due to excessive starch or dextrins.
- Deeper lauter cuts and more tannin extraction.
- Slower runoff and more turbid wort.
- Beer filterability and haze problems.

Enzyme Potential   For enzyme potential, we are looking for adequate production of diastase (alpha- and beta-amylase), which is required in the mash to convert starch into fermentable sugars. It is important to confirm that the activity of alpha- and beta-amylase has been preserved past the kilning process. Malt COA specifications involving enzyme potential include *diastatic power* and *alpha-amylase*.

Diastatic power derives from the combined action of the two major starch-converting enzymes, alpha- and beta-amylase. It is measured by the amount of starch hydrolyzed at $20\,°C$ in 30.0 minutes by the enzyme extracted from a sample of the malt. Diastatic power is measured either in degrees Lintner ($°L$), degrees ASBC (roughly equal to $°L$), or Windisch–Kolbach units ($°WK$). An approximate conversion is $°WK = 3.5 \times °L - 16$. Diastatic power is a measure of how effective the malt is at converting starch to sugar. The minimum diastatic power for malt to convert its own starch is about $40\,°L$. When base malts and other starch sources are used, brewers generally look for an average diastatic power of at least $50\,°L$.

Alpha-amylase activity can be measured independently of beta-amylase by using starch that has been pretreated with beta-amylase. As the alpha-amylase hydrolyzes the starch chains, they lose their ability to make a color in iodine. The unit of measure is the dextrinizing unit (DU), which is the amount of alpha-amylase that will hydrolyze starch at a rate of 1 g/h.

Today's barley malt contains ample enzymatic potential. Poor diastatic power is not usually a problem for most brewers. Only those brewers using large amounts of adjuncts need to be concerned. Typical enzymatic potential in barley malt can tolerate 30–40% adjunct additions to the mash with no issue. The major impact of enzyme potential for a brewer is *consistency*. Brewers should be able to repeat the same degree of **attenuation** for each brew given the same mashing protocol. Maintaining *consistency* in enzyme potential from lot to lot is important.

Color and Flavor   Color and flavor are indicators of kiln performance. Were the conditions of the kiln such that desirable color and flavor were achieved? Malt color is controlled during the kilning process by the severity of thermal treatment. As the temperature and treatment time increase, so does the concentration of Maillard products. Because malt is the main source of beer color, malt is graded for color.

Conventionally, measurement of malt color is based on the amount of blue light (**wavelength** of 430 nm) that a sample of wort produced from the malt will transmit (allow to pass through). The approximation is that beer is essentially beer-colored; all that matters is how light or dark it is. But this model is not accurate for all beer or wort. Some beer is more yellow and some more red, a topic discussed further in Chapter 13.

A sample of malt is milled and mashed following a standard procedure in which 50 g of malt is mashed to a final dilution of 450 g (malt plus water). This comes to 1.04 lb of malt per US gallon of water. The wort is filtered, and the color measured in the same way as for beer (see Chapter 13). The color of the test wort, ASBC or EBC, is reported as the malt color. An older system of color, called the Lovibond scale, gives similar numbers to the ASBC scale.

Malt flavor is a critical kilning outcome, but its inclusion in malt COAs is not standard. Part of the challenge in describing malt flavor has been on the development of a relevant method to describe malt-specific flavors. The ASBC Methods of Analysis has been recently updated to include Sensory Analysis 14: Hot Steep Malt Sensory Evaluation Method. This method can be easily adopted by any brewery. In brief, 50 g of coarse-ground malt is added to 400 mL of 65 °C (149 °F) distilled water in a thermos®. It is capped, shaken for 20 seconds, and steeped for 15 minutes. The "mash" is then filtered through fluted filter paper into a collection container with a vorlauf step for the first 100 mL of filtrate. Wort sensory analysis is to be conducted within four hours of filtration. This method is convenient and designed to limit starch modification, and thus sweetness, which can obscure subtle malt flavors. Malt flavor wheels using this method have been developed to further refine descriptive analysis of malt flavor.

4.3 HOPS

Hops (Figure 4.16) serve several functions in beer. One family of components provides a bitter taste to balance the sweetness of malt, and another provides aromas to give each style of beer its character. Hops have antibacterial activity that can retard beer spoilage, they help stabilize beer foam, and they can prevent excessive foaming during boiling.

The first historical mention of the use of hops in beer dates to 822 CE in monastery rules written by Adalard of Corbie (761–826), a cousin of Charlemagne, whereby a portion of the hops given to the abbey in tithes would be set aside for beer. The first cultivation of hops was recorded in Bavaria around 859 CE. In Europe over a period of hundreds of years, hops supplanted **gruit**, a mixture of herbs based on bog myrtle (*Myrica gale*), also called sweet gale. Today, hops are regarded as an essential feature of beer. In the United States, "malt beverages" are defined by law to be prepared with hops.

The Biology of Hops

The scientific name for hops, ***Humulus lupulus***, was first described by Carl von Linné, founder of modern taxonomy. "Humulus" is derived from the

Figure 4.16 Hops.

Slavic name for hops, and "lupulus" from the Latin word meaning "little wolf." This description may be attributed to Roman scientist Plinius (Pliny) the Elder who, in his landmark *Naturalis historia*, described a plant called "lupus salicarius" or wolf of the willows. Perhaps this early nomenclature can be attributed to physical attributes of the wild hop plants from which hops were collected.

Hops are a flowering, climbing bine of the *Cannabaceae* family. Unlike vines, which spread via tendrils, bines climb by wrapping in a clockwise manner around a surface. In addition to hops (*Humulus*), there are two other genera in this family: the hackberry tree (*Celtis*) and hemp (*Cannabis*). Hops are further divided into five subspecies (*H. lupulus* spp. *lupulus*, *H. lupulus* spp. *americanus*, *H. lupulus* spp. *neomexicanus*, *H. lupulus* spp. *lupuloides*, and *H. lupulus* spp. *japonicus*).

Hops are **dioecious**, meaning there are separate male and female plants. Flowers of the male plants produce pollen that is wind-dispersed to fertilize the female flowers. A female flower, also called a **burr**, is depicted in Figure 4.17. Plants in a commercial hop field are 100% female. By excluding males, the hop farmer ensures no mating takes place, thus maintaining the exact same cultivar in the entire field. Also, fertilized **strobiles** contain seeds that are rich in lipids, that could negatively affect foam stability.

The hop **cone**, which is the fruit of the plant, is the part used for brewing. Cones develop from the female flowers. Technically described as a strobile (Figure 4.18), the hop cone has a central strig that contains four bracteoles (inner leaves) and a pair of bracts (outer leaves) at each node, arranged in a structure that resembles a green pinecone (Figure 4.19). **Lupulin** glands emerge from the base of each bracteole and produce yellow lupulin oil. Lupulin glands contain hard and soft resins. The soft resins contain the **alpha** and **beta acids** that provide the bitterness for which hops are prized. There are

Figure 4.17   Female hop flowers (burrs).

Figure 4.18   *Humulus lupulus*: hop plant (female) with strobiles.

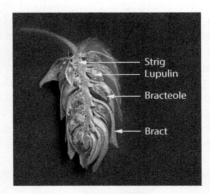

Figure 4.19   Hop strobile section.

over 300 compounds in the lupulin glands that are thought to contribute to hop flavor and aroma.

Hops are perennial plants that grow from rootstocks (rhizomes) each year. This allows female plants to be propagated by clipping and replanting the rhizomes from the rootstock. Hop shoots emerge in spring and are harvested in late summer or early fall. In commercial operations, hops grow on trellises up to 50 cm per week, reaching a height of 6 m (20 ft). Hops grow best in temperate climates with mild winters and require a long day (15 hours of sunlight) for optimal hop cone production. These growing conditions typically fall within the 35° and 55° latitude in the Northern and Southern Hemispheres.

The Pacific northwestern region of the United States and regions in Belgium, Germany, and the Czech Republic are prominent hop-producing regions. Southern Africa, Southern Australia, and New Zealand are also areas with emerging hop crops. Although the northeast region of the United States has an ideal temperate area for hop production, its humidity and rainfall promote disease and pest pressure. Powdery mildew and aphid infestations are more abundant in this area.

During harvest, small operations may require that hops be picked by hand, as was historically done. The bines bear prickly hairs, so pickers should wear gloves. Larger operations cut bines from the top, which then fall into trucks. Picking machines then separate the hops from stems, leaves, and other foreign material. The hops are dried in hop kilns at about 65 °C (150 °F) for 8–10 hours. They are then cooled and pressed into bales of about 90 kg (200 lb). Bales should always be stored cold to prevent oxidation and loss of quality.

Hops in storage are susceptible to spontaneous combustion. The risk is greatest when large quantities of hops are stacked up, when the hops are high in alpha acids, and when some of the hops are damp. A 2006 fire in Yakima, Washington, destroyed 4% of US hop production. Fortunately, there were no human casualties.

Hop Flavor Chemistry

There are hundreds of flavor components in hops; we will only cover the most prevalent compounds. Hops yield two broad classes of flavor compounds, bitter compounds and aroma compounds. These compounds are mainly found in lupulin.

Bitter Compounds   There are two families of hop acids, one of which is the alpha acids, sometimes called **humulones**. Figure 4.20 shows the three main alpha acids – humulone, cohumulone, and adhumulone – as well as a generic alpha acid in which the variable part of the molecule is represented by R. Bitterness is a taste; only dissolved materials can be tasted. In their original forms, compounds in hops are neither soluble nor bitter. To get bitterness,

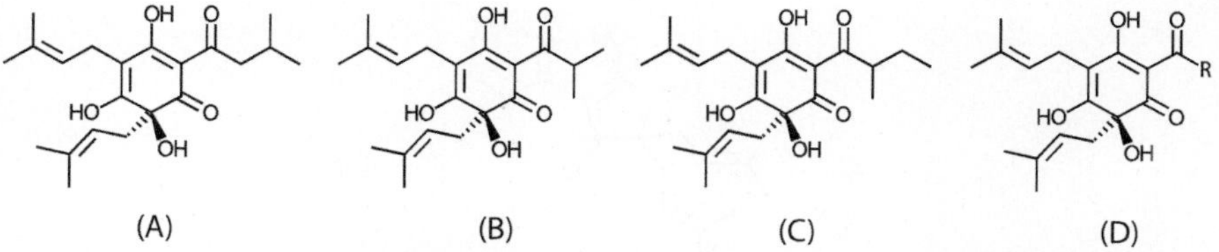

(A)	(B)	(C)	(D)

Figure 4.20   Alpha acids. (A) Humulone, (B) cohumulone, (C) adhumulone, and (D) generic alpha acid.

hop alpha acids found in the resin must undergo a reaction called **isomerization** to give **iso-alpha acids**. Hops typically range from 3.5 to 20% alpha acids by dry weight, depending on the variety, and about half that much beta acids.

The other family of hop acids is the beta acids, sometimes called **lupulones**. The beta acids include lupulone (shown in Figure 4.21), colupulone, and adlupulone, with the analogous differences in the side chain as the alpha acids. Unlike the alpha acids, the beta acids do not undergo isomerization. In their original form, they are largely flavor inactive, but they can produce bitter oxidation products, such as hulupone, shown in Figure 4.22. Hulupone has the same family of analogues as the alpha and beta acids, including cohulupone and adhulupone. The additional bittering partially compensates for loss of alpha acids during hop storage. Beta acids can be converted to bitter hop products by chemical processes, but not during brewing.

Compounds from hops are not bitter until they undergo a chemical change during boiling called isomerization. The isomerization products, called iso-alpha acids, also called **isohumulones** (shown in Figure 4.23), are bitter and moderately soluble. The flavor **threshold**, which is the minimum amount that can be tasted, for isohumulone is about 4 mg/L. Like the alpha and beta acids, the iso-alpha acids have corresponding analogues isocohumulone and isoadhumulone. Some brewers believe that isocohumulone provides undesirable harsh bitterness, but not everyone accepts this. Hop isomerization is covered in more detail in Chapter 8.

Humulinones (Figure 4.24) are another important class of hop bitter compounds. Structurally they are similar to iso-alpha acids, but they have an extra —OH group on the ring. Humulinones arise from oxidation of alpha

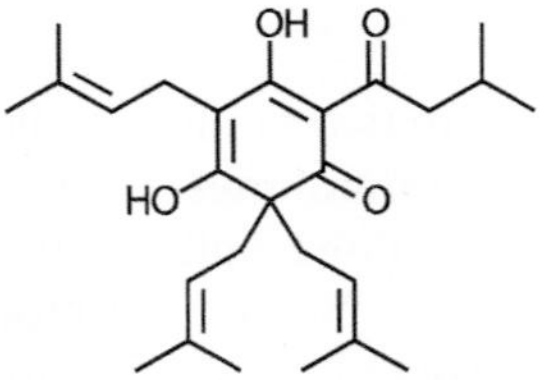

Figure 4.21   Lupulone, a beta acid.

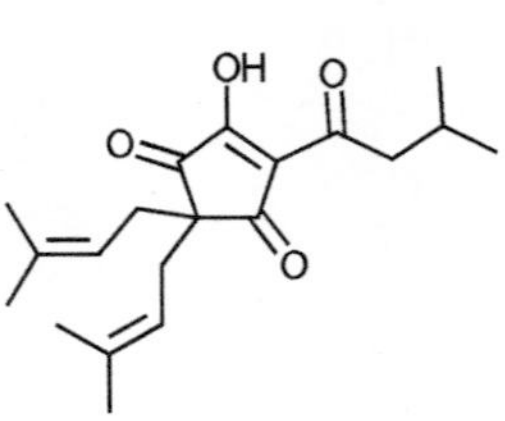

Figure 4.22   Hulupone, enol form.

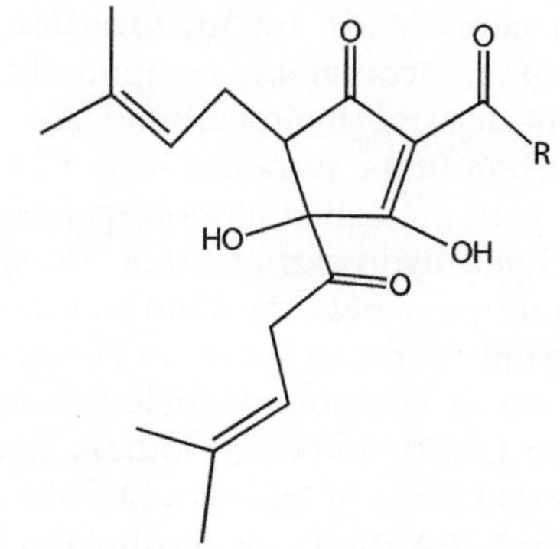

Figure 4.23   Generic iso-alpha acid.

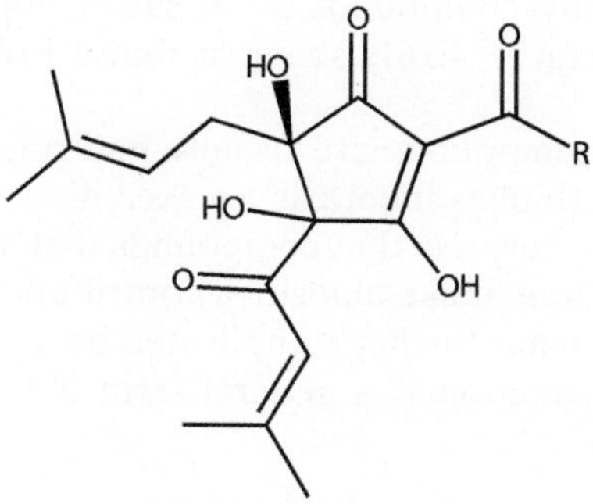

Figure 4.24   Humulinone.

acids; they do not require boiling to become bitter. They are about two-thirds as bitter as iso-alpha acids. They can compensate, to some extent, for the loss of alpha acids with storage, and they can be a significant source of bitterness in dry-hopped beers.

Aroma Compounds   Aroma compounds, which are heat sensitive and volatile, are in a fraction called hop oil. There may be over 400 compounds in hop oil, most of which have unknown molecular structures and make unknown contributions to beer flavor. Many of these compounds are present at barely detectable concentrations, but their flavor contributions could nonetheless be substantial. The standard dogma for aroma compounds seems to be that compounds that are abundant in "noble hops" (Hallertauer, Tettnang, Saaz, and Spalt) are pleasant and desirable and all others are harsh and objectionable. Experience suggests that the complex aroma profiles of the different hop strains defy simplified analysis in terms of good and evil compounds. The three main classes of aroma compounds are **hydrocarbons**, which contain only carbon and hydrogen; oxygen-containing compounds; and sulfur-containing compounds. Most plant-derived hydrocarbons are **terpenes**, compounds assembled from five-carbon units. Terpenes with 5 carbon atoms are called hemiterpenes, those with 10 are called monoterpenes, and those with 15 are called sesquiterpenes. Some hydrocarbons that are significant in most hop strains include the monoterpene beta-myrcene and the sesquiterpenes alpha-humulene and beta-caryophyllene, all shown in Figure 4.25.

Oxygen-containing aroma compounds include alcohols (have OH), carbonyl compounds (have C=O), and many others. Some of these have low volatility, and it is likely that some of them could make an aroma contribution even after boiling. Figure 4.26A shows the alcohol linalool, which contributes a floral aroma. A comparison with myrcene (Figure 4.25A) shows that linalool can be prepared by adding water (H—OH) across a double bond in myrcene. Of the carbonyl compounds, esters can be important sources of fruit and flower aromas. Figure 4.26B shows isobutyl isobutyrate, which has a pleasant fruit aroma.

Sulfur-containing compounds are usually bad-smelling. The example in Figure 4.27, S-methyl thiohexanoate, is an exception. It carries a pineapple aroma. It should be no surprise that compounds that smell bad on their own can, at low concentrations, make a positive contribution to beer flavor.

Hop flavor and aroma is further complicated by a process called **biotransformation**. Biotransformation is a general term for the conversion of one

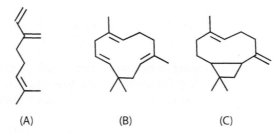

(A)   (B)   (C)

Figure 4.25   Hop terpenes. (A) Myrcene, (B) humulene, and (C) caryophyllene.

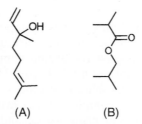

(A)   (B)

Figure 4.26   Oxygenates. (A) Linalool and (B) isobutyl isobutyrate.

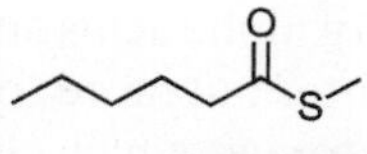

Figure 4.27 *S*-methyl thiohexanoate.

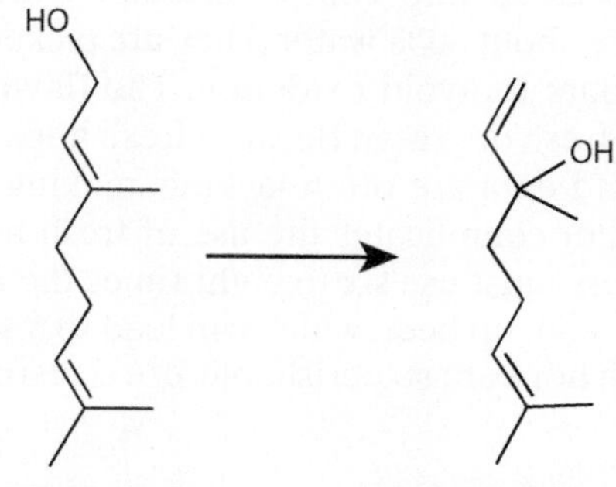

Geraniol   Linalool

Figure 4.28 Biotransformation of geraniol to linalool by yeast.

compound into another through biological enzymes. In the case of beer, the action of yeast can affect the final hop aroma. As an example of biotransformation, geraniol (rose-like) from hops has been shown to be transformed into linalool (citrus-like) by brewer's yeast (Figure 4.28). Brewer's yeast can also affect flavor through modification of glycosides. Hops contain glycosides, which consist of a sugar, like glucose, **covalently** attached by a glycosidic link to another compound, called (if it is not a sugar) the aglycone. In hops, many aroma-active compounds can serve as aglycones. Glycosides of linalool, geraniol, nerol, vanillin, benzaldehyde, and phenylacetate as well as other ketones, aldehydes, and alcohols have been discovered. Certain yeast strains express beta-glucosidases, enzymes capable of hydrolyzing the glycoside, which releases the aglycone. Glycosides are typically flavor neutral, but the aglycone released by hydrolysis may be flavor active.

Hops Processing and Products

One of the key issues in hopping beer is the efficiency of extraction of flavor compounds into the wort or beer. The low pH of beer makes extraction difficult. The percentage of the amount of a flavor compound extracted to the total amount in the hops is called the hop **utilization**. The term is mostly used for the extraction of hop bitter compounds. The type of hop product and details of the brewing process affect hop utilization.

Hop aging begins at harvest. As soon as the hop plant is removed from the ground, oxidation ensues. Hops do not improve with age; they are best used fresh. Oxidation drives the loss of alpha acids and the formation of humulinone and hulupone, which are less bitter but more soluble. Once baled, hops should be immediately placed in cold storage, which also mitigates fire risk. Advanced hop products, like extracts, are very shelf stable.

Fresh Hops   Unprocessed hop cones, variously called fresh hops, green hops, and wet hops, are about 80% water. They are picked fresh and should be used within 24–48 hours to avoid oxidation. The flavors imparted by fresh hops are described as fresh or grassy. Because fresh hops must be used quickly, analytical methods and data are often lacking, making consistency difficult. Excess moisture further complicates the use of fresh hops. Since fresh hops are 80% water, brewers must use six to eight times the amount of fresh hops as dry hops. The hops soak up beer, which can lead to a significant loss in beer volume. Because fresh hops are so perishable, use is restricted to a few batches at harvest time.

Leaf Hops   Dried hop cones, often compressed to save space, are called leaf, flower, or whole cone hops. After harvest, the hop cones are cleaned, layered on perforated shelves in a kiln, also called an **oast**, and dried in a stream of air in the temperature range of 50–65 °C until the water content reaches the desired level, typically about 10%. The details of the depth of layering, the air flow rate, and the heating and cooling program are critical to the character of the hops, especially to the aroma. Different strains of hops may undergo different drying regimes. After cooling, the cones are pressed into bales of as much as 100 kg. Fifty kilograms of hops is sometimes called a **zentner** (abbrev Znt). The **leaf hops** may be used directly for brewing, but most of the hops undergo additional treatment.

Leaf hops present technical issues in beer processing. They are the least stable processed form of hops in storage. Leaf hops are irregularly shaped; they cannot be depended upon to feed smoothly from hoppers. They must be added manually or blown about with air or carbon dioxide. Their low apparent bulk density (~0.15 kg/L) adds to the cost of transport and storage. The void space within the cones absorbs an objectionable volume of beer or wort during beer production. Because of their large particle size, the efficiency of extraction of desired compounds (utilization) is low. Leaf hops are not compatible with the whirlpool and must be removed by other apparatus. Only a few brewers use leaf hops.

Pellet Hops   Some of the technical issues that plague leaf hops are solved by pellet hops. Pellets originate as leaf hops that are broken up to a powder at low temperature, usually in a **hammer mill**. The milling produces a powder that can pass through a 140 mesh sieve (105 µm). Milling ruptures the lupulin glands. The powder may be blended with other varieties of hops, and it may be separated by a program of freezing and sieving to give a lupulin-rich fraction. Pellets prepared without separation are conventionally known as type 90 pellets, because the yield of pellets is about 90% of the leaf hops. Concentration of the lupulin yields type 45 pellets. The separation is not perfect, so type 45 pellets generally contain less than twice as much lupulin as type 90 pellets. Immediately after preparation, the powder is driven with rollers through holes, typically 6 mm diameter, to give cylindrical pellets. Figure 4.29 shows 5 mm diameter pellets. The pelleting operation generates heat, so cooling is sometimes applied to avoid damage to the flavor components of the hops. The pellets are packed in air-resistant bags, purged with carbon dioxide or nitrogen, and sealed either with the inert gas or under vacuum.

The **apparent bulk density** of pellets is three to four times that of leaf hops. When added to wort or beer, the pellets quickly return to the powder from which they were pressed. The rupture of lupulin glands and the small particle size enhances utilization from about 20% in leaf hops to 35% in pellets, an increase of 75%. Pellet hops take less space, absorb less beer, and weigh less than leaf hops, because they are more dense, and are better utilized, so less is needed. If the hops contain traces of contaminants, like pesticides, less hops means less contamination of the beer. The pellets usually roll smoothly and are easy to dispense mechanically. Because of the compression, pellet hops are less subject to degradation than leaf hops, so it is easier to maintain their freshness. Pellet hops should be stored at −20 °C (−4 °F) to prevent oxidation. Pellet hops stored at 0–4 °C are best used within one year. Storage at −20 °C will maintain freshness for up to three years. Hop pellets usually contain 80% or more insoluble material that must be removed from the wort after boiling, usually in a **whirlpool**.

Hop Extract Flavor compounds can be extracted into solvents and concentrated for use in brewing. The solvents in general use are liquid or **supercritical** carbon dioxide and ethanol. Supercritical carbon dioxide is CO_2

Figure 4.29   Hop pellets, 5 mm diameter.

whose temperature is higher than 31.1 °C and whose pressure is higher than 73 bar, the **critical point**. In practice, pressures are in the range of 150–300 bar (2175–4350 psia). Formerly hexane (C_6H_{14}) and other **organic** compounds were also used as solvents, but they have fallen into disuse. Today, most hop extract is made with supercritical carbon dioxide. The product is called CO_2 extract or pure resin extract (PRE). Hop extract is about a factor of four more concentrated than the original hops. It contains none of the insoluble plant matter of the leaf hops or pellets, so removal of spent hops in unnecessary. Hop extract can be used as a substitute for leaf or pellet hops, although some adjustment in processing may be needed because the relative amounts of bitter and aroma components may have changed. Hop extract is a thick syrup. It often comes in containers with a fixed amount of alpha acid. As with leaf hops and hop pellets, hop extract must be boiled to **isomerize** the alpha acids into soluble bitter compounds. Utilization of alpha acids is about the same as that from pellets, about 35%. Hop extract, stored cool in a sealed container, retains its potency for years, especially for bittering.

Stabilized and Isomerized Hop Pellets   Stabilized hop pellets are prepared by the mixing in of about 2% food-grade magnesium oxide to the hop powder before pelletization. The heat generated by pelletization converts most of the alpha acids to their magnesium salts. This improves storage stability and moderately improves alpha acid conversion during the boil.

Isomerized pellets have the alpha acids converted to the soluble, bitter iso-alpha acids during processing. Conversion is highly efficient (>96%), in contrast to isomerization during the boil. Iso-pellet utilization is 45–55%. For the same level of bittering, a smaller amount of isomerized pellets is needed compared with standard pellet hops. Pre-isomerization does nothing to increase the yield of aroma compounds, so less hops will yield less aroma. Isomerized pellets are prepared by adding about 1% food-grade magnesium oxide (a base) and then pressing them into pellets. The pellets are packaged under vacuum or inert gas and are then held for several days at 50 °C to complete the isomerization. The processing procedures for isomerized hop pellets are about the same as those for type 90 pellets, but isomerized pellets do not require refrigerated storage if aroma is not an issue.

Isomerized Kettle Extract (IKE and PIKE)   Hop extract can be pre-isomerized by treatment with magnesium oxide and heating to 100 °C. The products are magnesium salts of the iso-alpha acids, which are solid. The mixture is treated with an acid to convert the salts into liquid iso-alpha acids. The resulting mixture, called **isomerized kettle extract (IKE)** is similar to kettle extract, but the alpha acid utilization is 50–60%. IKE contains fractions that need to be removed by boiling, so it cannot be added later in the brewing process. If the IKE is neutralized with potassium hydroxide, the iso-alpha acids are converted to potassium salts, which are more water soluble than the free acids. The resulting product is called **PIKE**.

Isomerized Hop Extract   Isomerized kettle extract can be refined by manipulating the pH and temperature in several steps to put the iso-alpha acids in a different phase from other components. The refinement can be carried out before or after isomerization. The refined product is called isomerized hop extract. Because the sources of off-flavors and other processing issues have been removed, isomerized hop extract can be added to the fermenter or even during conditioning to adjust bitterness.

Hydrogenated Hop Products   Iso-alpha acids are subject to degradation by light. The degradation products can undergo a series of reactions, discussed in Chapter 12, leading to the formation of 3-methylbut-2-ene-1-thiol known as **MBT**. This compound creates a skunky off-flavor, called **lightstruck**, even at a concentration in the range of a few nanograms per liter (ppt). This is especially an issue if the beer is packaged in colorless or green bottles, as more light is transmitted through the glass as compared with brown bottles. If hydrogen is added across some of the double bonds in the iso-alpha acids, susceptibility to this **defect** is virtually eliminated. Iso-alpha acid molecules have six double bonds. Three of these, a C=O bond and two C=C bonds, form a conjugated system of alternating double and single bonds. Conjugation gives the bonds extra stability, making them harder to hydrogenate. The remaining double bonds, two C=C bonds and one C=O bond, shown in red in Figure 4.30, can be hydrogenated to yield the corresponding single bonds. The C=O bond can be hydrogenated chemically by the action of sodium

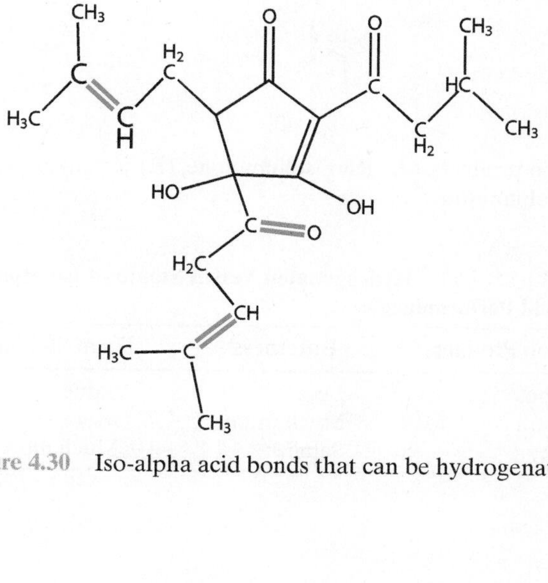

Figure 4.30   Iso-alpha acid bonds that can be hydrogenated.

borohydride (NaBH$_4$), yielding a product called rho isohumulone ("rho"), shown in Figure 4.31A. The two C=C bonds can be hydrogenated by direct metal-catalyzed addition of hydrogen, giving tetrahydroisohumulone ("tetra"), shown in Figure 4.31B. If both processes are carried out, the product is hexahydroisohumulone, shown in Figure 4.31C.

The hydrogenated products differ from untreated iso-alpha acids in more than their light stability. They contribute different levels of bitterness, and some of them enhance foam stability. Hexahydroisohumulone is sometimes used for foam stability even when light stability is not an issue. Table 4.11 compares the three hydrogenated iso-alpha acids with untreated alpha acid.

Analytical and Quality Control Procedures for Hops

Hops are dried and packaged at harvest time and may need to last for an entire year before use. The hops should be packaged in an inert atmosphere and kept sealed until used. They should be kept in a refrigerator or freezer. Any unused portion should be repackaged tightly and immediately returned to the cold storage. Opened hops should be used within a few weeks.

Physical Inspection   Hops, although only a small component of beer, make a large contribution to its flavor. Many brewers go to the growing regions at

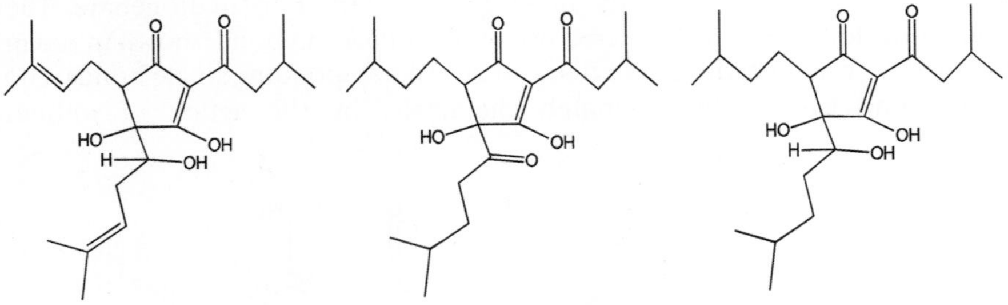

(A)   (B)   (C)

Figure 4.31 Hop products. (A) Rho isohumulone, (B) tetrahydroisohumulone, and (C) hexahydroisohumulone.

TABLE 4.11 Hydrogenated Versus Standard Iso-alpha Acid Performance

Hop Product	Bitterness	Foam Stability
Rho	Less	Same
Tetra	Much more	More
Hexa	Same	Much more

harvest time and inspect samples of the hops. They look at the hops, cut them open and look inside, crush them and smell them, make tea from them, etc. This is called *physical inspection*. A good lot of hops should have relatively low amounts of debris such as leaves, twigs, and seeds. The inspector should note the vibrancy of the color. Pellets should be crushable with finger pressure. Glossy-hard surfaces that resist crushing suggest excessive heating during pelletization. Further, hard pellets will not dissociate as easily when added to beer, potentially creating utilization problems.

In the hands of an experienced inspector, organoleptic evaluation can be very informative regarding the aroma quality. To evaluate dried hop cones, a handful of hops are rubbed vigorously between the palms to break the lupulin glands and warm the sample. During this process the feel of the hops can be noted, but most importantly, the aroma should be evaluated, noting any smell of oxidation. The shortcoming of this technique is that it tells nothing about bitterness, a key flavor attribute, because bitterness is not present in the raw hops, but it is developed by boiling.

Measurement of Alpha and Beta Acids   Hop suppliers invariably provide alpha and beta acid content. They can be measured several ways. One method for measurement of alpha and beta acids in hops is based on ultraviolet spectroscopy. A sample of hops is extracted and diluted. The absorbance of the solution is measured at wavelengths of 355, 325, and 275 nm. The results can be used to calculate the amounts of alpha and beta acids in the hop sample. Another method is conductometric titration with lead **acetate**. These methods are given in detail in ASBC Method Hops 6. Hop alpha and beta acids are more accurately measured using high performance liquid chromatography (HPLC) according to ASBC Method Hops 14. HPLC calibration standards prepared by the International Hop Standards Committee can be purchased from the ASBC.

Hop Storage Index   The absorbance values collected for measurement of alpha and beta acids can also be used to determine the hop storage index (HSI) according to ASBC Method Hops 12. The HSI provides a general overview on the "freshness" of the hops. The hop storage index is defined by $HSI = A275/A325$. An HSI of 0.2–0.25 indicates very fresh hops, 0.25–0.3 indicates fresh hops, 0.3–0.4 indicates slightly oxidized hops, and anything greater than 0.4 indicates high oxidation. While oxidation may negatively impact alpha acid utilization, slight oxidation may be considered beneficial for the creation of citrus and floral flavors. HSI and oxidative effects are affected by hop variety.

Total Essential Oils   The total oil fraction from hops is considered an important measure for hop quality and consistency from lot to lot. Suppliers usually provide the concentrations of total essential oil and of their constituents including myrcene (Figure 4.25A), humulene (Figure 4.25B), caryophyllene (Figure 4.25C), and farnesene, which is usually very low. Total essential oils in hops and hop pellets are determined by steam distillation (ASBC Hops 13) or by **gas chromatography** (ASBC Hops 17).

CHECK FOR UNDERSTANDING

Water

1. For the reaction $HCO_3^- + HPO_4^{2-} \rightarrow H_2CO_3 + H_2PO_4^-$, which is the acid and which is the base?

2. What is the concentration range in ppm for calcium ions in "moderately hard" water?

3. Express the hardness and alkalinity of Pilsen and Dublin water, as given in Table 4.4 above, in units of ppm as $CaCO_3$.

4. Go online and obtain your municipal water report. Comment on the quality of this water for beer production based on the sodium, hardness, and alkalinity.

5. Given the water profile below (in ppm), comment on each ion's contribution to beer production or flavor based on recommended concentrations. For which style of beer would this water be well suited?

Ca^{2+}	Mg^{2+}	Na^+	Cl^-	SO_4^{2-}	HCO_3^-
187	20	113	85	247	20

6. What are the differences among municipal, DI, and RO water?

7. How should all breweries treat municipal water for effective beer production?

8. You are interested in making an IPA with a very bright, firm bitterness as made famous by beer made in England at Burton-on-Trent. To approach the astringent bitterness of this region, you would like to do a single chemical addition to increase your sulfates to about 300 ppm. This process, informally called "Burtonizing water," simply requires the addition of gypsum. Gypsum is calcium sulfate dihydrate ($CaSO_4 \cdot 2H_2O$). You work for a brewery with the water profile given in problem 5.

Part I: Calculate the mass percent of sulfate in gypsum.

Part II: Calculate the amount (in grams) of gypsum required to raise the sulfate level to 300 ppm in 100 L of municipal water.

Part III: After addition of your gypsum in Part II, calculate the final Ca^{2+} concentration (in ppm) in this same 100 L batch of Burtonized water.

Malt

9. The drawings below represent end-on views of heads of barley. Choose the barley whose malt would be most suitable for a high-adjunct international lager. Explain your choice.

10. During the malting process, what is meant by the term *modification*? It could be argued that it is better to brew with an over-modified malt than an under-modified malt. Support this statement with an explanation of the science of malting and the implications of modification on the brewing process.

11. Organize the following malt types in order of decreasing diastatic power: chocolate, crystal, distiller's, Munich, pale ale, Pilsner, and Vienna. How does this compare to color?

12. Explain how to inspect new malt delivered to the brewery. What physical qualities can you immediately evaluate?

13. On a piece of paper, write three squares with the three steps of the malting process in order. Above each square, describe the process inputs (i.e. time, temperature, process conditions, etc.). Below each square, write the process outputs or checkpoints (i.e. final % moisture, CO_2, acrospires length, etc.)

14. Which carbohydrate in barley malt causes an undesirable increase in wort viscosity?

15. In your most recent shipment of pale 2-row malt, the COA (below) showed some significant differences from previous shipments. A. Explain how these differences would affect your brew if nothing on the process end was changed. Explain how you might adjust for these differences in the brewhouse given the new shipment.

Specification	Typical Measurement	New Shipment Measurement
Malt moisture	4.0	4.0
Extract (FG db)	80	70
Color (ASBC)	2.0	2.0
Total protein (T,%)	11.0	11.0
S/T ratio (%)	42	25
FAN	200	150
Beta-glucan (ppm)	100	150

Hops

16. Many sources incorrectly state that female hop flowers are used to flavor beer. Correct this statement.

17. What are some reasons for using hydrogenated hop extract?

18. What is the effect of boiling on the flavors provided by hops?

19. Your Director of Operations found an opened but resealed bag of hops in the cooler. It is at least three years old, but the Director would like you to use it in an upcoming collaboration beer. Describe why that is not a good idea. What scientific data could you present to support your argument?

20. What are humulinone and hulupone? How do they affect the brewing process?

BIBLIOGRAPHY

Almaguer C, Schönberger C, Gastl M, Arendt EK, Becker T. 2014. *Humulus lupulus* – a story that begs to be told. *J. Inst. Brew.* *120*(4):289–314. doi: 10.1002/jib.160. A detailed review article about hop biology, chemistry, and use in brewing.

American Society of Brewing Chemists. 2017. Hot steep malt sensory evaluation method. *Methods of Analysis*. Sensory-14.

Bamforth CW. 2002. *Standards of Brewing*. Brewers Publications. ISBN 978-0-937381-3.

Barth R. 2013. *The Chemistry of Beer: The Science in the Suds*. Wiley.

Cheetham NWH. 2011. *Introducing Biological Energetics*. Oxford University Press. ISBN 978-0-19-957593-0.

Combes RP. 1998. Hop products. *Brewer. 84*:29–35. Overview of the preparation and uses of hop products.

Eumann M, Schildbach S. 2012. Water sources and treatment in brewing. *J. Inst. Brew. 118*:12–21.

Hieronymus S. 2012. *For the Love of Hops*. Brewers Publications. ISBN 978-1-938469-01-5. Covers hop biology, processing, and use in brewing.

King A, Dickenson RJ. 2000. Biotransformation of monoterpene alcohols by *Saccharomyces cerevisiae, Torulaspora delbrueckii*, and *Kluyveromyces lactis*. *Yeast* *16*(10):499–506.

Kunze W. 1999. *Technology of Brewing and Malting*, International ed. VLB. ISBN 3-921690-39-0. Chap. 2.

Lewis ML, Young TW. 2001. *Brewing, 2nd ed*. Springer. ISBN 0-8342-1851-8. Chap 18. Fermentation biochemistry.

Mallett J. 2014. *Malt*. Brewers Publications. ISBN 978-1-938469-12-1.

Noonan GJ. 1996. *New Brewing Lager Beer*. Brewers Publications. ISBN 978-0-937381-82-3. p. 10.

Palmer J. 2017. *How to Brew, 4th ed*. Brewers Publications. ISBN 978-1-938469-35-0.

Palmer J, Kaminski C. 2013. *Water*. Brewers Publications. ISBN 978-0-937381-99-1. Comprehensive resource on water for brewing.

Sidor L, Peacock V. 2006. Hops and hop products. In Ockert K (editor). *Raw Materials and Brewhouse Operations*. Master Brewers Association of the Americas. ISBN 0-9770519-1-9. Chap. 4. p. 73–109.

Thomas D. 2014. *The Craft Maltsters' Handbook*. White Mule. ISBN 978-0-9910436-2-0.

Wahl R, Henius M. 1902. *American Handy Book of the Brewing, Malting, and Auxiliary Trades*. Wahl and Henius. Vol. *2*. p. 790.

Woodske D. 2012. *Hop Variety Handbook*. CreateSpace Independent Publishing Platform. ISBN 978-1475265057.

CHAPTER 5

GRAIN HANDLING AND MILLING

Malt is the second ingredient in **beer**, after water. Unlike water, malt is easily damaged during transfer and storage. Malt is susceptible to **spoilage** from **microbes**, mold, insects, and rodents. Malt can also be a source of dust that can be an inhalation hazard, a fire hazard, and a source of contamination to the **brewing** process. Malt may change hands several times on its way to the **brewery**. Once at the brewery it will be conveyed from the carrier's truck to storage to equipment to weigh out the **grain** bill to the mill and finally to the **mash** conversion vessel. The brewer must organize each step in the malt's journey to maintain malt quality and to minimize its hazards.

5.1 MALT STORAGE AND TRANSFER

Malt Receiving

Malt should always be inspected upon delivery before acceptance by the brewer. It is highly recommended that a contact person at the brewery be assigned the responsibility to develop and execute standard operating procedures regarding malt receiving.

Mastering Brewing Science: Quality and Production, First Edition.
Matthew Farber and Roger Barth.
© 2019 John Wiley & Sons, Inc. Published 2019 by John Wiley & Sons, Inc.

Malt inspection starts with the delivery truck. If receiving pallets or bags of malt, ask the truck driver if any other products were loaded in the same delivery. Bags are not impervious to **gas** and may admit unwanted odors. As each pallet is unloaded, check each for integrity; look for physical damage, staining, moisture, or foreign objects. Pay particular attention to infestation by insects, a topic covered in more detail in Section "Malt Infestations." The shipper (you) usually has the right to refuse acceptance while the carrier (the truck) is still there. Any malt that seems infested should go right back on the truck. If something looks damaged, stop immediately and note the problem on the **bill of lading**. Then follow up with the vendor's customer service representative. If receiving bulk malt from a truck or a train, samples from several locations in the shipment should be taken with a grain **trier**, as explained in Chapter 4. Each sample should be checked for quality and integrity. If it is not practical for the samples to be inspected before the delivery is unloaded, it is possible to keep the grain in a holding bin until it is accepted or rejected. In this case, there will need to be a clear, binding agreement on return of rejected grain for refund or credit, who pays for return shipment, and related issues.

Malt Storage

There are several ways to store malt; the appropriate method depends on the scale of the brewery. For a small craft brewery, malt storage and transfer may be all manual, with grain sacks and buckets transferred by hand or by **fork lift**. With larger-scale and more frequent brewing, malt may be stored in **super sacks** or a **silo** for bulk malt storage. This practice offers significant cost savings as malt suppliers often provide discounts for bulk shipments and may even offer financing options to support silo procurement.

Malt bags are delivered in 50 lb (22 kg) or 25 kg (55 lb) quantities on pallets that may contain up to 40 bags per pallet. Bag material may be double-walled paper or a more durable (and sometimes recyclable or reusable) woven polypropylene. Neither material is impervious to moisture or odor, so care must be taken to avoid each. Bags should be stored on pallets or industrial strength racks to keep the bags off the ground. Avoid high traffic areas as a snagged bag could lead to a malty mess. Ensure prolonged malt quality by storing your malt in cool, dry locations.

Large breweries may invest in silos and have the malt delivered "in bulk," that is, without packaging. The biggest consideration for purchasing a silo is the rate of malt usage. It is recommended that malt be stored in a silo for no more than six to eight weeks. Minimum recommended storage volumes are 22 000 kg (48 000 lb), which is also a manageable amount of malt for delivery via truck.

HOW TO OPEN A SEWN MALT BAG

Malt sacks are often sewn shut. It is better to remove the stitching than just to cut a hole in the bag or rip into the stitching, which can compromise the integrity of the bag and lead to polypropylene debris falling into the malt. Here is a reliable method to remove the stitching quickly and cleanly while opening a bag (Figure 5.1).

Step 1: Find the single loop stitching on one side of the bag. On the other side of the bag is double loop stitching.
Step 2: Face the side with the single stitch and grab the overhanging thread on the right side.
Step 3: Cut the double string closest to the bag.
Step 4: When you pull the single stitch string, the entire string will come off in one piece.

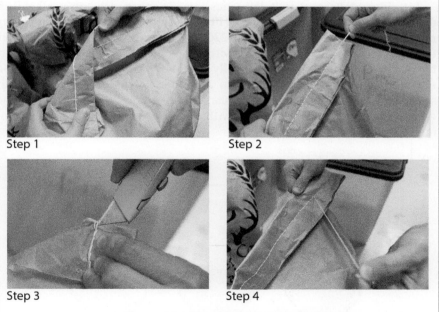

Figure 5.1   Opening a bag of grain.

You may wish to consider a silo if your consumption of a *particular type of malt* exceeds 2700 kg (6000 lb) a week.

Although silos are water tight, they are often single-walled vessels that are subject to environmental temperature fluctuations. During large warm and cool swings, the grain near the wall may release moisture, especially if any of the seeds can sprout. This moisture can lead to mold growth. Another area of concern for silos is the significant amount of dust. Dust represents an explosion hazard, so silos should be equipped with an air vent that feeds into a dust-collecting filter sock that is easy to replace from the ground level.

Malt silos may be filled pneumatically or mechanically and are often filled from the top. To ensure complete emptying, a bottom hopper angle (measured from one side of the hopper to the other) of 40–45° is recommended. However, even with such a hopper, uneven flow patterns called channeling can cause some malt to remain in the silo even after a new batch of malt has been loaded (Figure 5.2). For this reason, malt silos should be completely emptied once or twice a year to avoid accumulating stale malt. If more than one silo holds the same type of malt, one silo should be completely emptied before grain is drawn from another.

Intermediately sized malt deliveries come in **super sacks**, also called flexible intermediate bulk containers (**FIBCs**). Super sacks of malt typically range from 450 to 1000 kg (1000–2200 lb). They are reusable containers typically made of woven material with straps on each corner for transport by a fork lift. Malt super sacks usually have a 30–35 cm (12–14 in.) spout on the bottom for rapid emptying into a hopper. Racks designed for super sacks may have a large receiving shelf upon which the sack can sit. The opening then feeds into a hopper with an iris **valve** to control the flow of malt (Figure 5.3).

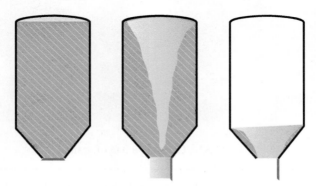

Figure 5.2   Channeling of malt in a silo during storage and emptying.

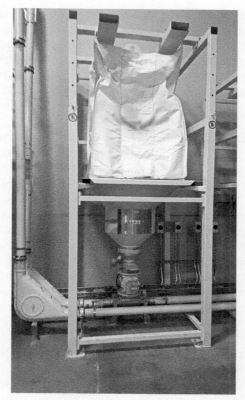

Figure 5.3 **Super sack on dispensing rack at Yards Brewing Company. Pipes to the left and below are parts of an aeromechanical conveyor.**

Malt Transfer

Malt Measurement The amount of each type of malt and other dry ingredients is a critical element of a beer style. The malt must be weighed carefully, especially the types that contribute intense or unusual colors or **flavors**. Small amounts of malt can be weighed out on conventional scales, but when hundreds or thousands of kilograms are used, the container from which or to which it is transferred is weighed before and after the transfer. The device that does the actual weighing is a **load cell**. A load cell is a piece of material or a hydraulic device that deforms slightly under the **weight** applied to it. The amount of deformation is measured with a strain gauge, typically a material whose **resistance** changes with extension, **compression**, or bending. Compression load cells can be installed under the feet of vessels that stand on the floor, or tension cells can be attached to the supports of those that hang from above. The load cells must be mounted so that the force passes through the cell in the proper direction. The load cells must be properly calibrated. If they are overloaded, they will give an erroneous reading, and they may not recover their accuracy if their elastic limits are exceeded. Abruptly dropping a large amount of material into a vessel equipped with load cells may permanently damage them. A load cell is usually accurate to some fraction of its full range, so a load cell designed for several **tonnes** may not give a precise reading for a few kilograms.

Malt Control　The speed of malt delivery by gravity feed from hoppers, silos, or super sacks is usually controlled by valves that can change the effective opening through which the malt flows. Usually the valves have gates or plates that enter the flow path at a right angle to the direction of flow. Valves that move parallel to the flow direction, like **butterfly valves**, experience wear from fighting the momentum of the grain, and their sealing **surfaces** are subject to damage from particles that are trapped when they close. **Gate valves** have a plate, sometimes described as a knife, that can be pushed into or pulled out of the opening to control grain flow (Figure 5.4). An iris valve is a variation on the gate valve in which several plates enter the flow path maintaining a roughly circular opening. A type of iris valve with fabric "plates" can be used to constrict the flow in a flexible tube like the spout of a super sack.

Conveying Systems　When only using 25 kg bags, malt transfer can be manual, but with larger volumes of malt comes the need for more efficient malt transport. The most cost-effective methods for transport are those with the shortest paths. Malt transportation from silo to mill to mash tun should be as short as possible while maintaining operator safety and grain integrity. Designing the brewery to have short paths can yield significant savings in the initial cost of conveying systems and in their operating costs.

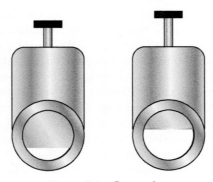

Figure 5.4　Gate valve.

Five types of conveyor systems are regularly used for moving malt and grain. Belt conveyors have a flexible belt that carries the material. Each end of the belt goes around a pulley; one pulley is connected to a motor that provides power, and the other pulley provides tension. The top of the belt is supported by a platform or by idler rollers that support the belt against the weight of the grain as shown in Figure 5.5. The idler rollers can be configured to form the belt into a trough, keeping the kernels from dropping off the edge of the belt. Belt conveyors can be up to 200 m long. They can be loaded or unloaded at any point. Enclosing them to control dust and contamination can be cumbersome and expensive. They do not easily accommodate turns, and they can raise or lower grain only through a small angle; otherwise the grain rolls down. Belt conveyors have many nip points that can injure workers if not properly covered with guards.

Screw conveyors, also called **augers** or **worms**, have a spiral band enclosed in a tube or trough, as shown in Figure 5.6. A motor turns the band, driving the grain along the tube toward the discharge point. Screw conveyors are generally limited to 20 m (65 ft). They can bend to a limited extent. The enclosure helps to control dust and contamination. They are most effective at horizontal transfer but can be used to lift grain at an angle of up to 30°. Capacity and energy efficiency fall off with lift angle.

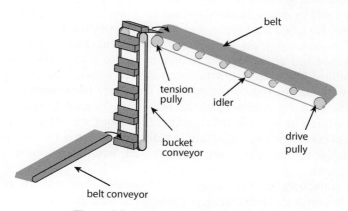

Figure 5.5   Belt and bucket conveyors.

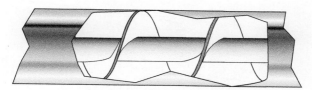

Figure 5.6   Screw conveyor.

The enclosed tube guards the major nip points during normal operation, but this may promote pinch points that cause malt breakage.

Bucket conveyors, also called *elevators* or *grain legs*, lift grain vertically in buckets attached to chains or a belt. The chains or belt goes around pulleys at the top and bottom with the drive shaft at the top. Bucket conveyors save space, because they are not limited to shallow lifting angles. They are generally loaded at the bottom as the bucket scoops malt out of a trough or gutter, and they are unloaded at the top. Large grain storage facilities, also called grain elevators, use bucket conveyors to raise the grain to the top of a tower from which it is distributed by gravity. The conveyor system is generally enclosed, providing protection against dust and contamination. It exerts little physical force on the malt that could cause breakage. It can be difficult to completely empty the bottom trough. A lower-level belt conveyor entering at one angle feeding a bucket conveyor that, in turn, feeds another belt conveyor exiting at a higher level in a different direction can be used to carry grain around a bend of any angle as shown in Figure 5.5.

Drag conveyors, also called *cable* or *chain conveyors*, consist of a tube with paddles, also called *flights*, moving the length of the tube. The paddles are moved by chains or cables. Drag conveyors can operate at any angle from horizontal to vertical. They can, with special mechanical gear, follow a curved path and go around (rounded) corners of large enough radius to guide the cable or chain. In the standard drag conveyor, the spaces between the paddles are filled with grain, and the motion of the cable or chain drags the paddles and the grain with them. The aeromechanical conveyor, also called a **cable** and **disk conveyor**, shown in Figure 5.7, is more common for grain handling. The paddles are quickly pulled by a cable. The grain is moved largely by the air moved by the paddles. The space between the paddles is about 20% grain by volume, and the rest is air. To load the conveyor, the grain is metered in to avoid overloading. The major advantage of the aeromechanical conveyor over the standard drag conveyor is that it causes less grain abrasion. Drag conveyors are enclosed, which cuts down dust and contamination.

Screw conveyors, bucket elevators, and tubular drag systems cannot be completely emptied. Therefore, it is essential to add specialty malt between

Figure 5.7   Aeromechanical conveyor.

additions of the main malt, so that any malt held up in the conveyor will not contribute unexpected flavors or colors to the next batch.

Pneumatic conveyors use air or a vacuum to move the grain through a pipe. The air is moved by a blower and it drives the grain with it. Pneumatic conveyors can operate at any angle and can negotiate narrower curves than drag conveyors. Pneumatic conveyors require equipment to separate the air from the grain at the delivery point. They use a good deal more energy than other conveying systems. A potential disadvantage of pneumatic conveyors is that the physical forces they exert on the malt may lead to breakage.

Malt Infestations

Pests, including insects and rodents, can cause expensive damage to stored malt. Inspections of malting facilities in Europe have found over 80 species of insects and mites. Those that represent the most risk are the grain weevil (*Calandra granaria*), the confused flour beetle (*Tribolium confusum*), the red flour beetle (*Tribolium castaneum*), and the saw-toothed grain beetle (*Oryzaephilus surinamensis*).

The grain weevil (3–4 mm) multiplies rapidly, laying some 100 eggs during a lifespan of seven to eight months. Adults lay eggs in a small hole they make on the surface of the seed. They seal the eggs in the hole with a gelatinous material that they secrete. When the larvae hatch, they burrow into the endosperm and finally emerge from the hole as adults. Such holes in malt are characteristic of grain weevil infestation, and once they are apparent, the malt has already been infested by at least one generation. As a result, damage can quickly escalate. A simple float test can help identify grain weevil-infested malt; grains that float on water are probably infested. The optimum temperature for grain weevil reproduction is 16–20 °C (61–68 °F). Lower temperatures do not kill the grain weevil, which accounts for its hardiness in surviving through the winter. It cannot multiply in malt with a moisture content lower than 10%.

The confused flour beetle (*T. confusum*) is more common in the northern United States, and the red flour beetle (*T. castaneum*) in the south. Both are around 4 mm long and commonly infest malt and flour. Unlike the grain weevil, they cannot survive the winter in unheated silos, and they do not burrow into the grain, so the larvae and adults can only feed on damaged grain or flour. The most common infestations are in bags or bins in the brewery where conditions favor breeding. With a lifespan of eight to nine months, an adult can lay up to 300 eggs. The life cycle from egg to adult takes about 90 days at 22 °C (71 °F) and 22 days at 30 °C (90 °F). They are resistant to high temperatures and can survive in malt with moisture of 5–7%, making them the most common insect infestation in brewer's malt.

Figure 5.8   Saw-toothed grain beetle infestation of malt flour.

The saw-toothed grain beetle (Figure 5.8) has a lifespan of 6–10 months. Like the flour beetle, it cannot burrow into grain and must feed on damaged malt or flour. It is extremely temperature tolerant, capable of surviving through the winter and withstanding temperatures of 40 °C (104 °F). Its life cycle takes about 72 days at 21 °C (70 °F) and 24 days at 27 °C (81 °F). It is less tolerant to low humidity and has not been found in malt below 9% moisture.

Because of the nature of farms, malthouses, and bulk transport and storage, bugs are unavoidable in malt. Prevention is better than treatment; thus brewers should inspect every shipment of malt and reject any infested malt. Malt stored within an acceptable moisture range (4–5%) will be resistant to infestation, and cool storage will prevent rapid insect breeding. Emptying malt storage silos several times a year helps to prevent infestations.

What if malt does become infested by insects? Title 21, Part 110.110 of the Code of Federal Regulations by the US Food and Drug Administration establishes maximum levels of natural or unavoidable contaminants of food that do not pose a health risk. Malt is not explicitly described. However, the "action level" for wheat is 32 or more insect-damaged kernels per 100 g. For wheat flour, the action level is 75 or more insect fragments per 50 g. This does not mean that the brewer should try to stay just under this level, but rather these are the limits for which the FDA considers the food unadulterated. Quality programs should be in place to ensure that insect damage in malt is kept to a minimum.

5.2 MILLING

Before the **starch** in grain can be converted to **sugar**, the grain must be crushed to expose the starch and release the enzymes. Milling breaches the waterproof **seed coat** and breaks the starch into small pieces, expanding the area of starch–water contact. Finer grinding provides more surface area of contact between the water and the starch, leading to more complete reaction of the starch. In most breweries, the bed of milled grain serves as a **filtration medium** during **wort separation** by the **lauter process**. If the particles are too small, the spaces between them will be narrow, and filtration will be unacceptably slow. Starch particles soften during mashing and are likely to form an impermeable paste. The **hulls** of the seeds help maintain permeability; damage to the hulls during milling should be avoided. Therefore, milling must be a balance between accessibility to the **endosperm** while leaving the hull intact. Some breweries separate wort by **mash filtration**. In this case the particle size can be very small. The hulls are not needed for filtration, but if they are pulverized, mashing conditions will need to be adjusted to avoid extraction of off-flavors.

Grain for brewing is prepared by either roller **mills** or **hammer mills**. The type of mill that is most suitable depends on other details of the brewing process, especially the method of wort separation. Most common to small breweries, roller mills crush the grain between two or more cylindrical rollers that pull the seeds through a narrow gap between them. In mills with multiple rollers, screens separate the finely milled material; the coarse grits are milled between another pair of rollers. The width of the gap is adjusted to give the desired particle size distribution to the milled grain. In a hammer mill, the grain is struck by rotating bars until the particles are small enough to escape through a screen. The resulting particles resemble flour. Mills used in the laboratory for testing are quite different from those used in production.

Roller Mill

The objective of roller milling is to break and expose the starchy endosperm of the seeds without excessive damage to the seed hull. Very fine grinding is undesirable because small particles provide more resistance to flow than larger particles. If wort separation is to be carried out by the lauter process, the wort will need to flow through a bed of **grist**. If the grist is too fine, flow will be impeded, causing processing problems. Excessive damage to the seed hulls should be avoided for two reasons. The hulls act as a processing aid during wort separation, helping to maintain grain bed **porosity** during lautering. Second, hulls contain **phenolic** compounds called **tannins** that, if extracted into the wort, give undesired **astringency** to the beer.

Small hull fragments have more surface exposed, so they are more susceptible to the extraction of phenolic compounds than larger pieces.

Dry Mill The starchy endosperm of grain is generally thick and brittle. When it passes through the mill, it shatters into small pieces. The hull is thin and somewhat flexible, so most hulls will emerge relatively intact. In some milling systems, the grain is "conditioned," by slight dampening with mist or steam to make the hull more flexible and less susceptible to damage. Dampening can provide higher **extract** yield and can make the **lauter process** faster.

The simplest type of roller mill has two rollers that roll toward one another. Each roller has grooves that help grip the malt. Damaging the grooves should be avoided as repairs or replacements are quite expensive. *The gap between the rollers is the critical process variable.* The gap is set to yield a compromise between high mashing efficiency provided by small particles and fast wort separation provided by large particles and undamaged hulls. A two-roller mill provides only one setting.

More control is possible with a four- or six-roller mill, as shown in Figure 5.9. The multiple-roller mill has two or three pairs of rollers. The gap is smaller between each succeeding pair. The output of the rollers goes through or across sieves, usually in pairs, driven either by vibration or by rotary beaters. Sieves separate the grist into fractions: flour, grits, and hulls. The multiple-roller mill provides more control over the particle size distribution of the grist, less damage to the hull, more efficient extraction in mashing, and the opportunity to separate the hulls from the starchy portion of the grist. Some brewers mash only the fine grists and then add the hulls to the lauter tun to improve flow. This reduces the risk of extracting astringent polyphenols from the hulls during the mash.

Dry milling is relatively simple and reliable. Maintenance issues with the mill are minor, and elaborate **cleaning** and sanitation are not needed. The grist can be stored in a **grist case** as needed before mashing, lending flexibility to the operation. The speed of the mill needs to be only enough to crush the grain in the time from one mash to the next. Storage of grist for more than 24 hours is not advised, because it promotes **lipoxygenase** activity in malt. Lipoxygenases are enzymes that cause oxygen to react with fats, giving rise to faster staling of beer. The disadvantage of dry milling is that it produces dust that can degrade the air quality of the brewery and can introduce microbial contamination. The dust is also a potential explosion hazard should any spark be generated.

Wet Mill Another type of roller mill is the wet mill. There are two major variations of wet milling. In *steep conditioning*, the grain is steeped in warm water in a vessel above the rollers, sometimes more than once. The steep

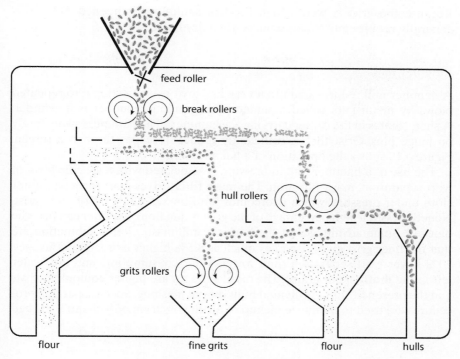

feed roller

break rollers

hull rollers

grits rollers

flour   fine grits   flour   hulls

Figure 5.9 Six-roller mill.

water is usually discarded. After milling, water is run through the steep tank and mill and into the mash tun. This method is seldom used today because of several issues. Some extract is lost with the steep water. Also, discarded steep water adds to the waste stream.

In *spray-steep conditioning*, the grain in the hopper is dry. The dry grain is metered into a conditioning compartment where it is sprayed with hot water. Excess water is removed and recirculated. The wet grain is metered onto the rollers and crushed. The water spray is adjusted to bring the moisture content of the hull to about 20–30%; the endosperm stays dry enough to fracture under pressure. The endosperm slides out of the relatively undamaged hull. Mashing actually starts at the mill, so the temperature of the steep water is set to give the correct mash-in temperature. Wet mills usually have two rollers. They do not use sieves, because it is not practical to sieve wet grist. For all wet mills, the grist must be mashed immediately after milling to avoid the growth of microbes. The mill must produce the required amount of grist quickly, because milling slowly, and accumulating wet grist, encourages the growth of microbes. Wet milling can give good hull integrity and high mashing and lautering efficiency. Milling dust is eliminated.

Because the grist is wet, a high level of sanitation is needed. Wet mills generally require more maintenance than dry mills.

Hammer Mill

A hammer mill reduces grain to a very fine grist by subjecting it to repeated blows by metal bars called hammers. The grinding chamber is a cylinder. A shaft rotates in the center carrying the hammers. The hammers swing freely on hinge pins. Grist that is sufficiently reduced escapes through a screen. Figure 5.10 shows the operation of a hammer mill.

The use of a hammer mill in brewing is associated with a specific form of wort separation, mash filtration. The **mash filter** drives wort through a filter cloth under pressure, which overcomes the flow resistance of the fine grist. Nonetheless, if a high percentage of the grist is fine flour, the filter can become blocked. Some advantages of the hammer mill/mash filter combination are that the finer grist gives rise to as much as 8.5% higher extraction efficiency, 30% lower wort separation time, lower water consumption, and less water left in the **draff** (spent grain). The disadvantages are higher equipment costs and the presence of finely divided hulls during mashing. Special sparging procedures are used to minimize pickup of off-flavors from polyphenols leached

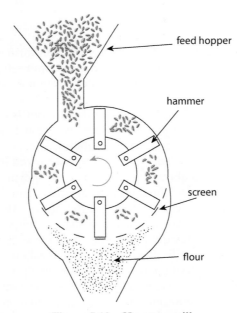

Figure 5.10   Hammer mill.

from the hulls. To lower the solubility of the polyphenols, the **sparge** water may be acidified, less sparge water may be used, and the sparge water temperature may be kept low.

MILLING SAFETY

A mill is a powerful, destructive mechanism. Every mill should be installed with an effective **lockout** device backed up by a lockout program that all workers understand and follow. The lockout device is an arrangement that allows a worker performing maintenance or repair on the mill to prevent anyone from mistakenly operating the mill before the worker is out of the potentially dangerous area. At a minimum, the lockout device should physically prevent the application of power with a locking device to which the worker has the key. Most lockout systems provide for multiple locks so that the device is secured until the last worker unlocks it. The lockout program is a system of procedures and worker training to make the lockout devices effective. The absence of a lockout program, or of an improperly trained worker bypassing it, can cause gruesome accidents.

A mill should contain a powerful magnet to pick up any bits of iron or **steel** in the grain before they can reach the rollers. Stray metal may damage the mill itself and, worse, can cause a spark. Malt dust is highly combustible; sparks must be suppressed by electrical grounding. The bearings of rotating parts, including the rollers and the sieve shakers, should be regularly maintained and monitored for overheating.

Mills should be located in a room separate from the brewhouse. This permits effective cleanup of dust after milling and prevents exposure of the brewhouse to airborne **bacteria**, as malt carries a variety of potential beer-spoiling bacteria. Ventilation of the mill should include a provision for trapping dust and preventing it from mixing with the air in the brewery. Grain dust is a serious inhalation hazard. A program of worker protection including **engineering controls** (like air filtration) as well as **personal protective equipment** (PPE) and worker training must be in place. Proper PPE includes respiratory masks that prevent dust inhalation. The recommended limit (National Institute of Occupational Safety and Health [NIOSH] and American Conference of Government Industrial Hygienists [ACGIH]) for grain dust is $4\,mg/m^3$. Grain dust is associated with lung problems like asthma and chronic bronchitis and lesser issues like eye and sinus irritation.

Laboratory Mills   Unmalted barley is milled in the laboratory with a knife mill. The **American Society of Brewing Chemists** (ASBC) recommends a Wiley intermediate model mill. A set of movable knives passes between stationary knives, chopping the sample to pass through an 18 mesh screen (1 mm opening).

Mills for malt analysis must produce a highly controllable and reproducible particle size distribution. The mill recommended by the ASBC and its European and British counterparts is the Buhler DLFU disk mill. This type of mill grinds the malt between a pair of abrasive disks. The particle size distribution is controlled by the spacing of the disks. When the mill is adjusted for a *fine grind*, 9–11% by mass is retained on a 30 mesh (0.6 mm) sieve. For a *course grind*, 74–76% is retained on a 30 mesh sieve. Analytical laboratory mills cost well over US$10 000. They are not part of the standard outfit of the average brewery laboratory.

Analytical and Quality Control Procedures for the Grist

Most production roller mill gaps are adjustable between 0 and 2.5 mm. Reasons for adjusting the roller gap are variability in malt size, differences in malt moisture, and the use of small grain adjuncts like rye. The mill gaps may also widen over time as vibrations and routine use cause slippage of the drive shaft. For this reason, the space between the roller mill gap is a critical process variable and should be checked as a critical **control point**. For six-roller mills, the break mill gap is typically 1.3–1.5 mm, the hull mill gap is 0.7–0.9 mm, and the grits mill gap is 0.30–0.35 mm. With a two-roller mill, a gap of 1.0–1.1 mm is recommended. Mill gap spacing can be checked with feeler gauges. Alternatively, a soft piece of metal, like solder, can be rolled through the gap, and its resulting thickness measured with a caliper.

Mill gap spacing is a critical control point because slippage can cause inefficiencies in the grist production. Too wide of a setting and uncrushed malt will pass into the mash, reducing the extract yield. Too narrow of a setting will provide too much flour and too little husk material, which may create lautering problems. Decisions to change the spacing will also depend on the condition of the malt (Table 5.1).

TABLE 5.1   Reasons to Tighten or Loosen the Malt Mill

Causes for Tightening	Causes for Loosening
Small or variable malt size	Uniform, plump malt
The use of small grain adjuncts	The use of plump grain adjuncts
Poorly modified malt	Well-modified, friable malt
High malt moisture	Low malt moisture
Uncrushed malt in grist	Too fine grist with little husk
Poor brewhouse yield	Long lauter times

Because of the implications of the grist in lauter performance and extract efficiency, the grist should be regularly monitored as a critical control point. Large six-roller mills may have sampling ports at various points in the grist pathway. These can be sliding drawers or half-cylinders that catch a sample of grist as it falls through the mill. The drawer can be slid out and its sample collected for analysis. Two-roller mills are not often equipped for sampling. A collection tool may be used to catch the grist as it drops into the mash. One of the disadvantages of wet milling is that evaluation of the particle size distribution by conventional means is not practical.

A preliminary evaluation of the grist should always be done by eye, specifically looking for:

- No intact malt kernels.
- The presence of fine grits with a low amount of flour.
- Split hulls with little to no adherent grits.

A more accurate method of measuring the grist is by particle size distribution. Particle size is measured with a series of sieves. Each sieve has a cylindrical frame with a standardized wire mesh soldered across the bottom. The frames come in different sizes. The ASBC methods call for 8 in. (203 mm) diameter sieves. A series of sieves with known openings in descending size is stacked with the largest openings on top. Usually a pan goes on the bottom, and a lid goes on top. A set of four sieves and a pan is shown in Figure 5.11. A weighed sample is introduced to the top sieve. Sometimes small rubber balls, typically 16 mm (5/8 in.), are added to the smaller sieves to help drive particles through. The stack of sieves is shaken for a set time in a standardized way, often with a mechanized shaker, although satisfactory results can be obtained by hand agitation. When the

Figure 5.11   Sieves from above and stacked.

shaking routine is finished, each sieve is weighed. Any material that is retained on a sieve is deemed to have a larger particle size than the sieve opening. The **mass fraction** or mass percent retained by each sieve represents the particle size distribution.

As shown in Table 5.2, there are at least three systems of sieves used in brewing. American brewers usually use the ASTM International system (formerly called the American Society for Testing and Materials). European brewers usually follow **European Brewing Convention (EBC)** procedures, which use the Pfungstadt series of sieves. Some laboratories use the Mitteleuropäische Brautechnische Analysenkommission (MEBAK) system (Central European commission for brewing analysis). The sieves relevant for brewing grist and the names of the fractions are given in Table 5.2. The sieves of different systems associated with a particular fraction do not always correspond closely in size. Some users also reduce the number of sieves used in the ASTM method to 10, 30, 60, 100, and the pan.

Table 5.3 describes typical particle size distribution percentages for the ASTM and Pfungstadt sieves. These numbers may vary based on several variables including grain bill, mill gap spacing, and malt condition, but each grist should be measured and recorded to enable standard acceptable ranges for

TABLE 5.2   **Sieve Systems Used in the Brewing Industry**

ASTM	Size	Pfungstadt	Size	MEBAK	Size	Fraction Retained
10	2.0 mm					Husk
14	1.4	1	1.270 mm	I	1.25 mm	Husk
18	1.0	2	1.010	II	1.00	Coarse grits
30	0.6	3	0.547	III	0.50	Fine grits 1
60	0.25	4	0.253	IV	0.25	Fine grits 2
100	0.15	5	0.152	V	0.125	Flour
Pan		Pan		Pan		Fine flour (dust)

TABLE 5.3   **Typical Particle Size Distribution Percentages for the Grist**

ASBC Sieve no.	Distribution (%)	EBC Sieve no.	Distribution (%)
10	13		
14	20	1	18
18	32	2	8
30	24	3	35
60	6	4	21
100	2	5	7
Pan	3	Pan	11

each brew. The pan fraction should be white flour, typically paler than the fraction retained on the smallest sieve. The pan fraction should not have a bitter taste or an astringent mouth feel. Problems here suggest excessive pulverization of the hulls, which could lead to excessive extraction of tannins and beta-glucans into the wort.

There is more to grist than particle size distribution. The physical appearance of the fractions can provide clues to milling and grain handling problems. Samples from the sieves should be spread on white paper and inspected. Color photographs will be helpful for identifying changes. The husk fraction should have the appearance of barely damaged husks. A 100 g sample of the husk fraction should be gently dropped into a 500 mL graduated cylinder without tapping or compression, just rolling from side to side to distribute the grist. The mass divided by the apparent volume is the apparent bulk density (ABD). A low ABD indicates relatively intact hulls. ABD should be monitored for consistency. Declining ABD over time may indicate a slipping mill gap that should be corrected to prevent troublesome lauters.

CHECK FOR UNDERSTANDING

1. Write a standard operating procedure for receiving bagged malt on pallets from a truck.

2. Discuss the various types of conveyors in terms of their suitability for transfer of malt from a dry mill to a grist case on the same floor in another room.

3. Discuss the advantages and disadvantages of wet milling compared with dry milling.

4. Give five safety items or procedures for dry mills and milling.

5. Under what circumstances would the use of a hammer mill be indicated?

6. How will downstream beer production be affected if the mill gap spacing is too wide?

CASE STUDY

A grist sample was collected from a six-roller mill whose roller gaps were 1.60, 0.90, and 0.35 mm. After separating the size fractions on a set of Pfungstadt sieves, the mass (g) of each fraction was weighed. Using the mass

TABLE 5.4 **Particle Size Analysis**

Sieve	Mass (g)	Mass Percent Targets	
		Min (%)	Max (%)
1	51.6	15	20
2	7.6	5	10
3	38.3	28	42
4	38.8	18	24
5	21.0	4	8
Pan	14.5	8	15
Total	171.8		

of the *total sample*, the mass percentage will need to be determined. The brewery has set minimum and maximum percentages that are acceptable for each sieve. The actual measurements and the target ranges are given in Table 5.4.

1. Determine the mass percent for each sample. Prepare a bar graph showing the measured mass percent for each sieve. Mark each bar with the target range.
2. Compare the measured mass percent with the target range, and note any deviations outside of the acceptable ranges.
3. What would you do to check or validate these measurements? What is your anticipated observation?
4. What could have caused these results?
5. How might these results impact brewing operations or beer quality?

BIBLIOGRAPHY

Briggs DE, Boulton CA, Brookes PA, Stevens R. 2004. *Brewing Science and Practice.* CRC. ISBN 0-5493-2547-1. Chap. 5

Fahy A, Spencer J, Dougherty J. 1999. Wort production. In McCabe JT (editor). *The Practical Brewer*, 3rd ed. Master Brewers Association of the Americas. p. 104.

Freeman JA. 1951. Pest infestation control in breweries and maltings. *J. Inst. Brew.* 50(5):326–337.

Helber J, Barr J, Bird T, Brynildson M, *et al.* 2003. Malt grist by manual sieve test. *J. Am. Soc. Brew. Chem.* 61(4):246–249.

Kunze W. 2014. *Technology Brewing and Malting*, 5th English ed. VLB. ISBN 978-3-921690-77-2.

Penn State Department of Entomology. 2001. Insect Advice from Extension. http://ento.psu.edu/extension/factsheets/pdf/

Wilkinson R. 2001. Brewhouse optimisation *Brewers' Guardian 130*(4):34–36.

CHAPTER 6

MASHING

Fermentable material in beer originates as **starch** from **malt**. This distinguishes **beer** from other **alcoholic beverages** that use **sugar** directly from fruit, honey, plant sap, and the like. Starch is insoluble and unfermentable; before the **yeast** can use it, it must be converted into sugar. In the **mashing** process, the **grist** is mixed with water and held at an optimal **temperature** for **enzymes** to convert insoluble starch to dissolved **carbohydrates**. The resulting liquid is called **wort**. The details of the mashing process, especially the temperature and **pH**, have a decisive influence on the character and **quality** of the wort and subsequently of the beer.

6.1 STARCH HYDROLYSIS

Before we discuss the conditions that affect mashing, we must understand the processes we seek to influence. The mashing process **hydrolyzes** starch to a **mixture** of fermentable sugars and unfermentable carbohydrates called **dextrins**. Starch consists of long chains of **glucose units condensed** together with all **glycosidic** links in the alpha (α, bent) configuration. Figure 3.8A shows a type of starch called **amylose**. The amylose **molecule** consists of

Mastering Brewing Science: Quality and Production, First Edition.
Matthew Farber and Roger Barth.
© 2019 John Wiley & Sons, Inc. Published 2019 by John Wiley & Sons, Inc.

hundreds of glucose units in one long chain with no branches. All glycosidic **bonds** are alpha(1→4). **Barley** amylose averages about 650 glucose units. There are two free ends; one end, called the **reducing end**, is a glucose with an **anomeric** carbon that has a free –OH group. The other end has no free anomeric –OH group. Another type of starch, called **amylopectin** (Figure 3.8B), contains alpha(1→4) chains with alpha(1→6) branches approximately every 30 glucose units. Barley amylopectin has an average of about 7000 glucose units per molecule and has hundreds of non-reducing (non-anomeric) ends. Figure 6.1 shows the details of an alpha(1→6) branch in which carbon 1, the anomeric carbon of one glucose unit, links to carbon 6, the carbon atom that hangs off the ring, of another glucose unit.

   Brewing yeast usually cannot use carbohydrates with more than three simple sugar units. They are not fermented, and they remain in the finished beer. The ability of yeast to use **trisaccharides** (**maltotriose**) is variable. Even some **disaccharides**, such as lactose, are not fermentable. Soluble carbohydrates with four or more glucose units are called dextrins. These soluble, non-fermentable carbohydrates raise the **final gravity** of the finished beer and contribute to the body. Beer with high levels of dextrins might be described as having a full mouth feel or body; one without might be described as dry or having a thin body.

   During mashing the starch converts to sugar in a reaction with water called hydrolysis, shown in Figure 6.2. With each hydrolysis reaction, a new reducing end is created. Starch will hydrolyze very slowly under ordinary conditions. To bring about the reaction at a reasonable speed, a **catalyst** is required. In the mashing process, enzymes from malt, such as amylase and limit dextrinase, catalyze the reaction. The process may be supplemented by additional

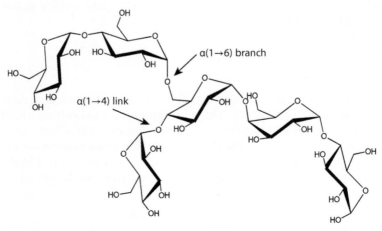

Figure 6.1   Amylopectin structure.

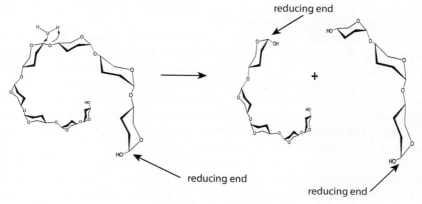

reducing end

reducing end

reducing end

Figure 6.2   Hydrolysis reaction.

sources of enzymes, usually purified from mold, although this practice is contrary to the ***Reinheitsgebot***. In brewing **sake**, a Japanese beer made from rice, the enzymes come from a mold called ***koji***, which is grown on the rice during a combined hydrolysis–fermentation step.

6.2   ENZYMES

Enzymes are biological catalysts made from proteins, usually incorporating other **ions** and molecules. Enzymes act upon one or more target **substrates** to make a product. Enzymes are unique among catalysts in their **selectivity** and **specificity**. Selectivity is the ability of a catalyst to speed up a reaction yielding a particular **product** among several possible reactions that could occur between the **reactants**. Specificity describes the ability of a catalyst to differentiate and interact with a particular target among a group of molecules that could undergo a similar reaction. **Alpha-amylase** catalyzes the hydrolysis of starch, but it is inert to similar reactions on lipids, proteins, or nucleic acids; it displays specificity. The products of alpha-amylase-catalyzed hydrolysis of starch contain a complex mixture of sugars and dextrins; alpha-amylase is not selective. **Beta-amylase** is also specific for starch, but the only sugar it makes is **maltose**, a disaccharide; beta-amylase is specific and selective.

Most enzymes are named after their target (*maltase* hydrolyzes maltose) or the reaction they catalyze, usually with the suffix "-ase". For example, **alcohol dehydrogenase** transfers hydrogen atoms from the –OH group in alcohol to **NAD⁺**, giving a carbonyl group and **NADH**. As discussed in Section "Reaction Rate" in Chapter 2, enzymes can work in either direction. Sometimes this results in enzyme names that seem backward. For example, in alcoholic fermentation the hydrogenation of acetaldehyde to give **ethanol** is catalyzed by alcohol dehydrogenase.

Enzyme specificity is governed by **intermolecular forces** between the enzyme and the substrate, its target. Often in the **tertiary structure** of an enzyme, there is a groove or cavity into which the substrate can fit. This groove or cavity is called the **binding site**. **R-groups** in or near the binding site stick to groups on the substrate via intermolecular forces (Section 2.5). The target substrate binds with higher affinity or more intermolecular forces than nontargets. The **bound** substrate is then positioned to engage R-groups on the enzyme in a location called the **active site** (Figure 6.5A). This is the precise location on the enzyme where the reaction occurs. This scenario is historically described as a "lock and key" model. Often the binding of the reactants leads to a change in the configuration of the enzyme that exposes the relevant **bonds** and brings them into position for reaction at the active site. This type of change is called an **induced fit**; it sometimes evokes images of opening and closing doors or gripping tongs. When the reaction is complete, the product is released, and the enzyme returns to its original form, ready to receive another substrate molecule. The enzyme is not consumed in the reaction.

Figure 6.3 illustrates the lock and key model for invertase, an enzyme that catalyzes the hydrolysis of sucrose ("inversion" because the rotation of plane-polarized light is reversed). The substrates are sucrose and water; the products

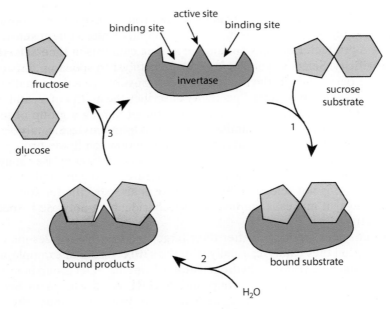

Figure 6.3   Lock and key model.

are glucose and fructose. Invertase is a highly specific enzyme; it cannot hydrolyze other disaccharides like maltose or lactose. In step 1, sucrose binds to the enzyme. Only sucrose can fit the binding site of invertase. In step 2, the bound sucrose reacts with water at the active site, giving bound products. In step 3, the products, glucose and fructose, are released; the enzyme resets, and it can then act on another sucrose molecule.

The speed (also called *rate* or *velocity*) of an enzyme-catalyzed reaction is proportional to the concentration of bound substrate. When there is a high concentration of substrate, nearly every enzyme molecule is occupied by substrate; adding more substrate will not increase the concentration of bound substrate. The rate under these conditions is at a maximum; increasing the substrate concentration will have no effect. When the concentration of substrate is low, the rate is close to a direct proportion with substrate concentration. Figure 6.4 models the effect of substrate concentration on rate over the entire range of substrate concentration. The maximum rate is marked V_{max}.

Barley alpha-amylase is one of the enzymes that hydrolyzes starch during mashing. It is a complex protein structure with 403 **amino acids**, 3 calcium ions (Ca^{2+}), and 153 water molecules, folded into a specific shape, as shown in Figure 6.5. The long tube is the backbone of the protein; in this model the R-groups and water molecules are omitted.

Some enzymes function as 100% protein, but many require additional components called cofactors. When a cofactor is required, the enzyme will function only when both the enzyme and the cofactor are present. Cofactors may be **organic** or inorganic, protein, small molecule, or ion. As can be seen in Figure 6.5, barley alpha-amylase requires three Ca^{2+} ions (indicated by gray spheres) that act as cofactors. Other metal ions including magnesium, zinc, and iron act as cofactors in other enzymes involved in the brewing process.

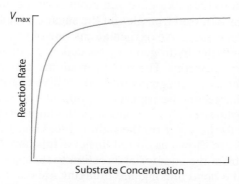

Figure 6.4 Rate of an enzymatic reaction.

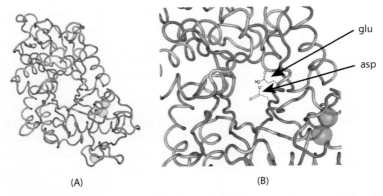

(A)   (B)

Figure 6.5   (A) Barley alpha-amylase. (B) Barley alpha-amylase detail showing glutamic acid and aspartate ion in the active site. *Source*: Drawing based on data from
Kadziola et al. (1994).

ALPHA-AMYLASE MECHANISM

Alpha-amylase increases the rate of hydrolysis of starch by as much as
a factor of 10^{15}, a quadrillion. Although the details of the **mechanism**
have not been unambiguously demonstrated, the scheme shown in
Figure 6.6 is accepted in broad outline. Figure 6.6A shows two sugar
units of a starch chain with a glycosidic link at the active site. The R's at
the right and left represent the rest of the starch chain; the E's at the
top and bottom represent the enzyme molecule. The group at the top
in Figure 6.6 (asp) is part of an aspartic acid R-group. The group at the
bottom (glu) is part of a glutamic acid. Figure 6.5 shows these groups in
the context of the active site. In Figure 6.6B the glutamic acid donates
its hydrogen ion to the linking oxygen atom, forming an **oxonium ion**.
The anomeric carbon (red triangle) on the sugar on the left is attached
to two highly electronegative oxygen atoms, one of which now shares a
pair of electrons with a hydrogen ion; this makes the oxygen atom even
more hungry for electrons. These oxygen atoms pull electrons away
from the anomeric carbon, giving it some positive charge so it **attracts**
the negative charge on the aspartate group at the top. Figure 6.6C
shows the aspartate forming a bond with the anomeric carbon while
the C—O bond to the sugar on the right is breaking. The forming and
breaking bonds are shown as dotted lines. As this happens, the hydrogen atom flips from the equatorial direction to the axial direction (red
arrow). When the bond making and breaking are complete, the starch

chain is broken. The part on the right is free and leaves; water (blue) enters. The part of the starch chain on the left is bound to the enzyme through the aspartate, as shown in Figure 6.6D. The water donates a hydrogen ion to the glutamate, leaving a **hydroxyl group**, shown in Figure 6.6E. The hydroxyl group attacks the anomeric carbon from below. Breaking and forming bonds are shown as dotted lines in Figure 6.6F. The hydrogen returns to the equatorial position (red arrow). In Figure 6.6G the hydrolysis is complete, the carbohydrate on the left is now free, and the enzyme has been restored to its pre-reaction state.

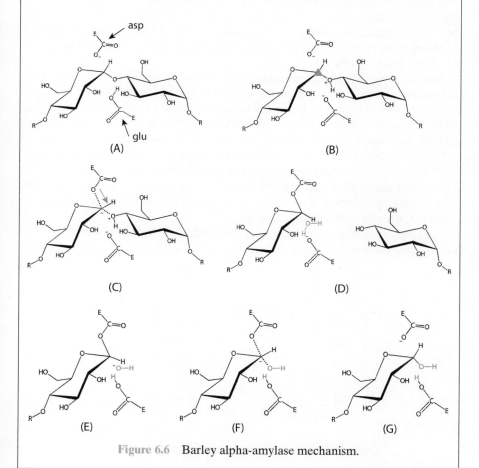

Figure 6.6   Barley alpha-amylase mechanism.

Non-catalytic reactions typically go faster as the temperature is increased. By contrast, the rates of enzyme-catalyzed reactions usually reach a maximum with temperature, but they slow down if the temperature is increased further, as shown in Figure 6.7. The reason for the loss of activity is that at sufficiently high temperature, the integrity of the three-dimensional structure of the enzyme begins to break down. The binding sites do not bind as well, and the R-groups in the active site lose the **geometry** they need to enhance the reaction. The protein becomes denatured. Depending on the enzyme and other conditions like pH, denaturation can be reversible, in which case activity is recovered when the temperature is lowered. More often, especially during the brewing process, denaturation is irreversible. At sufficiently high temperature, all enzymes experience irreversible denaturation. They unfold. Some organisms are adapted to extreme conditions like high temperature. Their enzymes are heat tolerant. The effects of temperature on two different enzymes are demonstrated in Figure 6.7. Heat-tolerant enzymes can be useful in certain biotechnological applications, like polymerase chain reaction (**PCR**).

Protein structure and enzyme activity are also affected by pH. Like optimal temperatures, enzymes have an optimal pH at which their activity is highest, decreasing at higher or lower pH values. Because pH affects the electrical charge on **acid** and **base groups**, interactions between different parts of the protein molecule and between the protein and the substrate are influenced by the surrounding pH. Figure 6.8 demonstrates the effects of pH on protein-hydrolyzing enzymes (**proteases**) isolated from two different organisms. The *Aspergillus niger* protease may be added during fermentation to remove haze-forming proteins. For this application, an enzyme with sufficient activity at pH 4.5, the pH of beer, should be used.

Enzyme conversion rates may decrease with time; the decrease is faster at higher temperature. We think of the net enzyme activity as the total amount

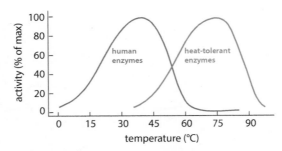

Figure 6.7   Effect of temperature on enzyme activity.

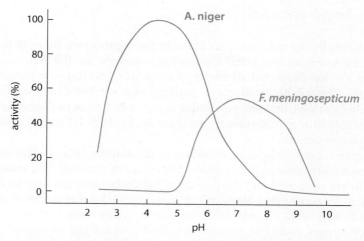

Figure 6.8   Effect of pH on proteases from two organisms. *Source*: After Edens et al. (2005).

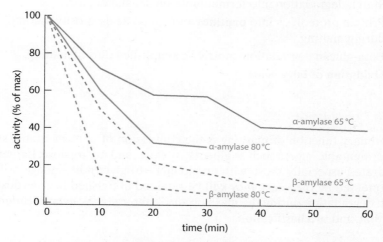

Figure 6.9   Alpha- and beta-amylase activity: time and temperature.

converted during the reaction. Figure 6.9 shows the activity as a percentage of the maximum during one hour at the specified temperature for the two main amylose enzymes. The mash temperature is rarely decreased; mash temperatures are only increased because activity that the enzyme had at the lower temperature is lost at a higher temperature. Lowering the temperature does not bring the activity back.

6.3   MASHING PROCESS

Mashing may be the most complex of the brewing processes. Some of the critical process variables that affect the mashing outcome are the composition of the grist, the **hardness** and **alkalinity** of the water, and the time/temperature treatment. The soluble products of mashing that are significant for brewing include fermentable and unfermentable sugars, dextrins, **beta-glucan**, proteins, other soluble nitrogen-containing compounds, lipids, polyphenols, and **flavor** and color compounds.

Mashing is essentially a continuation of the **malting** process. During malting, **germination** induces production of enzymes in the seed. The **endosperm cell walls** are degraded, but the process is halted before extensive hydrolysis of the starchy endosperm. Removal of water during **kilning** suspends enzyme activity. Rehydrating the malt in the mash reactivates some of these enzymes. The brewer uses these enzymes as tools to determine the sugar and protein profiles of the wort by varying the mashing parameters. Important enzymatic activities in the mash include:

- Starch degradation into fermentable sugars and dextrins.
- Protein proteolysis into **peptides** and amino acids; mostly accomplished during malting.
- Beta-glucan degradation; mostly accomplished during malting.
- **Oxidation** of **fatty acids**.

Starch Degradation

The primary function of mashing is the conversion of unfermentable starch into fermentable sugars such as glucose, maltose, and maltotriose. Larger carbohydrates, especially those with an alpha$(1\rightarrow6)$ glycosidic linkage, will not be fermented by the yeast; they will be retained in finished beer as dextrins. Starch degradation during the mash occurs in three stages: **gelatinization**, **liquefaction**, and **saccharification**.

Gelatinization   Gelatinization is the process through which crystalline starch granules take up hot water. Starch occurs in **grain** seeds tightly packed in granules; the starch chains are bound to one another by **hydrogen bonding**. During gelatinization, water molecules go between the starch molecules, replacing the starch–starch hydrogen bonding with starch–water hydrogen bonding. **Hydration** of starch causes the granules to swell and eventually to burst. During gelatinization, the **viscosity** of the mash increases because long starch molecules are released, and they restrict the motion of water. During liquefaction and saccharification the viscosity of the mash decreases, because

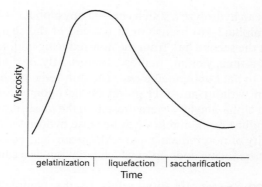

Figure 6.10   Change in viscosity over time during starch degradation in the mash.

TABLE 6.1   Gelatinization Temperatures of Cereal Grains Used in Brewing

Source	Gelatinization temperature
Barley	59–65 °C (138–149 °F)
Maize (corn)	70–80 °C (158–176 °F)
Oats	55–60 °C (131–140 °F)
Rice	70–80 °C (158–176 °F)
Rye	60–65 °C (140–149 °F)
Sorghum	70–80 °C (158–176 °F)
Wheat	52–54 °C (126–130 °F)

the long molecules are broken into smaller pieces. Figure 6.10 demonstrates the change in viscosity over time during the mash.

Barley has a gelatinization temperature around 59–65 °C (138–149 °F). Starchy **adjuncts**, such as rice and **maize** (**corn**), require higher temperatures for gelatinization, around 70–80 °C (158–176 °F). A comparison of grains used in brewing and their gelatinization temperatures is displayed in Table 6.1. As shown in Figure 6.9, barley enzymes are quickly destroyed at the elevated temperatures needed for gelatinization of adjuncts like corn and rice, so these adjuncts must be cooked in a separate vessel before they are added to the mash. Special considerations for the use of adjuncts will be discussed in Section 6.5.

Liquefaction   Liquefaction primarily refers to the reduction of viscosity in the mash as the result of alpha-amylase activity and the hydrolysis of starch into smaller chains. The two main mashing enzymes are alpha-amylase (α-amylase) and beta-amylase (β-amylase). These two enzymes work by related, but not identical, mechanisms. As discussed in Section 6.2, the big differences are in their selectivity and their stability in **solution**.

Alpha-amylase can hydrolyze a starch chain at any alpha$(1\rightarrow4)$ glycosidic link except near an alpha$(1\rightarrow6)$ branch or at the end of the chain. Beta-amylase operates only at the second link from the non-reducing end, so it removes two glucose units at a time, yielding maltose. Importantly, *beta-amylase can only cleave bonds from the non-reducing end*. The non-reducing end is the end of the starch chain without an −OH group on an anomeric carbon. If beta-amylase, but not alpha-amylase, were present in the mash, it would take a very long time to produce fermentable sugar because hydrolysis would be limited by the availability of non-reducing ends. Alpha-amylase activity is needed to create more non-reducing ends upon which beta-amylase can act.

Saccharification   Saccharification refers to the complete hydrolysis of starch into fermentable sugars: glucose, maltose, and maltotriose. The combined activity of alpha- and beta-amylase is required to fully hydrolyze all starch. In this system, alpha-amylase will cleave starch molecules at alpha$(1\rightarrow4)$ bonds once every 7–12 glucose units. This creates a new non-reducing end upon which beta-amylase can function. Due to the different lengths of the chains created by alpha-amylase cleavage, a different ratio of glucose, maltose, maltotriose, and dextrins may be produced. A typical wort sugar profile is shown in Table 6.2.

Variation in the mashing parameters (i.e. time, temperature, pH, **liquor**-to-grist ratio) can alter the activities of the amylase enzymes, influencing the wort carbohydrate profile.

Mash Temperature

Figure 6.11 models the effect of mashing temperature on the total wort solids, on the fermentable wort solids, and on the percentage of wort solids that are fermentable. This is a general trend whose details depend on the type of malt, the mashing time, the mash temperature profile, the capabilities of the yeast strain, and even the strain of barley from which the malt was made. Total

TABLE 6.2   Typical Carbohydrate Profile of a Standard Wort

Sugar	Percent
Maltose	50–60
Glucose	10–15
Maltotriose	10–15
Maltotetraose	2–10
Sucrose	1–5
Fructose	1–5
Dextrin	25

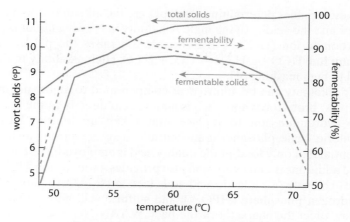

Figure 6.11   Extract and fermentability.

solids content increases with temperature by over 30% from 50 to 60 °C but only by another 5% from 60 to 70 °C. Fermentable sugar goes through a broad maximum from 52 to 68 °C and falls off at higher and lower temperatures. The **fermentability** is maximum at about 55 °C and falls off at a higher temperature, mostly because fermentable solid production is flat but total solids increases. In the range from 60 to 68 °C, increasing mashing temperature increases solids (**extract**) and decreases fermentability. Brewers can vary the mashing temperature to modify the character of the beer without much effect on the ethanol content (**alcohol by volume** [ABV]), which is governed by the amount of fermentable solids.

Mash pH

Enzymatic reactions, including mashing reactions, are generally affected by pH (introduced in Section 4.1). The generally recommended range for mash pH is 5.2–5.6, which falls between the optimal enzymatic pH for alpha- and beta-amylase. *The mash pH should be monitored and recorded for every brew at mash-in.* Higher pH tends to yield higher fermentability; lower pH favors higher extract (original gravity).

Studies of the details of the effect of pH on mashing outcomes are sparse and inconsistent. One reason for this is that it is not possible to adjust pH without introducing ions other than H_3O^+ and OH^-. Another technical issue is that the pH of the same material at mashing temperature is lower than that at room temperature where most measurements are made. For example, the pH of pure water at 25 °C is very close to 7, but at 70 °C it is 6.4. Many reports are vague about the conditions under which the pH was measured. Measuring pH at temperatures far from the laboratory temperature is difficult and prone to error.

The key variables determining mash pH are the alkalinity and hardness of the liquor and the acidity of the malt. Alkalinity in brewing liquor is provided by bicarbonate ion (HCO_3^-). Bicarbonate ion raises the pH by removing **hydronium ion**: $HCO_3^- + H_3O^+ \rightarrow CO_2 + 2H_2O$. High alkalinity in the liquor can lead to high mashing pH.

Water alkalinity can be adjusted or compensated to adjust the mash pH. The simplest brute-force approach is adding acid directly to the mash. Acids provide hydronium ion to replace that taken up by the bicarbonate. Hydrochloric, lactic, phosphoric, and sulfuric acids are most commonly used. It is essential that only food-grade quality acid is employed. The complication with acid additions is that every acid also provides a counterion. Hydrochloric acid adds chloride (Cl^-); lactic acid adds lactate ($C_3H_5O_3^-$); phosphoric acid adds hydrogen **phosphate** (HPO_4^{2-}); and sulfuric acid adds sulfate (SO_4^{2-}). Each may affect the character of the beer. WARNING: some of these acids are potentially corrosive and they can release heat on addition to water.

Malt provides acidity and **buffering** to the mash. Dark malt is more **acidic** than pale malt because the higher intensity kilning used to make dark malt removes **amines** by the **Maillard reaction**. Because amines are **basic**, their removal increases acidity. Mash acidity is also derived from the dissociation of barley malt phosphates: $H_3PO_4 \rightarrow H^+ + H_2PO_4^-$.

The water alkalinity can be lowered by **reverse osmosis**, **ion exchange**, or one of the other methods discussed in Chapter 4. Water hardness (calcium and magnesium ions) can, to some extent, neutralize alkalinity. Calcium ion and, to a lesser degree, magnesium ion react with phosphate in the malt to release hydronium ion: $3Ca^{2+} + 2HPO_4^{2-} + 2H_2O \rightarrow 2H_3O^+ + Ca_3(PO_4)_2$. Recent research suggests that this effect is not as great as had previously been thought.

Lowering of the mash pH can be accomplished by:

- Adding calcium in the form of gypsum or calcium chloride.
- Adding dark or **caramel malts**.
- Adding a small amount of **acidulated malt**.
- Sour mashing via **bacteria** addition.
- Adding food-grade acids.

6.4   MASH CONVERSION VESSEL DESIGN AND OPERATION

The mash vessel is the point at which barley malt is turned into brewer's wort. There are a variety of subtle design differences between different mashing vessels, but all operate with similar objectives, to obtain extract from the malt and to control the mash at specific temperatures that influence enzymatic activity. The types of enzymes involved and the impact on beer will be discussed.

Mashing In

Mashing begins with the addition of hot water, called hot liquor, to the grist, a process called **mashing in**. The water is usually prepared in advance in a vessel called a **hot liquor tank** (HLT). The optimal size of the HLT depends on the mash volume, the time between brews, and the use to which the hot water will be put. Hot liquor is required for **foundation water**, mashing, and **sparging**. The HLT may also serve as the source of water for rinsing and **cleaning**. If multiple batches are brewed on the same day, the water needs to be ready when the mash **tun** is clean from the previous batch. A general rule is for the HLT to have a capacity three times that of the mash volume. Small HLTs run the risk of insufficient amounts of hot liquor during production.

The grist from the mill is usually kept in a **grist case** until it is dispensed. The initial temperature of the liquor before addition to the grist is called the **strike temperature**. The temperature after mixing the liquor and grist is the **mash-in temperature**; it is always lower than the strike temperature, because the grist absorbs heat. Calculations for determining the strike temperature are discussed in the "MASH-IN AND DECOCTION TEMPERATURE" box.

Dry spots, clumping, and uneven temperatures should be avoided while mashing in. In small mashing vessels, the vessel is often filled with hot liquor at strike temperature, and the grist added through the top while mixing. *Dry grain should always be added to water*, because it can be difficult to mix if water is poured on top of the grist. There are several disadvantages to adding dry grist directly to hot water in the mash vessel. First, this method creates dust. Not only is this a safety hazard and a contamination threat, but it can lead to small losses in extract. Second, to reach the appropriate mash temperature, the water must be hotter than the mash target. The first grist to be added to the hot water experiences elevated temperatures that may deactivate malt enzymes, negatively influencing mash parameters. Finally, directly adding dry grist during mash can lead to dry clumps, called *dough balls*, that do not **hydrate** well.

Some systems place a grist case directly above the mashing vessel to contain dust during mash-in. Ideally, the grist case should be large enough to hold the entire malt bill for the brew, so that it can be added in a controlled and consistent manner. The grist case is connected to the mash vessel by a short **pipe** and is controlled by a slide **valve**, thus containing the dust. Figure 6.12 shows two grist cases. The smaller one on the left feeds the adjunct cooker; the larger one on the right feeds the mash conversion vessel.

It is best to mix the grist and brewing liquor while maintaining the same liquor-to-grist ratio the whole time. With proper control of the **strike water** temperature, the resulting mash-in temperature is held for the entire addition process. This can be accomplished with a **premasher**, also called a foremasher. These devices are typically installed in a pipe above the mash vessel, injecting water at mash-in temperature as the grist falls into the mash vessel, as shown

Figure 6.12   Grist cases at Yuengling Beer Company, Pottstown, PA.

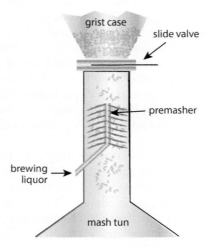

Figure 6.13   Premasher.

in Figure 6.13. To prevent clumping, the premasher may include mechanical mixing, or it may be designed with water jets that produce strong mixing. The **Steel's masher**, used for traditional British **ale**, has a horizontal screw **auger** followed by tines or rakes that mix the grist into the brewing liquor.

Prior to mash-in, a small amount of *foundation water*, at the appropriate strike temperature, should be added to the mash vessel. The volume of foundation water should at least be enough to cover 25–50 mm (1–2 in.) above the plate or mixer. Foundation water assists the spreading of the initial grist and is essential if using a false bottom for a combined mash/lauter tun to prevent clogs.

Liquor-to-Grist Ratio   The amount of water per unit of grist is the liquor-to-grist ratio, sometimes called the **mash thickness** (a misnomer, high liquor-to-grist is thin mash). Liquor-to-grist ratios can run as low as 1.6 L/kg (0.2 US gal/lb) for a very thick infusion that is difficult to mix. Or they can run as high as 5 L/kg (0.63 gal/lb), giving a thin **slurry** that is easy to **pump**. Mashing performance is significantly affected by mash thickness. A low liquor-to-grist ratio mash retains beta-amylase activity, resulting in higher wort fermentability.

The Iodine Check   The effectiveness of the mashing program in hydrolyzing starch can be quickly validated by an iodine test. Iodine is a light amber color, turning dark purple in the presence of starch. To run the test, a few mL of mash is placed on a white porcelain dish, and then a few drops of a solution that is about 0.25% in iodine and 0.5% in potassium iodide are added. If the solution turns purple, starch is still present. If the sample stays light brown, there is no starch. A negative test does not mean all hydrolysis reactions are completed. Iodine only interacts with long starch molecules, inserting itself into the helical shape. Large dextrins and long **polysaccharides** will not test positive, so this test is really only useful for monitoring early liquefaction. In general, an all-malt mash will produce a negative iodine test in 10 minutes or less. Monitoring the time at which the test becomes negative can be a useful method to monitor mash quality.

Mash Vessel Design

Once the grist has been added to the hot liquor in the mash vessel, it must be gently mixed. In small systems, and in days of yore, the mash was stirred with handheld oars. In moderate to large vessels, an agitator is built into the base of the unit as shown in Figure 6.14. The agitator should be set away from the walls and floor of the vessel to avoid crushing the hulls. Rotation speed will vary depending on vessel size, but the agitator should not exceed a **linear** speed of 2 m/s. Vigorous mixing applies **shear** forces that can damage the malt hulls, leading to polyphenol extraction and astringent beer. Slow but consistent mixing is essential to promote even temperatures, controlled enzymatic activity, and good beer quality.

The strike temperature of the hot liquor is chosen so that, after mixing, the mash reaches a target temperature. Afterward, it may be necessary to

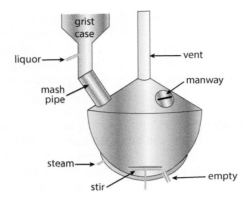

Figure 6.14   Mash conversion vessel.

maintain or to increase the temperature of the mash. Most modern mash vessels are equipped with a steam jacket. Direct fire heating may also be employed. In direct fire vessels, an agitator must operate continually to prevent scorching and burning of grain on the floor of the vessel.

Larger breweries use a separate **lauter** tun (discussed in Chapter 7) for wort separation, but smaller operations may use a combined mash and lauter tun. In this design, a false bottom keeps the malt hulls off the floor of the vessel and will permit **lautering**. If using a separate lauter tun, the entire mash is pumped into the lauter tun when mashing is complete.

Mashing Steps and Parameters

The critical process variables during mashing that determine wort quality are time, temperature, pH, and mash thickness. Regarding time and temperature, the mash will be held at a temperature for a specified period called a *step* or **rest**. Depending on the mashing regime, several rests may be performed. Table 6.3 provides a summary of typical mash steps and their effects.

The number of steps and the specific time and temperature of each step will depend upon the desired character of the wort. Most steps only require 15–20 minutes and need not exceed 30 minutes. A single step mash should persist for at least 60 minutes.

The lowest temperature step, the **acid rest**, was historically used to help lower the pH of the mash by promoting the activity of phytase, an enzyme that frees phosphoric acid from phytic acid (Figure 6.15). Phytase is most active between 35 and 52 °C and at a pH of 3–5. The resulting phosphates are necessary building blocks of DNA, RNA, and **ATP**, and phosphates can lower the pH of the mash to 4.8–5.2. Practical experience and well-**modified** malt used today indicate that mash pH is quickly and strongly buffered

TABLE 6.3   Effects of Temperature During Mash Rests

Temperature	Step (effect)
35–40 °C (95–104 °F)	Acid rest (ferulic acid release from hemicellulose; phosphate release via phytase)
45–50 °C (113–122 °F)	Protease rest (hydrolyzes proteins, increases FAN; decreases foam stability)
45–55 °C (113–131 °F)	β-Glucanase rest (hydrolyzes glucans; decreases viscosity)
60–65 °C (140–149 °F)	Maltose production rest (optimal β-amylase temp.)
66–70 °C (150–158 °F)	Saccharification rest (promotes α-amylase activity)
76–78 °C (168–172 °F)	Mash-out (decreases viscosity, deactivates enzymes)

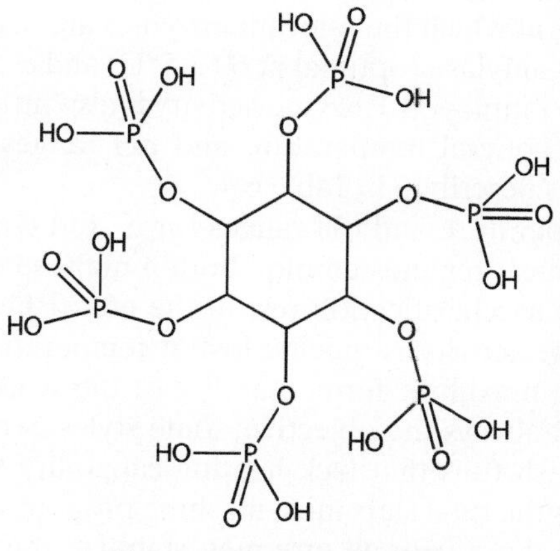

Figure 6.15   Phytic acid.

following mash-in. The buffering capability of the mash eliminates the need for the acid step in most brews. One exception is German wheat beers, where the rest is not to lower the pH, but to release ferulic acid. Free ferulic acid is important in wheat beers, because it is decarboxylated to 4-vinylguaiacol during fermentation (by POF+ yeast), which contributes to the characteristic phenolic flavor of the style.

Proteolytic enzymes are mostly deactivated during malting, but some brewers try to take advantage of any residual activity by including a protease or protein rest in their mash schedules. Protease hydrolyzes proteins from barley into smaller fragments, peptides, and amino acids. The protein rest can be used to increase **FAN** in the wort. A protein rest may be necessary to help

TABLE 6.4   Optimal Temperature and pH Ranges of Mash Enzymes

Enzyme	Optimal temperature	Optimal pH
β-Glucanases	40–60 °C (104–140 °F)	4.5–4.8
Proteases	45–50 °C (113–122 °F)	3.9–7.2
Lipoxygenase	45–55 °C (95–104 °F)	6.5–7.0
Phosphatase	50–53 °C (95–104 °F)	5.0
Limit dextrinase	55–65 °C (95–104 °F)	5.1
β-Amylase	60–65 °C (140–149 °F)	5.4–5.5
α-Amylase	70–74 °C (158–165 °F)	5.6–5.8

generate sufficient FAN when using adjuncts, but in all-malt beers, it is unnecessary and potentially detrimental to beer **foam**. If the protein rest is on the low end (45–50 °C), it may promote both protease and β-glucanase activity, while the higher end (50–55 °C) supports mostly β-glucanase.

The effects of each mash step on beer quality are ultimately guided by the optimal temperature at which the relevant enzymes are active. For this reason, a separate step for β-amylase (optimal at 60–65 °C) and α-amylase (optimal at 70–74 °C) is typically employed. Enzyme activity is also influenced by the mash pH. A summary of general temperature and pH ranges for broad enzyme groups in the mash is described in Table 6.4.

The optimal temperature and pH ranges for α- and β-amylase do not precisely align. Many mash regimes employ both a maltose rest in the 60–65 °C range followed by a saccharification rest in the 66–70 °C range. As shown in Figure 6.9, β-amylase activity is quickly lost at temperatures above 66 °C, so sequential steps can maximize fermentability of the wort. Nonetheless, high fermentability is not always the objective; some styles benefit from enhanced final gravity. For mash tuns that lack heating capability, this can be a virtue born of necessity. As the final step, most mashing programs involve a mash-out step above 72 °C to deactivate all enzymes, stabilize the wort, and decrease viscosity to accelerate lautering.

Mash Heating Programs

Because most enzymes irreversibly lose activity at temperatures that exceed their optimum range, but not at temperatures that are too low, mashing programs always increase rather than decrease the temperature. There are four general temperature programs often used for mashing. **Infusion mashing** makes no effort to change the mash temperature after mash-in. The target temperature is typically 65–67 °C (148–152 °F), which is within the narrow window of optimal activity for both alpha- and beta-amylase. Without temperature control, the temperature may slowly cool from the mash-in temperature as heat is lost to the surroundings. Infusion mashing is most suitable for well-modified malt.

In **decoction** mashing, a portion of the mash (typically one-third) is heated to boiling and returned to the mash tun to raise the temperature to the next step. Usually the mash is allowed to settle, and a portion from the denser phase is boiled. This avoids excessive loss of dissolved enzymes. The decoction method, popular for German-style brewing, dates from before **thermometers**; the mash boiling point established a dependable temperature. Decoction is suitable for malt that is not fully modified, because boiling helps gelatinize recalcitrant starch granules.

A variation of decoction is double mashing, suitable when high temperature gelatinizing adjuncts are used. The malt is mashed in at a low temperature. The adjunct, like rice or maize grits, is heated in another vessel. Usually some malt is included with the adjunct to help liquefy it and lower its viscosity. The adjunct mash is boiled to gelatinize the adjunct. The entire contents of the adjunct mash are then added to the malt mash, bringing it to the next temperature step. The most modern mashing programs involve continuous control of the temperature to take advantage of the properties of the enzymes in the mash.

MASH-IN AND DECOCTION TEMPERATURE

Strike Temperature

Our objective is to calculate the strike temperature, which is the water temperature needed to reach the desired mash-in temperature, accounting for the heat needed to heat the malt and the tank and the heat released by hydration of the malt. To set up the mash-in problem, we first write the energy conservation equation

$$Q_{water} + Q_{grist} + Q_{tank} + Q_{reaction} = 0$$

$Q_{reaction}$, called the "slaking heat," is energy released by chemical interaction of the dry grain with water. Because heat goes *out* of the reaction, this term is negative. For malt with 4% moisture, $Q_{reaction}$ is -40000 J/kg. For the first 3 Q's, Q = mass × specific heat × $(T_{init} - T)$, where T is the final mash-in temperature. We are looking for T_{init} of the water. The specific heat of water is 4184 J kg^{-1} °C^{-1}, and that of grist is 0.4 times that of water. The relevant equation derived from some algebra is

$$T_{strike} = T + \frac{0.4 M_{malt} T}{4184 M_{water}} - \frac{0.4 M_{malt} T_{malt}}{4184 M_{water}} + \frac{C_{tank} T}{4184 M_{water}} - \frac{C_{tank} T_{tank}}{4184 M_{water}} - \frac{40\,000 M_{malt}}{4184 M_{water}}$$

where T is the desired mash-in temperature. T_{strike} is the initial temperature of the water or the target strike temperature. The T's with subscripts

(continued)

are the initial temperatures of the malt and tank. The M's are masses in kg. C_{tank} is the **heat capacity** of the tank. The heat capacity of the tank is not often supplied by the manufacturer. Fortunately, the error introduced by ignoring it is not large. Even more fortunately, C_{tank} can be measured, as explained below. The calculation assumes that the supplies are at the indicated temperatures when they reach the mash tank.

Suppose we want to mash 3333 kg of grain with 10 000 kg of water. The desired mash-in temperature is 68.0 °C. The initial temperature of the tank and the malt is 22 °C. The heat capacity of the tank is 290 000 J/°C:

$$T_{strike} = 68.0 \,°C + \frac{0.4 \times 3333\,kg \times 68.0 \,°C}{10\,000\,kg} - \frac{0.4 \times 3333\,kg \times 22.0\,°C}{10\,000\,kg}$$

$$+ \frac{290\,000 \times 68.0 \,°C}{4184 \times 10\,000\,kg} - \frac{290\,000 \times 22.0 \,°C}{4184 \times 10\,000\,kg} - \frac{40\,000 \times 3333}{4184 \times 10\,000\,kg}$$

$$= 71.2 \,°C$$

This calculation is best managed in a computer spreadsheet.

Decoction Temperature

In a decoction, a portion of the wort is removed, boiled, and returned to the mash. The mash temperature after the boiling wort is added to the cool wort is given by the equation

$$T = \frac{T_{hot} M_{hot} + T_{cold} M_{cold}}{M_{hot} + M_{cold}}$$

Suppose 3333 kg of boiling wort at 100.2 °C is added to 6667 kg of wort at 35 °C. Plugging these numbers into the equation gives

$$T = \frac{100.2 \,°C \times 3333\,kg + 35 \,°C \times 6667\,kg}{10\,000\,kg}$$

$$= 48.0 \,°C$$

Measuring the Heat Capacity of a Tank

Meter water into the tank until the temperature sensor is covered. Record the amount of water. Cover the tank, and allow the tank and the water to equilibrate for an hour or so, with gentle stirring if available. Record the initial temperature (T_{cold}). Then meter in hot water at a known temperature (T_{hot}). Record the amount of hot water added and its temperature. Stir well and record the temperature (T). The heat capacity of the tank is calculated from the following equation:

$$C_{tank} = 4184 \frac{\left(M_{hot} \left[T - T_{hot} \right] + M_{cold} \left[T - T_{cold} \right] \right)}{T_{cold} - T}$$

6.5   THE USE OF ADJUNCTS

An adjunct is a source of fermentable material that is not malted grain. Adjuncts can provide starch that must undergo mashing, or they can supply sugar that can be added directly to the kettle or even to the fermenter (if **sterile**). Adjuncts are used as supplements to malted grain, seldom providing more than 30–40% of the fermentable sugar unless additional enzymes are added. Their major functions are to provide a crisp, light flavor and to enable high-gravity brewing. Some adjuncts increase fermentability to make beers with low **final extract**, like **light beer** and **malt liquor**. Adjuncts are also less expensive than malted grain, but the cost is somewhat offset because many starchy adjuncts require a separate vessel for gelatinization (see Table 6.1).

Starch Sources

Starchy adjuncts provide no starch-hydrolyzing enzymes. They depend on other components of the mash to convert their starch. If high concentrations of starchy adjuncts are used, it may be necessary to supplement the mash with enzymes from other sources, typically molds. The most common starchy adjuncts are maize (corn) and rice. Rice and maize contribute starch but are neutral in flavor and low in color and contribute minimal protein and **free amino nitrogen**. Maize, rice, and other grains are subjected to dry milling before they arrive at the **brewery** to remove the germ and bran, which are high in lipids that are undesirable in the brewing process. Maize milling yields a distribution of particle sizes from grits to flour. Rice milling yields whole rice grains, plus some broken grains that can be used in brewing. Starchy adjuncts can also be prepared from sorghum and millet. Additional milling is usually done at the brewery before the grain is cooked. Maize, rice, sorghum, and millet do not contain **gluten**, so they can form the basis for gluten-free beer. The labeling of food as gluten-free requires adherence to strict government regulations. Grains that can be malted, including barley, wheat, oats, rye, sorghum, and millet, can also be used as unmalted adjuncts. These adjuncts generally require adjustments to the milling and mashing processes because of their physical and chemical properties, such as high beta-glucan content, high protein, and lack of starch-hydrolyzing enzymes.

Neither rice nor maize will gelatinize fully at mashing temperature, so they must be cooked in a separate process before they are mixed with the malt mash. The milled grain is mixed with water at 40–50 °C (104–122 °F). Starch-degrading enzyme is provided, either by adding malt or mold-derived enzyme to the cooker. This prevents the mixture from setting up like pudding. The mixture is then heated and boiled for about 20 minutes. The boiling mixture is then added to the main malt mash to heat it to the saccharification temperature.

It is possible to gelatinize some starchy adjuncts during manufacture to avoid the cooking step. *Flaked* adjuncts are treated with hot air or steam and then pressed between hot rollers. Flaked barley, maize, rice, wheat, and oats are common. *Micronization* is a similar process, but the grain is treated with radiant heat instead of steam. Grains can also be *torrified*, a process of heating in hot air, causing the grains to expand like popcorn and become gelatinized. Torrified wheat, corn, and barley are typical. Because flaked, micronized, and torrified **cereals** are gelatinized in production, they can be added directly to the mash.

Sugar Sources

The addition of sugar adjuncts increases the initial gravity of the wort without requiring mashing. This is essential in **high-gravity** brewing, where concentrated beer is brewed from a high initial gravity wort, typically 18–20° **Plato**. It is difficult and highly inefficient to mash enough grain to make high-gravity wort. Instead, lower-gravity wort is supplemented with sugar. The high-gravity wort is then fermented and conditioned. The beer is diluted with deaerated, **carbonated** water to the correct ethanol content before release to customers.

Another application of sugar sources is in the brewing of very strong beer. Sugary adjuncts can be added during fermentation to avoid, to some extent, **osmotic stress** on the yeast that would result from pitching it into strong wort. Some sugar sources, like **candi sugar**, honey, molasses, and maple syrup, give special flavors to the beer. They can be quite expensive. Another flavorful adjunct is **malt extract**, which is prepared by mashing grain, separating the wort, and concentrating it by evaporation. Extract is available in syrup and powder form. Different types of malt yield different flavors of extract.

The most common sugar sources derive from wet milling of maize. Maize kernels are soaked in water containing sulfur dioxide. Starch is separated from the other components by a sequence of milling, **filtration**, flotation, and **centrifugation** in a device called a hydrocyclone. The starch emerges suspended in water and is hydrolyzed with acid or with enzymes to give a variety of products with different sugar profiles including **dextrose** corn syrup, refined liquid dextrose, high fructose corn syrup, and high maltose corn syrup. Evaporation can yield a solid called corn sugar (mostly glucose). A summary of common sugar adjuncts used in the brewery is provided in Table 6.5.

Sugar adjuncts are typically used at 5–15% of the total grain bill. They are best added to the boil kettle with ~10 minutes remaining to ensure sterilization. Sugar additions provide no FAN or zinc, which are essential nutrients for yeast health. Sugar adjuncts only contribute flavor if they were caramelized during manufacture, like molasses, brown sugar, and candi sugar. Molasses is a dark, viscous syrup obtained as a by-product of refining sugarcane. It contains about 85% solids and moderate amounts of calcium, magnesium, iron, and

TABLE 6.5 **Sugar Adjuncts**

Name	% Glucose	% Maltose	% Maltotriose	% Dextrins	Other
Dextrose (glucose)	100	0	0	0	0
Corn syrup	45	38	3	14	0
High maltose	10	60	0	30	0
Maltodextrin	0	1.5	3.5	95	0
Sucrose	0	0	0	0	100% sucrose
Invert sugar	50	0	0	0	50% fructose

phosphate. Molasses is around 50% fermentable. Brown sugar is typically 88% sucrose, 8% invert sugar (an equal mixture of glucose and fructose), and 4% molasses. Belgian candi sugar, in its finest form, is refined beet sugar that has been caramelized. Otherwise, it is caramelized invert sugar, potentially from alternative sugar sources.

Malt Extract   Malt extract is prepared by evaporation of beer wort to give a syrup (liquid malt extract [LME]) or powder (dry malt extract [DME]). Malt extract has no functioning enzymes, but it does contain free amino nitrogen and other yeast nutrients that are in the original wort. It gives color and flavor to the beer. Malt extract should be boiled like any wort. Malt extract is not as stable in storage as unmilled grain. It can be difficult to handle in large qualities because the syrup is very viscous and the powder absorbs water and steam forming sticky clumps. For the same amount of wort solids, malt extract is relatively expensive, so brewers who use it do so largely to boost the solids content of their own wort. Malt extract is not available in as many varieties as is malt. Nonetheless, it is possible to brew beer with malt extract as the only source of fermentables.

6.6   ENZYMES AND PROCESSING AIDS

Enzymes and processing aids are sometimes added to the mash, usually to overcome processing difficulties occasioned by undermodified malt, non-diastatic adjuncts, hulless adjuncts, or excessive beta-glucan.

Rice Hulls

Rice hulls are a processing aid that provide no extract or flavor, but they can be used to promote filterability during lautering, especially with glucan-rich adjuncts or those that lack hulls. Grain hulls prevent stuck sparges by increasing the permeability grain bed during lautering. They are typically used at 2–5% of the total grist. Rice hulls must be pre-wet and rinsed before use to remove dust that can compromise clarity and flavor. They can be rinsed with the sparging apparatus in the lauter tun before transfer of the mash.

Farbebier

Very dark wort is fermented to produce **Farbebier** (Ger: colored beer), which is sometimes used in small quantities to increase beer color, especially by brewers who comply with the *Reinheitsgebot*. Farbebier has a color of around 4000 °SRM, so the amount needed does not significantly affect the flavor or other properties of the beer.

Enzymes

Enzyme preparations in solid or solution form are available as brewing processing aids. They are usually extracted from cultures of bacteria or mold. Any enzymes used for brewing must be certified for food use. Nonetheless, their levels of activity and the virtually inevitable presence of other enzymes may not be standardized. Enzymes prepared by different suppliers using different **purification** protocols may behave quite differently. Brewers should use reliable suppliers that can deliver enzymes with consistency. Enzyme preparations must be stored cool and used fresh, or their activities may decline over time. Enzymes can be useful in many stages of brewing. Here we will discuss those that are added to the mash. Some commercially available enzymes are too stable for brewing use. Enzymes that are not deactivated by boiling or pasteurization may create problems in packaged beer.

Alpha-Amylase   Alpha-amylase hydrolyzes alpha(1→4) glycosidic bonds that are not too close to branches on a starch chain. If there is insufficient alpha-amylase activity during mashing, the wort fermentability will be low. Alpha-amylase from various species of the aspergillus mold has maximum activity around 60–65 °C (140–149 °F). This enzyme is not as stable as barley alpha-amylase, but it can be useful to provide activity when adjuncts dilute the malt. Some alpha-amylases from bacteria are more thermally stable. These are used to liquify the adjunct mash-in the **cereal cooker**.

Limit Dextrinase   **Limit dextrinase**, also called pullulanase, debranches starch by hydrolyzing alpha(1→6) glycosidic links. Acting alone, the enzyme makes a minimal contribution to fermentability, but it creates more substrates for alpha- and beta-amylase.

Glucoamylase   **Glucoamylase**, also called **amyloglucosidase**, hydrolyzes a single glucose unit from the non-reducing end of a starch chain. Depending on the microbial origin, it may or may not be able to hydrolyze alpha(1→6)

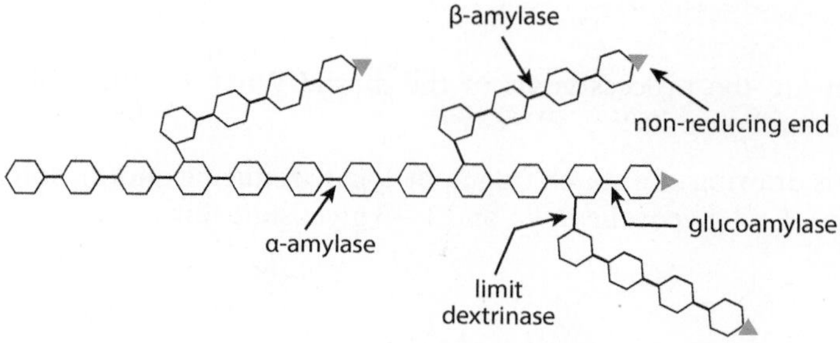

Figure 6.16   Targets for enzymes.

branch points. Glucoamylase is often added in the fermenter to make highly **attenuated** beer styles, like **light beer**, **malt liquor**, and brut IPAs. Use in the mash is possible, but it will hydrolyze maltose to glucose, potentially leading to problems due to **catabolite repression**. Figure 6.16 schematically shows target bonds for these starch hydrolysis enzymes.

Beta-Glucanase   Beta-glucan (Figure 3.11) is a **gum** that, when dissolved in wort, raises the viscosity, slowing down lautering and other filtration processes. It comes from degradation of grain cell walls. Slow filtration exposes the wort to possible oxidation and allows it to extract polyphenols from grain. Well-modified malt has little beta-glucan, but undermodified and unmalted grain can have enough to interfere with brewing. Beta-glucanase hydrolyzes the beta-glucan, quickly lowering the mash viscosity.

Pentosanase/Xylanase   Plant cell walls contain insoluble **cellulose**, plus a variety of potentially soluble carbohydrates, including beta-glucan and pentosans, which are polysaccharides containing pentoses. Two of the major pentoses are **xylose** and **arabinose**. The pentosan-degrading enzyme usually used in brewing is arabinoxylase. Used in conjunction with beta-glucanase, it can improve processability. The risk is that arabinoxylase can release ferulic acid from its ester linkages to the pentosans, allowing it to dissolve in the wort. Depending on the fermentation conditions, the ferulic acid is converted into 4-vinylguaiacol, a strongly flavored phenolic compound. Some beer styles, notably weizenbier, are expected to have 4-vinylguaiacol (Figure 2.20), but it is a **defect** in others.

Protease   Proteases hydrolyze proteins. In mashing, they are used to increase free amino nitrogen, especially in mashes with a high content of adjunct. Yeast needs free amino nitrogen to make nitrogen-containing compounds for their life functions. Protease should be used cautiously, because it can hydrolyze foam-positive protein, leading to well-fed yeast but deficient foam.

CHECK FOR UNDERSTANDING

1. What are the process goals of the mash? What are the major process inputs and outputs?

2. In this drawing of a starch fragment, locate reducing end(s), non-reducing end(s), 1→4 glycosidic links, and 1→6 glycosidic links.

3. In the drawing for problem 1, locate points of action for alpha-amylase, beta-amylase, limit dextrinase, and glucoamylase.

4. Describe what is meant by the *fermentability* of wort.

5. What is the relative percentage of most common carbohydrates in the mash?

6. What critical process variables of the mash would you change to create a dry beer?

7. You have a grain bill of 4000 kg with a liquor volume of 100 hL. Calculate the strike water temperature to achieve a mash-in temperature of 65 °C.

8. What is the purpose of a premasher?

9. Describe the best practices for mashing in.

10. What is the effect of temperature and pH on general enzyme function? Be as specific as possible by describing the effect on mashing enzymes.

11. Why is a slow gentle mash mixer recommended over vigorous stirring?

12. Design a mash program including time and temperatures to include a protein rest, maltose rest, saccharification rest, and mash-out step. Explain the purposes of each step by describing specific enzymatic activity promoted during each step.

13. Determine the amount of mash required in a single decoction step to raise the temperature of a mash from 55 to 65 °C assuming a boiling temperature of 100 °C.

14. Describe how the following mash conditions might (i) affect enzyme activity and (ii) influence the character of the finished beer: a thin mash at 4 kg/L, a mash pH of 5.8, a single step infusion mash at 68 °C (155 °F) for 60 minutes, and a mash-out at 76 °C (170 °F).

15. Why does rice need to be cooked in a separate vessel before use in a brewer's mash?

CASE STUDY

A brewery has a dark ale with an original extract of 14 °P and a final extract of 4 °P. Mash-in is at 69 °C (156 °F). Recently, the original extract has fallen to 13.5 °P, and the final extract has drifted up to 6 °P, which is well out of the specification. What are some likely causes for this problem in the mash? Propose a program to diagnose and fix the issue, preferably without excessive expense, downtime, and loss of good will.

BIBLIOGRAPHY

Bamforth CW. 2001. pH in brewing: an overview. *MBAA Tech. Q.* 38(4):1–9.

Barth R, Zaman R. 2015. Influence of strike water alkalinity and hardness on mash pH. *J. Am. Soc. Brew. Chem.* 73(3):240–242.

Briggs DE, Boulton CA, Brookes, PA, Stevens R. 2004. *Brewing Science and Practice.* CRC. ISBN 0-8493-2547-1. p. 46–49. Detailed discussion of added enzymes.

Edens L, Dekker P, van der Hoeven R, Deen F, de Roos A, Floris R. 2005. Extracellular prolyl endoprotease from *Aspergillus niger* and its use in the debittering of protein hydrolysates. *J. Agric. Food Chem.* 53(20):7950–7957.

Fix G. 1999. *Principles of Brewing Science.* Brewers Publications. ISBN 978-0-937381-74-8. Chap. 1.

Kadziola A, Abe J, Svensson B, Haser R. 1994. Crystal and molecular structure of barley alpha-amylase. *J. Mol. Biol.* 239(1):104–121.

Krottenhaler M, Back W, Zarnkow M. 2009. Wort production. In Eßlinger HM (editor). *Handbook of Brewing.* Wiley-VCH. ISBN 978-3-527-31674-8. Chap. 7.

Kunze W. 1999. *Technology of Brewing and Malting, International Edition.* VLB. ISBN 3-921-690-39-0. Chap. 3.

Lewis MJ, Robertson IC, Dankers SU. 1992. Proteolysis in the protein rest of mashing – an appraisal. *MBAA Tech. Q.* 29(4):117–121.

Sammartino M. 2015. Enzymes in brewing. *MBAA Tech. Q.* 52(3):156–164. Readable coverage of malt enzymes as well as added enzymes.

CHAPTER 7

WORT SEPARATION

After **mashing** is complete, the **mash tun** contains a **mixture** of dissolved **carbohydrate** and undissolved particles, including **grain hulls**, **fats**, and **protein**. During separation, the carbohydrate-rich **wort** is removed. It will be required for subsequent **fermentation**. The major goal of **wort separation** is to extract as much of the carbohydrates as possible from the grain bed while avoiding transfer of hulls or extraction of harsh **tannins**. There are several methods for wort separation; each is a variation on **filtration**. The wort is run through pores that are too small to admit the undissolved particles but large enough to allow a satisfactory flow rate for the wort. In **lautering**, the hulls of the **grist** serve as the **filter medium**. All wort separation methods involve two steps: (i) running off the *first wort* and (ii) **sparging** (rinsing) the grains to produce *second wort*.

7.1 WORT SEPARATION PROCESSES

Lautering (Ger: pure; clear) is the process of wort separation in which the spent grains serve as filter material. The process is completed with one of three types of equipment: a **lauter tun**, a combined mash/lauter tun, or a **mash filter**.

Mastering Brewing Science: Quality and Production, First Edition.
Matthew Farber and Roger Barth.
© 2019 John Wiley & Sons, Inc. Published 2019 by John Wiley & Sons, Inc.

The lauter tun and mash filter are separate vessels from the mash tun and require a low **shear pump** to transfer the entire contents of the mash.

Lauter Tun

Lautering is typically performed in a dedicated vessel called the lauter tun (Figure 7.1). It has a wide diameter to allow for a shallow grain bed depth and a wide flow area. This design increases flow by increasing the flow area (item 1 in Table 7.1) and decreasing the depth of the grain bed (item 2). A bed depth of 25–50 cm (10–20 in.) is common.

A few centimeters above the *true bottom* of the vessel is a **false bottom** with slotted plates to hold up the grain and to allow wort to pass. The false bottom supports the grain bed but does not perform the filtration. The design

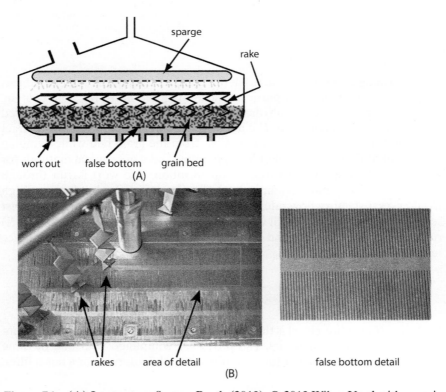

Figure 7.1   (A) Lauter tun. *Source*: Barth (2013). © 2013 Wiley. Used with permission. (B) Mash/lauter tun; Levante Brewing Company with detail of false bottom. *Source*: Photo: Naomi Hampson.

TABLE 7.1   **Application of Darcy's Equation to Flow Rate in the Lauter**

Increase Flow/Lower Pressure Drop	Decrease Flow/Increase Pressure Drop
Wide flow area	Narrow flow area
Shallow grain bed	Deep grain bed
Large grain particles	Finely ground grain particles
Low wort viscosity	High wort viscosity
Loose grain packing	Tight grain packing – collapsed grain bed
Higher temperature (lower viscosity)	Lower temperature (higher viscosity)
Low β-glucans (low viscosity)	High β-glucans (high viscosity)

of the false bottom is critical. It is designed to give minimal **pressure** drop if it is kept clear of deposits. The slots must be narrow enough to prevent the passage of spent grain but wide enough to prevent clogging. A common design element is to mill the underside of the slots slightly wider than the top to prevent blockage. With slot gaps around 0.7 by 80 mm, the total open area in a false bottom is about 10%.

The lauter tun is also equipped with rotating knives or rakes (also called cutting **units**) that slowly dig furrows in the grain bed to promote flow rate. The grain bed should only be raked if the flow rate is low or the differential pressure is high. Each rake arm is equipped with several knives that are designed to make channels efficiently without collapsing the grain bed. The height of the rake can be raised or lowered depending on the **turbidity** of the wort. If the wort is turbid, the rake should be raised. If the wort is clear, the rake should be lowered as close to the false bottom as possible. Some rakes include a pivoting plough-like crosspiece that assists spent grain removal.

Before transfer of the mash, hot water is added to the lauter tun, with enough volume to slightly cover the false bottom. This water, called **foundation water**, pre-warms the lauter tun and prevents air pockets under the false bottom. Because foundation water will dilute the mash, the minimum amount should be used. The entire contents of the mash tun are then pumped into the lauter tun from above the false bottom. Some particles may initially slip through the false bottom; therefore the early wort with particles is gently pumped from below the false bottom and returned above the false bottom, an operation called **vorlauf** [Ger: forerun]. The vorlauf sets up a bed of grain on the false bottom and is run until the wort is clear. It is this grain bed that serves as the **filtration medium**, trapping small particles and clarifying the wort (Figure 7.1).

Once the wort is clear, it is sent to the boil **kettle**. The wort is allowed to run through several **valves** in the true bottom, from which it is transferred to the kettle, sometimes via an intermediate vessel called a **prerun tank**. Runoff should take 90–120 minutes. The *first wort* is collected until the grain bed is just visible. At this point sparge water is sprayed onto the top to maximize wort extract. The wort collected after sparging is called *second wort*. Sparging is discussed in Section "Sparging."

Most lauter tuns have a grain-out port for the spent grain. When wort separation is complete, the port is opened, and the knives are positioned to drive the grain into a collection bin.

The **lauter process** has some disadvantages. These include:

- Complexity of the moving parts, including the knives and their raising/lowering gear, the sparge arm, and the grain-out system.
- Dilution of the wort with foundation water and sparge water limits the strength of the wort that can be efficiently obtained.
- Large requirement for floor space.

Mash/Lauter Tun

Small **breweries** and traditional British **ale** breweries use the same vessel for mashing and wort separation. The mash tun has a slotted false bottom like that of a lauter tun. The vessel is deeper and narrower than a lauter tun to conserve **heat**. The grain bed may approach 1–2 m deep, significantly deeper than a dedicated lauter tun. The traditional mash tun has no knives, so the wort must percolate through the entire bed. These design elements may increase lauter time as compared with a dedicated lauter tun. There is also no transfer of the mash to another vessel, so a subsequent brew cannot begin until the grain is removed.

Mash/lauter tuns are more susceptible to a *set* or *stuck* mash. If the wort is drawn off too quickly, pressure builds up that compacts the grain bed, closing the spaces through which the wort flows. When knives are unavailable, the bed must be elevated by pumping water from under the false bottom, a process termed **underletting**. This method completely disrupts the grain bed and requires that vorlauf be repeated.

Mash Filter

Mash filters are most common in large breweries where **extract** recovery efficiency is critical. Mash filters, also called filter presses, came into use for wort separation in 1901. The unit consists of:

- A fixed plate with accommodation for piping called the head.
- A movable plate, called the follower, connects to a screw or hydraulic ram to hold the plates together against the pressure of the feed pump.
- A stack of vertically oriented plates fitted with gaskets form chambers.

· Each chamber has a filter cloth supported by one wall of the chamber. Grooves or corrugations in the wall supporting the filter cloth allow the clear wort to drain.

The filtration process starts with pumping the mash into the chambers. Pressure from the pump drives the clear liquid through the filter cloths and into drains on the outlet sides of the chambers. After filtration, the pressure is released, and the plates are separated one at a time, allowing the solid to drop into a box below. The modern form of the mash filter, shown in Figure 7.2, was introduced by Meura Corporation in 1987. It is a conventional plate and frame filter with the addition of an elastic **membrane** that can expand under air or water pressure to drive the contents of the chamber against the filter cloth. This results in lower moisture in the **draff** (<70% compared with 85% for a lauter tun) and higher extract recovery. Some models have two kinds of plates, one with a membrane on each side and one with a filter on each side. Other models have identical plates, each with a filter on one side and a membrane on the other.

The filtration cycle starts with pumping mash into the bottoms of the chambers (Figure 7.2A). As the mash enters, the pressure drives wort through the filter pads to runoff channels behind them. When all wort is transferred, the mash valve is closed, and the membranes are inflated with compressed air, driving wort from the grain and through the filter cloths (Figure 7.2B). Hot water is pumped through the mash channels to sparge the grain as the air in the membranes is released (Figure 7.2C). The volume of sparge water admitted should match with the membrane fluid released to maintain the integrity of the filter cake. Sparge water remaining in the grain is pressed out by inflating the membranes again.

Because the wort is driven through the grain under pressure, maintaining high **permeability** is less important than it would be for gravity feed. Brewers take advantage of this by crushing the grain to small particles in a **hammer mill**. The small particles are mashed more efficiently, so less grain is needed for the same extract. The wort separation process is faster; no vorlauf is required. Mash filter manufacturers claim that they can process 14 batches a day. The mash filter is more efficient in water and energy use than the lauter tun. It produces a drier draff, saving transportation costs for spent grain. The pressure and the shallow grain bed provide less flow resistance to viscous wort, allowing the use of a greater variety of **malts** and **adjuncts** than is practical with other separation systems. For small breweries, the high capital for installation ordinarily does not match the savings in efficiency. Even small breweries in remote locations where the costs of malt, water, utilities, and draff disposal are high may benefit from mash filtration.

Nessie Filter   A novel mash filtration system, called "Nessie," was introduced by the Ziemann Holvrieka Company in 2016. The system relies on rotating

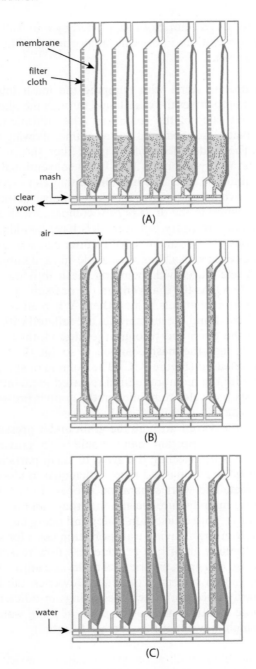

Figure 7.2 (A) Mash filtration, filling. (B) Mash filtration, compression. (C) Mash filtration, sparging.

circular porous sintered **stainless steel** filter wheels mounted vertically. Two wheels are connected at the rim to form a cylindrical vessel whose sides are the wheels. Mash is fed in between the wheels and clear wort comes out the sides. The rotation of the unit carries the grain to a chute. The demonstration system has four pairs of wheels through which the grain flows in sequence. Weak wort from the last two units is fed back to the previous units, and sparge water is added to the fourth unit. The Nessie system is designed for continuous operation. At the time of this writing, brewers have had little experience with this system.

Sparging

After the first wort is transferred to the boil kettle from the lauter tun, the grain is sprayed with hot 71–78 °C (160–172 °F) water to wash additional **sugar** from the grain, a process called sparging. The sparge water picks up additional extract from the grain bed and carries it to the kettle.

Batch sparging is a simple method where the wort is drained from the grain bed completely during the lauter phase. The lautering vessel is then refilled with water, the grain bed is re-established, and then it is drained again. Today, the first and second wort are usually combined, but in the past, the second runnings were used to make a low-alcohol beer called *small beer* or *table beer*. Continuous sparging (also called fly sparging) is a method of supplying a continuous sprinkle of water on the grain bed, such that water addition rate matches the wort drain-out rate. This practice is most common, as it maintains the filter bed. A thin layer of water is maintained over the grain bed to prevent excessive oxidation. Sparging must be monitored, because excessive sparging can dilute the wort and extract harsh-tasting tannins from the hulls.

There are several sparging devices including simple **spray balls**, spray nozzles, and sparge arms, which are rotating **pipes** with holes throughout. Even coverage with spray water is critical to avoid channeling, which leads to a loss in extract.

Figure 7.3A shows the grain bed before sparging. The spaces between the grains are filled with wort. In Figure 7.3B, sparge water is added at the top and wort is drawn from the bottom of the vessel. In this view, enough sparge water has been added to displace the wort in the upper portion of the bed. Sugary wort is still attached to the outside of the grains, illustrated by the grain border. Figure 7.3C shows the situation when most of the wort between the particles has been driven out. Additional water carries the entrained sugar, starting at the top (or wherever the sparge water enters).

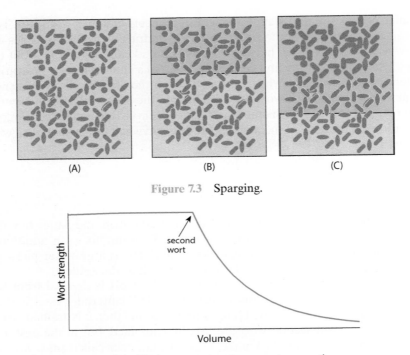

Figure 7.3 Sparging.

Figure 7.4 Extract recovery from wort separation.

As sparging continues, the sugar content of the wort decreases rapidly. In addition, the **pH** will rise as the sparge water displaces the **acidic** compounds from the malt. Sparging is usually terminated when the **specific gravity** of the runoff falls below 1.005 (2°P) or the pH exceeds 6. Subsequent runoff is generally of poor quality and will contain excessive concentrations of phenolic compounds or minerals extracted from the grain. Some breweries attempt to recover value from this weak wort by adding it to the mash-in water of the next batch.

The sparging water should be consistent with the mashing water. If any salt additions were made to the mash water, the same additions should be made to the sparge water. Sparge water benefits from pre-acidification to pH 6 to prevent extraction of tannins.

The general pattern of extract recovery during the wort separation is shown in Figure 7.4. The first wort collected during the lauter is high in extract. First **wort strength** is typically 4–6°P higher than the target beer. Because of sparging, the extract decreases in the second wort. The deeper the grain bed, the more slowly the wort strength declines; thus more sparging water is needed for extract recovery. It is advantageous to minimize the amount of sparge water required so that high-gravity wort is collected. Water can be added to lower the specific gravity if desired, but raising the gravity requires extensive boiling, which takes time, consumes energy, and influences the character of the beer. Another consideration is that the wort separation process is usually the slow step, so less sparging means faster brewing.

Depending on the beer and the recipe, wort separation is halted when:

- The target extract is recovered in the kettle.
- The target volume is obtained in the kettle.
- The runoff strength drops below 2 °P.
- The runoff pH exceeds 6.

7.2 FLOW AND PRESSURE

Differential Pressure

The principles of fluid flow are relevant to every aspect of beer production, but they are the key issues in wort separation. Differential pressure is the difference in pressure under the false bottom and the pressure above the grain bed. If the differential pressure is too high, the force compresses the grain bed and fills the slots of the false bottom, stopping flow and creating a stuck mash. Many lauter tuns are equipped with simple measurement tools called manometers (Figure 7.5), which measure the liquid pressure at two points. One is installed below the false bottom and one is installed above the grain bed. With **sight glasses** placed next to one another, the difference in liquid heights on the manometers indicates the differential pressure. Typically, 50–150 mm (2–6 in.) is an acceptable range during the lauter.

The pressure exerted by a column of liquid is given by

$$P = \rho g h$$

where ρ is the density of the liquid (in kg/m^3), g is the acceleration of gravity (9.8 m/s^2), and h is the height of the column (in m). One of the important

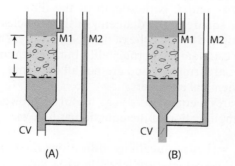

Figure 7.5 Differential pressure. M1 and M2, manometers; CV, control valve. (A) Valve closed and (B) valve open.

results of fluid dynamics is that the sum of all the pressure drops in the flow path is equal to the total pressure. In Figure 7.5A the valve is closed, stopping all flow. The liquid in both manometers comes up to the liquid level in the tank, showing no loss of pressure in the grain bed or the valve. The pressure below the false bottom is the full system pressure. The entire pressure drop occurs at the valve. In Figure 7.5B the valve is partially open, and the liquid flows through the system. The pressure due to gravity above the grain bed remains unchanged, but the flow through the bed causes a pressure drop, represented by the difference between M1 and M2. The grain bed and the control valve each provide part of the pressure drop. We can increase the pressure drop at the control valve by making its opening smaller. This will reduce the pressure drop at the grain bed, and it will lower the flow rate.

During brewing operations, the differential pressure should be monitored. Excessive pressure should be relieved by reducing flow either by slightly closing a valve or by reducing the speed of a pump. Never completely stop flow. Slight pressure is required to maintain the organization of the grain bed.

Darcy's Equation

The grain bed is an example of a packed bed. Another example is a filter, like **diatomaceous earth**. The packed bed resists the flow of wort requiring there to be differential pressure across the bed for fluid movement. The greater the resistance to flow, the greater the pressure needed to overcome it. Equivalently, the greater the flow rate, the greater the pressure differential across the bed. These concepts are captured in Darcy's equation:

$$Q = \frac{BA\Delta P}{\eta L}$$

where Q is the volume flow rate, B is the permeability, A is the area of the bed perpendicular to the flow direction, ΔP is the pressure drop at the bed, η is the viscosity of the wort, and L is the depth of the filter bed. Permeability is proportional to the square of the diameter of the particles that make up the filter medium multiplied by the cube of the bed **porosity**. There are equations that can be used to estimate the permeability, but application to beds of grain, where the particles are irregular and are not all the same size, is challenging work best left to chemical engineers.

Nonetheless, Darcy's equation is very useful to brewers in a qualitative way. Consider a simplification of the equation as follows:

Flow rate = (bed permeability × filtration area × pressure) /
(wort viscosity × bed depth)

Brewers can use Darcy's equation to determine how changes in lauter performance will influence runoff speed. For example, if the wort is drawn off more quickly, the differential pressure increases. As another example, consider what happens if a larger diameter lauter tun is employed. Assuming the same mash volume, the filtration area increases (A) and thus the flow rate increases. Some considerations of lauter equipment and processes on flow rate are described in Table 7.1.

7.3 LIPID REMOVAL

Brewing grain contains about 3% **lipid** (dry basis). Of this, about 90% is removed during wort separation. Wort turbidity is caused by an excess of long-chain fatty acids. Although some lipid in the wort is beneficial to yeast health and fermentation, lipids are generally undesirable. Excessive lipid in the fermenter wort leads to:

- Faster fermentation.
- More **vicinal diketones** (VDK) but faster VDK reduction.
- Lower esters.
- Increase in **amino acid** uptake by yeast.
- More **higher alcohols**.
- Less **foaming** during fermentation.
- Variable foam stability.

Because of its large role in lipid removal, wort separation has an important effect on wort quality. Consider a batch of grain containing 100 kg of lipid. If 88% of the lipid is removed during wort separation, 12 kg remains. If, by contrast, 92% of the lipid is removed, 8 kg remains. The less efficient lipid removal (by four percentage points) produces a wort with 50% more lipid.

The traditional mash/lauter process provides the most lipid removal, followed by the lauter tun. Mash filtration is generally less efficient for lipid removal. Some factors that enhance lipid removal include:

- Deep gain bed.
- Narrow openings in the false bottom.
- Slow runoff of wort.
- Slow rake speed (<3 m/min).
- Shallow raking (more than 5 cm above the false bottom).

7.4 MANAGEMENT OF SPENT GRAINS

Brewing generates about 20 kg of spent grain per **hectoliter** of beer (50 lb/US bbl). In small systems, the grain can be collected by manually shoveling or raking into collection tubs. Lauter tuns usually have sweep arms as part of the rake system that drives the draff into discharge pipes. Mash filters open plate by plate, releasing the draff to a bin below. For larger volumes, the draff is moved by auger, progressive cavity pump, or rotary screw with compressed air to a holding tank. With adequate storage, spent grain can be held for about seven days until significant **spoilage** occurs.

The water content of spent grain is 80% or more for combined mash/lauter tuns, around 75% for a dedicated lauter tun, and as low as 65% for a mash filter. These differences are quite significant in several ways. If the draff output is 20 kg/hL at 80% moisture, the dry content weighs 4 kg. At 65% moisture, the same 4 kg of dry material would come to 11.4 kg of wet draff. The difference of 8.6 kg is nearly 9 more **liters** of beer sold instead of discarded.

Most truck freight volumes for spent grain are around 22 600 kg (50 000 lbs) per truck. A rule of thumb on spent grain generation is about 18 kg/hL of beer (50 lb/bbl), so one truckload corresponds to about 1300 hL (1000 bbl). Draff for shipment must be kept below 80% moisture due to the risk of shifting **weight** distribution in transit. Spent yeast after fermentation may be mixed with the draff as long as moisture remains below 80%.

Spent grain has significant value as animal feed. Yeast and trub can be mixed with the draff, but no filter aids on other non-feed materials can be included. Some food safety regulations that apply to animal feed will need to be followed, including documentation practices. Using the grains for animal feed is almost always a much better option than disposing of it as waste.

Some considerations in designing for spent grain management include:

- How to collect draff after wort separation.
- How to transport the draff within the **brewery**. Minimal use of conveyors saves maintenance and operating costs. Draff that falls behind and under equipment will need to be **cleaned** up before it rots and makes an unsanitary situation.
- How to store the draff for shipment. Wet draff can be stored for no more than one week in a closed bin. Storage should accommodate somewhat more than a full truck load so that if the truck is late, there is still room to brew.
- Trucks need around 4 m (13 ft) overhead clearance from the draff storage bin.

CHECK FOR UNDERSTANDING

1. What is the purpose of wort separations? Describe the typical process inputs and outputs for each step of the process.

2. What some factors that limit the rate at which wort can be separated?

3. Compare and contrast the use of a dedicated lauter tun with a combined mash/lauter tun.

4. List the pros and cons of using a mash filter over a lauter tun.

5. Identify three factors about lauter tun design that influence the pressure drop across a bed of particles.

6. What is sparging? What are its benefits and risks?

7. How does the use of a mash filter influence the milling process?

8. Considering Darcy's equation, propose three changes the brewery could make to improve (speed up) the lautering process.

CASE STUDY

Case Study 1

You are the Director of Operations at a brewery with a 10 bbl brewing system. You have noticed an increase in total lauter time over the last several brews. Standard protocol takes about 1.5 hours, but the last batches have taken more than two hours. Explain your approach to identify, verify, and correct this increase in lauter time.

Case Study 2

You are the Brewing Manager at a brewhouse with an annual production volume of 100 000 bbl. Recently your sensory team noted an increase in lingering bitterness attributed to astringency in the finished beer. Explain which set(s) of quality control data you would review next in order to troubleshoot the problem. Selecting the two most likely scenarios, explain your approach to verify and correct this increase in astringency.

BIBLIOGRAPHY

Anness BJ, Reed RJR. 1985. Lipids in Wort. *J. Inst. Brew. 91*:313–317.
Barth R. 2013. *The Chemistry of Beer: The Science in the Suds*. Wiley.

Becher T, Ziller K, Wasmuht K, Gehrig K. 2017. A novel mash filtration process (part 1). *Brauwelt Int.* 35(3):191–194.

Kunze W. 1999. *Technology Malting and Brewing, International Edition.* VLB. ISBN 3-921690-39-0. p. 224–250.

Master Brewers Association of the Americas. Wet Spent Grains. https://www.mbaa.com/brewresources/TechTips/Pages/Wet-Spent-Grains—Tank-Suggestions-for-the-Craft-Brewer.aspx.

O'Rourke T. 2003. Mash separation systems. *Brewer Int.* 3(2):57–59.

CHAPTER 8

BOILING, WORT CLARIFICATION, AND CHILLING

After **wort** has been collected from the spent **grains**, the **sugar**-laden liquid is sent to the boil **kettle**. Here it will be boiled for at least 60–90 minutes. The most common methods of heating the boil kettle are by direct flame or by steam. After boiling, the wort may be clarified in a **whirlpool** tank. The hot wort is then chilled in a **heat exchanger** and oxygenated before entry into **fermentation**.

There are multiple process goals for boiling including:

- **Isomerization** of **hop alpha acids.**
- **Sterilization** of the wort.
- **Enzyme** inactivation.
- Formation and removal of **trub.**
- Removal of **volatile off-flavors.**
- Addition of wort clarifiers.
- Development of **flavor** and color.
- **Concentration** of the wort.

Mastering Brewing Science: Quality and Production, First Edition.
Matthew Farber and Roger Barth.
© 2019 John Wiley & Sons, Inc. Published 2019 by John Wiley & Sons, Inc.

Isomerization of **hop alpha acids** provides the characteristic bitterness of beer. In Section 4.3 we covered the chemistry of hops and the isomerization **reaction**. The high **temperatures** of boiling drive the formation of soluble, bitter **iso-alpha acids**. Boiling hops in wort also extracts other hop compounds; however many hop **aroma** compounds are lost with the steam. *Early addition hops* are those added for the length of the boil. Early addition hops promote bitterness. *Late addition hops* are those added with only 0–20 minutes of boil time remaining. Late addition hops promote flavor and aroma with less contribution to bitterness.

Sterilization of the wort is critical because microorganisms may have entered the production process at multiple points. While the **brewhouse** should be kept in **clean** condition, **malt** is laden with **bacteria** and wild **yeast** that may survive the temperatures of the **mash**. Elimination of possible contamination by these **microbes** during boiling helps to maintain consistency during fermentation.

Enzyme inactivation fixes the **carbohydrate** contents of the wort. While most malt enzymes are **denatured** at mash-out temperatures, boiling destroys any remaining activity. Fixing the carbohydrate content ensures a consistent ratio of fermentable sugar to **unfermentable dextrin**, yielding a consistent product.

Formation and removal of trub is necessary to remove hop debris, denatured proteins, and **precipitated lipids**. Many proteins are denatured during the boil and are unable to refold. **Hydrophobic** or nonpolar patches become exposed. The nonpolar sections of the protein **coagulate**, stick to lipids and nonpolar **polyphenols**, and drop out of **solution** as a precipitate. There are two types of break material formed during wort processing, **hot break** and **cold break**. Hot break is formed during the boil and is removed in the whirlpool while the wort is still hot. Cold break forms after the wort is cooled, typically in the **chiller**. It may or may not be removed before fermentation depending on the **brewery**.

Removal of volatile off-flavors is needed for compounds like dimethyl sulfide (DMS). The **molecule** has a very low aroma **threshold**, contributing a flavor of canned **corn**. DMS removal is covered further in Section "Dimethyl Sulfide."

Addition of wort clarifiers, also called kettle finings, improves long-term clarity of the finished beer by removing proteins and/or polyphenols. The high temperature and convection of the boil causes rapid solubility and dispersion of clarifying agents. The most common wort clarifying agent is **carrageenan**, also called **Irish moss**. Carrageenan is a negatively charged **polymer** that binds to positively charged proteins, which then precipitate with the trub.

Development of flavor and color is the result of increased **Maillard reactions**. Previously discussed in Section "Types of Malt," Chapter 4, in the context of flavor changes during malting, this same process occurs in boiling wort, as wort is rich in simple sugars and **amino acids**. The Maillard reactions produce molecules with caramel, toast, and biscuit flavors while also darkening the color of the beer by a few **SRMs**.

Concentration of the wort is achieved through evaporation. Over a period of 60 minutes, evaporation of 5–10% of the wort is common. As the boiling intensity increases, so does the evaporation. A gentle, rolling boil leads to a lower evaporation rate and better formation of hot break as a result of lower **shearing** forces.

8.1   HEAT TRANSFER

Principles

Boiling and chilling are **heat** transfers; heat moves from one medium to another. The general term for a device that transfers heat is a heat exchanger. Some examples of heat exchangers used in **brewing** include steam boilers, wort boilers, **energy** recovery equipment on boiling kettles, **wort chillers**, jacketed fermenters, and flash pasteurizers. Refrigeration **units** have two heat exchangers, one to accept heat on the cold side and one to release heat on the hot side. A heat exchanger has two materials, nearly always **fluids**, at different temperatures separated by a material that conducts heat.

Heat always moves from the higher temperature to the lower temperature:

$$\text{Heat rate} = -kA\frac{dT}{dx}$$

where the heat rate is the speed at which heat is transferred (power), k is the **thermal conductivity**, A is the area across which the heat travels, and dT/dx is the temperature gradient, the change in temperature over a given distance. The minus sign indicates that the heat flows in the direction opposite the temperature gradient, from hot to cold. This equation is only easy to use when the conducting material is uniform throughout and the temperature gradient is the same throughout, a situation that seldom arises in practice. The equation is more useful to show general trends. For example, heat flows faster when there is more **surface** for heat transfer, when the material through which heat flows is a good conductor, when the temperature difference across the conductor is greater, and when the conductor is thinner.

Heat transfer is complicated in boiling, because steam lowers the effective thermal conductivity. Table 8.1 gives thermal conductivities of materials of interest to brewers. Some highlights are the very high conductivity of copper and aluminum, the relatively low conductivity of carbon **steel** and **stainless steel**, and the insulating properties of water, steam, and carbon (representing, perhaps, the effect of burned-on coatings on wort heating surfaces).

TABLE 8.1   **Thermal Conductivities of Various Materials**

Material	Conductivity ($J s^{-1} m^{-1} K^{-1}$)
Air	0.024
Aluminum	205
Carbon	1.7
Copper	401
Ethylene glycol	0.25
Iron, cast	55
Steel, carbon	54
Steel, stainless	16
Water, ice	2.18
Water, liquid	0.58
Water, steam	0.016

Source: Data from *The Engineering Toolbox*. Provided to illustrate concepts only.

The high thermal conductivity of copper combined with its favorable material properties made it the material of choice for boiling kettles and heat exchanger **pipes** in the past. Even today, the boiling kettle is sometimes called a "copper." Today, stainless steel is commonly used for heat exchange surfaces, despite its relatively poor thermal conductivity. Its superior strength and stiffness allow thinner plates and tubes to be used, compensating to some extent for its poor thermal conductivity. Stainless steel is not wet by water as well as copper, leading to problems that we will discuss in the section on boiling technology. The decisive issue is that stainless steel can be cleaned with materials like sodium hydroxide (**caustic**) that would damage copper. Stainless steel is also less prone to **corrosion**. Aluminum is not suitable for direct contact with wort or beer because it is corroded by acids. Aluminum cans are protected by a chemical lining.

Heat Exchangers

The two most common types of heat exchangers in brewing are the shell and tube exchanger and the plate exchanger. The shell and tube exchanger contains tubes carrying one fluid surrounded by a shell carrying the other (Figure 8.1). The plate heat exchanger has a stack of plates with spaces between them. Fluid flows in alternating spaces, typically in opposite directions. The arrangement of the plates in the stack is shown schematically in Figure 8.2. Any heat exchanger is most efficient when the hot and cold fluids flow in opposite directions. In the case of the shell and tube exchanger in Figure 8.1, cold fluid entering the tubes at the left is warmed by lukewarm fluid in the shell. The already hot fluid in the tubes at the right needs the hottest fluid in the shell to give it more heat.

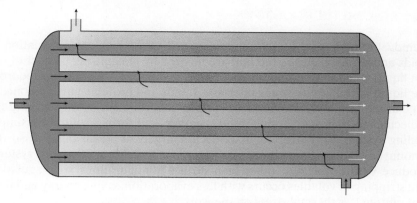

Figure 8.1   Shell and tube heat exchanger.

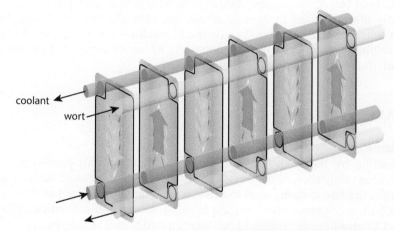

coolant

wort

Figure 8.2   Plate heat exchanger, expanded for clarity. *Source*: After Brigg et al. (2004).

The critical issue in all types of heat exchanger is that the fluids themselves are poor heat conductors. To get satisfactory heat exchange, the fluid must move in a way that brings all parts of it close to the heat exchange surface. This is accomplished by establishing **turbulent** flow in the fluid. Turbulent flow causes the fluid to mix well across the inside of the channel that carries it. The more turbulent the flow, the thinner the boundary layer, which is a stagnant layer against the channel walls. Heat will cross the boundary layer by conduction through the poorly conducting fluid, so a thinner boundary layer is good for efficient heat exchange. Turbulence is encouraged by high flow rates, high density, low viscosity, and features like projections or dimples in the wall of the channel.

8.2 BOILING TECHNOLOGY

Removal of undesired flavor compounds and to some extent coagulation of lipids and proteins during the boil are favored by maximizing contact of the hot liquid with the resulting steam. Volatile compounds can only escape from the liquid at the liquid–**vapor** interface. Lipids and proteins initially concentrate at the liquid surface, so more surface area offers more opportunity for molecules to find one another, bind together, and form a solid. In a simple boiling system, wort-steam contact occurs largely at steam bubbles, so the amount of contact depends on the vigor of boiling. More advanced systems produce wort droplets or other high-surface configurations, so coagulation and stripping of volatiles occurs with less evaporation, saving energy and time and putting less thermal stress on the wort.

Boiling is the largest consumer of energy in the brewhouse, which makes it a perennial target for improved efficiency. New proprietary systems are continually placed on the market as others are, with less fanfare, discontinued. Even a simple boiling system can be made much more efficient by the inclusion of energy recovery from the steam. As we cover some of these systems, we will deal with the basics of how they work.

SAFETY NOTES

Flame
Heating by flame should not be undertaken without provision for venting the firebox to the outside. This applies to flame used directly on the kettle and to flame used to raise steam. Flame produces toxic fumes, including colorless, odorless, and highly toxic carbon monoxide (CO). A CO monitor/alarm should be installed and tested regularly.

Boilover
Any boiling process is susceptible to **boilover**, where **foam** accumulates on top of the kettle, leading to overflow of boiling liquid. Boilovers are particularly robust with wort, because proteins and polyphenols can stabilize foam. Exposure to wort boilover can cause severe, extensive, life-threatening burns. The problem is exacerbated when the victim is on a narrow catwalk or ladder with no good escape route. A dedicated water hose to spray down a kettle should always be in easy reach of the brewer. In addition, the use of a level detector connected to a cutoff switch or valve to remove heat if the foam reaches an abnormal level is recommended. All standard operating procedures for the boil should involve safety considerations for a boilover. The details of the plan depend on the layout and procedures of your brewery.

Vacuum

Many brewery vessels are heated by steam jackets. Steam jackets of any size must have one or more valves to let air in and out. When steam is admitted, the air must be allowed to escape, or it will prevent the steam from reaching all parts of the jacket. When the steam is turned off, it will **condense**, leaving a vacuum in the jacket unless air is admitted. The vacuum can crush the vessel. Even without such a dramatic failure, the repeated push and pull will weaken the metal. The solution is a thermostatic air vent–vacuum breaker (TAVB). The TAVB is a **valve** that lets air out until the heat of the steam closes it. On the cooling cycle, it lets air in to prevent a vacuum from forming. Every jacketed vessel should have a properly installed and maintained TAVB.

About Steam

If a sample of water (or any other liquid) is sealed in an evacuated (no air) box, the water molecules have a broad range of energies whose average increases with increasing temperature. As the temperature increases, some molecules have enough energy to break free of the forces keeping them in the liquid. They evaporate and enter the vapor **phase**. Once in the vapor phase, some molecules will strike the surface of the liquid and become captured and condensed, rejoining the liquid phase. The rate (speed) of evaporation depends strongly on the temperature. The rate of condensation depends mostly on the concentration of molecules in the vapor, which is proportional to the pressure.

At the start, the pressure is low and evaporation dominates. As the pressure builds up, the rate of condensation rises and becomes equal to that of evaporation. The situation where a process and its reverse reach the same rate is called **equilibrium**. In the case of vapor–liquid equilibrium, the pressure reaches a steady value as long as the temperature does not change. The pressure at which this occurs is called the **vapor pressure**; it is strongly dependent on the temperature. Figure 8.3 shows pressure/temperature data for water plotted in the form of a **phase diagram**. Above and to the left of the line is the vapor region; the water exists in the form of **superheated** steam. Below and to the right is the liquid region. The pressures and temperatures on the line itself represent liquid in equilibrium with **saturated** steam. Selected vapor pressure data are provided in Table 8.2.

The **boiling point** of a substance is the temperature at which its vapor pressure is equal to the pressure upon it. Table 8.2 can be used to show that the boiling point of water at 1.99 bar is 120 °C. If liquid water at 140 °C were to be used to heat a vessel, the water and all its plumbing and **pumps** would have to

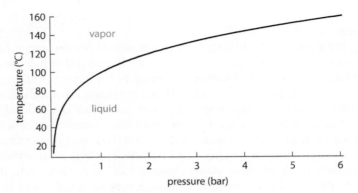

Figure 8.3   Phase diagram of water.

TABLE 8.2   **Vapor Pressure of Water**

Temperature (°C)	VP (bar)
10	0.01
20	0.02
30	0.04
40	0.07
50	0.12
60	0.2
70	0.31
80	0.47
90	0.7
100	1.01
110	1.43
120	1.99
140	3.62
160	6.18

Source: Data from Rumble (2017).

operate above 3.62 bar (37.8 psig). The boiling point of beer wort is higher than that of pure water at the same pressure by about 0.2 °C per degree Plato. Steam in breweries is often saturated at about 4 bar, giving a temperature of 144 °C. When saturated steam is used to provide heat, the steam temperature remains the same as long as the pressure is constant. Steam provides heat by condensing to water. At 4 bar, each kilogram of steam condensed releases about 2100 kJ of heat (960 **BTU**/lb). In practice, saturated steam is contaminated, to some extent, with a mist of liquid water. The portion of the steam that is in the liquid phase will not release heat; 4 bar steam that is 5% liquid will release only 1995 kJ/kg (0.95×2100 kJ/kg) of heat on condensation. The fraction of the steam that is vapor is called the **quality** of the steam.

The water produced by condensation of steam must be removed or it will interfere with steam flow. **Steam traps** are located at low points in the system. The trap opens in response to the presence of liquid water. The most common type of trap has a bulb that floats on the water, opening the valve when there is water in the trap and closing it when there is no water to float the bulb. The steam trap should be mounted over a drain or routed to a collection tank for reuse.

About Boiling

Boiling is not as simple as it seems. When heat is applied to a vessel, the material of the vessel and the rate of heat transfer influence the boiling process. Up to a point, the speed at which steam forms on a heated surface increases as the surface becomes hotter. At higher surface temperatures, the boiling rate decreases with increasing temperature. This paradoxical behavior occurs because there are two modes of boiling influenced by the wettability of the material. On a wettable surface or at low surface temperatures, a bubble forms, and then part of it breaks off and floats away, leaving a smaller bubble that grows as more steam is generated. This mode is called **nucleate boiling**. Nucleate boiling is desirable because heat transfer, and hence boiling rate, is rapid.

At higher surface temperatures or on non-wettable surfaces, the bubbles grow more rapidly than they can break off. The attached bubbles combine to form a film of steam. The film interferes with the flow of heat from the surface to the liquid, so boiling slows down. This mode is called **film boiling**. Here the surface of the vessel becomes significantly hotter than the liquid. This can cause scorching of wort. Materials like copper, which are wet by water, are less susceptible to film boiling.

Boiling Systems

Direct Fire Kettle   In direct fire boiling, the wort is heated by a fire under the kettle (Figure 8.4). The major advantage of direct fire is the simplicity and low cost of the equipment. Because of the uneven heating of the kettle surface, some caramelization of the wort is likely, especially at hot spots. The resulting flavor may be a **defect** or a desired style characteristic. Both the fire side and the wort side of the kettle are subject to accumulation of solid deposits, called **fouling**, which interferes with heat exchange, so the kettle must be cleaned often. The flame must not be lit until the entire bottom of the kettle is covered with wort; otherwise the wort will boil dry and char in places. For the same reason, the flame must be extinguished before the kettle is

Figure 8.4   Direct fire at Philadelphia Brewing Company.

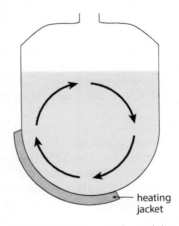

heating
jacket

Figure 8.5   Jacketed kettle. Convection mixing of wort shown.

empty. Direct fire is not energy efficient. A significant fraction of the heat produced bypasses the wort and goes up the flue. Direct fire becomes less practical for large kettles, because volume of wort outpaces the external area of the kettle. Direct fire is unusual in kettles above 35 hL (30 bbl).

Jacketed Kettle In a jacketed kettle, a hot fluid—usually steam—is conducted through an enclosed space on the outside of the kettle and brought into contact with the bottom and sides of the kettle, as shown in Figure 8.5. In this example, the heating is applied asymmetrically, which enhances wort mixing by convection (curved arrows). The temperature of the fluid in the jacket must be higher than the target temperature for the wort. Steam is the most common method of kettle heating in breweries. Steam operates at pressures of 2–4 bar absolute (15–45 psig), making the steam plumbing relatively easy. However, it does require substantial capital investment for double-walled jackets, piping installation, and a boiler facility. Steam requirements across the entire brewery should be taken into consideration, because large changes in demand can cause inefficient heating. Jacketed kettles have a limitation of about 235 hL (200 US bbl).

Internally Heated Kettle   Wort can be boiled by a coil of tubes inside the kettle that carries steam or even combustion products from a flame. These systems have fallen out of favor because of limitations on how much tubing surface can be accommodated in the kettle and especially because cleaning and maintenance can be complicated.

An internal **calandria** is a shell and tube heat exchanger mounted inside the kettle. The unique feature is that the tubes are open at the upper end or both ends to allow wort to flow in and out. Steam is provided to the shell to boil the wort. The wort is propelled upward by the convection current created by steam flow, in which case both ends of the tubes are open, or it can be pumped. Intermediate and combined systems using mechanical pumping as well as steam-driven pumping are in use. Usually, the output of the calandria shoots up to a deflector plate that breaks up the liquid into droplets, as shown in Figure 8.6. The extra surface on the droplets facilitates removal of volatiles at a moderate rate of evaporation. Some calandrias have an auxiliary output below the liquid surface to provide mixing. The heat exchanger tubes provide extra surface for heat exchange, to some extent overcoming the size limitation of direct fired and jacked kettles.

External Calandria   Putting the heat exchanger outside of the kettle removes the limitation on heat transfer surface. A pump cycles the wort from

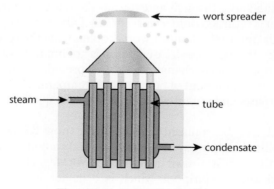

Figure 8.6   Internal calandria.

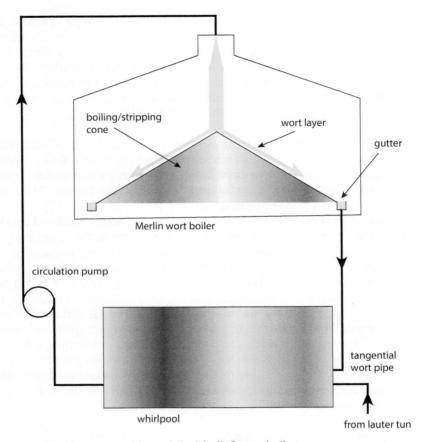

boiling/stripping cone

wort layer

gutter

Merlin wort boiler

circulation pump

tangential wort pipe

whirlpool

from lauter tun

Figure 8.7 Merlin® wort boiler.

the bottom of the kettle through an external shell and tube heat exchanger. The heated wort is returned onto a wort spreader above the liquid level in the kettle. As with the internal calandria, wort droplets from the spreader enhance boiling efficiency. Because the size of the unit is not limited by the dimensions of the kettle, more heat exchange surface is provided. More surface can exchange the required amount of heat at lower temperature, reducing fouling of exchange surfaces and potentially enhancing wort quality.

Merlin   The Merlin®, introduced by Steinecker (now Krones) in 1999, is a unique wort boiling/stripping system whose central element is a steam-heated cone that boils the wort, as shown in Figure 8.7. Wort from the wort separation unit is pumped into a whirlpool, which is a device for wort clarification explained in Section "Hot Wort Clarification." A circulation pump delivers the wort to the top of the conical heater, from which it flows under gravity in a thin layer to a gutter at the base of the Merlin vessel. Wort from the gutter is fed by gravity to a tangential pipe back into the whirlpool, causing slow whirlpool action. After a rest of about 10 minutes, circulation is resumed briefly to remove any remaining volatiles, and then the wort is chilled. It is claimed that the Merlin system can function satisfactorily with as little as 4% evaporation.

Direct Steam Injection   Steam at a temperature higher than the boiling point can be added directly to wort to boil it. The energy of the steam is used to break up the wort into droplets, facilitating the removal of undesired volatile compounds. Standard brewery steam, called **process steam**, is not suitable for **direct steam injection**. Instead, **culinary steam** must be prepared, either in a separate boiler or by treatment of process steam. Steam injection avoids the problem of fouling of the heat exchange surface. The major disadvantage of steam injection is dilution of the wort by condensation of the injected steam.

Continuous Boiling   The continuous wort boiling process involves heating the wort at high pressure to about 135 °C for a specified time without boiling and then lowering the pressure to allow boiling to occur. The key advantage is that because the process is continuous, heat from the later steps can be used in the earlier steps. Figure 8.8 illustrates one version of the process. Wort from the **lauter tun** or **mash filter** is heated in three stages to about 135 °C under pressure, so it remains liquid. The temperature is maintained in a holding tube for about three minutes. The pressure is released in two stages in flash vessels in which part of the wort boils into steam. The steam from the second, cooler flash vessel preheats the incoming wort. The steam from the first, hotter flash vessel provides additional heat. The third heating stage uses fresh steam from the boiler to heat the wort to the final temperature of 135 °C. The condensate from the three heat exchangers can be used to heat water for other operations. The process only works in a brewery that is in continuous operation. To maintain continuity, it is recommended that several wort separators provide feed and several whirlpools accept the output of the boiling unit. Although the concepts behind continuous boiling were developed many years ago, few breweries use it.

Energy Recovery

Boiling is the single largest consumer of energy in the brewery; it consumes about 30% of the energy expended for wort production. Nearly every brewery makes some effort to reduce the cost and environmental impact of the burning of fuel for wort boiling. The most straightforward way to conserve

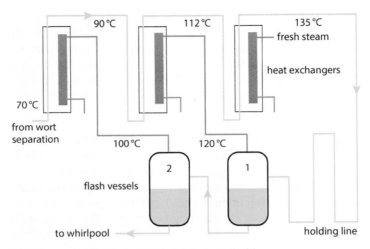

Figure 8.8   Continuous boiling.

energy is not to use it at all. Many of the boiling systems described above are designed to use less energy by reducing the amount of evaporation while maintaining good wort quality. Nonetheless, even a very efficient boiling system evaporates at least 4% of the wort. Evaporating 1 L of water from beer wort at 1 atm pressure requires 2.3 MJ of heat.

The simplest recovery scheme to recapture the energy of evaporation is to condense the vapor from the kettle and use the heat to warm water. This can be accomplished by spraying cool water into the steam. Water is heated to about 90 °C (194 °F) in this way, but the presence of volatile flavor compounds from the boil can limit its use.

Direct contact of steam with water is avoided by using a heat exchanger. The hot water prepared in this way can be stored in an insulated tank for general heating tasks, such as heating water for mashing, cleaning, and rinsing. A tall, narrow tank will supply hotter water from the top (90–95 °C = 194–201 °F) and cooler water (70–80 °C = 158–176 °F) from the bottom.

The usefulness of energy is greater at higher temperature. For example, if the temperature of the steam is increased to 110 °C (230 °F), it can assist with wort boiling. Table 8.2 shows that saturated steam at 110 °C has a pressure of 1.43 bar (6.2 psig). By compressing the steam from atmospheric pressure, which is 1.01 bar (0 psig) at sea level, to 1.43 bar, the temperature will increase to 110 °C. Most of the energy to accomplish this temperature rise comes from condensation of some of the steam, while some comes from the mechanical work of compression. When the compression is provided by a mechanical compressor, typically powered by an electric motor, the process is called *mechanical vapor recompression* (MVR). The compression can also be provided by a steam jet, a process called *thermal vapor recompression* (TVR). The compressor used for MVR is noisy and a likely source of maintenance issues. The steam jet for TVR presents fewer such issues but generates extra condensate. If there is a use for this extra hot water, TVR can be advantageous.

Hot Wort Clarification

After boiling, the wort has a heavy load of solid particles. Some of this material is spent hops, and some is material precipitated from the wort during boiling, called hot break or trub. The total wet mass of solids is typically in the range of 1–1.5 kg/hL (2.5–4 lb/bbl) of wort but varies depending on the strength of the wort and the hopping rate. On a dry basis, the hot break has about 50–70% protein; 15–20% polyphenols, including insoluble hop resins; and a few percent of other materials, including fatty acids. It is desirable to remove this material to avoid excessive solids in the chiller, filtration issues after fermentation, and the persistence of **haze**-forming materials. Before clarification, the wort should be protected from shear forces that will disrupt the hot break particles, making them difficult to remove.

Hop Back   When **leaf hops** are used, the hops and hot break can be filtered out in a device variously called a **hop back**, **hop jack**, or **montejus** (formerly there were distinctions among these). The hop back is a vessel with a slotted false bottom or a mesh screen. The hot wort with hops is pumped into the vessel where the hops are caught on the false bottom or screen and serve to filter out the hot break particles. Wort entrained in the solid may be recovered by sparging, or it may be wrung out by driving the hops against a screen with an **auger** or belt. Many brewers are reluctant to recover wort in this way because of potential quality issues. Some brewers add additional **aroma hops** to the hop back to give the beer a hoppy aroma. The wort is then chilled before there is any opportunity for the aroma compounds to vaporize out.

Whirlpool   The whirlpool is the most common method for wort clarification. It is most suitable when hops are introduced in the form of pellets, powders, or liquid products rather than as **leaf hops**. The whirlpool is a separate cylindrical vessel, typically with a flat bottom or a slight slope toward the outlet. Optimal whirlpool geometry is a height-to-diameter ratio of 1 : 1. A pipe enters the vessel from the side at about 30° from the tangent to the wall, as shown in Figure 8.9. When the wort is introduced through this tangential pipe, it rotates in the vessel. The momentum provided by filling the whirlpool may be enough to keep the wort rotating, or it may be necessary to use a pump to recirculate the wort to keep it moving. The rotation of the wort drives the trub and hop

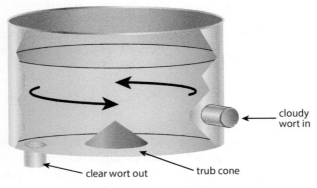

Figure 8.9   Whirlpool.

particles toward the outer wall from which they settle to the bottom and then collect in a cone-shaped pile at the center. Once all the wort is in the tank, it is allowed to swirl for 10–20 minutes. After this time, the clear wort is pumped out from an outlet at the bottom. Some tanks are designed with three or four sequential outlets along the side of the tank, with the last outlet on the bottom. Sequential outlets permit precise removal of wort while maintaining the trub pile. By slowing the runoff rates near the trub pile, the integrity of the pile is promoted, which may lead to higher wort recovery.

Whirlpools should be designed to have no projections or dead legs that obstruct flow. Sometimes, when there are issues with eddy formation, flat circular rings, called Denk rings, are installed. The rings are about three-quarters of the inside diameter of the vessel and 25–60 cm (10–24 in.) above its floor. The rings suppress secondary eddies that interfere with the flow pattern in the whirlpool.

Small breweries may use the boil kettle as a whirlpool. As one would expect, a dual-purpose vessel is not as effective as individual single-purpose vessels. When the kettle is used as a whirlpool, it is not available for the next batch of beer, so the brewing cycle is longer. The kettle is not optimized as a whirlpool, so it may lack tangential inlets and optimal outlets. Furthermore, the presence of **thermometer** probes, dip tubes, and other obstructions may create eddies.

Wort Stripping   Sometime an extra step, called *wort stripping*, is added after wort clarification to remove any remaining undesired volatile compounds. One configuration is to run the wort down a column lined with beads, which causes the wort to spread in a thin layer. Steam is run up the column to remove the volatile compounds. Wort stripping is not appropriate for beer styles in which hop aroma is desired.

8.3 BOILING CHEMISTRY

Hop Acid Isomerization

A major function of wort boiling is to **isomerize** alpha acids, which are precursors to the hop bittering compounds in beer. Three alpha acids account for about 98% of these compounds in hops; they differ only in one side chain. The three main compounds are shown in Figure 8.10 along with a generic alpha acid in which the variable portion is shown as **R**. Before isomerization, the alpha acids are insoluble and are not flavor active. During boiling the alpha acids are slowly converted (isomerized) to iso-alpha acids, shown in Figure 8.11. **Bonds** on asymmetric carbon atoms are represented as a solid wedge emerging toward the viewer and a dashed wedge receding away from the viewer.

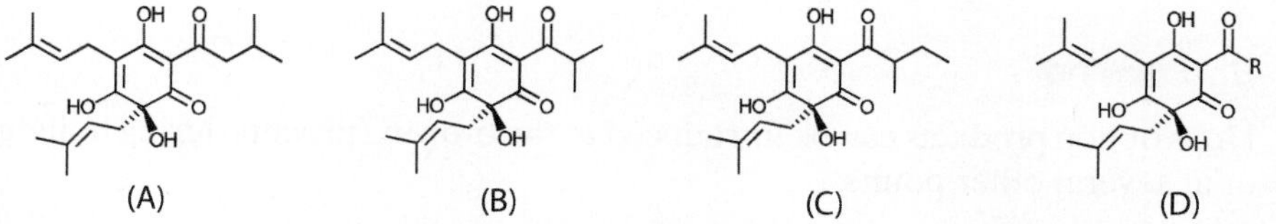

Figure 8.10 Alpha acids. (A) Humulone. (B) Cohumulone. (C) Adhumulone. (D) Generic alpha acid.

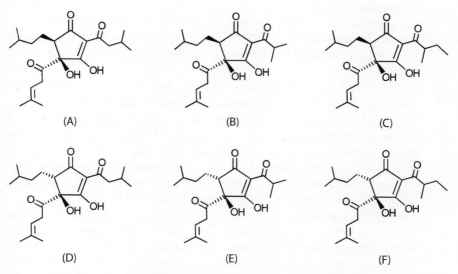

Figure 8.11 Iso-alpha acids. (A) *cis*-Isohumulone. (B) *cis*-Isocohumulone. (C) *cis*-Isoadhumulone. (D) *trans*-Isohumulone. (E) *trans*-Isocohumulone. (F) *trans*-Isoadhumulone.

Isomerization introduces an additional asymmetric carbon atom, so there are six iso-alpha acids. The absolute configurations were established quite recently; older literature may not be correct.

The isomerization reaction is first order, that is, its rate (speed) is directly proportional to the amount of alpha acid present. Other factors that affect the rate are temperature (faster at higher temperature), pH (faster at higher pH), and the concentration of metal ions like Mg^{2+} (faster at higher concentration). The rate is lower when the carbohydrate content of the wort is higher. Under normal wort boiling conditions with hop pellets, no more than about 30% of the available alpha acids are isomerized and soluble. The amount of iso-alpha acids produced is less than the amount of alpha acids consumed. This is because once the iso-alpha acids are formed, they can undergo degradation to non-bitter compounds.

Hop Dosing

Hops or hop products can be introduced to the brewing process during boiling or at several other points:

- *First wort hopping.* The hops are put into the kettle before the wort from the separation process is added. This is anecdotally said to give smooth bitterness.
- *Boiling hops.* Hops are added to the kettle when boiling begins. This is the usual way to provide bitterness to beer. If the hops are added manually, the heat should be cut off during addition to minimize the risk of a boilover. This method and first wort hopping retain little aroma from the hops.
- *Late kettle addition.* Hops are added part way through the boiling process to decrease the loss of aroma compounds in the steam.
- *Hop back or whirlpool addition.* Hops are added to the hop back (for leaf hops) or the whirlpool (for pellet hops). When the hot wort is processed in the vessel, aroma compounds picked up by the wort have no opportunity to escape with the steam, so they are retained at least until fermentation. They may still contribute bitterness through extraction of hulupones and humulinones (Figures 4.22 and 4.24).
- *Dry hopping.* Hops are added to the fermenter at some point in the fermentation process. Equipment to perform the **dry hopping** operation without introducing air is called a **hop doser**, but is also known under a variety of trade names mostly based on projectile weapons (hop rocket, **hop cannon**, hop gun, hop bazooka, etc.). These devices permit purging of the hops with carbon dioxide. They then use pressure to push the hops into the fermenter. Dry hopping greatly enhances hop aroma, especially

when the hops are added after most of the fermentation process is complete, minimizing loss of aroma in the carbon dioxide. **Isohumulone** decreases during dry hopping because it sticks to the hop solids. However, loss of bittering from isohumulone can be compensated by release of humulinone (Figure 4.24), which dissolves readily from the added hops. The result can be gain or loss of bitterness from dry hopping, depending on the balance between the iso-alpha acids lost and the humulinone released.

HOP CREEP: OVER-ATTENUATION FROM DRY HOPPING

Dry hopping can cause unexpected refermentation in a fermenter or packaged beer. Historically this was observed as **hop creep**, whereby an otherwise terminal gravity beer demonstrated an additional decline in specific gravity after dry hopping. Hops, like many plants, contain small quantities of **starch hydrolysis** enzymes, including alpha- and **beta-amylase** and smaller amounts of **limit dextrinase** and **amyloglucosidase**. These enzymes, if not deactivated by boiling, can convert unfermentable dextrins in packaged beer to fermentable sugars. If a dry-hopped beer is packaged before accommodating the dry hop creep, it may lead to dangerous carbon dioxide pressure and out-of-specification alcohol concentration.

Laboratory studies have shown a 2 °P increase in attenuation of finished beer to which hops and yeast have been added. This additional attenuation could increase **carbonation** by 4 volumes and alcohol content by 1% by volume. The pressure caused by increased carbonation could lead to dangerous failure in the packaging.

Dimethyl Sulfide

Dimethyl sulfide (DMS: CH_3-S-CH_3), although noticeable in most **lager** styles, is regarded as an off-flavor, especially when present in excess. The details of DMS production are covered in Chapter 12. Boiling can lower the DMS concentration by a process called steam distillation. Because of its volatility, the DMS is carried off with the steam. To keep DMS to a low level, condensate from the DMS-containing steam must be removed; it should not drip back into the kettle. Exhaust vents from the kettle should be designed with condensation collection traps. At boiling temperature, DMS is continuously made from precursors in the malt. Unless the wort is cooled quickly after boiling, more DMS will be made in the hot wort. DMS can be made from a different precursor during fermentation, so vigorous boiling and rapid cooling may not be the entire solution. Excessive DMS may also be removed by carbon dioxide released during vigorous fermentation. Slow fermentation or unhealthy yeast may give rise to DMS in beer.

8.4   CHILLING

Chillers

Before yeast is added to start fermentation, the wort must be chilled to a compatible temperature, which can be as low as 8 °C (46 °F) for lager yeast. In the past, hot wort was put into a broad shallow vessel called a **coolship**, where it released its heat to the surrounding air. The method is seldom used today, except to produce special flavor effects. The two disadvantages are potential for contamination from microorganisms in the air and the amount of time and space required. Also, there is little potential to recover energy from wort cooled in a coolship.

Nearly all commercial breweries use some form of fluid-cooled heat exchanger, generically called a chiller, to chill wort. The most common type of chiller is the plate heat exchanger (Figure 8.2), sometimes called a **Paraflow**. The plate heat exchanger takes little space, is enclosed and isolated from environmental microbes, is easy to **sanitize**, can cool a large volume of wort quickly, and can recover a good fraction of heat from the wort in the form of hot water (70 °C = 158 °F or more). The efficiency of heat exchange is greater when the difference between the temperature of the wort and that of the coolant is larger.

More rapid cooling and a lower final temperature can be attained when two chillers are used in series. The first chiller uses ordinary water as a coolant, and the second uses a refrigerated coolant, which can be chilled water, water with ethylene or propylene **glycol**, or alcohol to prevent freezing, or refrigeration fluids such as hydrochlorofluorocarbons (HCFCs). Two-stage cooling can reduce the load on the refrigeration unit and save energy. The flow capacity and final temperature for a chiller depend on the number and size of the cooling plates, the inlet temperature of the coolant, and the rate of flow of the coolant and that of the wort. Another significant factor is the buildup of material on the plates. If the cooling capacity of the chiller is insufficient, wort will have to be run more slowly, so some of it will have to wait for a long time before chilling. This could result in an increase in DMS in the beer. The chiller is a potential source of microbial contamination; it should be regularly monitored for cleanliness. The chiller should be taken apart for inspection and gasket replacement on a regular schedule.

Cold Break

About 40–350 mg/L of solid material, called cold break, precipitates into the wort on chilling, rendering it cloudy. Cold break is about half protein, and most of the rest is polyphenols and carbohydrates. The solid is very fine and does not settle well. It can be removed by filtration or by blowing small air bubbles through the wort to make a layer of foam, to which much of the cold break adheres. Most brewers do not attempt to remove cold break. Some reasons to omit this step are as follows:

- Indications that the presence of cold break can be beneficial to fermentation.
- An additional process on chilled wort risks introduction of microbial contamination.
- Injection of air can deplete **antioxidant** properties of the beer.
- An additional step adds time and expense.

Cold break can be measured with an **Imhoff cone**. These long, transparent cones contain volumetric gradations to quantify sedimentation. Cold trub levels are often reported in mg/L or mL/L and should be used to check consistency from each brew. Imhoff cones can also be used to measure lauter **turbidity**, hot break, and yeast **flocculation**.

Oxygenation

Dissolved oxygen (DO) is an important nutrient for yeast. The yeast cells do not use the oxygen for **respiration**; rather they need it to synthesize new **plasma membranes** during cell division. Oxygen is added directly after chilling because the solubility of **gas** increases as the liquid temperature decreases. The usual target for oxygen is 7–18 mg/L (5–13 mL oxygen/L at 0 °C and 1 atm).

Oxygen is typically added to wort in-line as it exits the chiller. Once fermentation begins, no more oxygen should be added. Oxygen can be added directly to wort as food-grade oxygen. Air can be used, but the partial pressure and hence the solubility of oxygen from air is only 21% of that from pure oxygen at the same pressure. The solubility of oxygen in beer wort is about 20% less than that in pure water at the same temperature and pressure. Equipment for wort aeration includes ceramic or stainless steel sintered porous stones or a **Venturi device**. Dissolved oxygen (DO) in wort can be monitored with a selective membrane **electrode**.

All equipment that is or could be exposed to pure oxygen under pressure must be rated for oxygen service. The system must be designed and maintained to avoid dangerous reactions. The **regulator** inlet must be free of dirt, dust, tape, and solvents. The tank valve should be opened slowly to avoid a dangerous oxygen hammer, which can result in steel and **brass** becoming combustible.

CHECK FOR UNDERSTANDING

1. What is the purpose of the boiling step? Describe the critical process inputs and outputs for the procedure.

2. Identify six places in a brewery where heat exchange takes place.

3. Compare and contrast direct fire systems with steam-driven calandrias.

4. What is the role of turbulence in heat exchange?

5. Of the materials listed in Table 8.1, copper and aluminum have the highest thermal conductivities. Why is stainless steel favored in brewery heat exchangers?

6. As the heating power to a boiler is increased, the boiling rate reaches a maximum and then declines. Explain this phenomenon.

7. Describe trub. What is it? How is it formed? Why is it removed? How is it removed?

8. What is an external calandria?

9. Describe essential design elements of a whirlpool.

10. Explain the influence of dry hopping on perceived bitterness in finished beer.

11. Discuss the importance of wort aeration. Consider what the effects of under-oxygenation and over-oxygenation might be on yeast performance.

CASE STUDY

You are the Regional Director for Craft Brewery Acquisitions at a major brewery organization. One of the flagship brands of a recently acquired brewery, an American light lager, was moved to production at another facility. You have since received many consumer complaints about this beer, which detail a disappointing change in flavor. The complaints began after the production location was moved. To follow up, you conducted a sensory evaluation of the original beer from the original location (Beer A) and the new beer from the new location (Beer B). Beer A is noted as having significant canned corn flavor, which complements the lager style. This flavor is completely absent in Beer B; it is not true to type. Explain the potential reasons for this change in flavor upon moving production to another location. Propose the next course of action to identify, verify, and correct this change at the new location.

BIBLIOGRAPHY

Anness BJ, Bamforth CW. 1982. Dimethyl sulphide—a review. *J. Inst. Brew.* *88*:244–252.

Bamforth CW. 2014. Dimethyl sulfide—significance, origins, and control. *J. Am. Soc. Brew. Chem.* *72*(3):165–168. An accessible update to the 1982 Anness review.

Brigg DE, Boulton CA, Brookes PA, Stevens R. 2004. *Brewing Science and Practice.* CRC. ISBN 0-8493-2547-1.

The Engineering Toolbox. www.engineeringtoolbox.com.

Holmes CP, Hense W, Donnelly D, Cook DJ. 2014. Impacts of steam injection technology on volatile formation and stripping during wort boiling. *MBAA Tech. Q.* *51*(2):33–41. doi:10.1094/TQ-51-2-0514-01.

Kapral D. 2016. Applications and importance of thermostatic air vents and vacuum breakers in the brewery: vacuum crushed vessels are unusable. *MBAA Tech. Q.* *53*(3):137–139. doi:10.1094/TQ-53-3-0723-01.

Kattein U, Herrmann M. 2009. Comparison of different wort-boiling systems and the quality of their worts and resulting beers. *MBAA Tech. Q.* *46*(4):1–6. doi:10.1094/TQ-46-4-1012-01.

Kirkpatrick KR, Shellhammer TH. 2018. Evidence of dextrin hydrolyzing enzymes in cascade hops (*Humulus lupulus*). *J. Agric. Food Chem.* *66*(34):9121–9126.

Kollnberger P. 1984. Wort boiling systems – new developments. *MBAA Tech. Q.* *21*(3):124–130.

MBAA. Master Brewers Brewery Safety. https://www.mbaa.com/brewresources/brewsafety/Pages/default.aspx.

Ockert K. 2014. Raw materials and brewhouse operations. *MBAA Practical Handbook for the Specialty Brewer, Vol. 1.* Master Brewers Association of the Americas. ISBN 0-9770519-1-9.

O'Rourke T. 2002. The process of wort boiling. *Brew. Int.* (June):26–28. Brief descriptions and advantages and disadvantages of several boiling systems.

Rehberger A, Luther GE. 1999. Wort boiling. In McCabe JT (editor). *The Practical Brewer, 3rd ed.* Master Brewers Association of the Americas. Chap. VII. p. 165–199.

Rumble JR. 2017. *Handbook of Chemistry and Physics, 98th ed.* CRC Press. ISBN 9781498784559.

Steenackers B, De Cooman L, De Vos D. 2014. Chemical transformations of characteristic hop secondary metabolites in relation to beer properties and the brewing process: a review. *Food Chem.* *172*:742–756. About hop bitter compounds.

Urban J, Dahlberg CJ, Carroll BJ, Kaminsky W. 2013. Absolute configuration of beer's bitter compounds. *Angew. Chem. Int. Ed.* *52*:1553–1555. doi:10.1002/anie.201208450.

Weinzierl M, Niedaner H, Stippler K, Wasmuht K, Englmann J. 2000. Merlin – a new wort boiling system. *MBAA Tech. Q.* *37*(3):383–391.

Willaert R. 2001. Wort boiling today. *Cerevisia 26*:217–230.

CHAPTER 9

FERMENTATION

After **wort** is made in the **brewhouse**, it is oxygenated and sent to a **fermenter** where **yeast** is added for **fermentation**. This begins the cold side of the **brewing** process. It is often said that brewers make **wort**; yeast makes beer. The brewer's job is to provide the yeast with the most favorable environment for making good beer. The process by which yeast converts the **sugar** in wort to ethanol and carbon dioxide is **alcoholic fermentation**. **Ethanol** and carbon dioxide are primary **flavor** compounds in virtually every style of beer. Fermentation also yields many flavor-active, minor products that can have a significant effect on the character of the beer, discussed further in Chapter 12.

In this chapter, we will discuss the basic process of fermentation, including the underlying chemistry. We will discuss fermentation equipment and design, process monitoring and control, and best practices for yeast management in the **brewery**.

9.1 FERMENTATION PROCESS

Yeast, when first **pitched** into wort, undergoes a classic microbiological growth curve consisting of a **lag phase**, an **exponential phase** or logarithmic (log) phase, a **stationary phase**, and finally, a **decline phase**. This growth curve

Mastering Brewing Science: Quality and Production, First Edition.
Matthew Farber and Roger Barth.
© 2019 John Wiley & Sons, Inc. Published 2019 by John Wiley & Sons, Inc.

is demonstrated in Figure 9.1. In classic microbiological growth curves, these four phases represent the life cycle of microorganisms in a liquid **culture**. During the lag phase, the cell count does not change. During the logarithmic (log) or exponential phase, cell division occurs, resulting in a rapid increase in total cell count. During the stationary phase, the rate of cell division is the same as the rate of cell death; thus total cell count reaches a plateau. Finally, during the decline phase, cell death prevails, and the total cell count declines.

If we think of beer fermentation as a microbiological culture system, the fermentation curves are slightly different, particularly because we start with a high number of yeast cells and maintain ideal conditions to prevent cell death. Furthermore, we often measure only the yeast in suspension and not the total cell count. The amount of yeast in suspension during fermentation is a useful tool to monitor fermentation. An illustration of a typical fermentation curve for **ales** is shown in Figure 9.2.

Lag Phase

During fermentation, the lag phase can last from 1 to 15 hours for properly pitched wort or up to 48 hours for underpitched wort. During this phase, the yeast cells acclimate to the environment, assimilate or absorb nutrients, and prepare for cellular division. There is little to no fermentation during the lag phase.

All-malt wort is typically an adequate source of the necessary **amino acids**, vitamins, and minerals required for normal yeast cellular physiology. Important vitamins include riboflavin, biotin, and inositol. Important minerals include sodium, phosphorus, sulfur, copper, iron, zinc, and potassium. All these nutrients are required to produce new **proteins** and **enzymes** that will

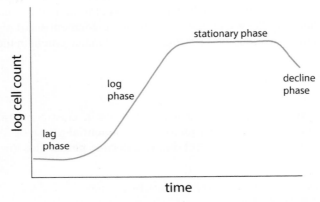

Figure 9.1   Classical microbiological growth curve.

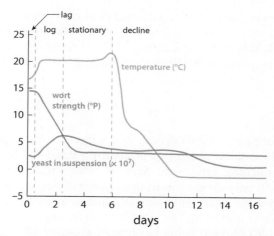

Figure 9.2   Typical trends over time during fermentation of an ale.

help the yeast cells grow, divide, and conduct alcoholic fermentation. Without enough nutrients or number of cells, the fermentation may be delayed or sluggish or lead to off-flavors.

Of the most essential minerals that yeast require, zinc is one that may be insufficient in wort. Yeast cells require 0.1–0.2 ppm of zinc, which they cannot produce themselves. Zinc is supplied by malt, but much of it is **bound** in an inaccessible form. In addition, some of the zinc adsorbs upon (sticks to the **surface** of) **trub** and is removed in the whirlpool. Zinc is a cofactor for yeast **alcohol dehydrogenase**, an essential enzyme required for alcoholic fermentation. With insufficient zinc in wort, fermentation may stall or be sluggish. Zinc is also required for **flocculation**, which is the clumping together of yeast cells; it facilitates better yeast harvesting and beer clarity.

Zinc is a common additive in commercial yeast nutrients or it may be supplied by addition of zinc sulfate heptahydrate ($ZnSO_4 \cdot 7H_2O$) or zinc chloride ($ZnCl_2$). Because zinc is adsorbed by trub, zinc salts should be added directly to the fermenter. Zinc addition must be carefully controlled; too much zinc can be detrimental to yeast health, leading to decreased fermentation rates and even cell death. Because the ***Reinheitsgebot*** forbids the addition of zinc salts, some German breweries provide zinc by placing a zinc **electrode** in the lauter tun.

As the cells grow and prepare for cellular division during the lag phase, the presence of oxygen in wort is critical. An oxygen concentration of 8–15 ppm by **weight** is recommended in wort before fermentation. The yeast cells need the oxygen to produce **sterols** and **unsaturated fatty acids**, which are essential components of the **cell membrane**. Oxygen-deficient wort leads to insufficient nutrient uptake due to poor membrane integrity and thus reduced **viability** and cell growth. Wort with a high solids content characteristic of

high-gravity brewing may require a second dose of oxygen 12–18 hours into fermentation to prevent stalling because of **osmotic stress** and a need for more cell growth. Too much oxygen may lead to excessive cell multiplication, leading to consumption of wort sugar for yeast growth rather than ethanol production.

Growth Phase

Once the cells are prepared to divide, they will enter the exponential (log) phase of growth. It takes about two hours for a yeast cell to divide. At the recommended pitching rates (see Section "Pitch Rate and Addition of Yeast"), yeast will only divide two to three times during a normal fermentation. During this time, yeast will assimilate amino acids, small **peptides**, and sugar in a particular hierarchy. The classification of amino acid assimilation is shown in Table 9.1. Amino acids have been grouped into four categories based on their rates of assimilation by yeast, though some variation may exist between strains.

During cellular division and early fermentation, many key flavor **molecules** are produced from metabolic processes (see Chapter 12). Fermentation begins during the growth phase. Sugars are then assimilated in a particular order, which is important for fermentation performance. Figure 9.3 shows the concentrations of the major sugars during fermentation. The removal of sugar by fermentation is called **attenuation**. Sucrose is taken up rapidly and is completely attenuated in 10 hours. Fructose increases at first, because it is released by the hydrolysis of sucrose. When the sucrose is depleted, **glucose** and fructose decline rapidly and are completely attenuated after 24 hours. **Maltose** is assimilated slowly, and **maltotriose** even more slowly, and **dextrins** (soluble **carbohydrates** with four or more sugar units) are not fermented at all.

Maltose and maltotriose are carried into the cell by proteins in the cell membrane called permeases. The activities of these transporters are sensitive

TABLE 9.1   Amino Acids Grouped Assimilation Rate by Brewer's Yeast

Group A Fast	Group B Intermediate	Group C Slow	Group D No assimilation
Arginine	Histidine	Alanine	Proline
Asparagine	Isoleucine	Glycine	
Aspartic acid	Leucine	Phenylalanine	
Glutamic acid	Methionine	Tryptophan	
Glutamine	Valine	Tyrosine	
Lysine			
Serine			
Threonine			

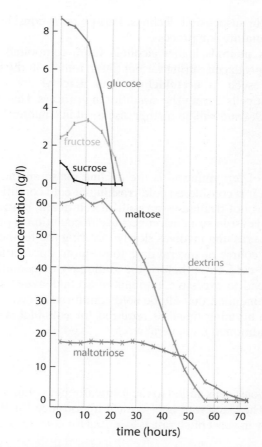

Figure 9.3 Typical trends for carbohydrate assimilation by yeast during ale fermentation. Note the expanded scale for glucose, fructose, and sucrose. *Source*: After Boulton and Quain (2006).

to glucose concentration. In a process known as **catabolite repression**, if an excessive amount of glucose is in the wort, maltose and maltotriose will not be utilized and assimilation will be inhibited. This is not usually a problem in all-malt wort, but when high concentrations of sugar **adjuncts** are used, uptake may be inhibited, resulting in a high **final gravity** (FG).

Stationary Phase

Fermentation continues and is completed in the stationary phase of fermentation. Although the total number of yeast cells remains the same throughout fermentation, yeast cells flocculate or **sediment** during this phase. Thus, the number of cells in suspension declines. This phase is considered part of the **conditioning** or maturation process.

Certain **esters**, **phenols**, higher alcohols, sulfur compounds, and other flavor compounds produced during fermentation remain in the beer. Other flavor compounds such as **acetaldehyde** and **diacetyl** are reabsorbed and reduced by yeast cells during the conditioning process. This concept is illustrated in Figure 9.4 and will be further discussed in Chapter 12.

Decline Phase

The final stage of fermentation is the decline phase; in the context of the brewing process, it is considered cold **crashing** or cold conditioning. The total cell number does not decline as a result of cell death, but rather there is a large decline of yeast in suspension because flocculation is promoted during this phase. Cold crashing involves slowly lowering the temperature to −1 °C (30 °F) over the course of several days to promote flocculation. Moving too quickly into the cold crash, before the stationary or conditioning phase is complete, will lead to excessive amounts of acetaldehyde or diacetyl in the beer. It is recommended that ale be cold conditioned for two to seven days with longer conditioning times as required for **colloidal stability**. **Lager** is usually cold conditioned for several weeks.

Pitch Rate and Addition of Yeast

The number of yeast cells added to the fermenter per volume of beer is called the **pitch rate**. The pitch rate is of critical importance for the flavor of the beer; consistency in yeast pitch rate is essential for consistency in beer quality.

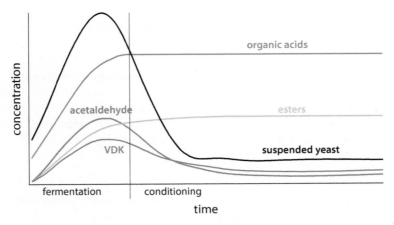

Figure 9.4   Flavors from fermentation.

In general, underpitching wort will promote flavors associated with cellular metabolism, like esters. On the other hand, overpitching wort will negatively affect yeast health, which is especially important if the yeast will be reused for subsequent batches. Fewer yeast cells in an underpitch means longer lag times, slower fermentation rates, and the potential for inadequate attenuation. In contrast, overpitching leads to a shorter lag time, faster fermentations, but the potential for a thin body. Overpitching puts stress on the yeast, which can lead to poor **foam** quality. Stress can cause yeast cells to secrete a **protease** enzyme that **hydrolyzes** foam-positive proteins and interferes with foam quality. A summary of the general effects of pitch rate on beer quality is displayed in Table 9.2. As would be expected, these effects depend upon yeast strain and wort character. Optimal pitching rates must be determined by the brewer and kept consistent during production.

How much yeast **slurry** should be added to the fermenter? How is pitch rate determined? There are several methods for determining the amount of yeast to add to a fermentation. The method employed depends on the technical capabilities and size of the brewery. The easiest method is addition by weight or by volume. It is recommended that 0.4–1.2 kg of yeast slurry per **hectoliter** (1–3 lb per barrel) is sufficient for fermentation. While this method will make beer, it is inconsistent because the number of yeast cells in a mass or volume of slurry is variable. Pitching based entirely on weight or volume of slurry yields inconsistent fermentations.

Pitching by weight or by volume can be made more accurate by estimating percent solids of the slurry. A rough estimate of the cell count in a yeast slurry is 25 million cells/mL for each percent solids, so a slurry with 40% solids would contain about 1 billion (10^9) cells/mL. To determine the percent solids, mix the yeast well to distribute the cells evenly in suspension. Collect a specific volume of the resuspended slurry in a graduated container. Allow the slurry to settle overnight at 0–4 °C or centrifuge and then determine the

TABLE 9.2   **Pitch Rate Effects on Beer Quality**

Underpitch	Overpitch
More yeast growth	Less yeast growth
Longer lag phase	Short lag phase
Longer fermentation	Shorter fermentation
Slower diacetyl reduction	Faster diacetyl reduction
Increase in esters	Decrease in esters
Decrease in fusel alcohols	Increase in fusel alcohols
Increase in sulfur compounds	Increase in autolysis
Under attenuation risk	Optimal attenuation
Full body	Thin body
Better foam retention	Poor foam retention

percentage of solids compared to the total volume. This method may vary by strain or by **trub** amount. To correct for trub volume, add 1 mL of a 50% **solution** of sodium hydroxide (**caustic**) to 50 mL of yeast slurry. The caustic will dissolve the trub, leaving only yeast after settling or centrifugation. WARNING: Sodium hydroxide is dangerously corrosive; it must be prepared, handled, and disposed of only by properly trained persons.

The most accurate and best method for determining the amount of yeast in a slurry is to count the cells. One way to do this is with a **hemocytometer**. A hemocytometer is a modified **microscope** slide with a well of a fixed depth and a grid etched in as shown in Figure 9.5. To obtain an accurate cell count, one should determine the number of cells in a square 1 mm on an edge (1 mm^2). For yeast, the center square is typically used. If the yeast culture is dense, counting the entire 1 mm^2 space can be challenging; thus it is standard practice to count five different 0.2 mm^2 smaller squares within the center square (indicated in Figure 9.5 as the magnified box in the upper right). The total number of cells in five boxes is then multiplied by 5 to approximate the

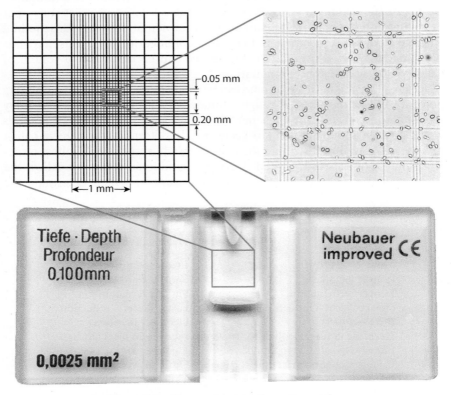

Figure 9.5   Hemocytometer for cell counting.

cell count in the total $1\,mm^2$ space. This number can be used to infer the number of cells per milliliter (see "Determining Cell Count in a Slurry" box for more details).

DETERMINING CELL COUNT IN A SLURRY

To count yeast, $10\,\mu L$ of a diluted slurry is added to the hemocytometer chamber. Using a compound light microscope, a 40× **objective** is used to visualize the grid. For best accuracy, five squares are counted within the central $1\,mm^2$ grid, typically the four in the corners and one in the center. The number of cells within each $0.2\,mm$ square should be around 20–70 cells. If the number is more than 100, then further dilution is required. The total of for the five squares is multiplied by 5 to estimate the total cell number within the 25 squares in the $1\,mm^2$ grid. The chamber is $0.1\,mm$ deep, so the volume is $1\,mm^2 \times 0.1\,mm = 0.1\,mm^3 = 10^{-4}\,mL$. To calculate the cell number per milliliter, use the following equation:

$$\text{Cells per milliliter} = (\text{Total number of cells in } 1\,mm^2 \text{ grid}) \times (\text{Dilution factor}) \times (10^4)$$

The dilution factor depends on how the slurry was prepared prior to counting. If the slurry was diluted 1:100, this number is 100. If the slurry was undiluted, it is 1. The 10^4 unit is the conversion factor to go from the grid volume of $0.1\,mm^3$ to $1\,mL$. To calculate the total cell number in a slurry, the number of cells per milliliter is multiplied by the total slurry volume in milliliters.

To practice, assume you have prepared a yeast sample for counting by diluting $2\,mL$ of slurry to a final volume of $100\,mL$ for a dilution factor of $100/2 = 50$. You pipetted $10\,\mu L$ into the hemocytometer. The following cell counts were recorded in five $0.2\,mm^2$ grids: 31, 21, 22, 28, 22. The sum of these numbers, 124, is multiplied by 5 to estimate the cell count in $1\,mm^2$: $124 \times 5 = 620$ cells in $1\,mm^2$. Cells per milliliter are determined by the following: $620 \text{ cells} \times 50 \times 10^4 = 3.1 \times 10^8 \text{ cells/mL}$. If the sample came from a $40\,L$ slurry, then the total cell count is $(3.1 \times 10^8 \text{ cells/mL}) \times 40\,L \times (10^3 \text{ mL/L}) = 1.24 \times 10^{13} \text{ cells}$.

With many yeast strains, counting cells in the slurry as-is can be difficult due to high cell density and/or extensive flocculation. If the yeast is too dense, a 1:10 or 1:100 dilution is often required. This is best accomplished with 1× **phosphate**-buffered saline (**PBS**). Water can be used if PBS is not available. If the yeast cells are highly flocculent, the sample should be diluted 1:10 with 0.5% sulfuric **acid** or 0.5 M EDTA **pH** 7.0 and slowly stirred with a stir bar for several minutes. Pipetting and dilution must be done precisely to ensure accurate cell counts.

Cell counting can be used to determine the amount of yeast, but exactly how much yeast should be added to wort? The industry standard is to pitch 1×10^6 (1 million) cells/mL/°P of the target wort for ale and 1.5×10^6 cells/mL/°P for lager. This is more of a general recommendation than an inviolable standard. Many ale strains provide improved flavor when underpitched (7.5×10^5/mL/°P). Conversely, overpitching (1.5×10^6/mL/°P) will lead to a faster fermentation. For high-gravity brews (higher than 16°P), it has been demonstrated that significant drops in viability occur after only 24 hours at normal pitching rates; thus pitching rate is increased depending on the **wort strength** (°P).

CALCULATING PITCH VOLUME

The total number of viable cells required to pitch a batch of wort is

$$\text{Viable cells required} = (\text{Desired pitch rate in cells}/\text{mL}) \times (\text{Wort volume in mL})$$

Suppose we are pitching 10 bbl of 12 °P wort at the industry standard for ale of 10^6 cells/mL/°P. The desired pitch rate is 12 °P × 10^6 cells/mL/°P. The wort volume is 10 bbl × 117.4 L/bbl × 1000 mL/L = 1.174×10^6 mL. The cells required come to 1.174×10^6 mL × 12 × 10^6 cells/mL = 1.35×10^{13} cells. We will pitch with the slurry whose cells we counted in the previous box to be 3.1×10^8 cells/mL. We calculate the volume of slurry needed from mL slurry = cells needed/cells/mL = 1.35×10^{13} cells/3.1×10^8 cells/mL = 4.35×10^4 mL = 43.5 L.

Inclusion of Viability
Viability is measured as a percentage. During cell counting on a hemocytometer, the total number of cells and the number of dead cells (**stained** blue with **methylene blue**) are recorded. Percent viability is determined by

$$\% \text{ Viability} = \frac{\text{Total number of cells} - \text{Dead cells}}{\text{Total cells}} \times 100\%$$

In the previous box we counted 124 cells. Suppose that six of them stained blue with methylene blue. The percent viability is $(124 - 6)/124 \times 100\% = 95\%$.

The percent viability can be used to determine the total number of viable cells in a slurry or to determine how much more of a slurry to add. To get the pitch volume including only viable cells, we divide by the fractional viability: 0.95. So we should pitch 43.5 L/0.95 = 45.8 L.

Viability and Vitality

The first criterion for satisfactory fermentation performance is live yeast cells. Yeast cells that display signs of life are said to be viable. Live and dead yeast cells look the same under the microscope, so **indicators** of life and death must be applied to make the distinction. Methylene blue and methylene violet are perhaps the most frequently used indicators for viability. These dyes are readily absorbed by all yeast cells, living or dead, staining them blue. Viable cells quickly reduce the dye, turning it colorless. Under the microscope, viable cells are colorless; dead cells stain blue. Combined with cell counting, this provides a quick way to estimate viability.

A cell can be hanging on to life, but still be in poor condition. The distinction between strong, healthy cells and weak, vulnerable cells is **vitality**. But what is the difference between viability and vitality? Is there any need to differentiate between the two in brewing? In truth, the difference can be subtle, but we like to think of viability as "capability of surviving" and vitality as "metabolic state characteristic of the ability to thrive under stress." A yeast cell might be alive (viable), but is it also capable of thriving in the presence of stress (vital)?

Measuring viability with methylene blue, or the similar methylene violet, is common practice because it is cheap and easy. Nonetheless, it has drawbacks. First, when yeast viability is reduced below 90%, the method grossly overestimates viability. Second, actively dividing cells reduce the dye more slowly, which can lead to misinterpretation. In general, if yeast viability is consistently less than 90%, an alternative method of measurement should be considered. Other membrane-permeable dyes can only enter the cell when the membrane is compromised, i.e. dead. One such dye, propidium iodide, binds to DNA and fluoresces. Slurry pH is also an effective and easy tool to examine overall quality of brewing yeast. As yeast viability declines, the pH increases. This is a quick tool for measuring the quality of yeast slurries in which the pH of the barm beer is determined.

Capacitance, which is the ability to store electric charge, serves as another viability indicator. An intact viable cell acts as a small capacitor, because charge is stored by the **plasma membrane**. If the cell is dead, the membrane is disrupted and cannot store charge; hence it makes no contribution to the capacitance. A capacitance probe produces a radio-frequency electric field and measures the electrical impedance (**resistance** to high-frequency **current**).

The impedance is inversely proportional to the capacitance. The capacitance is directly proportional to the number of live cells in the sample chamber. A benefit to using capacitance is that trub, other solids, and **gas** do not influence the measurement.

Replication capacity measures the ability of yeast cells to divide. The yeast is diluted and spread on an **agar** medium in a **Petri dish**. Viable yeast cells multiply to form colonies that grow to become visible. The colonies are counted, and the count is multiplied by the dilution factor to obtain the total viable number of cells. This method is severely limited by time as it takes 24–48 hours to see yeast colonies. The test can be run more quickly under a microscope, because it takes less time for a colony to become recognizable under magnification. Microscopic examination of a sample of yeast cells to determine the fraction that are budding, called the **budding index**, can also provide viability information.

Several automated cell-counting systems have been developed to help the user standardize and expedite cell counting. Devices such as the Cellometer® (Nexcelom, Lawrence, MA), the Countstar® (ABER Instruments, Aberystwyth, West Wales), and the Nucleocounter® (ChemoMetec, Allerød, Denmark) use cell imaging systems to calculate cell number. They may also include reagents and methods for including viability and vitality measurements.

Rate of oxygen uptake, sugar uptake, carbon dioxide release, and ethanol release in laboratory fermentations are effective methods for measuring vitality. Vitality can also be measured by the quantification of intracellular carbohydrates. **Glycogen** is a storage **polysaccharide** that is an indicator of vitality. It is quickly depleted in early fermentation but is then replenished shortly thereafter. Trehalose, a disaccharide of glucose, is another indicator of vitality. This sugar protects the membrane during stressful conditions like **dehydration** and freezing. Intracellular pH is another measurement of vitality. Healthy yeast cells drive out hydrogen ions. More active yeast cells have a pH around 6.5, while less vital yeast will be around pH 4–5. Intracellular pH can be measured with fluorescent pH indicators.

For highest accuracy, cell counting should be combined with measurement of cell viability; dead cells should not be included in the count when determining pitch rates. For example, if a brewer pitched yeast that was only 80% viable, then the actual number of productive yeast cells added would be short by 20%. The count of viable cells can be used to refine measurements based on slurry weight or volume to allow the faster, more convenient slurry methods to yield reliable pitch rates.

The Importance of Temperature

Temperature directly affects yeast performance. Temperature should be precisely controlled for each stage of fermentation. The optimal temperature depends upon the yeast strain, the beer style, and the characteristics of the fermentation. In general, if fermentation is too cool, yeast performance will be slow with the potential for premature flocculation. If fermentation is too hot, yeast will divide and ferment more quickly but at the cost of a significant increase in off-flavors like **fusel alcohol** and acetaldehyde. Centuries of brewing practice have identified the typical temperature ranges for the various stages of fermentation, which are provided in Table 9.3 for the production of ales and lagers. These ranges represent typical numbers and may vary depending on the beer and the yeast.

In the laboratory, yeast grows best at 30–32 °C (86–90 °F). For most brewer's yeast **propagations**, 25–28 °C (77–82 °F) is recommended. For beer production, ale fermentations are conducted around 15–22 °C (60–72 °F) and lager fermentations at 8–14 °C (46–57 °F). All fermentations are pitched at a slightly cooler temperature than primary fermentation to permit the yeast to acclimate and to accommodate a slight temperature rise from heat generated during early fermentation.

For an ale fermentation, the wort might be pitched at 15 °C (58 °F) with a primary fermentation at 18 °C (64 °F). Beer **specific gravity** or **density** should be checked and recorded daily to monitor fermentation progress. Primary fermentation typically lasts 4–10 days. When the fermentation is 2–3 °P from final attenuation, the temperature may be raised 2–3 °C (5–8 °F) for two to three days to facilitate the reduction of diacetyl. This step is called the diacetyl rest. After the diacetyl rest, the beer is slowly cooled to 0 °C (32 °F). The **green beer** is then conditioned at this temperature for zero to seven days depending on the style or the need.

For a lager fermentation, the wort might be pitched at 7 °C (45 °F) and allowed to ferment for 5–14 days at 9 °C (48 °F). After primary fermentation, the beer may be subjected to a diacetyl rest at 11 °C (52 °F) for two to three days. Then the beer is slowly cooled at a rate not exceeding 2 °C (4 °F) per hour, until 4 °C (39 °F) is reached. The beer is conditioned at this temperature for two to six weeks, a process called lagering [Ger: warehouse].

Selection of an optimal fermentation temperature depends on the yeast strain employed and the flavor characteristics desired. Across multiple yeast

TABLE 9.3   Typical Temperature Ranges for Modern Fermentations

	Ale		Lager	
Stage	Minimum	Maximum	Minimum	Maximum
At pitching	15 °C (58 °F)	22 °C (72 °F)	6 °C (42 °F)	11 °C (52 °F)
At primary fermentation	16 °C (60 °F)	25 °C (77 °F)	8 °C (48 °F)	14 °C (57 °F)
Cold conditioning	–1 °C (30 °F)	6 °C (43 °F)	–1 °C (30 °F)	6 °C (43 °F)

strains there are large variations in fermentation temperature ranges. For example, certain Scottish ale strains prefer cooler temperatures in the 16 °C range, whereas some Saison strains benefit from warmer temperatures above 22 °C. In fact, some brewers permit a free temperature rise during fermentation with Saison strains, as some have been reported to produce excellent flavors at temperatures as high as 35 °C (95 °F). Yeast providers usually recommend fermentation temperatures for their individual yeast strains.

Throughout fermentation, yeast releases heat as a result of metabolic activity. This causes the temperature to increase, especially within the cone of the fermenter where yeast collects. Without proper temperature control, the temperature of the fermentation will continue to rise with potential detrimental effects on beer and yeast quality. Typical characteristics of fermentations conducted too warm include:

- Fast attenuation.
- Poor yeast viability and autolysis.
- High fusel alcohols.
- High acetaldehyde.

The target pitching temperature is typically 2–3 °C cooler than fermentation temperature. After pitching, the temperature is raised to the fermentation target over a period of 18–36 hours. Most flavor compounds are produced in the first 72 hours of fermentation, illustrating the importance of temperature control from the outset of fermentation. Near the end of fermentation, before final gravity has been reached, the temperature may be raised another 2–3 °C over one to two days to help promote final attenuation and expedite acetaldehyde and diacetyl reduction during conditioning. The expected temperature changes for an ale fermentation are shown in Figure 9.2.

Once flavor maturation has been attained, the beer is cold crashed to as low −1 °C (30 °F). Because of the presence of ethanol, which has a freezing point of −114 °C (−173 °F), beer has a freezing point below −2 °C (28 °F). The fermenter should be cooled no faster than 1–2 °C/h. Rapid cooling causes thermal stress that can reduce yeast viability by more than 10%. Cold conditioning ultimately promotes yeast removal from beer due to flocculation, and it helps to remove colloids, giving improved clarity after filtration.

Krausen, Brandhefe, and the Use of Antifoam

As yeast releases carbon dioxide, foam called **krausen** is generated in the fermenting beer. Depending on the stage of fermentation, more CO_2 and thus more foam is produced. It is recommended that the fermenter be filled only to 70% capacity to provide **headspace** for krausen.

Traditionally, fermentation progress was tracked by the appearance of the krausen. In the earliest stages of fermentation, as the yeast cells divide, a fine layer of bubbles or foam may appear. *Low krausen* indicates the onset of vigorous fermentation and is marked by fine bubbles with light brown caps. *High krausen* indicates peak fermentation with the maximum foam height and the largest foam bubbles. As fermentation slows, the krausen collapses, leaving dark brown foam caps. And finally, with completion of fermentation, the krausen collapses completely, leaving a dirty brown layer. **The krausen ring**, or **brandhefe**, is often left behind on the fermenter wall as the high krausen collapses. This layer of cold break containing lipids, protein, and yeast can be very difficult to **clean** after it dries. After the fermenter **CIP** cycle, the effectiveness of removing the brandhefe should be visually inspected.

Some traditional brewing practices require physical removal of the brandhefe during fermentation. If left to fall back into finished beer, **autolyzed** yeast (yeast cells that have been damaged by their own enzymes) in the brandhefe can cause an **astringent** bitterness. Some brewers permit foam to gush out of the airlock during fermentation to avoid a buildup of the brandhefe. This practice has the disadvantage of loss in final beer volume and loss of beer foam-positive proteins; it is not recommended.

The brewer can protect the foam quality of the finished beer by adding an antifoam during fermentation, eliminating the krausen. This seems counterintuitive, but by adding antifoam during fermentation, more foam-active proteins are retained in the finished beer, thus improving its foam. Also important is that the use of antifoam allows the fermenter headspace to be more completely filled, so more beer can be made. By avoiding krausen loss in a blow-off tube, more yeast is retained for reuse.

The most commonly used antifoam products are food-grade silicone-based materials. The *Reinheitsgebot* forbids them, and the US Food and Drug Administration (FDA) has strict rules on their use. For silicone-based antifoams, the FDA requires that no more than 10 ppm of silicone can be present in finished beer. Silicone-based antifoam sticks to yeast during fermentation; thus most will be removed from finished beer with appropriate removal of yeast. When silicone is used, it would be prudent to take extra care to remove the yeast effectivity. Practices that wring extra beer from yeast should be tested carefully for their effect on the finished beer. Strict adherence to manufacturer recommended concentrations is essential to avoid carryover into finished beer. Alternative antifoam products have been successfully developed that use natural oils and **hop** products. Vegetable oil products have the added benefit of improving yeast health while potentially reducing wort oxygenation needs.

Tracking Fermentation

In modern fermentations, fermentation progress is monitored by a beer property that varies with solids content. The solids content in beer is mostly carbohydrate, much of which is consumed during fermentation. We will draw very clear distinctions among the measurement (specific gravity, **refractive index**), the interpretation (solids content, degrees **Plato**), and the significance to the fermentation process (**degree of fermentation, real attenuation**). We will call attention to the suitability of various sensors as **in-line, on-line**, and **off-line** analyzers. In-line analyzers are mounted in the **pipes** or vessels to provide continuous monitoring of the process. On-line sensors are installed near the process so that a sample can be diverted to the sensor for analysis. The analysis can be automated or under operator control. Off-line analysis requires that a sample be collected and carried to a laboratory.

Solids and Ethanol Content   The **extract**, or solids content, of wort or beer is the fraction or percentage of the sample that consists of dissolved solids. The extract in wort before fermentation is the **original extract (OE)**; that after fermentation is the real extract (RE) also called the **final extract**. Wort is a **mixture** of various carbohydrates, proteins, and inorganic material. Beer has all of these plus ethanol and small amounts of other volatile **organic** components. The usual measurements of extract and alcohol make simplifying assumptions to give results at an appropriate level of detail without consuming excessive laboratory time and expense. For extract, the sample is compared with a reference sample containing only water and sucrose. A sample of wort that gives the same measurement (specific gravity, refractive index, etc.) as a solution of 11% sucrose in water is considered to have an OE of 11%, or 11 °P, irrespective of the exact composition of the solids actually present. For beer, the volatile and nonvolatile components must be selectivity analyzed. A sample of beer whose volatile components when dissolved in the same mass as the original beer give the same measurement as that of 5% ethanol by weight would be considered 5% **alcohol by weight (ABW)**. For example, suppose we **distill** 100 g of beer and separately collect the distillate (the part that boils) and the residue (the part that does not boil). We add water to each fraction to bring each to 100 g. If the refractive index of the distillate is the same as that of 5% ethanol by weight in water, the beer is considered 5% ABW, irrespective of the nature of the components actually present. If the specific gravity of the residue is the same as that of 4% sucrose by weight in water, the RE of the beer would be considered 4 °P.

Density/Specific Gravity   Density is the mass of a sample divided by its volume. Specific gravity is the mass of a sample divided by the mass of an identical volume of water. Alternatively, the specific gravity can be calculated from the density of the sample divided by the density of water. For example, a specific gravity of 1.050 means that the substance is 5% heavier than an equal volume of water. The density, in units of g/mL or kg/L, is numerically close to the specific gravity. Specific gravity is measured with a calibrated **hydrometer** that displaces an amount of water whose mass is equal to the mass of the hydrometer. High precision hydrometers that can accurately measure to the thousandth are most suitable for commercial breweries. Before use, the hydrometer should be clean and dry. The specific gravity is read where the bottom of the liquid **meniscus** crosses the hydrometer scale, as demonstrated by the dashed red line in Figure 9.6.

The specific gravity of wort before fermentation is called the **original gravity (OG)** or initial gravity, and the specific gravity of the wort after fermentation is called the final gravity (FG). A brewer should keep daily logs of density or specific gravity to monitor fermentation progress. Once the specific gravity is stable for three days, it is considered the FG. OG and FG are used to estimate **alcohol by volume**. The calculation is complicated, because the density varies with both solids content and alcohol content. Equations in

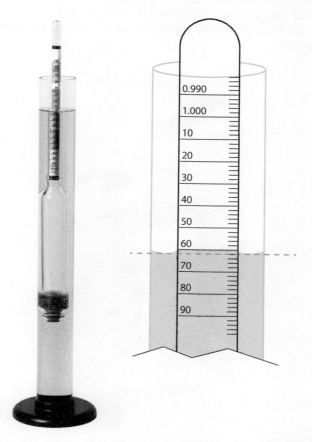

Figure 9.6   Hydrometer showing specific gravity of 1.062. *Source*: Photo © 123rf.com. Roger Siljander.

Chapter 15 relate estimated alcohol content, **real degree of fermentation** (RDF), and **real extract** (RE) to OG and FG, but they may not be reliable enough for labeling and taxation.

Density or specific gravity can be determined to high precision off-line by weighing a known volume of sample at a controlled temperature in a **pycnometer**, a glass vessel that can be reproducibly filled with an exact volume. This method takes about 30 minutes in the hands of a skilled operator; it is too slow for most beer production and has been generally abandoned by the industry. A more rapid and advanced method involves the oscillating density meter, or **densitometer**, which consists of a U-shaped tube filled with sample. The tube is made to vibrate and the vibrational period (time per cycle) is measured. The vibrational period is proportional to the square root of mass of the tube plus sample. The mass and volume of the tube are known, so the density can be calculated. Devices based on this principle often take the form of in-line analyzers or of on-line portable handheld units (Figure 9.7).

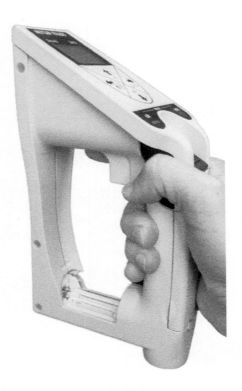

Figure 9.7   Handheld densitometer.

Certain wort or beer parameters can affect density measurements and may cause errors in interpretation. For example, beer must be decarbonated before an accurate measurement can be made. Also, density is affected by temperature, so unless temperature compensation is included in a device, steps should be taken to ensure the sample is read at 20 °C (68 °F). Finally, ethanol is less dense than water, so this needs to be considered for density measurements during and after fermentation. Further discussion on the effects of ethanol on density is given below.

Refractometer   Refractive index, also called *index of refraction*, is the ratio of the speed of light in a vacuum to that in the sample; it can be measured with a device called a **refractometer**. When light enters a low index (fast light) material, like a sample of beer wort, from a high index material like a glass prism, the light bends toward the high index material. If the angle of the light entering the low index sample is flat enough, all the light can be bent out of the sample. The angle at which this happens is called the critical angle; it can be used to determine the index of refraction of the sample. The refractive index of water varies in a predictable way when sucrose is added. The higher the sugar concentration, the slower light travels in it, so the higher the refractive index. As with density, the presence of alcohol can introduce a substantial discrepancy in the measurement of solids content, so much so that a refractometer is only recommended for OE measurements. Refractometers can be used as in-line devices, if precautions are taken to deal with the presence of ethanol, gases, and light scattering by undissolved solids.

Degree of Fermentation/Attenuation   The fraction of the mass of pre-fermentation dissolved solids lost in fermentation is the real degree of fermentation (RDF). For example, if the wort after boiling contained 100 kg of dissolved solids and the fully fermented beer contains 35 kg of dissolved solids, the amount fermented is 65 kg, and the RDF would be 65%. One complication is that the mass of the beer is lower than that of the wort because some of the sugar is converted to carbon dioxide and some to yeast, neither of which is part of the beer. For example, suppose 1000.0 kg of wort with OE of 16 °P yields 943.8 kg of beer with RE of 5.42 °P. The mass of wort solids is 1000 kg × 0.16 = 160 kg. The mass of beer solids is 943.8 kg × 0.0542 = 51.2 kg. The RDF is given by

$$RDF = \frac{\text{Mass wort solids} - \text{Mass beer solids}}{\text{Mass wort solids}} \times 100\%$$

which comes to 68%. The real attenuation, which is the percentage decrease in the concentration of dissolved solid (degrees Plato), is not the exactly the same as the RDF. In the example above, the OE is 16 °P and the RE is 5.42 °P. The equation below gives the real attenuation:

$$RA = \frac{OE - RE}{OE} \times 100\%$$

which comes to 66%. The difference between the RDF of 68% and the RA of 66% is not huge, but it is quite significant.

In the previous example, the RA was 5.42 °P. Although a 5.42% solution of sucrose in pure water would have a specific gravity of 1.023, the ethanol in the beer (ABV = 7.0%) lowers the specific gravity to 1.008. So far, we have been dealing with straightforward reality. We now leave all of that behind and see what happens if we were to calculate the solids content of the beer from the specific gravity *as though the ethanol were not there*. The solids content corresponding to a specific gravity of 1.008 is 2.05%, which is called the **apparent extract** (AE). The degree of fermentation based on this imaginary solids content is 87%, called the **apparent attenuation**:

$$AA = \frac{OE - AE}{OE} \times 100\%$$

To take one more step back from reality, we can use the approximation that the extract is approximately proportional to the specific gravity minus one. We will call the attenuation calculated from specific gravities the *approximate apparent attenuation*:

$$AAA = \frac{OG - FG}{OG - 1} \times 100\%$$

Chapter 15 has more detail about the relationships among specific gravity, solids content, alcohol content, and degree of fermentation.

There is a valid way to determine the real (not apparent) degree of fermentation. Take a sample of beer and decarbonate it by stirring. Weigh the sample or a representative portion, enough to insert a hydrometer. Boil the sample carefully until its volume is less than two-thirds of the volume before boiling. Cool the sample. Add purified water to bring the mass back to its original value. This is the water and the solids, but the alcohol has been boiled off. Now determine the specific gravity at 20 °C and look up the solids content on a Plato table. This is the final or real extract. Of course, this is strictly an off-line method.

After fermentation, how do we know if all available extract was fermented? We need to know the **attenuation limit**, which is measured by a **forced fermentation** test. A 200 mL sample of wort is taken before fermentation. The wort is then overpitched with 15 g of fresh yeast slurry or 2 g of rehydrated dry yeast. Using a stir bar or gentle shaking to accelerate fermentation at 20 °C, the final gravity should be reached in 48 hours. The attenuation limit of a given wort can be influenced by a variety of factors including yeast performance and changes in wort **fermentability** due to changes in process. The forced fermentation test is also useful when troubleshooting poor attenuation. Results will help determine whether the problem comes from yeast performance (forced fermentation test is lower than the observed FG) or wort quality (the forced fermentation test is higher than usual due to elevated unfermentable sugars as a result of inconsistent **mashing**).

9.2   FERMENTATION REACTIONS

The biological purpose of fermentation in yeast is to capture chemical **energy** in the form of **adenosine triphosphate (ATP)** from fermentable sugar. In contrast to **respiration**, fermentation uses no oxygen. The net **reaction** for alcoholic fermentation is $C_6H_{12}O_6 + 2ADP + 2H_3PO_4 \rightarrow 2CH_3CH_2OH + 2CO_2 + 2ATP + 2H_2O$. This reaction occurs in 12 steps, each with a specific enzyme. The chemical details of fermentation are shown in Figure 9.8. Many of the compounds have acid properties, so the charge and number of hydrogen atoms depend on the pH. To make it easier to follow the reactions, we show all acids in the **acidic** (with H^+ bound) form, hence pyruvic acid ($C_3H_4O_3$) rather than pyruvate ion ($C_3H_3O_3^-$). The first ten steps are called **glycolysis**, or the Embden–Meyerhof–Parnas pathway. Some form of glycolysis is carried out in every living cell. Steps 11 and 12 are the ethanol synthesis reactions; their biological purpose is to regenerate **NAD⁺** from **NADH**. In addition to the main products, fermentation potentially produces small quantities of many compounds that can have a major influence on the flavor of the beer.

From the point of view of the yeast cells, the important product of fermentation is ATP. All living cells need a constant supply of ATP for their life functions. For the brewer and the drinker, the key product is ethanol, the **psychoactive** ingredient in **alcoholic beverages**.

Each step of the glycolysis pathway is **catalyzed** by an enzyme, mostly named after the reaction or its reverse. For example, reaction 12 is catalyzed by *alcohol dehydrogenase*, even though the actual reaction, hydrogenation of acetaldehyde, is the reverse. Under conditions of plenty of ethanol and NAD^+, the same enzyme would catalyze dehydrogenation instead of hydrogenation. Eleven other enzymes are involved.

In step 1, one of the —OH groups on a sugar (glucose is shown, but other sugars can also serve) is **condensed** with a phosphate group that comes from the hydrolysis of ATP. The hydrolysis consumes a water molecule, and the condensation releases one.

In step 2, the glucose phosphate molecule rearranges to give fructose phosphate. This makes the molecule more symmetric. The oxygen is in the

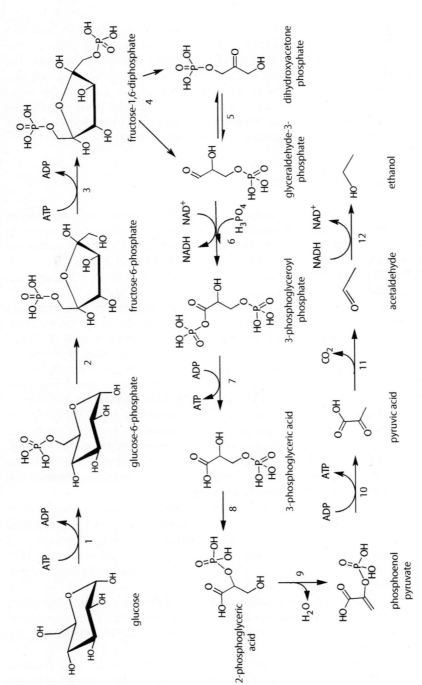

Figure 9.8 Glycolysis and ethanol synthesis.

middle of the five-member ring, and a one-carbon side chain hangs off the ring on each side of the ring oxygen.

Step 3 is similar to step 1. Another —OH group is phosphorylated by ATP on the corresponding carbon to the right of the ring oxygen. Fructose diphosphate (sometimes called fructose bisphosphate) is essentially identical across a line going through the ring oxygen and cutting through the **bond** opposite it. This is where the molecules will come apart. Up to this point, far from making ATP, the pathway has used up two molecules of ATP.

In step 4, the molecule breaks down the middle, leaving two 3-carbon fragments, each with one —OH group, one phosphate group, and one carbonyl (C=O) group. Two bonds are broken, and two bonds are made; the bonds made are the second bonds in the C=O double bonds. The two products are sugar phosphates, but one is an **aldose** (C=O on an end carbon) and the other is a **ketose** (C=O on the middle carbon).

In step 5, the two sugar phosphates interconvert. Because step 6 uses only the aldose, the net effect is to convert the ketose (dihydroxyacetone phosphate) to the aldose (glyceraldehyde phosphate). From this point on, we will follow only one of the two molecules produced in step 4, but keep in mind that each glucose molecule gives rise to *two* of the molecules shown.

Step 6 is an oxidation reaction in which one of the hydrogen atoms on phosphoric acid (H_3PO_4) and one on the end carbon of the aldose react with NAD^+ to give NADH and a hydrogen ion. The resulting compound, phosphoglyceroyl phosphate, is an anhydride, that is, a condensation compound of two acids. Figure 9.9 shows that if we were to hydrolyze off the top phosphate, the products would be phosphoric acid and a compound with a —COOH group (highlighted), a **carboxylic acid** called 3-phosphoglyceric acid.

In step 7, **adenosine diphosphate** (**ADP**) is phosphorylated to ATP by phosphoglyceroyl phosphate. This is an ATP payoff step. Because it happens twice for each glucose molecule that originally reacted, the ATP molecules consumed in steps 1 and 3 are replaced.

In step 8, the phosphate migrates from the end to the middle carbon atom of the three-carbon chain. This prepares the way for the production of a high-energy compound in step 9.

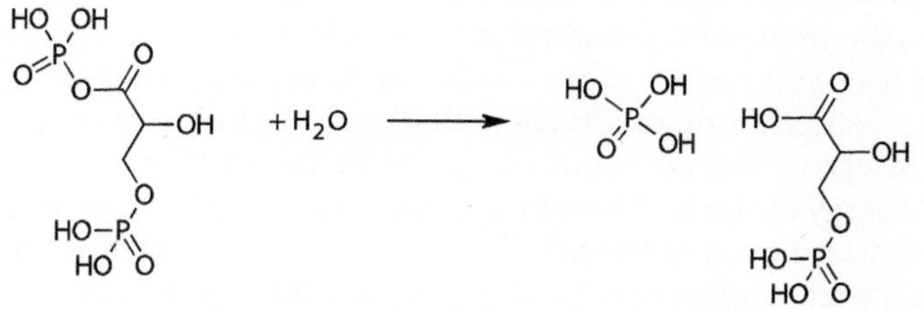

Figure 9.9   Hydration of phosphoglyceroyl phosphate.

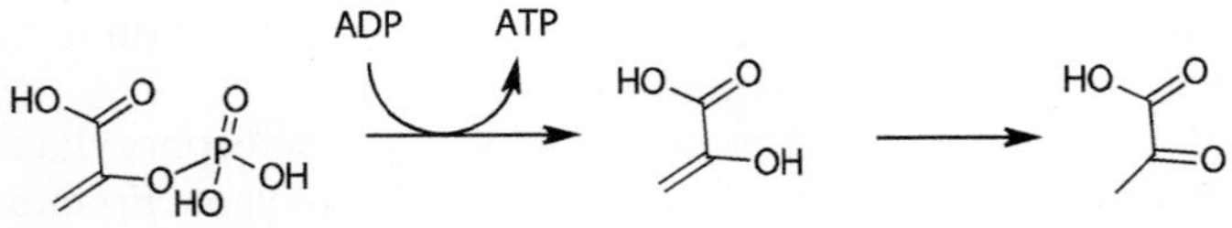

Figure 9.10   Formation of pyruvic acid.

In step 9, water is eliminated to give phosphoenol pyruvate, a high-energy compound.

Step 10 is the phosphorylation of a second ADP to give ATP. We can understand this reaction as a combination of two reactions. The phosphate group is hydrolyzed, leaving an —OH group. A compound with an —OH group on a carbon that has a double bond is an **enol** (EEN-all). Enols are usually unstable; the —OH hydrogen shifts to the other end of the double bond, and the second bond in C=C shifts to form a C=O double bond, giving pyruvic acid as shown in Figure 9.10.

This completes the glycolysis process proper. Two ATPs were hydrolyzed to ADP (one per 3-carbon unit) and four ADPs were condensed to ATP (two per 3-carbon unit). The net production of ATP is two molecules per molecule of glucose. During glycolysis, two molecules of NAD^+ per glucose were reduced to NADH in step 6. The process cannot continue unless the NAD^+ is regenerated. The next two steps, which are the ethanol synthesis reactions, regenerate the NAD^+.

Step 11 releases carbon dioxide to give acetaldehyde. In step 12, NADH reduces acetaldehyde to ethanol, an essential component of alcoholic beverages. This regenerates the NAD^+ that was consumed in step 6.

9.3   ENERGY AND ATP

Energy

Energy can be modeled as an imaginary **fluid** that enters a system to perform a non-spontaneous function and then exits when the system does something spontaneous. A key feature is that energy is not created or destroyed, but it can be degraded by being converted into a less usable form. One way to store energy for use in electrical systems is to use electrical energy to **pump** water to a high reservoir. When more electricity is needed, the water, pulled by gravity, is released through a turbine that generates electricity. The water in the high reservoir has potential energy by virtue of its position. It has less potential energy at the bottom of the turbine race, but some of the decreased energy was transferred to the turbine and then to the electrical grid. The falling of the water is spontaneous. It is coupled through the electrical grid to non-spontaneous processes like pumping beer. None of these transfers are perfect; some of the energy is cast off as waste heat.

Every living yeast cell constantly uses energy. When a compound is made that has more energy within its molecules than the starting materials, energy must be supplied. Energy is used to drive reactions that otherwise would go in the reverse direction and to move materials in the direction opposite to the direction that they would otherwise flow. In a mechanical system, we might store energy in a spring. We can recover the energy by having the spring do work as it relaxes to its unstressed position. Energy can be stored in chemical compounds in a similar way, but the underlying **mechanism** is the interactions of charged bodies. If we use energy to drive like charges together or pull unlike charges apart, that energy can be recovered by letting the charges return to their normal positions. In the case of yeast cells during fermentation, this energy is provided by the breakdown of sugar in the glycolysis reaction. Many chemical reactions are accompanied by release of energy. If nothing is done to capture and channel the energy, the energy will be released to the surroundings as heat. To use the energy to drive a desired process, we need to couple it to some device like a turbine, or a spring, or an electrochemical cell. In living cells, the coupling is done by energetic molecules. The energetic molecules quickly release their energy to support the life processes of the yeast. Ultimately, all the energy released by the fermentation process is degraded to heat. If all of this energy were trapped in the fermentation vessel, the temperature would increase beyond acceptable limits. For this reason, cooling must be provided to control the temperature of fermentation vessels, and specifically the yeast slurry, as will be discussed in Section 9.6.

ATP

In most sequences, the molecule that connects the process that releases energy and the one that uses energy is adenosine triphosphate, ATP. The hydrolysis reaction of ATP to ADP is shown in Figure 9.11. This reaction releases energy in a way that can be coupled to drive a non-spontaneous process. Often the phosphoric acid (PA) gets attached to another molecule, as in Figure 9.8, reactions 1 and 3. When energy is available, it can be used to condense ADP and phosphoric acid to make ATP. ATP is highly energetic and unstable. ATP that is not immediately needed is used to make intermediate storage compounds that last longer, like sugar or starch.

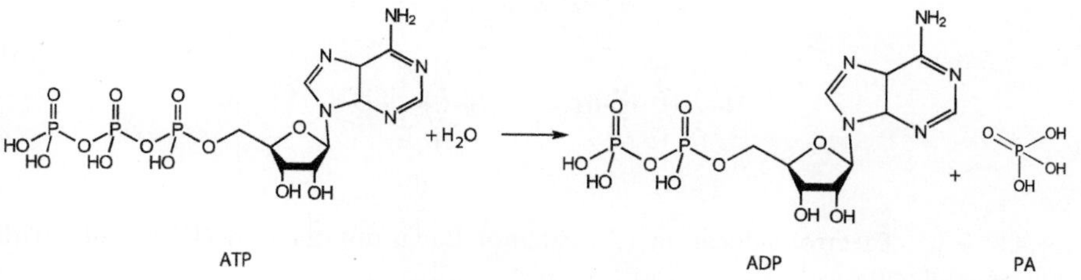

Figure 9.11   ATP hydrolysis.

9.4   OXIDATION, REDUCTION, AND NAD

Reactions involving the gain or loss of **electrons** by atoms are called **oxidation–reduction** reactions. The atom that loses electrons is said to be oxidized, and the one that gains electrons is reduced. Often the *molecules* whose atoms are oxidized or reduced are themselves said to have been oxidized or reduced, as long as oxidation and reduction occur on different molecules. If an atom is oxidized, another atom must be reduced; the total number of electrons does not change. A reaction that is sometimes used to remove iron from brewing water is $2Fe^{2+} + Cl_2 \rightarrow 2Fe^{3+} + 2Cl^-$. It is easy to see that each Fe^{2+} ion loses an electron and each chlorine atom gains an electron.

There are many important reactions in which it is not so easy to follow the loss and gain of electrons. To deal with these cases, we use a system of allocating electrons called **oxidation numbers**. The oxidation number is a fictive charge that an atom would have if all unshared electrons were assigned to that atom and all shared electrons were assigned to the more **electronegative** atom sharing them. The sum of the oxidation numbers on all atoms in an ion or molecule adds up to the charge on that ion or molecule, so for the reaction above, the oxidation numbers are +2 for Fe^{2+}, zero for each chlorine atom in Cl_2, +3 for Fe^{3+}, and −1 for Cl^-. For more complicated cases, we need a process for determining the oxidation number. We will apply this process to ethanol step by step. The first step is to draw the **Lewis structure** showing all **valence electrons**, shared and unshared, as dots, as shown in Figure 9.12A. The electronegativities of the atoms involved can be found in Figure 2.3: H = 2.2; C = 2.6; O = 3.4. Next, starting with the most electronegative atom, oxygen in this case, draw a circle around each atom that allocates all unshared electrons to the atom and the shared electrons to the more electronegative atom of the bond. If there is a bond between two atoms of the same electronegativity, allocate half of them to each atom. Do the same for the remaining atoms, in order of decreasing electronegativity until all valence electrons have been allocated. Ethanol has a total of 20 valence electrons, as shown in Figure 9.12A. Figure 9.12B shows the electron allocation for oxidation numbers. The oxygen atom gets its two nonbonding pairs, both electrons in the bond with carbon, and both electrons in the bond with the hydrogen for a total allocation of eight electrons. The carbon atom on the right gets neither of the two

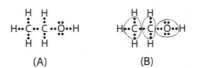

(A)   (B)

Figure 9.12   Electron allocation. (A) Ethanol, Lewis dot diagram. (B) Ethanol with electron allocations.

electrons in the bond with oxygen, all four of the electrons in the bonds with the two hydrogens, and one of the two electrons in the bond with the carbon on the left for a total allocation of five electrons. The carbon on the left gets the remaining seven electrons.

To determine the oxidation number for each atom, subtract the number of electrons allocated from the **periodic table main group** number. For the six hydrogens, $1-0 = +1$. For the carbon on the left, $4-7 = -3$. For the carbon on the right, $4-5 = -1$. For the oxygen, $6-8 = -2$. The sum of the oxidation numbers is $(6 \times +1) - 3 - 1 - 2 = 0$; ethanol is uncharged. The allocation for acetaldehyde is shown in Figure 9.13.

In acetaldehyde, the oxidation numbers are the same as in ethanol except for the right carbon, which has an oxidation number of $4-3 = +1$. In the alcohol synthesis reaction (step 12 in Figure 9.8), the reactant (starting material) is acetaldehyde, and the product is ethanol. The oxidation number of the corresponding carbon atom goes from $+1$ to -1 in the course of the reaction. Acetaldehyde is reduced by two electrons. The electrons to reduce the acetaldehyde come from a complicated molecule called nicotinamide adenosine dinucleotide (NAD), which exists in two forms, NADH and NAD^+. Figure 9.14 shows NADH being oxidized to NAD^+. The electrons released by this reaction reduce another molecule, acetaldehyde in this case. NAD is a cofactor in many cell processes involving oxidation or reduction. The part of the molecule that releases or accepts electrons is the nitrogen-containing nicotinamide ring shown at the top of Figure 9.14 with oxidation numbers highlighted. Figure 9.15 shows the electron allocations from which the oxidation numbers were calculated. The circles are color-coded according to the number of allocated electrons. Oxidation numbers of two of the carbon atoms in this ring change under oxidation. The carbon atom opposite the nitrogen goes from oxidation number $= -2$ in NADH to oxidation number $= -1$ in NAD^+. The carbon atom to the right of the nitrogen goes from oxidation number $= 0$ to oxidation number $= +1$. The net change in oxidation number is $+2$; two electrons are released. Acetaldehyde gains these two electrons in converting to ethanol. In alcoholic fermentation, the purpose of the alcohol synthesis reaction is to oxidize NADH to regenerate NAD^+, which is needed for the oxidation reaction in step 6 of Figure 9.8 as well as for other cell processes. If oxygen were involved, NADH would be oxidized in a different pathway in which the electrons are ultimately taken up by oxygen and additional ATP is generated.

Figure 9.13   Electron allocation for acetaldehyde.

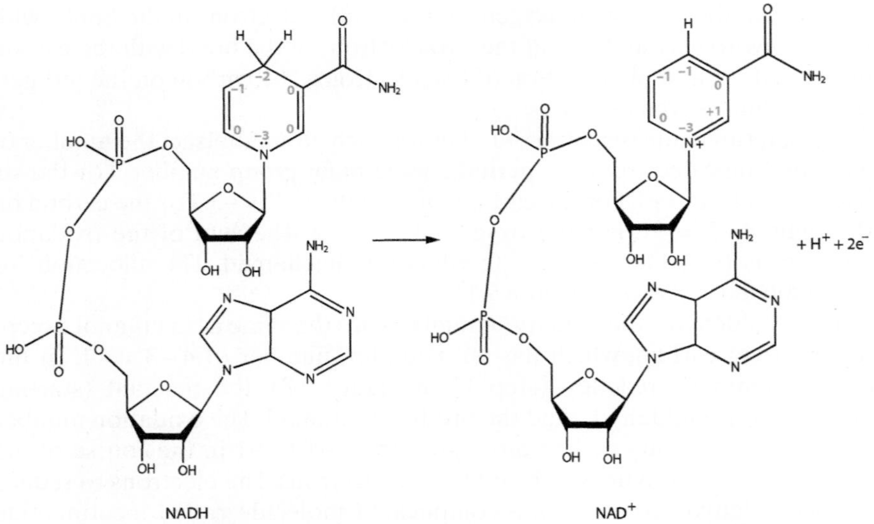

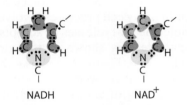

Figure 9.14   NADH oxidation to NAD⁺. Oxidation numbers highlighted.

Figure 9.15   Electron allocation in NADH and NAD⁺.

9.5   FERMENTATION EQUIPMENT

Beer is generally classified as either ale, which is fermented at 18–22 °C (64–72 °F), or as lager, which is fermented at 8–15 °C (46–59 °F). Most ale is fermented with a species of yeast called *Saccharomyces cerevisiae*, which tends to rise to the top with CO_2 bubbles during fermentation. **Lager beer** is fermented with *Saccharomyces pastorianus* (formerly called *Saccharomyces carlsbergensis*) yeast, which sinks to the bottom. Eventually all yeast sinks, but some ale fermenters take advantage of the rising tendency of ale yeast. After wort is prepared, chilled, and oxygenated in the brewhouse, it is pumped into a fermentation vessel, or fermenter. Fermenters have been made in a range of sizes, shapes, and materials. Commercial fermenters are usually in the range of 1200 L (12 hL = 10 US bbl)) to 600 000 L (6000 hL = 5100 bbl). They are usually made of stainless steel. Fermenters larger than about 2000 hL must be made on-site, which adds to their cost. The major objectives of the fermenter are as follows:

- Safely vent carbon dioxide produced by fermentation.
- Protect the beer from contamination by unwanted **microbes**.
- Carry away heat generated by the fermentation process.
- Allow the yeast to be recovered from the beer.

Once yeast is added to the wort, the mixture is regarded as fermenting beer. When fermentation is complete, the beer may require additional conditioning for maturation or carbonation. Beer should always be free from microbiological contamination, which would prevent the collection and reuse of yeast in subsequent fermentations.

Cylindroconical Fermenters

The most popular fermenter configuration is the jacketed stainless steel **cylindroconical** vessel (CCV), shown in Figure 9.16. The upper part of the vessel is cylindrical, and the lower part forms a cone, usually with sides sloping about 60° from horizontal. A cooling jacket surrounds parts of the vessel to carry away the heat generated by fermentation. In some systems there are two or three independently controlled jackets, usually one for the cone and one or two for the cylinder. It is highly recommended that for CCVs larger than 10 hL, a dedicated jacket for the cone, distinct from the jacket for the cylinder, be used to more precisely control the heat generated by yeast during fermentation. This concept is further discussed in Section 9.6.

The wort and yeast are typically added to a CCV through a fill/empty line. This line connects to an inner **racking** arm that is placed above the expected level of the yeast, allowing the beer to be collected without disturbing the yeast. Often the racking arm is configured in an "L" shape, allowing rotation of the arm. In some systems this pipe enters from the bottom port. On the top of the fermenter is a vent pipe that allows the carbon dioxide to escape, as well as plumbing for a CIP system. At the bottom of the cone is a valve that is used to collect yeast or to drain the tank during cleaning. The **zwickel** is a port to provide samples for quality measurement. Figure 9.16 is simplified. In working systems all lines are connected to the CIP system. There are sensors in the fermenter and in its jacket to monitor temperature, pressure, and possibly specific gravity or some other indication of the progress of fermentation.

An often overlooked but essential component of CCVs and other closed vessels is a **pressure relief valve** (PRV). Pressure relief valves are typically connected to the top of a fermenter through a tri-clamp connection. Pressure

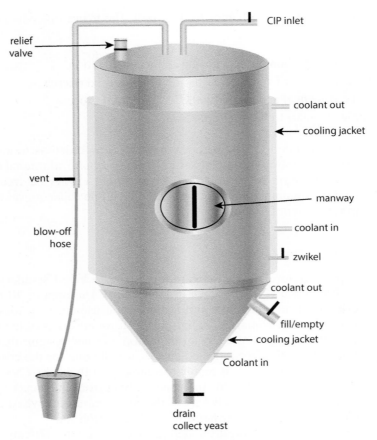

Figure 9.16   Cylindroconical fermenter.

changes are frequent in a CCV, and if pressure changes in a closed vessel, the tank may implode, damaging the tank, or explode, an extreme safety risk. Pressure relief valves for fermenters must be able to vent in either direction. The pressure relief valve prevents vacuum failure by allowing air to enter the tank if the tank pressure gets too low or for gases to escape from the tank if excessive pressure builds up. It is important that pressure relief valves be regularly inspected and subject to CIP cleaning, as clogged or damaged pressure relief valves may result in failure. Fermenter pressure relief valves are normally mounted at the top of the fermenter where they can only be reached by a ladder. They are bulky and heavy. If provision to make servicing them safer and more convenient can be built into the fermenter design, they are more likely to get regular inspection and maintenance.

Vacuum in a CCV only occurs when the tank is unvented, resulting in a closed system where valves were either mistakenly closed or blocked. The reduction in pressure may result from one of the following scenarios:

- Pumping or draining liquid from the tank.
- A hot CIP step followed by a cold rinse.
- A caustic (NaOH) wash after fermentation without removing the carbon dioxide. The carbon dioxide reacts according to $NaOH + CO_2 \rightarrow NaHCO_3$. If the tank is sealed, this reaction can remove gas, leaving a vacuum. Pressures less than 0.2 bar below atmosphere (0.8 bar absolute = 3 **psi** of vacuum) can crush a fermenter. Even if the tank is not sealed, the loss of caustic to this reaction will interfere with CIP effectiveness.

After fermentation, the yeast sediments and falls into the cone. This allows the yeast to be harvested with minimal loss of salable beer. Because the system is closed, the risk of contamination of the beer or the yeast is minimized. Independently controlled cooling jackets allow the temperature to be made cooler on top to promote mixing by convection or cooler on the bottom to promote sedimentation of the yeast. The uncluttered interior of the CCV makes cleaning and sanitation with standard CIP equipment easy and inexpensive. For best sanitation, the interior **surfaces** should have a smooth finish, especially the cone area.

Open Fermenters

There are still breweries that use open vats, especially for fermentation of ale. The ale yeast and resulting carbon dioxide form a layer on top of the beer, providing some protection from contamination. In some open fermenters there is a provision to collect some of the yeast during fermentation for use in subsequent batches. The Yorkshire square-style open fermenter, shown in Figure 9.17, has two chambers separated by a deck. The deck has an opening called the *manhole* for the yeast and one or more tubes, called *organ pipes*, that extend to near the bottom of the lower chamber to carry beer down. Often there is a pump to recirculate beer from the lower chamber to the upper chamber to rouse (stir) the yeast. Temperature control is maintained by a cooling jacket or by coolant pipes in the lower chamber. The vessel is filled until the lower chamber is full, and the upper chamber has a few centimeters (1 in) of beer. During fermentation, ale yeast rises through the manhole and collects in the upper chamber. Beer returns through the organ pipes. Near the end of fermentation, the yeast in the upper chamber is collected. Yorkshire squares are still used in the British Isles to make traditional ales, although some of them are actually circular rather than square to facilitate cleaning.

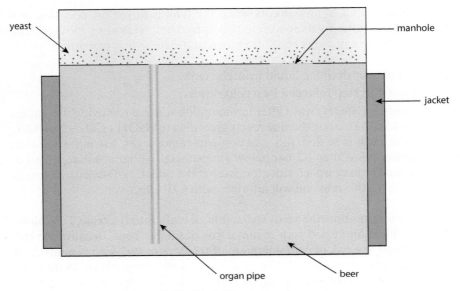

Figure 9.17   Yorkshire square fermenter.

There are several disadvantages to open fermentation. Onerous precautions, including special ventilation, are needed to minimize the risk of contamination by unwanted microbes. Many systems are difficult to clean and sanitize. The depth of the fermenter is limited to around 2 m (80 in.). Beer losses are greater than in cylindroconical fermenters, especially in the solids that sink to the bottom. Carbon dioxide from fermentation is vented directly into the fermentation room with no provision for recovery or control. Nonetheless, brewers who use open fermentation credit it with providing unique flavors.

Fermentation Cooling Requirement

Alcoholic fermentation releases about 560 kJ of heat per kilogram of sugar fermented (240 **BTU**/lb). This comes to about 5.6 kJ/°P/L. At a maximum ale fermentation rate of 0.33 °P/h, the cooling output **power** would be about 50 W/hL. This is the rate at which heat must be transferred through the walls of the fermenter to the cooling jacket and hence to the refrigeration system. Lager beer ferments slower, so it requires less cooling power.

Some methods of brewing require that the beer be quickly chilled from the final fermentation temperature to around 0 °C, a process called cold crashing. Cold crashing causes yeast to flocculate and fall to the bottom of the fermenter. Although crashing is part of conditioning rather than fermentation, we consider its cooling needs here, because it affects the design of the fermentation system. To crash 1 hL of beer from 18 to 0 °C in 12 hours requires 175 W of cooling power, not counting the cooling required for the fermenter vessel, its jackets, the coolant, and the cooling lines. This dwarfs the cooling requirement for fermentation temperature control; it must be considered in the initial design.

9.6 TEMPERATURE MONITORING AND CONTROL

Temperature

Temperature is the most important fermentation variable. At warmer fermentation temperatures, more fusel alcohols and esters are produced, affecting the character of the beer. Low temperatures can make the yeast sediment prematurely, stalling fermentation before all sugar is consumed. Temperature is usually controlled with a cooling loop, shown in Figure 9.18. Chilled coolant is pumped from a refrigeration unit, around a closed loop, through the cooling jackets and other items to be cooled, before returning to the chiller. Solenoid or **air-actuated valves** direct coolant through the fermenter cooling jacket or coil or block it if the temperature is low enough. The valves are opened and closed by signals from a **temperature controller** connected to sensors in the fermenters. A bypass valve opens when the pressure in the coolant inlet line exceeds a preset value, giving a return path for coolant when the cooling requirement is low and protecting the equipment from excessive pressure.

In most breweries, the coolant is a mixture of propylene **glycol** (PG) and water informally called "glycol." It is essential that the PG is food grade. The PG concentration is adjusted to give a mixture whose freezing point is lower

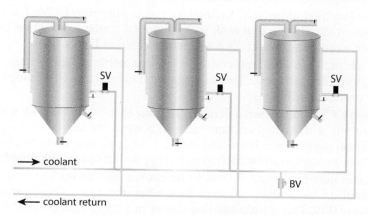

Figure 9.18   Cooling loop. SV, solenoid or air actuated valve. BV, bypass valve.

TABLE 9.4   Propylene Glycol Solution Freezing Points

PG Mass %	Freezing Point °C	Freezing Point °F	Specific Gravity
0	0	32	1.000
10	−3	26	1.008
20	−8	18	1.017
30	−14	7	1.026
40	−22	−8	1.034
50	−34	−29	1.041

Source: Adapted from EngineeringToolbox.com

than any temperature encountered in the system. This is a good deal lower than the lowest process set point; the coolant must be colder than the fermenter temperature, and allowance must be made for some temperature rise in the lines. Freezing points and specific gravities (useful for checking the PG concentration) are given in Table 9.4. PG is not a refrigerant; it is a secondary coolant that must give up its heat to a refrigeration system. Efficiency is lost in the exchange. PG is not as good a heat transfer medium as water, so too high a concentration will detract from cooling performance. The PG concentration should be high enough to give a freezing point well below the lowest temperature encountered by the coolant to avoid formation of slush.

There are some common considerations when designing and installing coolant lines:

- Place your chiller as close to the vessels as possible to minimize restriction in the coolant piping.
- Ensure proper sizing of your coolant lines to avoid excessive flow resistance.
- Use the "first in/last out" method for connecting your CCVs. From your coolant supply, the first tank to be supplied coolant should be the last tank to enter the return line, as demonstrated in Figure 9.18.
- Place pressure and temperature gauges in coolant supply lines where they can be easily read. This will allow you to better catch cooling system or pressure problems.
- Each CCV should have a dedicated sensor to measure product temperature and a dedicated coolant valve to control the supply of coolant. A manual control valve bypassing the solenoid or air-actuated valve is highly recommended to allow coolant to be supplied in case of solenoid failure.
- If you may be adding vessels to be cooled, like additional fermenters, install an expandable coolant loop after the last CCV, as shown in Figure 9.19. This simple solution allows more CCVs to be added in the future without interrupting cooling to the existing units.

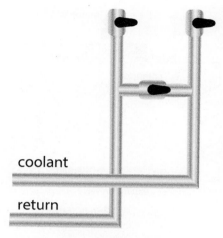

coolant

return

Figure 9.19   Expandable coolant loop.

Refrigeration

Breweries usually have a central refrigeration unit, also called a chiller or primary coolant system that provides cooling to the **wort chiller**, fermenters, conditioning tanks, and sometimes the beer storage and serving areas. The job of the primary coolant system is to extract heat at low temperature and release the heat to a higher temperature. Heat moves spontaneously from high to low temperature, so the non-spontaneous process must be coupled to a spontaneous process, normally the burning of fuel at the electrical power station.

Refrigerators in breweries almost always operate with a liquid–**vapor** cycle, shown in Figure 9.20. The refrigerant is a working fluid that can exist in the liquid and vapor state in the range of pressure and temperature required by the unit. The fluid absorbs heat when it vaporizes and releases heat when it condenses. Warm vapor at low pressure enters the compressor. The compressor is driven by a motor, doing work on (putting energy into) the fluid, increasing its pressure and temperature. The hot compressed fluid passes through a heat exchanger called the **condenser**, where it releases heat either to the atmosphere or to cooling water. The condenser coils get hot; they are often placed outdoors. The cooled vapor, still at high pressure, condenses to liquid, releasing more heat. The high-pressure liquid passes through an expansion valve, where the pressure is lowered, causing some of liquid to vaporize, absorbing heat, and going to lower temperature. The low-pressure liquid–vapor mixture passes through a heat exchanger called the evaporator. The liquid–vapor mixture continues to vaporize and to absorb heat from whatever is to be chilled. Then it enters the compressor to start the cycle

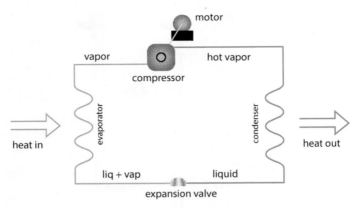

Figure 9.20   Refrigerator.

again. The heat released by the system is equal to the heat absorbed by the evaporator plus the energy provided by the compressor motor, plus any waste heat, including the heat released by the motor. Because of the large amount of heat released, it can be advantageous to locate the primary coolant system outside the brewery. It is easy to forget about the refrigeration unit. It often operates out of doors and out of sight. But when it breaks, the brewery cannot make beer. Beer in process may even be ruined. Monthly maintenance is recommended.

Smaller breweries usually use indirect cooling, provided by a secondary coolant system. The primary coolant system cools a coolant, often water and propylene glycol, that is pumped around to provide cooling to the equipment in the brewery. This is convenient; the brewers do not have to deal with refrigerant plumbing, but it is not efficient, as energy transfer between the systems is not perfect. Some large breweries use direct refrigeration in which the refrigerant itself, rather than a coolant, flows through the cooling lines.

Refrigerants are classed as CFCs (containing carbon, fluorine, and chlorine), HCFCs (also including hydrogen), HFCs (hydrogen, fluorine, and carbon), **hydrocarbons**, ammonia, carbon dioxide, and some others. The CFCs are nontoxic, are essentially non-flammable, and have good refrigeration properties, but are now banned because they deplete the ozone layer and have a high potential for global warming. Replacements vary in their desirability as refrigerants. For example, ammonia has good refrigerant properties, but it is toxic and flammable. It is used in large systems, including some in breweries. Common refrigerants in primary coolant systems include R-404A (HFC blend), R-134A (HFC), R-401A (HCFC), and R-717 (ammonia).

Temperature Control

Temperature is regulated by a system of temperature sensors, heaters, or, in this case, coolers and a temperature controller to control the application of heating or cooling in response to the measurement at the sensor. The simplest control system is the on–off controller. When the temperature is too high, cooling is applied. When the temperature is too low, cooling is turned off (same concepts apply to heating, but in reverse). On–off control is not usually very precise; the process temperature tends to overshoot on heat-up and undershoot on cool-down. A more sophisticated control scheme is proportional control. When the temperature is within a *proportional band*, the cooling power increases with temperature. Because the cooling power is less as the temperature reaches and goes below the set point (desired temperature), undershoot and overshoot are less; control is better. The proportional integral (PI) and proportional integral differential (PID) control add additional layers of sophistication to the control scheme.

Adjusting cooling or heating power is usually a matter of cycling between zero power and full power. For example, if 30% cooling power is needed, the solenoid valve on the coolant line would be open 30% of the time and closed 70%. This is usually easier to manage than devising a system to open and close the valves partway.

Temperature Sensors   The resistance temperature detector (**RTD**) is a metal, usually platinum, resister whose resistance increases as temperature increases. The metal takes the form of a fine wire or a thin film. Platinum RTDs are rugged and have a **linear** response (a graph of resistance against temperature is a straight line) and a wide temperature range. They are manufactured to recognized standards, making them interchangeable (for units that follow the standard). The sensor element is relativity bulky, making RTDs difficult to use for small samples, like **well plates** or Petri dishes. Platinum RTDs can be used up to 960 °C (1760 °F).

The **thermistor** is a ceramic or polymer whose resistance (usually) decreases as temperature increases. Thermistors can be made very small. In most parts of the temperature range, they have a greater resistance change with temperature than an RTD. The biggest disadvantage is that the change in resistance is not at all linear with temperature. Another disadvantage is that their construction is not fully standardized, complicating matching and interchange. The thermistor material is fragile; it must be packaged for protection. The temperature range for a thermistor is limited, typically up to 130 °C (266 °F).

A **thermocouple** is a junction of dissimilar metals, typically in the form of two wires spot-welded together. The thermocouple produces an **electrical potential** in the range of tens of microvolts per degree celcius. Standardized pairs of alloys are used in various temperature ranges. The thermocouple junction can be made quite small. The thermocouple pairs that are most often used in breweries are iron/copper-nickel (iron/constantan, Type J) and nickel-chromium/nickel-aluminum-manganese-silicon (chromel/alumel, Type K). Type J gives a larger output, but the iron wire is subject to **corrosion**. In the brewery temperature range of −5 to 120 °C (23–248 °F), the copper/copper-nickel (copper/constantan, Type T) gives the best measurement performance.

One complication of thermocouples is the reference junction. At some point in the circuit, the thermocouple wires must be attached to ordinary conductors. This is called the *reference junction*; the reference junction is actually two junctions. In Figure 9.21 the two thermocouple alloys are A and B. They join at the measuring junction, which is situated at the point where the temperature is to be measured. The connections between the A wire and the B wire to copper wires are the reference junctions. The **voltage** output is based on the temperature difference between the measuring junction and the reference junction. Standard calibrations assume a reference junction at 0 °C. Maintaining zero-degree reference junctions is not usually convenient, so correction circuits called reference junction compensators are usually used. Although a thermocouple does not require power, the reference junction compensator does. A type K thermocouple can measure to 1260 °C (2300 °F).

A *silicon bandgap sensor* is a semiconductor diode used as a temperature sensor. When current flows in the forward direction, there is a voltage across the diode that is directly proportional to the absolute (Kelvin) temperature. At 25 °C (298 K) the voltage is about 600–700 mV, so the sensitivity is about 2 mV/°C with high linearity. Silicon bandgap sensors, also called *p–n* or *diode sensors*, are not yet in regular temperature control use, but they have advantages for measuring low to moderate temperatures that may soon become decisive.

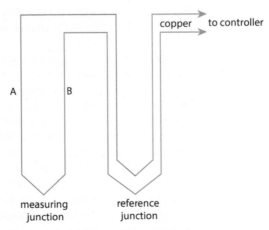

Figure 9.21 Thermocouple circuit.

9.7 YEAST HANDLING AND REPITCHING

It can be highly cost effective to reuse yeast from one fermentation into another. Repitching, historically called *backslopping*, is the process of adding yeast collected after fermentation is complete to another fermenter filled with fresh wort. As discussed previously, the standard pitching rate is 1×10^6 viable cells/mL/°P. Collecting yeast from a fermenter is called **cropping**.

Transferring Yeast for Reuse

There are several mechanisms for transferring yeast from one fermenter to another depending on the capabilities of the brewery. Some brewers may simply crop the yeast into a bucket from the racking arm or the drawoff port on the bottom of the cone. The yeast is then poured into a second fermenter through the **manway**. While this method allows easy access to the yeast for sampling, it is never a recommended procedure because of the risk of contamination. This method can be improved by using a **yeast brink**.

A simple yeast brink can be made from a modified half-barrel **keg** with a tri-clamp port on the top and on the side near the bottom of the keg. A half-barrel brink can hold about 45 L (12 US gal) of slurry, which is typically enough yeast to pitch 25–30 hL (20–25 US barrels) of wort. Each tri-clamp should be outfitted with valves for controlled transfer. Yeast is added or removed through a transfer arm on the bottom port. Sanitary transfer from brink to fermenter is controlled by applying pressure to the top of the brink. It is essential that each brink be equipped with a pressure relief valve to prevent hazardous pressure buildup. The use of a brink permits easy homogenization of the slurry, easy access for sampling, and easy transportation around the brewery.

Yeast quality will quickly plummet after storage in a brink. For this reason, harvested yeast should always be stored cold and for no more than 72 hours before use. While good viability may persist longer than 72 hours, glycogen storage reserves quickly decline, leading to long-term vitality issues. A simple quality check of stored yeast slurry is checking the pH of the **barm beer**. As yeast quality declines, the pH increases. Yeast slurry with a pH greater than 4.9 should be discarded.

Another practice for yeast cropping is cone-to-cone transfer. Here the yeast slurry is pushed from one fermenter cone to another through a transfer hose. Movement may be facilitated by gravity, pumping, or pressure on the supplying cone. This method is only recommended when precise volumes can be measured via an in-line flow meter or by measurement of weight by load cells on the fermenters. Accurate mass flow meters mounted on **skids** (ABER Instruments®) are also suitable. Mass of slurry can be correlated with cell count in the lab to more accurately estimate cell count. Integrating measurement with pump control permits exclusion of yeast with solids content that is too high (trub) or too low (end of slurry). The low-solids slurry can be redirected to the next stage of processing. The disadvantage of cone-to-cone transfer for small breweries is the management of yeast strains and time; there must be a tight schedule between the end of one fermentation and the start of another.

Not all yeast can be reused for all beers. As a significant amount of barm beer will carry over into the subsequent fermentation, yeast should only be reused for similar beer styles. For example, it is not advised to reuse an English ale strain from a porter in a fresh batch of an English pale ale. The porter barm beer will contribute flavor and color changes to the no-longer pale ale. As another example, it may not be possible to harvest yeast from an IPA with significant hop additions in the fermenter because the yeast would be mixed with copious amounts of hops. Yeast that has been exposed to high alcohol or high specific gravity is in poor condition because of ethanol toxicity or **osmotic stress**. Such yeast is not suitable for repitching.

When repitching, yeast should be transferred to the fermenter after a portion of the wort is added. This process facilitates mixing. For example, 5 hL of wort is transferred to a 30 hL fermenter, and then the yeast is pitched. The rest of the wort is then added, resuspending and rousing the yeast.

After the yeast is pitched and after all the wort is added to the fermenter, the yeast should be counted to ensure consistency in pitch rate. Total yeast in suspension may be counted immediately after filling. Another approach is to wait two to four hours before sampling to permit the trub, hop resins, and dead yeast to settle before counting the yeast in suspension. Either approach is sufficient, but the same method should be used consistently.

Cropping Yeast Heterogeneity

Not all yeast is created equal in the cone. Because large, more dense particles will settle first, the first material to settle after filling a fermenter is the trub. Trub is darker than yeast; a **sight glass** clamped in place between the fermenter's transfer port and a T-**pipe fitting** with valves at either downstream port in the transfer line, as illustrated in Figure 9.22, allows the trub to be dumped before the yeast is transferred. The valves can be used to direct beer to a hose routed to a drain or collection tank for trub or to another hose to the next process point for the yeast slurry. The sight glass provides manual identification of the trub, which is a lumpy, dark brown material, as compared with the yeast slurry, which is a creamy off-white color. When pulling yeast from the cone, flow rate is important. If the rate is too high, channeling will occur, resulting in a loss of yield as a significant amount of slurry remains on the fermenter cone walls. Any yeast left behind in unfiltered beer may lead to flavors attributed to autolysis. In addition, high velocities create mechanical stress on the yeast due to turbulence and shear forces. It is recommended that

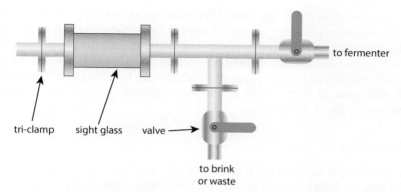

tri-clamp   sight glass   valve →

to fermenter

to brink
or waste

Figure 9.22   Transfer of yeast slurry from cone-to-cone is facilitated by a sight glass and a T-pipe fitting with valves.

the fluid velocity not exceed 1 m/s (3.3 ft/s) for yeast transfer. An even flow rate to avoid channeling is also important to ensure consistency in the yeast crop. There can be heterogeneity in yeast quality depending on the depth in the cone. Not surprisingly, slurry density and yeast cell number are highest at the bottom of the cone. It is interesting that alcohol percentage has also been found to be highest at the bottom. Heterogeneity in the fermenter cone suggests that yeast cropping practices should be standardized within the brewery to get yeast from the same level each time yeast is cropped.

What is the best time to crop yeast during fermentation? Yeast cropping after cold crashing is a common method for yeast harvesting, because cold conditioning increases flocculation and settling of yeast in suspension. Cropping at this step maximizes yeast yield, but it has the potential drawback of lower viability. Cold conditioning introduces thermal stress on the yeast and has been shown to reduce viability by about 8%. For this reason, if enough yeast for subsequent brews can be cropped after terminal gravity and before cold crashing, this is an ideal time. Yeast held in the fermenter cone for more than two days should be discarded rather than cropped to prevent off-flavors from autolysis.

Drauflassen or "Topping Up"

Drauflassen is a German brewing term for the process of "topping up" or adding wort to an already fermenting tank. Brewers may not produce enough yeast for a fermentation due to propagation or production limitations. This is also a common practice in brewing operations where fermenter capacity is larger than the brewhouse size. When topping up a fermenter, it is important to initially fill the tank with cooled, oxygenated wort as normal. A standard pitching rate should also be used. Then after approximately 24 hours or during low krausen, an equal volume of cooled, oxygenated wort is added. Such addition maintains the low krausen and will continue to promote cellular growth as needed for the fermentation. It is essential that the second wort addition be oxygenated to support yeast growth. It is also essential that the wort temperature is the same as the fermentation temperature to prevent temperature shock to the yeast.

Stratification

The practice of topping up a fermentation may lead to a condition called **stratification**. Two layers of liquid, the low krausen fermenting beer and the freshly added wort, do not mix; they form distinct separated zones, each with its own temperature and composition. Separation occurs because of differences in density as a result of different sugar and ethanol concentrations in the two zones. In this scenario, the dense, sugary freshly added wort enters at the bottom, while the partly fermented low-density beer is pushed to the top. Stratification causes problems in flavor and consistency.

Stratification is noted as a sudden temperature drop measured in the lower zone when the two zones mix. A model observation that might indicate stratification is shown in Figure 9.23. The temperature drop was the result of the top fermentation chilling beyond that of the bottom fermentation where the temperature sensor was located. Addition of sampling ports and temperature sensors at both the top and bottom of the fermenter shed light on the problem. How or precisely when coalescence occurs is less understood.

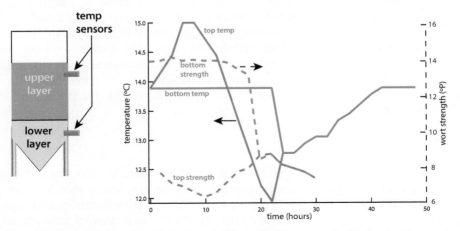

Figure 9.23 Stratification after second filling of a fermenter. Solid line temperature (scale on left). Dashed line wort strength (scale on right).

In summary, fermenter stratification may be caused by:

- Unequal temperatures between the low krausen beer and the freshly added wort.
- Large deviation in density between the low krausen beer and the freshly added wort as a result of permitting fermentation to persist too long before the second addition of wort.
- Inadequate mixing of wort additions due to low velocity pumps or inlets.

Yeast Generation Number

In contrast to laboratory or wild yeast strains, brewer's yeast is stuck in an asexual life cycle of replicative **budding**. If there are sufficient nutrients and space for cell division, yeast will continue in its life cycle. But given limitations, yeast will exit into a vegetative state, called the stationary or G0 stage. Once enough resources are available, the yeast cell will reenter the cellular division cycle. The asexual yeast cell cycle is illustrated in Figure 9.24.

As yeast cells divide, the parent cell forms bud scar in the cell wall. In this manner, the age or reproductive history of a yeast cell can be determined by the number of bud scars on the cell surface. Each time the cell divides, the original cell, called the *mother cell*, produces a new yeast termed the *daughter* or *virgin cell*. Because of this cycle, every yeast culture or slurry contains 50% daughter cells. This concept is modeled in Figure 9.25.

Because all yeast populations contain 50% new cells, then yeast can be used *ad inifinitum*, or again and again forever, right? Not quite. Yeast are repeatedly subjected to stress from the presence of alcohol and carbon

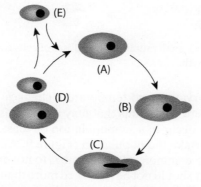

Figure 9.24 Yeast cell cycle. (A) Cell grows. (B) Bud emerges. (C) Genetic material and organelles split. (D) Daughter cell separates. (E) If nutrients are insufficient the cell enters dormancy. When conditions become favorable the cell returns to the cycle.

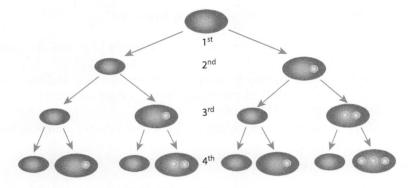

Figure 9.25   As a yeast population divides, it will contain 50% new, virgin cells. Four generations shown.

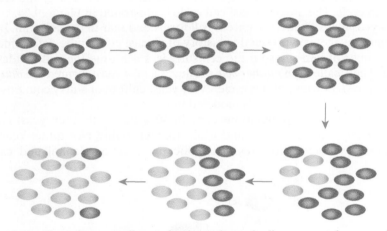

Figure 9.26   Over time a small mutation in a subset of cells can overtake a population due to exponential growth.

dioxide and from starvation and temperature changes. Stress has the potential to introduce genetic mutations that may affect yeast performance. While a mutation may only occur in a subpopulation of yeast, over time the effects can escalate and overtake the population (Figure 9.26).

For this reason, the common rule of thumb is to not exceed ten generations, meaning a yeast slurry should not be repitched more than ten times. Monitoring of yeast quality in the slurry can help reveal any potential problems because of high generation number. Flavor may be impacted, but other qualities are often affected first. Slower fermentation rates, reduced flocculation, declines in viability, and delayed diacetyl reduction are all metrics that may indicate poor yeast performance as a result of age. While a maximum of 10 generation numbers is generally recommended, ultimately the brewer should keep careful track of any such deviations in performance to help inform recommended yeast generation number. In practice, generation number may be as low as 4 or as high as 30. Generation number is strain dependent.

9.8 YEAST PROPAGATION

In the previous section, we discussed how yeast can be repitched from batch to batch. But once the yeast reaches the recommended generation number, how does a brewer get new yeast? Commercial yeast suppliers can provide large slurry volumes from 0.5 to 100 hL. If purchasing yeast from a commercial supplier, it is highly recommended to keep track of quality by verifying cell count and viability, as yeast may be subject to unseen stresses during shipment. Of course, with the convenience of obtaining a pitchable volume of yeast slurry comes an increase in price. With sufficient resources and management, in-house yeast propagation is relatively straightforward, but breweries that use many yeast strains may find that using commercial suppliers makes good business sense. Commercial suppliers can handle private yeast strains under exclusivity agreements.

Yeast propagation has a different goal than wort fermentation. The goal of propagation is to make yeast; the goal of fermentation is to make good beer. Stirring and aeration are good for making yeast, but bad for making beer. The first pure yeast culture system was established by Emil Christian Hansen in 1883 at the Carlsberg Laboratory. He was the first to culture individual yeast cells, which he subsequently grew into a slurry in a sugar solution. The concepts of his discovery remain today and are divided into three main phases: cell culture, laboratory propagation, and commercial propagation. In-house propagation requires the utmost attention to **aseptic** technique and **sterile** or sanitary conditions to prevent microbial contamination of the yeast. It requires a laboratory isolated from the brewery. Positive pressure in the laboratory space is ideal to prevent backdrafts into the lab from the brewery floor.

Cell Culture

A pure culture of yeast is used to seed the propagation. Normally yeast colonies can be obtained from an **agar** plate or **slant**. A slant is a test tube with agar and a growth medium that has been allowed to gel at an angle, as shown in Figure 9.27. When stored at 4 °C (34 °F), most colonies will be viable for up to one month, at which point the cultures should be regenerated by re-**streaking**

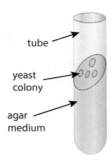

Figure 9.27   Yeast culture on slant.

on fresh medium. Yeast extract–peptone–**dextrose** (YPD) medium is the most commonly used yeast medium for cell culture on agar.

Yeast cultures may be ordered directly from a national cell culture bank or university, or they can be maintained in-house, where the yeast is resuspended in a 10–20% **glycerol** solution and frozen at −80 °C (−112 °F) in a special freezer. Storage in liquid nitrogen is also possible but requires routine replacement of the liquid nitrogen. Cultures stored at −80 °C can survive for three to five years before the stocks should be regenerated. Common freezers that hold at −20 °C (−4 °F) should not be used, because the refrigeration cycle can cause temperature changes with a negative effect on long-term viability.

Laboratory Propagation

A single colony is used to inoculate 10 mL of culture medium, which is incubated for 24–48 hours at 25–30 °C (77–86 °F) with shaking at 180 rpm. YPD broth is recommended for this first step. From this initial culture, the yeast is propagated by an increase in volume by a factor of 10. The 10 mL culture is used to inoculate 100 mL of 8–12 °P wort, which is again cultured with shaking or stirring for 24–48 hours at 25–30 °C (77–86 °F). Once the culture reaches maximum density, it is ready for transfer to the next step. During laboratory propagations, it is important to use a low-gravity wort to avoid excessive stress on the yeast. If a shaker is not available, a stir plate can be used, but not one that gets hot. All laboratory-scale cultures should be incubated at 25–30 °C (77–86 °F) with shaking at 180 rpm. The 100 mL culture is then used to inoculate 1 L of wort, incubated at the settings described above. Figure 9.28 illustrates the proper steps for laboratory propagation. Shaking or stirring is recommended during all propagations because increased aeration promotes cell growth and viability. Stirred cultures produce four times more yeast than static cultures. All transfer steps should be conducted in a laminar flow hood to prevent any contamination.

It is at this point that laboratory propagation reaches its limit. A 1–10 L step is technically challenging due to limitations in space. Some breweries

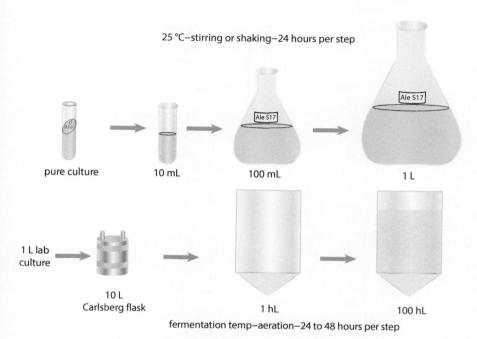

25 °C–stirring or shaking–24 hours per step

pure culture   10 mL   100 mL   1 L

1 L lab culture

10 L Carlsberg flask   1 hL   100 hL

fermentation temp–aeration–24 to 48 hours per step

Figure 9.28   Laboratory propagation of yeast.

employ a Carlsberg flask (Figure 9.29). The Carlsberg flask has a capacity of 20 L. Wort from the brewhouse can be **sterilized** and cooled in the flask. An aeration port allows oxygenation of the cooled wort with sterile air or oxygen. The laboratory culture is then injected into the flask through a sterile port. When the cells reach maximum density, the culture can be transferred to a commercial propagation system.

Commercial Propagation

If the brewery has no dedicated propagation equipment, a sanitized fermenter can be used. In this method, sterile, oxygenated wort is added to the yeast slurry, increasing the volume by 10–25 times for each step. The yeast is allowed to grow for 24–48 hours. It is often not necessary to cold crash the propagation. Instead, the entire yeast culture can be transferred to the fermenter. It is recommended that the volume of wort in the fermenter be no more than 20 times that of the culture. This method can produce 80–100 million cells/mL.

A commercial propagation system can produce up to 200 million cells/mL, yielding enough yeast to pitch into 40 times the volume of wort. For example, yeast produced in a 10 hL propagation system may generate enough yeast to

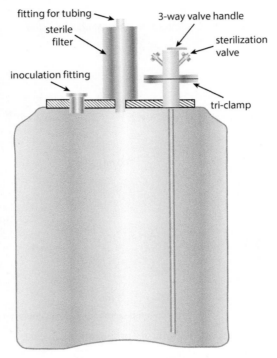

fitting for tubing

sterile
filter

inoculation fitting

3-way valve handle

sterilization
valve

tri-clamp

Figure 9.29   Carlsberg flask.

pitch into a 400 hL fermenter but should always be verified with cell counting. As discussed with laboratory propagation, wort aeration and resuspension through shaking or stirring is recommended to increase biomass or cell number. For this reason, many commercial propagation systems include ports for aeration and an agitator plate for gentle stirring.

Single-vessel, double-vessel, or continuous fed-batch systems may be used in commercial propagation plants. All systems should have excellent CIP capabilities and internal sterilization by steam. For single-vessel systems (Figure 9.30), a 5 hL tank permits sterilization of wort, cooling, aeration, and sterile transfer of yeast into the system from the Carlsberg flask. Once the fermentation reaches high krausen, the entire contents are used to pitch a 25–30 hL (20–25 US bbl) fermenter. This method requires a laboratory propagation for each new batch of yeast.

A double-vessel system includes one 5 hL vessel and one 30 hL vessel. Here, a 5 hL propagation can be performed in tank 1, as described above, while 25 hL of wort is prepared in tank 2 (Figure 9.31). When the first tank reaches high krausen, its contents are transferred to the sterile wort in tank 2, the larger tank. Meanwhile, the first tank is cleaned and sterilized. When the

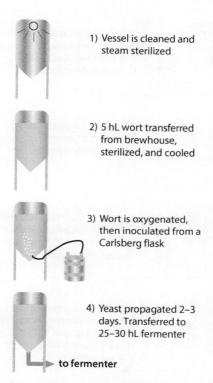

1) Vessel is cleaned and steam sterilized

2) 5 hL wort transferred from brewhouse, sterilized, and cooled

3) Wort is oxygenated, then inoculated from a Carlsberg flask

4) Yeast propagated 2–3 days. Transferred to 25–30 hL fermenter

to fermenter

Figure 9.30   Single vessel commercial yeast propagation in the brewery.

second tank reaches high krausen, 25 hL of yeast is available for fermenters, while 5 hL is transferred back to the first tank. Now the 5 hL tank is used to hold the yeast until another 25 hL of wort can be prepared in tank 2. This method permits continual reuse of the yeast in the propagation system and only requires a laboratory propagation for the first pitch. A continuous fed-batch system also includes two tanks, but one is used to sterilize and store wort, while the other is used solely for propagation. Using a pump, sensors to monitor fermentation, and a programmable logic controller (PLC) system, wort is dosed into the propagation system with continuous aeration. As the yeast cell number increases, so does the wort volume. Once the culture reaches a critical mass, some of the yeast is transferred out of the tank for fermentation and the continual culture loop continues.

Active Dry Yeast

Active dry yeast represents a convenient solution for brewers with no propagation systems. Under proper conditions, active dry yeast can be stored for up

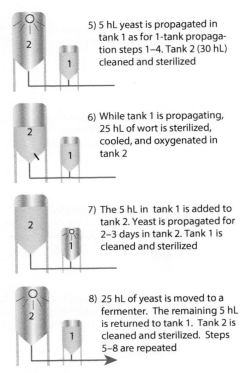

5) 5 hL yeast is propagated in tank 1 as for 1-tank propagation steps 1–4. Tank 2 (30 hL) cleaned and sterilized

6) While tank 1 is propagating, 25 hL of wort is sterilized, cooled, and oxygenated in tank 2

7) The 5 hL in tank 1 is added to tank 2. Yeast is propagated for 2–3 days in tank 2. Tank 1 is cleaned and sterilized

8) 25 hL of yeast is moved to a fermenter. The remaining 5 hL is returned to tank 1. Tank 2 is cleaned and sterilized. Steps 5–8 are repeated

Figure 9.31 Double vessel commercial yeast propagation in the brewery.

to three years with minimal effects on fermentation performance or beer flavor. Because the active dry yeast can be stored for extended periods, different strains are more easily employed in the brewery, as they are available on demand with cell counts necessary for healthy fermentations. The availability and speed with which a new yeast strain can be used in the brewery overcomes the time and resource management necessary for yeast propagation systems. Active dry yeast can also be used for secondary fermentation in bottles or kegs.

Typical cell counts for dry yeast are 5 billion cells/g, typically at 93–95% dry matter. Specific numbers are strain and supplier specific. Dry yeast is usually vacuum packed in 500 g containers. Oxygen and humidity negatively affect the viability of dry yeast, so once the packet is opened, all yeast should be used as quickly as possible. Storage of sealed packets is best at 4 °C (34 °F).

Another factor required to promote active dry yeast viability is rehydration methods. Improper yeast rehydration may lead to a reduction in viability, ultimately leading to performance problems such as longer diacetyl rests, stuck fermentations, longer lag phases, longer time to final gravity, and poor utilization of maltotriose. For best rehydration, yeast should be sprinkled onto the surface of 10 times its weight in sterilized brewing water in a sanitized vessel with the water at a temperature of 30–35 °C (86–95 °F). The active dry yeast should be left undisturbed for 15 minutes and then gently stirred to resuspend the yeast completely. Best practice is to then bring the yeast to the same temperature as the target wort to reduce thermal stress. A common misconception is that active dry yeast can be simply sprinkled onto wort. **Brewing liquor**, rather than wort, distilled water, or **reverse osmosis** water, should be used to resuspend active dry yeast to avoid loss of viability as a result of osmotic stress.

CHECK FOR UNDERSTANDING

1. Why do yeast cells carry out fermentation?

2. Draw a standard microbiological growth curve and label the lag, log, stationary, and decline phases. On the graph, overlay the trends observed when measuring yeast cells in suspension of a typical ale fermentation. Describe the differences between a traditional microbiological growth curve and yeast during fermentation.

3. What is the distinction between viability and vitality? How can they be distinguished in practice?

4. Why are cell counting and cell viability measurements important in brewing operations?

5. Discuss the importance of pitch rates in brewing a consistent, high-quality beer.

6. How many cells do you need to ferment 20 barrels of a 12 °P wort?

7. You have 225 billion yeast cells in 1 L of culture at 97% viability. How much slurry do you need if your fermentation requires 200 billion cells?

8. You prepared a sample for counting by diluting 1:10 in a methylene blue solution. You then counted the five chambers as described above and measured 20, 18, 22, 17, and 23 cells at 100% viability. How many cells per milliliter do you have?

9. In the question above, if you have a 2 L slurry, how many total cells do you have available?

10. You prepared a sample for counting by diluting 1:10 in a 0.5 M EDTA solution and then 1:10 in a methylene blue solution. You then counted the five chambers as described above and measured 12, 14, 22, 13, and 17 cells at 96% viability.

a. How many cells per milliliter do you have?

b. If you have a 1 bbl slurry, how many cells total do you have?

c. If you are going to ferment 10 bbls of a 10 °P wort, how many cells do you need?

d. Determine how much slurry is needed for this fermentation. How might underpitching affect fermentation performance?

11. In Figure 9.2, why does the temperature decline abruptly at six days?

12. Describe the temperatures changes that must occur during a fermentation. What is the purpose of each step?

13. State two advantages of using an antifoaming agent during fermentation?

14. Why does the barley plant provide the seed with starch instead of ATP?

15. What do yeast cells require from wort other than sugar for a healthy fermentation?

16. Describe the best practices for ensuring optimal yeast cell counts. Which steps are prone to creating errors in accuracy?

17. Describe the process of topping up. How can this lead to problems in the brewing process?

18. How would you determine how many times a yeast could be repitched before replacing it from a supplier?

19. You met a brewer who bragged that the brewery was on their 324th generation of a yeast strain. First, explain how this could be possible. Second, explain the drawbacks of this approach.

20. You have been asked by your Director of Operations to investigate the cost of implementing an in-house yeast propagation system. Start by explaining the equipment needs and process flow for yeast propagation in a brewery with 20 bbl fermenters.

21. Discuss the best practices of maintaining yeast in a brink before reuse.

CASE STUDY

Case Study 1

Your sensory team has noticed higher amounts of fusel alcohols in one of the brewery's flagship beers. Create a plan to identify potential areas of concern in the brewing process which could be responsible. Propose a plan to evaluate and solve the issue.

Case Study 2

Your brewery makes a dry-hopped pale ale with an OG of 15 °P and a FG of 2.0 °P. Recently your FG has been stopping at 5.0 °P. Troubleshoot this problem. Suggest potential reasons for this stall and how you will identify, evaluate, and solve the problem.

BIBLIOGRAPHY

American Society of Brewing Chemists. *Methods of Analysis*. Table 1: Extract in Wort or Beer. doi: 10.1094/ASBCMOA-TableWortBeerBrewingSugars.

Barth R. 2015. The Role of Alcoholic Fermentation in the Rise of Biochemistry. In Barth R, Benvenuto MA (editors). *Ethanol and Education*. Oxford. ISBN 978-0-8412-3059-0.

Boulton C, Quain D. 2006. *Brewing Yeast and Fermentation*. Blackwell Science. ISBN 978-1-4051-5268-6.

Carey D, Grossman K. 2014. Fermentation and Cellar Operations. In Ockert K (editor). *MBAA Practical Handbook for the Specialty Brewer, Vol. 2*. Master Brewers Association of the Americas. ISBN 0-9770519-2-7.

Casey GP, Ingledew WM. 1983. High-gravity brewing: influence of pitching rate and wort on early yeast viability. *J. Am. Soc. Brew. Chem. 41*: 148–152.

Cutaia AJ, Reid AJ, Speers RA. 2009. Examination of the relationships between original, real and apparent extracts, and alcohol in pilot plant and commercially produced beers. *J. Inst. Brew. 115*:318–327.

Fix G. 1999. *Principles of Brewing Science*. Brewers Publications. ISBN 0-937-381-74-8.

Gosselin Y, Van Nedervelde L, Boeykens A, Janssens P. 2017. Shelf life and consistency of active dry yeast for breweries. Proceedings ASBC Annual Meeting. Tech Session 4, paper 11.

Jones M, Pierce JS. 1964. Absorption of amino acids from wort by yeasts. *J. Inst. Brew. Dist. 70*(4):307–315.

Kapral D. 2008. Stratified fermentation: causes and corrective actions. *MBAA TQ 45*(2):115–120.

Kunze W. 2014. *Technology Brewing & Malting. 5th ed*. VLB Berlin. ISBN 978-3-921690-77-2.

Layfield JB, Sheppard JD. 2015. What brewers should know about viability, vitality, and overall brewing fitness: a mini-review. *MBAA TQ 52*(3):132–140.

O'Conner-Cox E, Mochaba FM, Lodolo EJ, Axcell B. 1997. Methylene blue staining: use at your own risk. *MBAA TQ 34*(1):306–312.

Pasteur L. 1876. *Studies on Fermentation*. Translation by Faulkner F, Robb DC. Reprinted 2005. BeerBooks.com. ISBN 0-9662084-2-0.

Powell CD, Quain DE, Smart KA. 2004. The impact of sedimentation on cone yeast heterogeneity. *J. Am. Soc. Brew. Chem. 62*(1): 8–17.

Stokes C. 1999. In-line Analytical Instrumentation. In McCabe JT (editor). *The Practical Brewer, 3rd ed*. Master Brewers Association.

Troester K. 2013. Step Up Your Starters. Presentation at the National Homebrewers Conference. Philadelphia, PA.

Van den Berg S, Van Landeschoot A. 2003. Practical use of dried yeast in the brewery industry. *Cerevisia 28*(3):25–30.

Van Zandycke SM, Fischborn T, Peterson D, Oliver G, Powell CD. 2011. The use of dry yeast for bottle conditioning. *MBAA TQ 48*(1):32–37.

White C, Zainasheff J. 2010. *Yeast: The Practical Guide to Beer Fermentation*. Brewers Publications. ISBN 0-937-381-96-9.

CHAPTER 10

CONDITIONING

In ancient times, beer was often consumed during active **fermentation**. Today, several operations are typically performed before the beer is ready for packaging or serving. These operations are collectively called **conditioning**, otherwise known as *beer stabilization*. Before conditioning, the beer is called **ruh beer** or **green beer**. When green beer reaches final **attenuation** in the **fermenter**, the beer will be cold conditioned in the same tank for further **flavor** maturation and stability. This is called **secondary fermentation**. A small **brewpub** might then transfer the beer into a **bright beer tank** for **carbonation** and serving. Packaging **breweries** may subject the beer to **centrifugation** and **filtration** to prolong flavor and **colloidal** and microbiological stability.

Beer conditioning increases the shelf life of the beer. Shelf life is the period of time a beer remains salable, defined primarily by the brewer's acceptable flavor profile. Shelf life may range from weeks for certain **hop**-forward beers to months for most common **ales**. If steps are taken to reduce oxygen, ensure sound **brewing** practice, and secure good raw materials; proper beer conditioning will further promote shelf stability. Not all aspects of conditioning are practiced by all breweries. Small brewers may only conduct warm and cold maturation in the fermenter followed by carbonation in a bright beer tank. More advanced conditioning steps are not always required unless beer is

Mastering Brewing Science: Quality and Production, First Edition.
Matthew Farber and Roger Barth.
© 2019 John Wiley & Sons, Inc. Published 2019 by John Wiley & Sons, Inc.

packaged for sale outside the brewery. The longer the beer has to last after production, the greater the need for elaborate conditioning protocols.

Specifically, beer conditioning serves the following purposes:

- Flavor maturation and stability.
- Reduction of **haze** forming compounds.
- Clarification of the beer before service or packaging.
- Stabilization of **foam**.
- Reduce risk of **spoilage** microorganisms.
- Ensure appropriate carbonation levels.

10.1 WARM CONDITIONING

Secondary fermentation often takes place in the same **cylindroconical** vessel as fermentation, aptly named a unitank. Only if space permits, or if extensive **lagering** is required, will green beer be transferred to another vessel for conditioning. In Figure 9.2 we discussed some of the temperature changes that occur during secondary fermentation. These include an increase in temperature for warm conditioning. This step aids in flavor maturation through the reduction of **acetaldehyde** and **diacetyl**, which are produced by **yeast** during fermentation (Figure 9.4). The other temperature change is cold conditioning, which facilitates yeast removal and colloidal stability. Once these steps are complete, beer is sent to a bright beer tank for carbonation and temporary storage before serving or packaging. Bright beer tanks are not typically used for conditioning; they are holding tanks.

Diacetyl Reduction

The major off-flavor compounds in ruh beer are diacetyl (chemical name: 2,3-butanedione) and 2,3-pentanedione, collectively called **vicinal diketones** (**VDKs**), shown in Figure 10.1.

The VDKs have C=O groups on adjacent carbon **atoms**. They provide a buttery flavor characteristic of artificially flavored popcorn. The aroma **threshold** of diacetyl is 0.15 ppm; that of pentanedione is 1 ppm. One of the major objectives of ruh beer processing is to minimize the VDK off-flavor. It can be difficult to analytically distinguish diacetyl and pentanedione, so they are collectively discussed as VDKs, often used interchangeably with the term diacetyl. Here when we mention diacetyl, we specifically refer to 2,3-butane-dione, which is the focus of our discussion.

Diacetyl arises during fermentation as a by-product of a pathway used by yeast to produce the **amino acids** valine and leucine; pentanedione arises

diacetyl   2,3-pentanedione

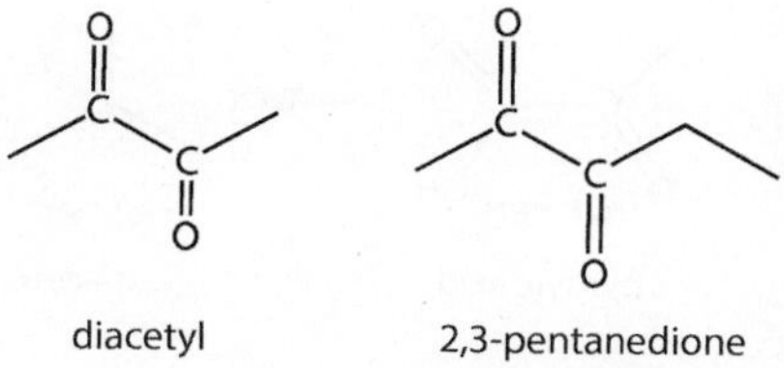

Figure 10.1   Vicinal diketones.

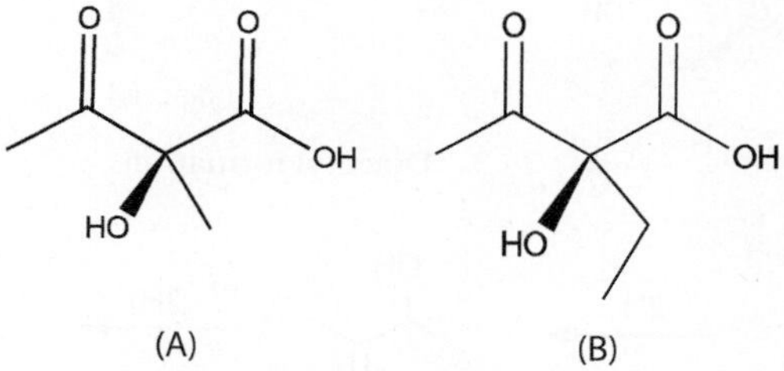

(A)   (B)

Figure 10.2   VDK precursors. (A) Alpha-acetolactic acid. (B) Alpha-acetohydroxybutyric acid.

from a pathway to isoleucine (Figure 3.16). When the yeast produces these amino acids, intermediates are produced that cannot be quickly processed to the next step in the pathway, so they are secreted from the yeast cell and into the beer. During fermentation, diacetyl is made from alpha-acetolactic acid, and 2,3-pentanedione is made from alpha-acetohydroxybuteric acid (AHBA), shown in Figure 10.2, by nonbiological oxidation reactions. The process for diacetyl is shown in Figure 10.3. Acetaldehyde and pyruvic acid, both products of glycolysis, form alpha-acetolactic acid (officially (S)-2-hydroxy-2-methyl-3-oxobutanoic acid). This intermediate transfers two hydrogen atoms to an oxidizing agent (Q in Figure 10.3) and releases carbon dioxide to give diacetyl. The yeast quickly hydrogenates the C=O groups to −OH groups, yielding low flavor alcohols, as shown in Figure 10.4. The 2H shown over the arrow represents addition of hydrogen atoms from NADH. The flavor problem arises when the beer is separated from the yeast while the alpha-acetolactic acid precursor is still present. Diacetyl is still produced, but it is not removed because the reaction shown in Figure 10.4 requires live yeast. When the yeast is removed, the precursor compounds remaining in the beer continue to be oxidized, but there is no yeast to hydrogenate the resulting VDK.

Diacetyl is more of a concern in **lager beer**, which is fermented below 15 °C, than in ale. The higher ale fermentation temperature gives more rapid conversion of precursors to VDK during fermentation and conditioning, so the sources of VDK are generally depleted within a few days. By contrast, the

acetaldehyde   pyruvic acid   α-acetolactic acid

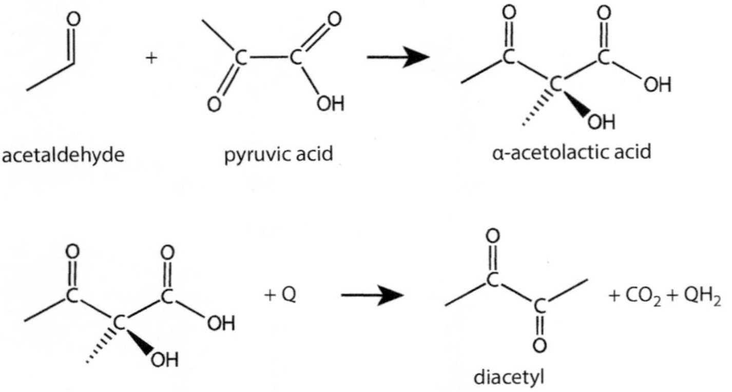

diacetyl

Figure 10.3   Diacetyl formation.

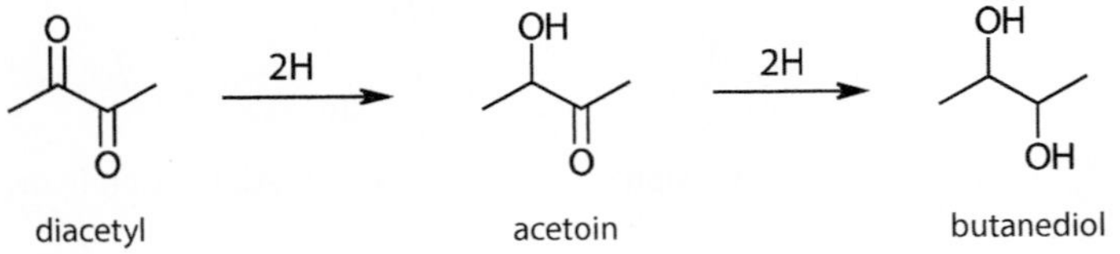

diacetyl   acetoin   butanediol

Figure 10.4   Diacetyl reduction.

lower lager fermentation temperature gives much slower depletion of VDK precursors, which can give rise to VDK above the flavor threshold after fermentation. To fully deplete VDK precursors in lager beer, the beer is held for a prolonged period in contact with yeast at low temperature. This prolonged conditioning gives lager beer its name [*lager*: Ger. warehouse]. Three to four weeks of lagering is standard. Some brewers raise the temperature of the beer at the end of primary fermentation to increase the rate of VDK precursor conversion. This is called a *diacetyl rest*. The diacetyl rest can enhance the effectiveness of lagering, but the higher temperature after sugar is depleted can stress the yeast and decrease its viability.

The lager process can be bypassed altogether by treating the beer with *immobilized yeast*, which is yeast cells attached to beads. The process starts with removal of the fermentation yeast by centrifugation. The clear beer is heated briefly to 60–90 °C to convert all precursor to VDK. The beer is chilled and run through a bed of immobilized yeast. Removal of VDK can take as little as 20 hours. The use of immobilized yeast is not a common practice.

Other Sources of Diacetyl in Beer   Diacetyl can increase as a result of **hop creep**. After **dry hopping**, the reactivation of fermentation may lead to a second wave of diacetyl production. Enough time is needed after dry hopping

α-acetolactic acid   acetoin

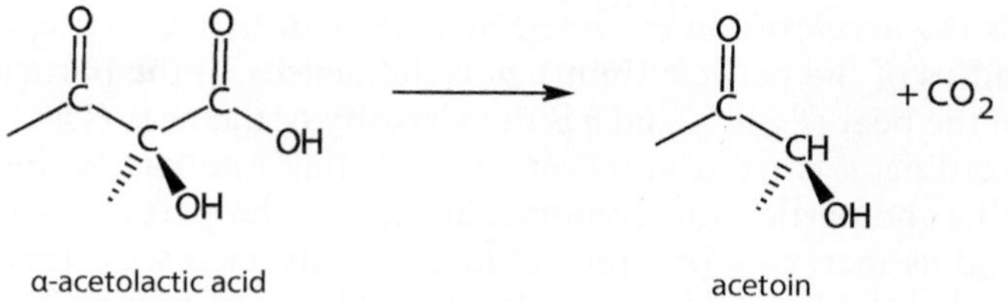

Figure 10.5   Enzymatic treatment for VDK control.

for diacetyl reduction. Diacetyl can also be produced as a result of **bacterial** infection. Spoilage bacteria do not produce 2,3-pentanedione. The typical ratio of diacetyl to 2,3-pentanedione is 3:1 to 5:1. A higher ratio suggests bacterial contamination.

The production and reduction of diacetyl is yeast strain dependent. Production also depends on the amount of valine present in the **wort**. As valine is assimilated, diacetyl production decreases, but if wort is deficient in or depleted of valine, then diacetyl production increases. All-**malt** wort usually has enough valine, but extensive use of **adjuncts** may warrant caution.

Enzymatic VDK Control   Recently, a bacterial **enzyme**, **alpha-acetolactate** decarboxylase (α-ALDC), has become available. This enzyme is added at the beginning of fermentation at a rate of 1–1.5 g/hL of wort. It directly **catalyzes** the conversion of alpha-acetolactic acid, the precursor for diacetyl, to acetoin and carbon dioxide according to the reaction shown in Figure 10.5. Use of α-ALDC bypasses the spontaneous production of diacetyl and the need for yeast as the means for reduction.

10.2   CLARIFICATION

A current trend is to produce hazy or unfiltered beer that requires minimal processing. But most styles of beer are expected to be served clear and **bright**. Yeast and other solid particles are removed. Materials that can react to give **haze** particles during the expected shelf life of the beer must be excluded. Clarification will promote the long-term flavor and colloidal stability of beer in packaging.

Beer is eventually clarified by gravity, but during production, clarity is typically induced by the addition of **fining** agents and/or by mechanical means. Clarification is governed by Stokes' law where the velocity of a particle settling in liquid is determined by the equation

$$v = \frac{2r^2 \left(\rho_p - \rho_b\right) a}{9\eta}$$

where a is the acceleration provided by gravity or by a centrifuge (in m/s^2), r is the radius of the particle (in m), ρ_p is the **density** of the particle, ρ_b is the density of the beer (kg/m^3), and η is the **viscosity** of the beer (Pa·s). For gravitational settling, a is fixed at 9.8 m/s. The settling rate can be increased by encouraging the particles to combine, increasing the particle radius. As the particle radius increases (r^2), the settling velocity increases. This is accomplished during the brewing process by the addition of finings. The acceleration can be increased with a **centrifuge**. The centrifugal acceleration is given by $a = 4\pi^2 f^2 R$, where R is the distance from the center of rotation and f is the spinning rate in rotations per second (RPM/60). The use of a centrifuge increases the settling rate and clarifies the beer faster. Higher beer viscosity (η) or density (ρ_b) makes settling slower.

Yeast Removal

Yeast can be removed by **sedimentation**, centrifugation, or filtration, all of which are enhanced by addition of finings. After fermentation, the beer is chilled, sometimes to near the freezing point. At the low temperature, most strains of yeast clump together and sink to the bottom of the vessel, a process called sedimentation. If the vessel has a cylindricoconical form, the yeast can be harvested for use in another fermentation. Sedimentation tends to be slow and, for some strains of yeast, incomplete.

Sedimentation is highly dependent upon the ability of the yeast to **flocculate**. The tendency of a yeast to flocculate is a defining feature of yeast strains. This information is typically provided by the yeast supplier. Flocculation typically occurs during the **stationary phase** of fermentation as cell division ceases. As the cells stick together, their sedimentation rate increases, and they sink to the bottom of the **CCV**. Highly **flocculant** yeast improves downstream clarification processes, since there is less material that needs to be removed.

Changes in flocculation character may be caused by:

- Yeast health – Genetic drift or mutation may reduce the ability of yeast to flocculate.
- Calcium ions in wort – Flocculation is calcium dependent.
- Premature yeast flocculation (PYF) – PYF may be caused by fungal contamination of barley used for malting.
- Contamination by non-flocculant yeasts or bacteria.

Centrifugation

Sedimentation is usually followed by additional clarification steps. The modern method of choice for secondary yeast removal is the centrifuge. Beer is fed in at the center where it enters a system of rapidly spinning (>5000 RPM) bowls. The bowls accelerate the beer, greatly increasing the settling velocity. The yeast cells are ejected at the edges, and the clarified beer emerges at the center. The yeast is not contaminated with any filtration material and can be added to the **brewer's grains** for use as cattle feed. Centrifugation uses high-speed moving parts operating at close tolerance. Good maintenance is essential. The unit is noisy, and it may cause an increase in beer temperature. Poorly adjusted units may also increase the amount of dissolved oxygen in the finished beer. If the centrifuge were to be operated at the very high velocity that would be needed to completely remove haze particles from the beer, the mechanical forces could damage the yeast cells, increasing haze. For this reason, centrifuged beer usually requires additional filtration to achieve the desired brightness.

Finings

The sedimentation of yeast is greatly enhanced by the addition of a fining called **isinglass**. Isinglass is a **protein**-containing material derived from the swim bladders of certain tropical river fish. Isinglass molecules take a positive charge at beer pH. This allows them to trap yeast cells, which are negatively charged, causing them to **coagulate**. Isinglass is frequently used for **cask** conditioning. Isinglass may not be acceptable to **vegan** consumers and those who observe kosher laws.

The main source of haze in properly brewed beer, that is, free of **microbes**, **starch** particles, and processing aids, is the particles formed from the interaction of **polyphenols** and certain protein fractions. If either of these can be removed, the tendency of the beer to become hazy can be greatly reduced. Both polyphenols and proteins have some benefits, so their removal can have side effects. Excessive protein removal can reduce **head** retention (foam stability); excessive polyphenol removal can affect flavor stability.

The family of polyphenols most implicated in haze formation are proanthocyanidins. An example called procyanidin B3 is shown in Figure 10.6. The molecule is a dimer, the two parts of which are color-coded in the figure. Proteins that are haze active generally have a high content of the amino acid proline, shown in Figure 10.7. Proline is unique among amino acids in having the amino group as part of a ring. The presence of proline in a protein introduces a kink in the chain.

Most of the finings used in beer production are targeted to polyphenols or to haze-active proteins. The finings are added after yeast removal. The finings and the molecules they bind are usually removed by a filtration process. Specific finings and enzymes to assist in clarification are discussed in detail in Section 13.4.

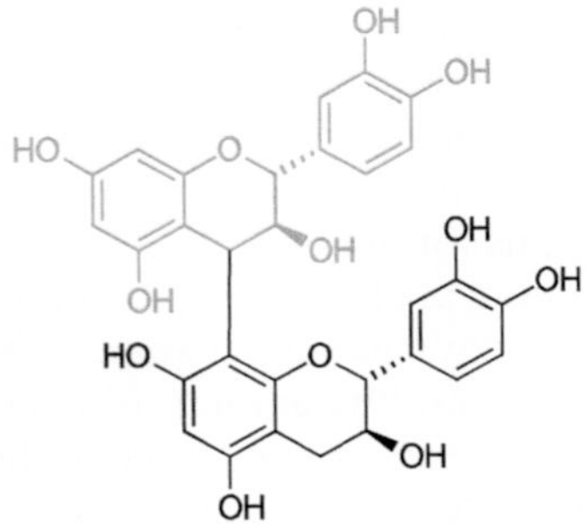

Figure 10.6   Polyphenol: procyanidin B3.

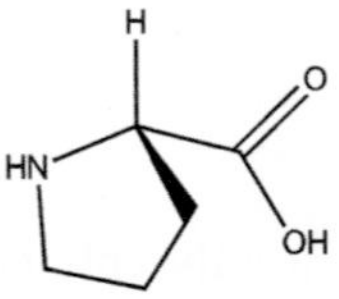

Figure 10.7   Proline.

Filtration

Filtration involves driving the beer through a porous material called the filtration medium. The medium traps particles that are larger than its pores. Filtration is used to remove particles at many stages of the brewing process, including processing of water, processing of syrup adjuncts, separation of wort, removal of yeast, removal of finings, clarification before packaging, and removal of microbes. The filtration process can be *depth filtration*, used in most brewery processes, or *membrane filtration*, used mostly for removal of microbes as an alternative to **pasteurization**. In depth filtration, the thickness of the filtration layer is much greater than the diameter of the particles targeted for removal. The particles thread their way through filter medium following a complex path with many opportunities to get trapped. One depth filtration method is to use a porous powder supported on a screen or cloth. The supported powder is called the **precoat**. Sometimes a precoat of finer particles is laid down first, followed by a top layer of coarser particles. This arrangement increases filtration capacity. Sometimes additional filtration medium powder, called **admix** or **body feed**, is added to the beer before it arrives at the filter. This increases filtration efficiency and capacity. Pressure drop in a filter is governed by **Darcy's law**, discussed in Section 7.2.

Membrane filters are thin sheets or tubes with pores of a specific size. Each trapped particle blocks a pore, so filtering heavy sediment with a fine-pore filter will lead to clogged filters and hazardous pressure buildup. Pressure gauges should be present upstream and downstream of all filter systems. The

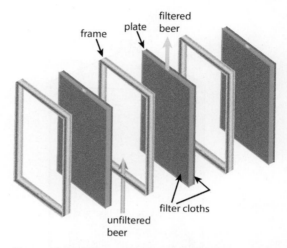

Figure 10.8   Plate and frame filter exploded view.

pressure upstream of the filter is called *back pressure*. The back pressure should be monitored throughout the filtration process. Any excessive increase in pressure indicates potential blockage of the filter. For this reason, filter size should be matched to the required filtration capacity. Membrane filters are usually used for final filtration when the concentration of particles to be removed is low. **Microbial filtration** is often performed with a series of filters, the last of which is a membrane filter with 0.45 μm or smaller pores.

Plate and Frame Filter/Sheet Filter   The plate and frame filter (Figure 10.8) is a less specialized version of the **mash** filter, shown in Figure 7.2. Filtration occurs on filter cloths attached and supported by plates. Each plate (except the ones on the ends) is a box with filter cloths on each side. Any beer that enters the plate must go through a filter cloth. The plates are separated by frames, which are spacers made of metal or plastic. The unfiltered beer enters the frames and crosses into the plates through the filter cloths. Usually the actual filtration occurs on filtration medium, typically **diatomaceous earth** (DE) or **perlite** coated on the filter cloths. A special type of plate filter is used with sheet filters. For sheet filtration, the frame is reduced to a spacer gasket. The filter cloth is a composite material made from **cellulose** fibers, and filtration media like DE or perlite with binders made of food-grade resins. Sometimes finings are included in the sheet filter composition.

Candle Filter   A candle filter (Figure 10.9) is a pressure vessel with parallel perforated tubes inside, called candles. The tubes serve as the support for the filter medium. A **slurry** of medium is fed in at the bottom of the vessel and

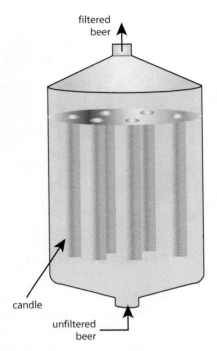

filtered
beer

candle

unfiltered
beer

Figure 10.9 Candle filter. In practice there are many more candles.

through the candles. The perforations trap medium on the outside of the candles, forming a precoat. The beer, which may be mixed with filtration medium (admix), follows the same path. Clarified beer emerges from the inside of the candles and flows out of the device.

Pressure-Leaf Filter   A pressure-leaf filter (Figure 10.10) consists of an array of plates, called leaves. Each leaf is covered on one or both sides by a fine screen. The screens are supported by coarse mesh to provide stiffening. The leaf is sealed at the edges. The inside of each leaf leads to a tube that is connected to a **manifold**. If the leaves are mounted vertically, there is usually a bar across the tops of the leaves to provide mechanical support. Horizontally mounted leaves are supported at the center by the manifold **pipe**. Before each run, the leaves are precoated with filter medium. Beer, possibly mixed with filtration medium (body feed), is fed into the filtration vessel. The beer is driven under pressure through the medium coating the leaves, through the screen, and into the tube to the manifold. The filter cake (solid that stays on the filter) can be removed by spraying or shaking. In some units with horizontally mounted leaves, the manifold pipe can be rotated by a motor, spinning off the filter cake. This system provides a fast cleanout cycle, but the rotating seals that allow the shaft to turn must be maintained meticulously to avoid leaks. The terms

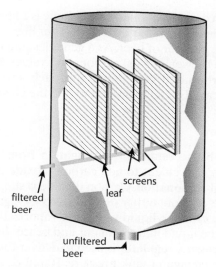

Figure 10.10   Pressure-leaf filter.

"vertical" and "horizontal" are often loosely applied either to the mounting direction of the filter leaves or to the configuration of the filtration vessel.

Filtration Media   Diatomaceous earth (DE), also called **kieselguhr**, filtration is very effective in removing particles from beer, including yeast and finings. DE is a porous mineral derived from skeletons of marine algae. The material is milled to give the correct particle size and sometimes acid-washed to remove undesired materials, like iron. DE can be used as a precoat and body feed, or it can be included in the composition of sheet filters. The DE traps particles in its pores. The major disadvantage of DE is that it is toxic and cancer causing when inhaled. Its exposure limit is $0.05\,\mathrm{mg/m^3}$, so precautions must be taken for safe handling. After yeast removal, the used DE–yeast **mixture** has no value and must be disposed of safely. Nonetheless, DE is generally recognized as a very effective filtration medium.

Perlite is **hydrated** volcanic **glass** that is expanded with heat and then milled. It is safer than DE, having a recommended exposure limit of $5\,\mathrm{mg/m^3}$ (100 times that of DE). Perlite, lacking the complex pore structure of DE, is less effective than DE at removing small particles from beer.

10.3   CARBONATION

Beer is usually served with dissolved carbon dioxide (CO_2) at a level of about 0.5% by **weight**. The flavor threshold for carbon dioxide is 0.1%, so carbon dioxide makes a major contribution to beer flavor. American brewers measure the amount of CO_2 in beer as *volumes*. One volume is 1 **liter** of CO_2, measured at 0 °C and 1 atmosphere (1.013 bar or 14.7 psia) per liter of beer. One volume is equal to 1.96 g CO_2 per liter of beer. As a general standard, open to stylistic interpretation of the brewer, normal carbonation values are 1.5–2.6 volumes, although highly **carbonated** beers may approach 4.0 volumes, requiring special bottles with caged corks to accommodate the elevated pressures.

Forced Carbonation

Although it is possible to retain the carbonation from fermentation, most brewers find it more convenient to add carbon dioxide after clarification, a process called **forced carbonation**. Carbon dioxide is supplied in cylinders as a liquid at a pressure of about 60 bar (870 psia). As CO_2 **gas** is drawn off from the **headspace** of the cylinder, liquid **vaporizes**. Vaporization absorbs heat, so the temperature of the liquid in the cylinder and hence the pressure decrease during delivery. A pressure **regulator** is used to deliver the gas at a controlled pressure despite the change in tank pressure. Rapid or excessive release of CO_2 from a cylinder may cause the regulator to freeze, blocking release.

Carbonation Volume

The carbonation volume in beer increases with pressure and decreases with temperature. For this reason, beer is always carbonated cold. The equation below relates the pressure in absolute bar at the liquid surface (P) to the carbonation in volumes (V) and the temperature in °C (t) for "average beer." To convert to atmospheres, multiply by 0.9869. To convert to psia, multiply by 14.50:

$$P = \frac{6167V}{\exp\left(\dfrac{2505}{t+273.15}\right)}$$

If you are not coding this equation into a computer, you may find it convenient to use Table 10.1. This table can be used to approximate the headspace pressure required to reach a certain carbonation volume at a specific temperature. Neither the equation nor the table should be considered exact. The solubility parameters of carbon dioxide in beer vary with the composition of the beer.

Equipment and Procedures

Carbon dioxide can be introduced into the beer by driving it through a porous "stone" made of particles of **stainless steel** or ceramic. The stone releases the gas in the form of small bubbles that offer a large surface of contact with the

TABLE 10.1 Pressures for Various Temperatures and Carbonation Volumes

Temperature (°C)	−2	0	2	4	6	8	10	12
Volumes CO$_2$	Absolute Pressure (bar)							
1.0	0.60	0.64	0.69	0.73	0.78	0.83	0.89	0.94
1.2	0.72	0.77	0.82	0.88	0.94	1.00	1.06	1.13
1.4	0.84	0.90	0.96	1.03	1.09	1.17	1.24	1.32
1.6	0.96	1.03	1.10	1.17	1.25	1.33	1.42	1.51
1.8	1.08	1.15	1.23	1.32	1.41	1.50	1.60	1.70
2.0	1.20	1.28	1.37	1.46	1.56	1.67	1.77	1.89
2.2	1.32	1.41	1.51	1.61	1.72	1.83	1.95	2.08
2.4	1.44	1.54	1.65	1.76	1.88	2.00	2.13	2.27
2.6	1.56	1.67	1.78	1.90	2.03	2.17	2.31	2.45
2.8	1.68	1.80	1.92	2.05	2.19	2.33	2.48	2.64
3.0	1.80	1.92	2.06	2.20	2.34	2.50	2.66	2.83
3.2	1.92	2.05	2.19	2.34	2.50	2.67	2.84	3.02
3.4	2.04	2.18	2.33	2.49	2.66	2.83	3.02	3.21
3.6	2.16	2.31	2.47	2.64	2.81	3.00	3.19	3.40
3.8	2.28	2.44	2.61	2.78	2.97	3.16	3.37	3.59
4.0	2.40	2.57	2.74	2.93	3.13	3.33	3.55	3.78
4.2	2.52	2.69	2.88	3.08	3.28	3.50	3.73	3.96
4.4	2.64	2.82	3.02	3.22	3.44	3.66	3.90	4.15
4.6	2.76	2.95	3.15	3.37	3.59	3.83	4.08	4.34

liquid, making the dissolving process faster. Sometimes membranes in the form of hollow tubes are used to introduce the CO_2. The membrane avoids some complications of two-phase flow. CO_2 can be introduced into a batch in the bright beer tank, or it can be dosed into the beer as it flows to the packaging unit.

Carbonation pressure in the bright beer tank is ideally measured in the headspace at the top of the tank. In practice, many brewers are not equipped for this, so they measure the pressure at the CO_2 tank regulator. The regulator pressure is always higher than the headspace pressure because of pressure loss in the carbonation stone and in other restrictions. Flow through a carbonation stone is influenced by capillary action. When the stone is in water or beer, the liquid is drawn into the pores. Pressure is needed to drive the liquid out. Smaller pores have a higher capillary pressure. If the pressure driving the gas through the stone is less than the capillary pressure, beer will be drawn into the stone and no gas will flow. To estimate the carbonation pressure, the brewer will measure the capillary pressure by placing the stone in a bucket of beer and adjusting the flow to give a curtain of small bubbles. At the onset of the bubbles, the pressure on the gauge will be the capillary pressure plus the variable resistance from friction in the stone and in any other places like a control valve in the line.

The hydrostatic pressure (in bar) exerted by the beer in the tank is about 0.1486 times the height of the beer (in meters) measured from the CO_2 inlet to the beer surface. In psia, this comes to 0.6567 times the height difference in feet. The pressure at the regulator is approximately the sum of the capillary pressure, the hydrostatic pressure, and the carbonation pressure. Suppose the target carbonation is 2.4 volumes and the temperature in the bright beer tank is 4 °C. The capillary pressure at the stone is measured at 0.35 bar. The carbonation stone is 0.5 m from the bottom of the tank, and the tank is filled to 3.0 m from the bottom. The carbonation pressure from Table 10.1 is 1.90 bar. The hydrostatic pressure is $(3\,m - 0.5\,m) \times 0.1486\,bar/m = 0.37\,bar$. The required delivery pressure is the sum of these three pressures: 1.90 bar + 0.35 bar + 0.37 bar = 2.62 bar. When the pressure at the surface of the bright tank is 1.90 bar, the pressure resisting flow will balance the applied pressure, and flow will stop. This method cannot be relied upon to give highly consistent carbonation.

It is ideal if the rate of introduction of gas is slow enough that the gas dissolves before the bubbles reach the surface. Excessive breakout of gas can lead to removal of flavor compounds and foaming that makes the beer hard to pump and package. Also, foam formation will use up foam-positive proteins that are needed for **head retention**. These proteins do not fully redissolve when the foam settles, so visible particles called **bits** can be left in the beer. Nonetheless, the deliberate purging of beer with gas is sometimes attempted to deal with issues like excessive oxygen or **volatile** off-flavors. It is a desperate measure, best avoided.

Nitrogen

In **nitrogenated** beer (also called nitrogenized beer), some of the carbon dioxide is replaced with nitrogen (N_2) to provide very stable foam with smaller bubbles. The creamier head is preferred in some styles by some drinkers. The foam character arises because nitrogen is only 64% as dense as carbon dioxide, and it is much less soluble. The effects of nitrogenation on beer foam are discussed in Chapter 13. On the negative side, nitrogen does not provide the characteristic tartness and prickly sensation of carbonic acid. Nitrogen is considered less appropriate for hop-forward beers because it does not seem to carry aroma as well as carbon dioxide. However, because there is less carbonic acid bite, which can obscure flavors, it may be argued that nitrogenated beer is more flavorful.

Nitrogen may be introduced in beer using the same techniques as for carbonation but using N_2 or a N_2/CO_2 mixture instead of pure CO_2. The standard N_2/CO_2 mixture for nitrogenated beer is about 75% nitrogen by volume (3 : 1 ratio). Because nitrogen is less soluble than CO_2, it requires greater headspace pressure to get the gas into **solution**. Liquid nitrogen is sometimes added to cans and bottles just before closing. Beer with nitrogen requires special techniques for draft delivery, discussed in Chapter 11.

10.4   BEER AGING

Aged Beer Categories

Aging is the extended storage of beer to allow characteristic flavors to develop. Aging is often accomplished in wooden **barrels** or in glass bottles after packaging. Few brewers have the resources to commit their permanent vessels for the three months or more required for aging. There are two general categories of aged beer. The first category is strong, conventionally brewed beers. These are sometimes aged, often in their bottles, to smooth out harsh flavors arising from ethanol and higher alcohols. Some examples are barley **wine** and imperial stout. The second category is beer that is exposed to or inoculated with microbes such as bacteria or wild yeast. These beers require a prolonged, multi-step aging process before release. Sometimes they are blended to help overcome the uncertainty inherent in microbial aging. Some examples are **coolship** ale, Flemish oud ale, and **lambic** ale.

The fermentation and aging of beers of the second category tend to occur in a sequence of phases in which different groups of fermentive organisms dominate. The sequence depends on the details of the wort processing, the nature of the microbes and how they get into the beer, fermentation temperature, season of the year, and even the geographic location of the brewery. In a typical case, the dominant organisms during aging and fermentation could be:

- *Enterobacteriaceae*: Consume sugar and make lactic acid.
- *Saccharomyces* and ***Brettanomyces*** species: Perform the main fermentation, consume sugar, and make ethanol and carbon dioxide.
- Lactic acid bacteria: Consume sugar and make lactic acid.
- **Acetic acid** bacteria: Consume ethanol and oxygen and make acetic acid.

These organisms are tolerant to oxygen; they can grow in casks and barrels that allow oxygen to enter. Acetic acid bacteria require oxygen; they get energy by reacting ethanol with oxygen. Apart from *Saccharomyces cerevisiae* (ale yeast), these yeasts and bacteria would be regarded as spoilage organisms in conventional beer. Brewers who intend to brew microbially aged beer in addition to conventional beer need to take precautions to avoid cross-contamination.

Barrels

Barrels are often used to hold beer during parts of the fermentation and aging processes. Some breweries that specialize in aged beer have large permanent wooden vessels called **foeders** [FOOD er]. Most aging barrels are made of American or European white oak. The barrel is built of narrow planks called **staves**. A groove called the **croze** is cut near the ends of the staves to accommodate the heads, which are the flat planks that make up the barrel top and bottom. The staves are bent with heat to make the barrel wider in the middle than at the ends. The staves are held with the long edges in contact under compression provided by hoops (see Figure 10.11). The inside of the barrel is usually subjected to a heat treatment that can vary from a light toasting to charring in a gas flame. The **bung** hole is drilled in one of the staves to let the product in and out. Wood shrinks on drying and expends when it absorbs moisture. Barrels that have become dry must be rehydrated or they will leak. Brewers often use barrels that have been used for other beverages. In the United States, barrels from Bourbon whiskey are most popular. United States standards of identity require that Bourbon be aged in new charred oak barrels. Because the barrel is new only once, it then becomes available for purchase by brewers (and **wine** and **spirits** makers) at an estimated wholesale price of $60–70 for a standard 200 L barrel plus the cost of shipping, which can be very significant.

The barrel exchanges material with the beer inside as well as with the surroundings. Traces of the previous contents of the barrel can flavor the beer. Oxygen leaks in slowly, affecting the beer directly by providing reactive oxygen species and by supporting the growth of oxygen-dependent microbes like acetic acid bacteria. Growth of acetic acid bacteria can be limited by ensuring the barrel is "topped up," meaning the barrel is completely filled to the bung. Over time, it may be necessary to top up with fresh beer due to loss from evaporation.

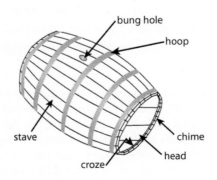

Figure 10.11 Barrel.

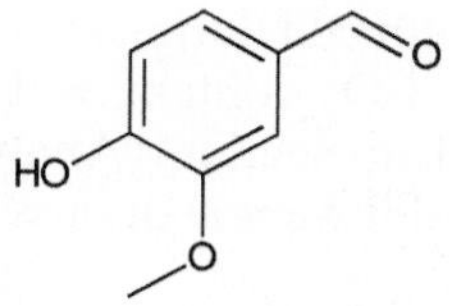

Figure 10.12   Vanillin.

During aging, volatile substances in the beer can escape, changing the flavor. The pores in the wood give shelter to microbes that produce flavor-active substances. The wood itself releases many flavor compounds such as vanillin (Figure 10.12), which has a vanilla flavor. The type of wood, its growing conditions, the seasoning of the wood, and the heat treatments to which it was exposed during barrel construction all affect the types and quantities of flavor compounds.

Wood is not as easily maintained as stainless steel; it is subject to mold and rot. If struck or dropped, it will crack rather than dent. A typical barrel has 50 or more joints that can spring leaks. The first step in barrel maintenance is to plan a place for the barrels. Ideally, they should be stacked on racks allowing individual access. The barrel cellar should be dry enough to discourage mold but not so dry that the wood shrinks. Around 75% relative humidity is recommended. Real installations usually deviate from ideality. One issue that should not be subject to compromise is identification of the contents of the barrel. The labels need to make the contents easily identifiable by the person sent to get the barrel, who may not even have been around when the barrel was filled months or years ago. Irrespective of what is on the label, it must remain attached and readable for the entire duration of the aging process.

Barrels are typically cleaned with steam or hot water. Their hoops need occasional adjustment to keep the staves tight. Small leaks through the wood can sometimes be repaired by the application of wax. Larger leaks can be repaired by driving in pegs. An installation with more than a few dozen barrels should employ a trained person to supervise their handling and maintenance.

Samples can be taken from barrels with a **sterile** pipette through the bung. When barrels are stacked, access may be difficult. Vinnie Cilurzo, brewer and owner of Russian River Brewing, has standardized the sampling of beer through a small port (2.8 mm = 7/64 in.) on the head stoppered with a stainless steel nail. The nail can be easily removed and resealed as sampling is required.

10.5 ANALYTICAL AND QUALITY CONTROL PROCEDURES FOR CONDITIONING

After conditioning, beer should be finished and ready for consumption. The most informative and universally applied quality control procedure for conditioned beer is tasting. This will be covered in more detail in Chapter 12.

Although carbonation is introduced during conditioning, it is normally tested after packaging. Measuring CO_2 content will be covered in Chapter 11. Analysis of haze is covered in Section "Analysis and Quality Control of Haze," Chapter 13. At this point we will discuss procedures for evaluation of ethanol and diacetyl content.

Beer contains varying amounts of carbon dioxide that can interfere with most analytical results. Before any analysis is undertaken, the beer must be effectively degassed. Many degassing methods involve agitation of the beer using techniques ranging from pouring it back and forth between two beakers to the use of ultrasound. Other methods involve filtration and the use of membrane technology. Some methods are specifically recommended for certain analytical methods. In any case, the same degassing procedure should always be used in your laboratory for a particular analysis to avoid the introduction of uncontrolled variables.

Ethanol

The classical and very time-consuming method for analysis of ethanol in beer is **distillation**. A measured sample of degassed beer is diluted with water and put into a laboratory still. The beer is distilled into an ice-cold receiver (to suppress evaporation of the ethanol) until about two-thirds of the beer–water mixture has been distilled. The distillate (liquid condensed after distillation) is essentially ethanol and water. Purified water is added to the distillate until the mass or volume is equal to the original mass or volume of the beer sample. The adjusted distillate should now have the same ethanol concentration as the original beer, but the only other **component** should be water. The mixture is then analyzed by measuring its specific gravity, refractive index, or some other property that varies in a predictable way with ethanol–water composition.

A more modern method is **gas chromatography** (Figure 10.13). The beer is degassed, an internal standard (typically 1-propanol) is added, the sample is quantitatively diluted, and then injected directly into the gas chromatograph. The sample mixture is vaporized at the injection port. A carrier gas (usually helium or nitrogen) carries the sample through a tube, called the **column**, that contains or is coated with a material that **attracts** and slows the flow of the components of the sample to different extents. The components reach the end of the column at different times depending on how strongly they stick to the column material. A **detector** at the end of the column responds to the **organic** components, including ethanol, as they arrive, providing an electrical signal to a data system.

One disadvantage of direct injection of beer is that upon drying in the injector, solids from the beer will deposit, requiring frequent maintenance. Instead of injecting beer directly, it is possible to put beer samples into sealed vials fitted with septa (rubber disks that allow a needle to enter). The vials are kept at a controlled temperature. An automatic system pulls a sample from

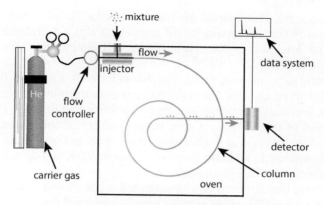

Figure 10.13   Gas chromatograph.

the headspace (air space over the liquid) in the vial and injects it into the carrier stream. The system will perform with less maintenance because the headspace sample does not contain the nonvolatile components of the beer; they stay in the vial. Gas chromatography is reliable, but it requires expensive equipment that takes a good deal of space, including space for potentially three gas cylinders, one for carrier and one each for fuel and oxygen for the detector. If hydrogen is used as the carrier (there is no reason that it should not be), then only two cylinders are required.

Several commercial instruments measure ethanol, mostly by **near infrared** (NIR) spectroscopy. NIR is light in the range of 780–2500 nm. Mathematical methods are applied to the **transmittance** to compute the concentrations of components of mixtures. Another instrument uses heat generated by catalytic combustion of the ethanol in excess air. These units are costly, but they do not take much space or consume expensive supplies.

Low ethanol levels (<0.5 ABV) like those in "nonalcoholic beer" can be determined by adding NAD$^+$, a biological oxidizing agent, and ethanol dehydrogenase, an enzyme. The ethanol reacts with the NAD$^+$ to give NADH and acetaldehyde according to the reaction $CH_3CH_2OH + NAD^+ \rightarrow CH_3CHO + NADH + H^+$. The absorbance at 340 nm is a measure of the amount of NADH, hence the amount of ethanol in the original sample.

Diacetyl

Rapid, non-quantitative methods for detecting diacetyl rely on sensory perception. The method is typically conducted for a relatively quick go/no-go decision. Two 200 mL samples should be collected in flasks that can be lightly covered (but not sealed). One sample is left undisturbed at room temperature. One sample is heated at 60 °C (140 °F) for 20 minutes and is then returned to room temperature. The samples are then sniffed for the presence of diacetyl. Heating causes the conversion of all remaining alpha-acetolactic acid into diacetyl. The detection of diacetyl in the heated sample indicates that more conditioning is needed.

For quantitative measurement, the brewer needs to know the sum of VDKs and their precursors. There are two colorimetric methods: one involves distillation, and one is a purge and trap (bubbling gas through sample and condensation of the VDK in a cold trap). Neither of these methods have generated much traction in the brewing community, as they require extensive time for preparation. In addition, they only measure VDK and do not distinguish between diacetyl and 2,3-pentanedione.

The preferred method involves headspace gas chromatography. The sample, a solution of copper(II) sulfate ($CuSO_4$), and a measured dose of 2,3-hexanedione (a VDK that does not occur in beer) as an internal standard are put into a sealed vial capped with a septum (rubber disk that can accept a sampling needle). The sample is held at 60 °C long enough for the copper ion to oxidize VDK precursors to VDK. An autosampler inserts a needle into the septum of each vial, takes a sample from the vapor at the top of the vial (the headspace), and injects it into the gas chromatograph. This method is amenable to automation, and it can distinguish between diacetyl and 2,3-pentanedione. It has the disadvantages mentioned for gas chromatography under ethanol analysis, plus the additional issue of requiring an electron capture detector, which contains a radioactive electron source. In the Unites States, electron capture detectors operate under licensing requirements from the Nuclear Regulatory Commission (NRC). For most breweries, the equipment supplier provides the license, but the user is restricted in how the unit is serviced and disposed of.

CHECK FOR UNDERSTANDING

1. Identify three objectives of beer conditioning.

2. Why is beer heated before testing for diacetyl?

3. Name five finings and discuss how they are used.

4. What types of compounds are responsible for haze? How can they be removed?

5. Give brief descriptions of three types of filters.

6. What are the roles of precoat and body feed in filtration?

7. Ten hectoliters of beer at 4 °C is to be force-carbonated to 2.6 volumes. The beer already contains 1.2 volumes from fermentation. What will be the final pressure? What mass of CO_2 will need to be added?

8. What are the three sources of pressure in a bright beer tank?
9. Explain the two categories of aged beer.
10. What are some compounds that would be expected in barrel-aged beer?
11. Make a diagram of a gas chromatograph.

CASE STUDY

You work at a 25 hL (20 bbl) brewery. Standard conditioning processes involve cold conditioning in the fermenter, centrifugation, plate and frame filtration, and carbonation in a bright beer tank. During the last filter run, you noticed an increase in the back pressure on the plate and frame filter. It was managed, but then the lab results indicated higher turbidity outside of the acceptable limits. Explain how you will identify, isolate, and resolve the problem.

BIBLIOGRAPHY

Barth R. 2013. The science of carbonation. *Zymurgy 36*(6):65–69.

Barth R. 2018. Beer Conditioning, Aging and Spoilage. In Bordiga M (editor). *Post-Fermentation and -Distillation Technology*. CRC Press. ISBN 978-1-4987-7869-5. Chap. 6.

Cantwell D, Bouckaert P. 2016. *Wood and Beer: A Brewer's Guide*. Brewers Publications. ISBN 978-1-938469-21-3.

Constant MD, Collier JE. 1991. Alternative techniques for beer decarbonation. *J. Am. Soc. Brew. Chem. 51*(1):29–35.

Lewis MJ, Young TW. 2001. *Brewing, 2nd ed.* Springer. ISBN 978-0-306-47274-9.

Lindemann B. 2009. Filtration and Stabilization. In Eßlinger HM (editor). *Handbook of Brewing*. Wiley-VCH. ISBN 978-3-527-31674-8.

Patino H. 1999. Overview of Cellar Operations. In McCabe JT (editor). *The Practical Brewer, 3rd ed.* MBAA.

Siebert KJ. 1999. Effects of protein-polyphenol interactions on beverage haze, stabilization, and analysis. *J. Agric. Food Chem. 47*(2):353–362.

Speers RA, MacIntosh AJ. 2013. Carbon dioxide solubility in beer. *J. Am. Soc. Brew. Chem. 71*(4):242–247.

Verachtert H, Derdelinckx G. 2014. Belgian acidic beers: daily reminiscences of the past. *Cerevisia 38*:121–128.

Work H. 2014. *Wood, Whiskey, and Wine: A History of Barrels*. Reaktion. ISBN 978-178023-356-7.

CHAPTER 11

PACKAGING AND SERVING

The days when a customer took **beer** home in a galvanized **steel** bucket are long gone. Modern beer is served through an elaborate dispensing system or sold in secure and attractive packages. After **conditioning**, beer is stored in the **bright beer tank**. It is cold, **carbonated**, and ready to drink. Small **breweries** may have several bright beer tanks located close to the bar to which the **draft** system is directly connected. Alternatively, the beer is sent from the bright beer tanks to a *large pack* system or a *small pack* system. Large pack refers to large serving containers like **kegs**, while small pack refers to single-serving containers like bottles or cans.

Packaging permits storage and transport of high-**quality** beer, ensuring consistent carbonation and protection from light, while providing an access point for dispense. Quality packaging is immensely important; once the beer leaves the **brewery**, it is beyond the brewer's control.

Mastering Brewing Science: Quality and Production, First Edition.
Matthew Farber and Roger Barth.
© 2019 John Wiley & Sons, Inc. Published 2019 by John Wiley & Sons, Inc.

11.1   KEGS AND CASKS

Kegs

The keg is the most common vessel for holding beer to be served by the glass. It maintains quality and carbonation of the beer by protecting it from air and light while also permitting quick and easy dispense. Modern kegs all use the Sankey system described in the box "KEG DETAILS." Older keg systems, like Golden Gate and Hoff-Stevens, have been largely abandoned. A Sankey keg is a cylindrical **pressure** vessel with a single hole about 5 cm (2 in.) in diameter at the center of the top. A short tube called the **neck** extends the opening outside the keg. The neck accommodates a dip tube, called a **spear**, extending to the bottom of the keg, as shown in Figure 11.1. The spear has concentric **valves** to accommodate **gas** and liquid. Most kegs are made of **stainless steel**. They can be made with aluminum with a protective coating inside, but aluminum kegs are more susceptible to theft for the salvage value of the metal. Table 11.1 gives dimensions and capacity for standard kegs. A full US half-**barrel** keg weighs 73 kg (160 lb).

The spear may be held secured to the neck by a threaded system or a drop-in system. The drop-in system is more common in American kegs.

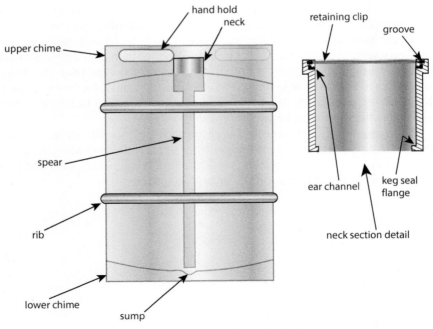

Figure 11.1   Keg with neck detail.

TABLE 11.1 **Typical Keg Dimensions**

Capacity	United States			Europe
	58 L (15.5 gal)	30 L (7.75 gal)	20 L (5.5 gal)	50 L
Name	Half barrel	Quarter barrel	Sixth barrel	
Height	594 mm (23⅜ in.)	594 mm (23⅜ in.)	594 mm (23⅜ in.)	532 mm
Diameter (incl ribs)	410 mm (16⅛ in.)	283 mm (11⅛ in.)	235 mm (9¼ in.)	408 mm

A drop-in spear is attached to the keg neck by lugs called "ears" on the spear body that are held by channels in the neck with a press and twist action. As a safety measure, in case the lugs work free of the channels, a spring-metal retaining ring, also called a circlip, fits into a groove in the keg neck above the top of the spear, preventing the spear from being ejected by pressure in the keg. Ejection of the spear is dangerous; it can cause injury to anyone in its path. The usual reason for spear ejection is tampering with the spear by customers, perhaps to perform an amateur refill of the keg, or improper servicing of the keg by workers lacking proper training, tools, or parts. Everyone who could be involved in filling or servicing kegs should be trained in the proper procedures and precautions. When working with returned kegs, extreme caution should be taken. The neck and top of the spear should be examined for gouges indicating screwdriver damage. The retaining ring should be examined for damage suggesting prying. Kegs showing evidence of tampering or improper servicing should be depressurized very carefully so that if the spear is ejected, it will cause no injury. Workers should be instructed never to put their heads immediately over the keg neck. Retaining clips are not to be reused. A new high-quality clip must be installed each time the spear is removed. The Internet features many dangerous procedures for working with kegs. Get legitimate training from an authoritative source.

The spear is attached to the beer dispense system with a keg **coupling**, a device that provides gas (carbon dioxide, sometimes mixed with nitrogen) to the headspace of the keg to drive the beer into the beer line, which is connected to another **fitting** on the keg coupling. The coupling attaches to the outside part of the spear, and a lever on it drives a probe into the spear that opens the gas and beer valves. A similar fitting is used to fill the keg with the connection reversed; one opening lets beer in and the other lets gas out of the keg. When used for beer service, kegs must stand upright.

Kegs are **cleaned** upside down. Cleaning **solutions** and/or steam is supplied from the dispense valve, flowing up the spear and spraying onto the "bottom" and sides of the keg. During cleaning the **CIP** solution should be pulsed at high and low velocities. This change in flow is also required to adequately cover the spear itself with cleaning chemical. The CIP chemicals or condensate is removed by the carbonation valve near the neck. After washing, the keg can then be **sterilized** with steam and evacuated and pressurized with carbon dioxide.

There are many types of keg couplings; they differ in the way the coupling connects to the spear. Each type of coupling is compatible only with a matching type of spear. The most popular coupling in the United States is the type D although the type G is also employed; many European beers come in type S kegs. It is most convenient if all the kegs used by a brewery have the same type of coupling. Filling and cleaning equipment use fittings based on the coupling configuration.

KEG DETAILS

The top of the Sankey keg has a hole about 50 mm in diameter. A short tube called the neck is welded to the hole. The spear, which fits into the neck, has the mating **surfaces** for the specific type of coupling for that keg. The top of the spear tube is surmounted by the spear body, which contains the concentric valves for beer and for gas. The spear body bears, via the keg seal, against a flange at the bottom of the neck, and it is secured at the top by threads or ears, so it cannot move up or down. Figure 11.2A shows the spear assembly in the closed position. The **ball valve** is pushed against the gas seal from below by a spring in the spear tube. A powerful spring outside the tube bears against a locking disk below pushing the entire spear tube, along with the gas seal, into a seat in the spear body. As the figure shows, the beer valve and the gas valve share the same polymer seal, but different surfaces of it couple with the seat of each valve.

When the keg is placed into service in a dispensing system, a keg coupling is installed. As shown in Figure 11.2B, a probe in the coupling pushes down on the beer valve ball, opening the beer valve. A seal on the probe engages the gas seal and drives the spear tube down against its spring, breaking the contact between the gas seal and the valve seat, opening the gas valve. The probe seal separates the gas from the liquid. Gas enters the coupling from the side, passes through the space created by the movement of the spear tube, and enters the **headspace** of the keg through openings in the spear body. The pressure of the gas drives the beer up the spear and into the probe, which is connected to the beer dispense lines. When the lever on the coupling is raised, the probe is withdrawn, and the springs close the beer and gas valves.

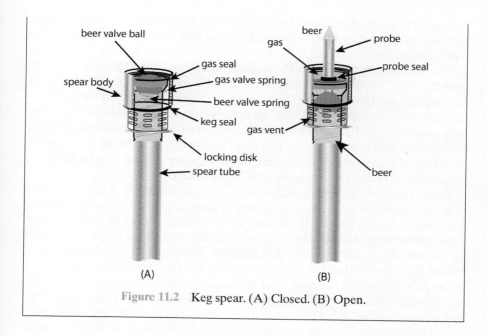

Figure 11.2   Keg spear. (A) Closed. (B) Open.

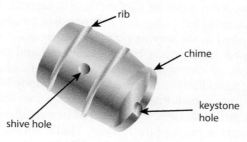

Figure 11.3   Cask.

Casks

Casks are modeled after the wooden barrels from which they originate; some are still in use. Wooden barrels are discussed in Chapter 10. The most common capacity for casks is the firkin (9 imperial gal = 40.91 L); a two-firkin cask is called a kilderkin; a half-firkin is a pin. Casks have no internal plumbing; they are just barrel-shaped vessels with two openings, the **shive** and the **keystone** holes, as shown in Figure 11.3. The ends of the cask are protected by rails, called **chimes**. Casks are usually made of stainless steel or passivated and epoxy-coated aluminum, but plastic casks are beginning to appear. The openings are named after the fittings that they accommodate. The shive hole, or bung hole, is on a curved side of the cask. The cask is filled and cleaned through the shive hole. It is closed with a shive, which is a stopper that has a provision for driving a hole, called the tut, through its center. The tut releases carbon dioxide and admits air. It can be fully or partly blocked with a peg, called the **spile**. Another opening on a flat surface of the cask is called the keystone hole. The keystone hole is placed so that when the shive hole is at the top, the keystone hole is near the bottom of the cask. The keystone hole is closed with a keystone, a fitting with provision for driving in a **tap** or faucet to connect the cask to a dispense system or to deliver beer directly to a glass. The shive and keystone are replaced each time the cask is filled; they are often disposable. They can be made of plastic, wood, or metal, sometimes with rubber parts.

Casks are used to prepare and serve cask-conditioned **ale**. After **fermentation**, the ale settles until the yeast count is correct, usually between 0.25 and 3 million cells/mL. The ale is run into casks through the shive hole. Priming sugar and a fining, traditionally isinglass or gelatin, are added to the beer either just before or just after it is put into the cask. The beer is left to ferment for a time at the brewery, then the shive is driven in, and the cask is delivered at once to the **pub**. On arrival the cask is set up, ideally horizontally, with the shive up and the keystone down on a rack called a **stillage**, or on wooden chocks. The cask is left undisturbed to allow the finings to clarify the beer. At the suitable time, a porous spile is driven through the shive, allowing carbon dioxide to vent at a controlled rate. After a day or two, the cask is tapped; a spigot is driven through a thin place in the keystone with a sharp rap of a mallet. Cask ale is not supposed to be pushed to the serving point by gas pressure; it is either fed by gravity or is **pumped**, traditionally by bar-mounted hand pumps. As beer is taken from the cask, air enters; thus the cask must be finished before the air spoils the beer. For casks that are not expected to be emptied within a few days, it is permissible to use a **cask breather**, which is a valve that admits carbon dioxide instead of air. Cask ale is a British tradition requiring a high level of cooperation between the pub and the brewery. Had it not been for a consumer movement called the **Campaign for Real Ale (CAMRA)**, cask ale would be a shadow of its current status. In addition to CAMRA, there is a professional organization called Cask Marque that provides inspection of and training about cask ale and other aspects of beer production and service.

Preparing and Filling Kegs and Casks

Kegs are usually filled with beer that has been subjected to a brief high temperature treatment called **flash pasteurization**, to be discussed in detail in Chapter 14. Many craft brewers avoid pasteurization; they clarify the beer by some combination of **sedimentation**, **centrifugation**, and **filtration** and package it without pasteurization. Casks are filled with unpasteurized, unfiltered beer. Before a keg or cask is filled, it must be inspected and cleaned. The cleaning process is invariably mechanized (although not necessarily automated) in the case of kegs. There are mechanized systems for casks, which can save time and supplies for breweries that have the cask volume to justify them. The preparation process for kegs follows a series of steps:

1. The keg is inspected for visible damage. The spear body is checked for tightness and evidence of tampering. The keg is tested to make sure it can hold pressure. It is unsafe, as well as a waste of money, time, and beer, to attempt to fill a keg whose condition is questionable.

2. Labels and marking on the outside of the keg are removed. The outside is scrubbed with hot **detergent**, often recirculated. The detergent is rinsed off with a spray of hot water.

3. The keg is mounted upside down on a keg **racking** machine. The machine has a fitting that functions like a keg coupling, opening the gas and beer valves in the spear.

4. The keg is emptied. This process is sometimes called de-**ullage**. Compressed air is driven into the beer port and up the spear. Any beer remaining in the keg exits at the gas port.

5. A series of washings (**caustic** and acid) and rinsings are performed. The cleaning solutions are driven into the beer port up the spear, so they spray the bottom of the keg. Liquids are driven out the gas port by compressed air.

6. The keg is sanitized with steam or a chemical **sanitizer**.

7. The keg is purged with carbon dioxide and then filled with CO_2 under pressure.

8. Beer is driven into the keg against the CO_2 pressure.

9. The keg is dismounted and turned right side up. A plastic cover is put on the spear opening. The keg is labeled with the type of beer, the best by date, and other required information. A barcode is often used for identification. This helps with proper tracking of beer lots for consumer complaints or potential recalls. The filled kegs are placed on **pallets** and wrapped with plastic for shipment.

The process is similar for casks, but casks have two openings, so liquids can go in one and out the other. Also, casks are not usually pressure-tested, so sterilization with steam is not possible.

11.2 GAS LAWS

Beer is dispensed under gas pressure. To understand the issues involved in dispensing beer, we need to look at the universal behavior of gases. Many of the same issues that apply to packaging beer also apply to beer **foam**, covered in Chapter 13.

Ideal Gas Law

The **molecules** of a gas are so far apart that they barely interact, except for collisions. In many ways they behave like featureless balls flying about the container; one molecule acts much like another. Under moderate conditions of pressure and temperature, a gas follows a universal equation, called the **ideal gas law** (IGL) that relates pressure, volume, amount (moles), and temperature. Some consequences of the IGL are that in a fixed-volume system, like a closed fermenter or a steam jacket, if the temperature goes up or down, the pressure follows. If the amount of material in the gas phase goes up or down, the pressure follows. As an example, if the steam to the heating jacket of a vessel is cut with no provision for letting air in or out, when the steam condenses (less gas), the pressure will drop, and the resulting vacuum could damage the vessel.

The mathematical formula of the IGL is $pV = nRT$. The temperature, T, must be the Kelvin (absolute) temperature, which is the temperature in degrees Celsius plus 273.15 K. R is the gas constant, whose value depends on the units. The pressure must be absolute pressure (see Section "Direct Draw Systems" for an explanation of gauge and absolute pressure). If the pressure is given in atmospheres and the volume in **liters**, the value of the gas constant is $0.082\,06\,L\,atm\,K^{-1}\,mol^{-1}$. If the pressure is in bar and the volume in hectoliters, $R = 8.3145 \times 10^{-4}\ hL\,bar\,K^{-1}\,mol^{-1}$. If the pressure is in pascals and the volume in cubic meters (pure **SI** units), $R = 8.3145\,J\,K^{-1}\,mol^{-1}$. The IGL has limitations that must not be overlooked. It works for gases only. It will not work for liquids or solids in the presence of their **vapors** or gases in the presence of liquids in which they dissolve. Carbon dioxide is delivered as a liquid. The cylinder pressure does not follow the IGL at all.

IGL EXAMPLE

200 mg of liquid nitrogen (molar mass = 28 g/mol) is vaporized at 25 °C in a sealed can whose headspace volume is 20 mL. Calculate the pressure, assuming the solubility of nitrogen in the liquid is practically zero.

$$p = \frac{nRT}{V}$$

$n = (0.200\,g)/(28\,g/mol) = 0.0071\,mol;\ T = 25 + 273\,K = 298\,K$
$p = (0.0071\,mol)(0.082\,06\,L\,atm\,K^{-1}\,mol^{-1})(298\,K)/(0.015\,L)$
$p = 11.6\,atm$. Of course, the can will fail before all the nitrogen vaporizes.

IGL EXAMPLE

Carbonation is specified by "volumes," that is, liters of CO_2 (measured at $0\,°C$ and $1\,atm$) per liter of beer. Calculate the mass of CO_2 (molar mass = $44\,g/mol$) in $355\,mL$ ($12\,oz$) of beer that is carbonated to 2.7 volumes.

Solution:

$V = (2.7\,L\ CO_2/L\ beer) \cdot 0.355\,L\ beer = 0.96\,L\ CO_2$

$$n = \frac{PV}{RT}$$

$n = (1\,atm)(0.96\,L)/(0.082\,06\,L\,atm\,K^{-1}\,mol^{-1} \cdot 273.15\,K)$

$n = 0.043\,mol;\ mass = 44\,g/mol \cdot 0.043\,mol = 1.9\,g$

Dalton's Law

In a gas **mixture**, each component of the mixture acts independently of the others, unless they react. We can use the IGL to calculate a partial pressure for each gas as though it alone occupied the entire volume. The total pressure of the mixture, which is what the pressure gauge shows, is the sum of the partial pressures. This is Dalton's law. An important consequence of Dalton's law is that in a gas mixture, the volume fraction is equal to the mole fraction. For example, 100 total moles of a mixture of 75% nitrogen and 25% carbon dioxide (by volume) will have 75 mol of nitrogen and 25 mol of carbon dioxide. For many purposes, a gas at a certain **partial pressure** in a mixture exhibits the same behavior as if it were unmixed at the partial pressure.

DALTON'S LAW EXAMPLE

Calculate the mass percent of nitrogen in "beer gas" that is 75% N_2 and 25% CO_2 by volume.

Solution:

Take $100\,mol$ total, which comes to $75\,mol\ N_2$ and $25\,mol\ CO_2$.

Mass $N_2 = (75\,mol) \cdot (28\,g/mol) = 2100\,g$

(*continued*)

Mass CO_2 = $(25\,mol) \cdot (44\,g/mol) = 1100\,g$

Total mass = $2100 + 1100\,g = 3200\,g$

Mass % N_2 = $(2100\,g/3200\,g) \cdot 100\% = 66\%$

Mass % CO_2 = $(1100\,g/3200\,g) \cdot 100\% = 34\%$

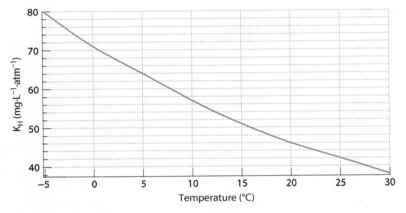

Figure 11.4   Henry's law constant for oxygen.

Henry's Law

Henry's law states that the solubility of a gas in a liquid is in direct proportion to the partial pressure of the gas. Mathematically, $c = kp$, where c is the concentration, p is the absolute pressure, and k depends on the gas, the liquid, and the temperature. The solubility of a gas in a liquid always decreases as the temperature increases. The numerical value of k depends on the units chosen for concentration and pressure. If the concentration is known, Henry's law can be used to determine the pressure. Figure 11.4 shows the Henry's law constant for oxygen in water, where the partial pressure of oxygen is measured in atmospheres and the concentration comes out in mg/L. Table 10.1 provides carbon dioxide pressures for different carbonation levels at different temperatures.

HENRY'S LAW EXAMPLE

See package oxygen example in Section "Measurement of Beer Packaging Quality."

11.3 KEG DISPENSE SYSTEM

The dispense system is the apparatus that gets the beer from the keg to the customer's glass. The beer must arrive with the correct carbonation, at the correct temperature, with the correct amount of **head** (foam), and with no deterioration of its **flavor**. Even the most skillfully **brewed** beer can lose its appeal if the dispense system is not doing its job. The essential parts of a dispense system are the chilling system, compressed CO_2, the keg coupling, the beer line, and the beer faucet. Dispense systems are divided into temporary, direct draw, and long draw draft systems.

Temporary Draft Systems

Temporary draft systems are used at events like parties, beer festivals, and the like. Hand pumps permit draft dispense direct from a keg. From a **coupler**, the beer is transferred through a vinyl hose to a plastic picnic faucet. Gas pressure for transfer is generated through a hand pump that drives air into the keg. The major disadvantage to hand pumps is that the air causes the quality of the beer to deteriorate quickly; hands pumps are only recommended when the beer will be consumed in a day. An improvement on this design is the integration of a small, single-use CO_2 cartridge that primes the keg.

Another temporary system, superior to the hand pump, is the **jockey box**, commonly used at festivals and other temporary locations. A standard coupler is attached to the keg and to a small, portable CO_2 tank. The beer line is directed through a cooler in which a plate chiller or copper coil is placed in ice. Faucets for dispense are attached to the cooler wall. The large surface area of the chiller allows the beer to be cooled in the line during hot summer months or long events. The cooling capacity of a cold plate is limited; it should be used for beer that comes from the keg at less than $13\,°C$ ($55\,°F$). Pressure is controlled with a regulator on the CO_2 cylinder. Due to the flow resistance of the chiller, higher pressure (1.3–2 bar, gauge = 20–30 psig) is often necessary for satisfactory dispense. For this reason, jockey boxes are not permanent fixtures, as elevated carbonation levels in the keg would result. It is recommended that the beer line be set up before ice is added to the jockey box to prevent excessive chilling, hence excessive carbonation of the beer.

Direct Draw Systems

In a direct draw system (Figure 11.5), the beer is driven by carbon dioxide at the pressure that corresponds to the partial pressure of carbon dioxide at the keg temperature at the correct degree of carbonation. Direct draw is used only when short draft line distances are possible. A common example of

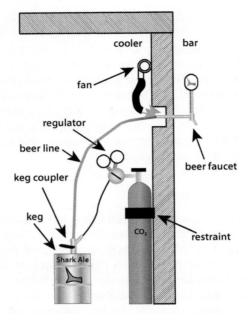

Figure 11.5   Direct draw system for a tap room.

short draw systems in craft breweries or taprooms is the placement of a walk-in cooler behind the bar. The faucets are then installed with their shanks through the wall of the cooler to be easily accessible by the bartender. In direct draw systems, only 1–1.5 m (4–5 ft) of 5 mm (3/16 in) line can be used for a balanced system. Longer lines will suffer from pressure loss and slow dispense rates.

Because of the short distance from the keg to the faucet, the pressure drop driving the flow is minimal, so the pressure at the faucet will be essentially the same as the pressure at the keg. If this condition is not satisfied, carbon dioxide will escape from the beer into the line, causing foaming and undercarbonation. Table 10.1 has carbonation pressures at various temperatures and degrees of carbonation. The complication is that gas pressure gauges nearly always read in gauge pressure, which is the pressure difference between the local pressure of the atmosphere and that of the gas. At sea level, the atmospheric pressure is usually close to 1.01 bar (14.6 psi). The pressure falls off with elevation at a rate of 0.113 bar per 1000 m (1 psi/2000 ft). The local pressure must be subtracted from the absolute pressure to get the gauge pressure.

In direct draw systems the service pressure may be too high, causing the beer to flow too quickly. This can be corrected by attaching a length of narrow bore tubing, called a choker, to restrict the flow rate. Unless the faucet is installed on an outer wall of the refrigerator, the beer line will need to be kept cool. A flow of air from the refrigerator may be adequate, depending on the layout. Otherwise a **python** line will be needed, as described below under long draw systems.

ELEVATION EFFECT ON GAUGE PRESSURE

Suppose beer is to be served at $2\,°C$ ($36\,°F$) with a carbonation level of 2.5 volumes in Boulder, Colorado (elevation $1655\,m = 5430\,ft$). At sea level, the gauge pressure would be $1.71 - 1.01 = 0.70\,bar$ ($10.2\,psig$). At $1655\,m$, the gauge pressure would be $0.70 + 0.113\,bar/km \times 1.655\,km = 0.89\,bar$ ($12.9\,psig$).

Long Draw Systems

If the distance or elevation change from the keg to the faucet introduces a significant pressure change, the dispense system must be designed as long draw. Figure 11.6 shows a simplified long draw dispense system with gas mixing. In this system, the primary regulator feeds a **manifold** with secondary

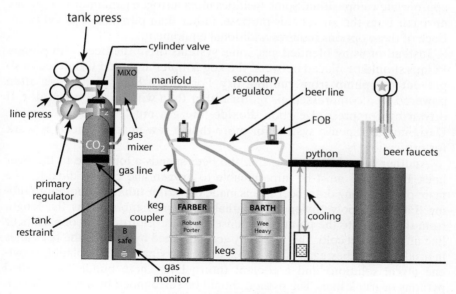

Figure 11.6 Long draw dispense system.

regulators, allowing different pressures for different kegs. The beers lines are connected to foam-on-beer detectors (**FOB** detectors) to shut off the connection to the beer lines when the keg runs out. The FOB detectors keep the lines from filling with foam, saving beer. The FOB is a float ball valve that seals when empty. The beer lines, from the cooler to the tap, are kept cool by a python arrangement, described below.

In a long draw system, extra pressure is needed to drive the beer. If all the pressure were provided by carbon dioxide, the beer would become overcarbonated, so a gas mixture is used. To avoid overcarbonating the beer, nitrogen, an inert gas, is mixed with the carbon dioxide. The carbon dioxide maintains the correct level of carbonation, and the nitrogen provides pressure but does not carbonate the beer. Suppose, for example, that the required carbonation pressure is 1.71 bar absolute, but we need 2.44 bar absolute to drive the beer through the system at the correct flow rate (usually $60\,mL/s = 2\,fl\,oz/s$). The extra pressure, 0.71 bar, would be made up with nitrogen. Dalton's law holds that the pressure fraction is equal to the volume fraction, so the mixture would be $1.71/2.44 = 0.70 = 70\%\ CO_2$. This is not the same as the mixtures typically used to deliver **nitrogenated** beer. The mixtures may be provided in a single cylinder but are often mixed from two cylinders with a gas blender mounted in-line.

Because of limitations mentioned below about compressed gases, it may be convenient to mix pure carbon dioxide and nitrogen on-site to produce the appropriate **composition**. Some facilities use a nitrogen generator to separate nitrogen from the air for this purpose, rather than purchase it in cylinders. Each of these options requires additional equipment.

Instead of using blended gas, some systems drive the beer with pumps. Pumps should be placed between the keg coupler and the FOB detector to prevent the pump from running dry. For safety, beer pumps are often powered by a compressed air motor rather than with an electric motor. If driven by compressed carbon dioxide, the gas must be vented outside. Designing a dispense system for more than a few nearby faucets is work for professionals.

Another long draw issue is that the beer spends a long time in the beer lines. The tubing must be impermeable to oxygen and carbon dioxide. The main part of a long draw beer line is made of **barrier tubing**, smoothbore tubing lined with polyethylene terephthalate. Barrier tubing runs the length from the cooler to the point of dispense. Also, the entire length of each beer line must be kept cold. This is usually accomplished by running the beer lines in insulated bundles, each of which has a line carrying chilled coolant (propylene **glycol** solution) and a coolant return line. These bundles are called pythons or trunk lines. The python should be surrounded by a vapor barrier, especially at any open ends, to avoid condensation.

ABOUT FAUCETS

Beer faucets are the final part of the dispense system. The two major types are rear sealing, shown in Figure 11.7, and front sealing. In all faucets, the beer flows when the dispense lever is pulled forward (to the left in Figure 11.7). The dispense lever rotates around a ball that is positioned with the ball washer. Above the ball washer is a friction washer that tightens the dispense lever to keep it from flopping about. Pressure is maintained on the washers by a nut called the bonnet (not shown). In the rear-sealing valve, the dispense lever engages the valve shaft, which has the valve seal washer at its back end (on the right in the figure). When the bartender pulls the dispense lever forward, the valve shaft is pushed back, pushing the washer off the valve seat and admitting beer, as

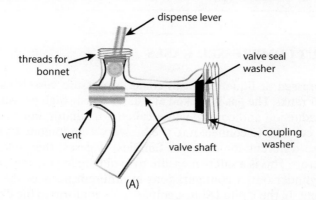

(A)

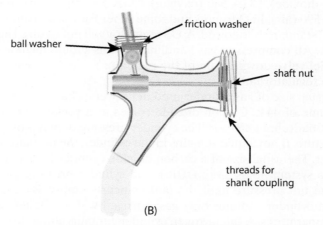

(B)

Figure 11.7   Beer faucet. (A) Closed. (B) Open.

(continued)

shown in Figure 11.7B. Front-sealing faucets have no valve shaft. The dispense lever has a sealing element that covers the faucet opening when the lever is pushed back.

The faucet mates with a stout threaded fitting called a shank (not shown) that seals with the coupling washer. The shank is usually firmly attached to a panel or tower and enables a connection between the faucet and the beer line. Shanks come in different lengths; a shank should be matched to the width of the wall through which it will pass. A coupling nut (not shown) screws to outside threads on the faucet, sealing the faucet to the shank. The shank and the faucet have a key arrangement to keep the faucet from rotating when it is attached. The back end of the shank has a hose fitting or other means to connect it to the beer line. Faucets should be removed and cleaned regularly to prevent loss of beer quality from **bacteria** and mold.

ABOUT COMPRESSED GASES

Compressed or liquefied gases come in pressure vessels called cylinders or tanks. The gas is packed at a dangerously high pressure. A pressure reduction and control valve, called a regulator, must be attached to the cylinder to connect it to the brewery's equipment. Cylinder outlets are different for different families of gases; they take different regulators. This is a safety measure to avoid mixing incompatible gases. The cylinder outlet configurations in different parts of the world are different. In the United States, connections conform to the Compressed Gas Association (CGA). Typical CGA designations are CGA 320 (carbon dioxide), CGA 540 (oxygen), CGA 580 (nitrogen), and CGA 510/346/590 (air). In Germany and some other European countries, the DIN 477 standard is followed. A US regulator will not fit on a European gas tank. All compressed gas handling systems should include a **pressure relief valve** downstream of the regulator to vent the gas safely in case the regulator fails.

The usual size of gas cylinder used in breweries has a nominal internal volume of 44 L. Carbon dioxide comes as a pressurized liquid at about 60 bar (about 850 psi). The cylinder pressure depends only on the temperature. If any liquid remains in the cylinder, the pressure will still be 60 bar. The usual size of a carbon dioxide cylinder for a permanent dispense system holds 23 kg (50 lb). The cylinder pressure gauge will show the full pressure until the tank is nearly empty. By contrast, a nitrogen cylinder contains only gas; nitrogen will not liquefy at ordinary temperatures. A full nitrogen cylinder contains about 7.6 kg (17 lb) at a pressure of about 150 bar (2200 psi). The internal pressure is proportional to the amount of gas remaining in the cylinder.

Gas mixtures of carbon dioxide and nitrogen are restricted to pressures low enough to prevent the carbon dioxide from becoming a liquid. If liquid forms in the cylinder, the release of gas becomes a distillation process. Nitrogen will be enriched in the gas phase and the gas composition will change as gas is withdrawn. To avoid this, the CO_2 partial pressure should not exceed 60 bar. For a mixture of 70% carbon dioxide and 30% nitrogen by volume, the total pressure must be kept below 78 bar. This is about half the pressure of a pure nitrogen cylinder, hence half the gas by volume.

Compressed Gas Safety

Safety considerations for gas cylinders are essential for the workplace. Gas cylinders contain a large amount of energy that can cause personal injury and property damage if improperly released. In addition, carbon dioxide is toxic at concentrations above 1%, and nitrogen can displace air, leading to an oxygen-deficient atmosphere. The following guidelines are recommendations for safe handling practices regarding compressed gas.

Gas cylinders should always be secured with restraints, like straps or chains. Falling cylinders have caused personal injury by landing on an employee. Also, if the valve strikes something and breaks off, the cylinders can rocket across a room, causing major damage. To prevent damage when not in use, cylinders should be secured with the supplied metal cap called a valve protector. The valve protector should always be in place when the tank is stored or moved and whenever it is not in service and attached to a regulator. A cylinder is most vulnerable when a regulator is in place. A fall could damage the regulator and abruptly release the gas with potentially deadly consequences. When the cylinder is to be moved, the regulator must be removed and the valve protector installed. A specialized hand truck with provision for securing the cylinder must be used. Rolling the tank for more than a few centimeters is an unsafe practice.

Brewery and taproom carbon dioxide sources should always be food grade. "Food-grade" designation is not about the purity of the gas but rather the cleanliness of the tank. A new tank of food-grade gas should have a shrink-wrap plastic cover on the tank outlet to prevent contamination. This must be removed before the regulator is installed. All cylinders are to be used with regulators. Although the pressure provided by the tank may change with contents and temperature, the regulator will

(continued)

maintain a constant pressure at the regulator outlet. The cylinder valve must never be opened without a regulator in place. The regulator attaches to threads (outside or inside, depending on the type of gas) on the tank outlet. The threads drive the mating surfaces of the regulator and the tank outlet together, but they do not make the seal. Some regulators seal by direct metal-to-metal contact of a convex gland with a concave cup. Some have flat mating surfaces and use a plastic washer to make the seal. The threads on the cylinder outlet and the regulator inlet are never part of the seal. Thread compound and thread tape (Teflon®) must never be used on cylinder threads; they can get into the regulator mechanism. Because a regulator is a mechanical device, it is subject to failure. There must be a pressure relief valve not far downstream to vent the gas safely should this happen.

Gas monitors must be in place in enclosed spaces in which compressed gas is used. For toxic gases, like carbon dioxide, the monitor should measure the concentration of the gas and should activate an alarm when the concentration exceeds the limit. For inert gases, the hazard is displacement of oxygen. Oxygen must be monitored to ensure a viable atmosphere. The alarm should be activated when the concentration of oxygen is below the acceptable limit. The monitors must be regularly calibrated with known mixtures to ensure that they are properly working. In-line gas monitors are also available. When placed in an easily visible place between the primary regulator and the secondary regulator(s), they can be useful to check for leaks or open gas valves.

ABOUT NITROGEN GENERATORS

Nitrogen generators for beer bars operate on either of two principles. In a pressure swing adsorption (PSA) system, clean, oil-free, dry compressed air passes at 4–8 bar over particles of carbon molecular sieve (CMS). CMS is prepared by heating polymers under conditions that give carbon with a network of well-controlled pores. The pore size is chosen so the smaller molecules, including oxygen, can enter the pores and stick to the internal surface. The larger nitrogen molecules stay outside and pass through the bed of particles. Selectivity is not perfect, but after multiple encounters with the CMS, good nitrogen purity can be attained. In time, the pores fill with oxygen and the efficiency diminishes. The CMS is regenerated by lowering the gas pressure, allowing the gas inside the pores to be released. During regeneration, the gas is discarded through a noise suppressor. Meanwhile, a second vessel is brought up to high pressure to provide a continuous stream of purified nitrogen. Each vessel is on stream for a few seconds to several minutes, and then it is regenerated. Most systems have a buffer tank to hold compressed nitrogen to level out flow variations. The other type of nitrogen generator relies on membranes that are selectivity permeable to oxygen. The membrane takes the form of a bundle of tubes. Clean, dry, oil-free air enters the tube under pressure. The oxygen escapes through the wall of the tube and the nitrogen is retained to exit at the end.

Either type of generator requires a robust, oil-free compressor and a manifold to clean and dry the inlet air. Some units are integrated with a compressor and gas mixing manifold that can be connected to a carbon dioxide tank. The output of the unit is the appropriate mixture of carbon dioxide and nitrogen. Nitrogen generators can routinely generate nitrogen of 99.7% purity, which is suitable when the nitrogen is to be used only to maintain dispense pressure. For nitrogenated beer dispense, where nitrogen dissolves in the beer, a higher purity may be desirable to minimize oxygen pickup. Generators with enhanced purity are available.

Nitrogen generated in-house is less expensive than nitrogen delivered in cylinders. Less handling of heavy, bulky nitrogen tanks saves labor and improves worker safety. About half of the gas cylinders used for beer dispense can be eliminated if nitrogen is generated on-site, even more if the gas had been supplied premixed or if nitrogenated beer is served. Nitrogen generators can be purchased or leased.

Nitrogenated Beer   Nitrogenated beer can provide a long-lasting head of very tiny bubbles that is suitable for some styles, primarily stout. The beer has low carbonation, 1.1–1.7 vol, and a tiny amount of dissolved nitrogen, typically around 60 mg/L, which comes to 0.05 vol. Nitrogenated beer is driven from the keg at about 2.5 bar gauge pressure (35 psig) with 25% CO_2 and 75% nitrogen. The faucet for nitrogenated beer has an insert that forces the beer through holes that provide about 1.4 bar (20 psi) of pressure drop and introduce turbulence that makes the beer foam. For reasons that we will discuss in Chapter 13, the resulting foam is composed of very small bubbles enriched in nitrogen.

Nitrogenated beer gas should never be used to dispense standard beer. Looking back at the example for the long draw system operating at 2.44 bar absolute with 70% CO_2, if 25% CO_2 were used instead, the CO_2 partial pressure would be 0.61 bar. This is less than half of the 1.71 bar required to keep the beer carbonated. If the total pressure were increased to give a CO_2 partial pressure of 1.71 bar, it would come to 6.84 bar absolute, or 5.83 bar gauge

(85 psig). This is enough pressure to blow off the hoses or to knock the glass out of the bartender's hand if the hoses stay on.

Canned or bottled nitrogenated beers use a widget to raise foam. The widget is typically a small plastic hollow ball with a small opening (0.6 mm). When a beer is bottled or canned, the gas builds up pressure inside of the package. Gas and beer will enter the widget until pressure equilibrium is reached. When the package is opened, the pressure in the container quickly drops, but the pressure in the widget is forced through the small opening. This turbulent jet of gas and beer from the small widget opening creates a thick head of foam.

11.4 BOTTLES

Glass bottles are viewed as prestigious packaging by consumers. They are mechanically stiff, making them easy to handle without damage. On the other hand, they are brittle and shatter into sharp shards that represent a safety concern. Several product recalls have been required because of potential glass inclusion from breakage on the packing line. Like other ceramics, glass is stronger under compression than tension. Unfortunately, the pressure inside a beer bottle keeps it under tension, so the walls must be thick. Bottles are heavy, and they take a lot of space to ship, store, and stock. Labels or plastic sleeves are often applied, adding a step to the packaging process.

The most common size for a beer bottle in the United States is 355 mL (12 oz). European bottles are often 330 mL (11.2 oz). Large format bottles, called "bombers," are also routine at 651 mL (22 oz) in the United States or 500 mL (16.9 oz) in Europe. The usual 12 oz long-neck bottle, shown in Figure 11.8, is 229 mm (9.02 in.) high and 61.3 mm in diameter and weighs 170 g (6.0 oz). The bottom of the bottle has an indentation called the punt that strengthens the bottle against the internal pressure. The bottles are sealed with a **crown cap** that crimps around a raised bead called the **finish** at the top of the bottle. The standard cap has 21 puckers called teeth. The inner surface that contacts the beer has a **polymer** liner. The seal is made with a plastic (PVC) washer in the cap, as shown in Figure 11.9. The cap is applied by bending down the rim to grip the bottle. The seals on bottle caps are slowly permeated by oxygen, which can cause the beer to go stale. There are caps whose linings absorb oxygen, potentially prolonging beer stability.

Plastic has been the material of choice for carbonated soft drink bottles but is uncommon for beer. Polyethylene terephthlate (PET) bottles are used in some European markets but have yet to take root in the United States. There are technical issues including the ability of plastic to retain carbonation, to exclude oxygen, and to withstand pasteurization. Improvements in plastic technology are ongoing. Some major brands are introducing plastic bottles, especially in large sizes. It is likely that the market share of plastic will grow as technical issues are resolved and consumers become accustomed to the new material.

Figure 11.8   Bottle.

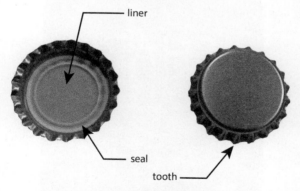

Figure 11.9   Crown cap.

11.5   CANS

About 60% of beer is packed in aluminum cans and 30% in bottles, and 10% is served on draft. The market share of cans has been increasing at the expense of bottles. Craft beer has been following this trend more slowly; about 20% of craft beer is packaged in cans, 50% in bottles, and 30% on draft with the can segment rising. It is interesting that newer, smaller breweries are packaging in cans at a much higher rate than larger, established breweries. This is likely the result of companies who manufacture canning lines with smaller footprints and lower prices.

In addition, cans have several important advantages for beer packaging. They are light, about 30 g (1 avoirdupois ounce); they admit no light or oxygen; their large opening allows rapid filling; they have a shape that can be packed and shipped efficiently; and, if enough are ordered, they can be decorated with labeling specified by the brewer, obviating the need for a labeling process (Figure 11.10). Alternatively, a sticker or shrink-wrap label can be applied to blank cans. Because a can is smaller and lighter than a bottle, more cans can be shipped and stored in the same space as bottles, reducing freight costs and maximizing warehouse storage. The disadvantages of cans include a negative (but fading) consumer perception compared with other packaging options, concern among some consumers about the epoxy lining material, and the susceptibility of empty cans to damage in the packaging process. Small breweries may find it difficult to buy and store enough cans to satisfy a minimum order.

Modern cans are made of an alloy containing about 1% each of magnesium and manganese in aluminum. The inside of the can has an epoxy coating

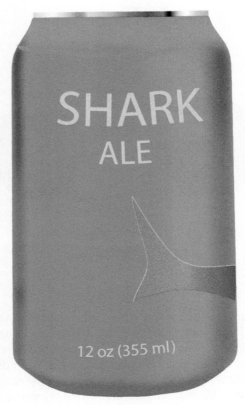

Figure 11.10   Can.

to keep the beer from making direct contact with the metal and **corroding** it. The can comes in two parts: the body, which includes the sides and bottom of the can, and the top, which has a scored opening with a riveted pull tab that allows the consumer to open the can without a tool. Empty can bodies come flared at the top to form a flange. The lid edge is bent down to form a hook. Inside the hook is a thin coating of a latex-based sealing compound.

11.6 FILLING BOTTLES AND CANS

The bottle and can filling operation in a brewery is easily the most mechanically complex, breakdown-prone, and exasperating activity in the brewery. It represents yet another site for beer loss and an essential area for quality control to ensure the best product possible will enter the market. Before the bottles or cans arrive at the filling machine, they must be removed from their pallets; bottles must be unpacked from their boxes; and the bottles or cans must be rinsed and drained. The filling operation must be carried out at high speed and precision and in a way that prevents entry of oxygen. Packaged beer should have no more than 0.2 mg of oxygen per liter (200 parts per billion). Most brewers try for about half of this. If the beer is **flash pasteurized** or sterile filtered, the filling operation must prevent the entry of microbes. Bottle and can packaging follows similar concepts with slightly different technical processes due to differences in the package.

The following are the typical steps of a packaging system in order:

1. Depalletize
2. Rinse
3. Fill
4. Pasteurize
5. Label
6. Pack
7. Palletize.

Depalletizing/Uncrating

Bottles arrive at the brewery on pallets that are moved about with **fork lifts**. Most countries have regulations requiring that all employees who will operate fork lifts be trained by an accredited organization to operate them safely.

New, non-returnable bottles arrive at the brewery in cardboard trays stacked on pallets. The pallets are loaded onto a **conveyor** and carried to the depalletizing machine. The machine grips an entire layer of trays and places them on a conveyor to the uncrating machine. Alternatively, the entire row of bottles is pushed forward onto a conveyor, and the entire pallet is then raised into position to transfer the next row. The cardboard trays are often removed by an automated suction cup device to be saved and returned to the glass supplier. Cans are arranged on pallets in layers separated by sheets of cardboard.

Some facilities receive bottles prepacked in branded boxes. This requires an uncrating machine to remove the bottles from the box. The uncrating machine has an array of grippers that pick up all the bottles in a tray. The gripper lifts the bottles clear of the trays and moves them from a conveyor to the bottle sprayer.

Rinsing

After depalletizing, the package must be rinsed free from any dust or debris. Can bodies from the depalletizer are herded along the conveyor between rails until they travel single file. At this point, the conveyor takes the form of a wire cage constraining the can. The cage twists to turn the cans upside down. The insides of the cans are rinsed with a spray of water. Another twist of the cage brings the washed cans to the upright positions for filling.

Bottles from the uncrating machine are herded into lanes and conveyed to a row of devices that grip each bottle and invert it. For new bottles, a jet of clean water, possibly containing a sanitizer, removes any dust. For returned bottles, a much more elaborate cleaning procedure with dedicated equipment is needed. The process includes label removal, soaking, hot chemical cleaning using a high-pressure jet, and several stages of rinsing. This practice is not common in the United States.

Both returned and new empty bottles may be subjected to computerized inspection to look for **defects** in or damage to the glass. The rejection rate for returned bottles is typically around 2%. Potential defects in glass bottles include:

- Split finishes with vertical cracks around the seam.
- Blisters caused by bubbles or air pockets in the glass.
- Inclusions of foreign material.
- "Bird swings" or glass webs, which are threads of glass across the interior.
- Variations in glass thickness caused by irregular glass distribution.

Filling

Fillers for bottles or cans can take the form of a rotating carousel with multiple fill heads, parallel in-line units each with a single head, or manual units in which the packages are mounted and removed by an operator. An overview of the filling operation is shown in Table 11.2. Bottles are usually lifted

TABLE 11.2   Filling Sequence

	Bottles	Cans
1	Center and seal bottle to filler head	Center can at filler head
2	Evacuate air from bottle (some systems)	
3	Purge bottle with CO_2	Purge can with CO_2
4	Repeat steps 2 and 3 (some systems)	Seal can flange to filler head (pressure from above)
5	Fill bottle with CO_2 to equalize pressure	Fill can with CO_2 to equalize pressure
6	Fill bottle with beer; beer pushes out CO_2	Fill can with beer; beer pushes out CO_2
7	Any bubbles allowed to rise (settling)	Any bubbles allowed to rise (settling)
8	Lower CO_2 pressure to atmospheric (**snifting**)	Lower CO_2 pressure to atmospheric (snifting)
9	Jet with deaerated water to foam	Blow CO_2 across the top
10	Cap the bottle	Cover the can and **seam** on the lid
11	Rinse any beer or foam from the bottle, dry	Rinse any beer or foam from the can, dry

into filling position, rather than the filler heads moving to the bottles. For can filling, the head often comes down over the can to minimize movement of the more fragile aluminum can. Most filling systems use counterpressure for filling, whereby the container is first pressurized to the same pressure as the beer to avoid foaming and loss of carbonation. Beer is then transferred only by gravity with the headspace gas in the bottle transferred back to the headspace in the filler.

Bottle fillers may deliver the beer through a tube that goes all the way to the bottom of the bottle, called a *long-tube filler*, or through a short tube with a deflector that directs the beer down the walls of the bottle, called a *short-tube filler*. These devices are designed to regulate fill height. For long-tube fillers, the filling of the bottle stops once the bottle is filled completely. Removal of the long tube then displaces a certain volume of beer in the package. The displaced volume by the tube is exactly the headspace volume in the bottle. A short-tube filler stops filling when the beer reaches the vent tube, blocking the transfer of CO_2 back to the tank. Without the ability to displace the gas, the beer stops filling. The vent tube is placed at the height required for final fill height.

Cans are filled through multiple short tubes to take advantage of their wider opening for rapid filling. Some can fillers include a simple ball float valve to regulate the fill height. When the ball floats on the rising beer, it seals the gas transfer valve, stopping the transfer of liquid. Mechanically raising or

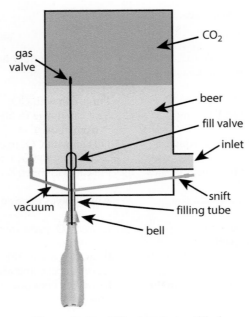

gas
valve

CO₂

beer

fill valve

inlet

snift

filling tube

vacuum

bell

Figure 11.11   Filler head, simplified.

lowering the float valve regulates the fill height. Specified volume filling can be by accomplished through direct volume control by delivering the beer from a straight-sided doser that has a constant inside area, so the level of the beer is a true measure of the volume. A level sensor in the doser controls the start and stop of the fill. Another method is to deliver beer at a defined flow rate. The fill is then regulated by flow time. Figure 11.11 shows the functional parts of a bottle filler head.

Bottles and cans must be accurately filled to avoid shorting the customer and for accurate government reporting. In advanced fillers, a sensor is used to scan the exact fill height of a package. Inappropriately filled containers are rejected and pushed out of the line. In smaller or more manual fillers, fill height is simply checked by **weight** through random sampling.

At the end of the fill step, the counterpressure is released, an operation called snifting. Immediately after filling, the beer is at risk for entry of air from the surrounding environment. To remove air from the neck and to prevent oxygen pickup by the beer, a jet of deaerated water is shot into the opening. Such agitation causes immediate CO_2 nucleation and foaming. The foam displaces the air with CO_2. At this point the package is immediately sealed before excessive beer is lost to foam.

The capper or **seamer** can either be part of the filling unit, or it can be located very close, so the package is closed within a few seconds of filling. Crown caps are supplied in bulk. The capper must organize them to face the correct direction. The cap is placed on the bottle, and a cone-shaped bell crimps it against the finish of the bottle (see Figure 11.8). Can lids come in stacks with the lids oriented in the same direction. The lid is usually dropped onto the can as it leaves the filler. The full can with its lid then moves into a seamer that seals the lid with two operations, one to wind the can body flange and the lid hook around one another and the other to flatten and seal the seam. Details of the seaming process are given in the box "SEAMING A CAN." Caps and lids are not ordinarily sanitized at the brewery. They must be handled with precautions to prevent contamination. Cappers and seamers require cleaning and sanitation.

SEAMING A CAN

After a can is filled, the lid is sealed in a two-operation process called seaming, shown in Figure 11.12. The lid is placed on the full can, which is brought to a platform on the seaming machine. The platform raises the can into the seaming head. A disk-shaped tool, called the chuck,

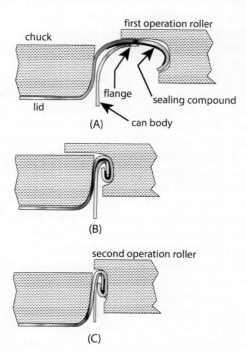

Figure 11.12 Seaming a can. (A) Section view of sealing surfaces of can in seaming head. (B) First seaming operation. (C) Second seaming operation.

(continued)

engages the lid and serves as the anvil for the seaming process. Another disk-shaped tool, called the first operation roller, approaches the closure from the side. The roller has a rounded groove that engages the can body flange and lid hook, as shown in Figure 11.12A. In the first seaming operation, the roller presses the flange and hook into a loose spiral, as shown in Figure 11.12B. The can and the roller spin so the entire circumference of the can is subjected to the first operation. The first operation roller is withdrawn, and the second operation roller approaches. This roller has a groove with a shallower profile than the first roller. It presses the seam flat, completing the seal, as shown in Figure 11.12C. The roller and chuck are withdrawn, and the platform lowers to discharge the sealed can.

Labeling

After the beer is sealed in its primary packaging, the bottles or cans may be subjected to a pasteurization tunnel. Pasteurization is covered in detail in Chapter 14. After pasteurization, bottles will need labels. In addition, all packaging needs batch identification and date (packaging or expiration date). This is usually printed with an ink jet printer directly on the neck of the bottle or the bottom of the can.

Labels must stand up to handling and condensation. They must be applied accurately to give a professional appearance. Labelers are mechanical devices that must repeatedly and accurately perform a complex sequence of actions. The two main types of labelers are glue applied and pressure sensitive. Glue-applied labelers are more complex, because they must handle glue as well as labels and bottles. The labeling process for glued labels is less expensive than that for pressure-sensitive labels. Also, there are more choices in the type of adhesive to use. Glued labels can be applied to cold or wet bottles, with the correct type of glue. In many operations, the bottles are dried quickly by **air knives** before labeling to remove post-filling rinse water. Pressure-sensitive labels can give refined effects, for example, the label can be printed on a transparent background giving the impression of printing directly on the glass.

A typical rotary, glue-applied labeler has two rotary tables: the machine table and the container table. The machine table has the label magazine, the glue roller, and several paddles called gluing pallets. When the motion of the table brings a pallet to the glue roller, glue is applied. Usually the glue is applied in a pattern of lines. The pallet with glue moves to the magazine, which applies a label to it. The pallet with the label now applied continues to travel until it reaches the gripper cylinder, which transfers the glued label from the pallet to a bottle on the container table. The gripper cylinder has fingers that pick up the label through indentations in the pallet. It then carries the label to the bottle. The bottle moves to the next station where rollers or brushes press it to the bottle. Some labelers use hot-melt glue applied directly to the bottle. In this case, the bottle itself acts as the pallet, accepting glue from the glue roller and labels from the magazine. Some labelers have a provision to add batch information to each label. Label application should be checked throughout the packing process to ensure consistency.

Cans are usually preprinted with labels by the bulk supplier. The major drawback for small brewers is that the typical minimum order is about 200000 cans, all with the same labeling. This comes to about 7000 hL of beer. For a moderate-size craft brewery, this can be more than 10% of annual output. There are secondary suppliers that can provide shorter labeling runs, at a significant increase in price. Smaller canning operations use unlabeled cans with labels applied as described above or even potentially by hand. Unlabeled packaging provides more flexibility to use the package across multiple brands.

Secondary/Tertiary Packaging

Packaged beer is shipped in boxes or cases. The cases are usually printed with the design theme of the brewery and of the beer brand. Often the bottles or cans are restrained in units of four or six within the case. Packaging for the packages is *secondary packaging*. The standard case for bottles holds 24 units. Beer case blanks come flattened with only the side seam glued. They are erected and glued at the bottom, usually with hot-melt. A packer with gripper heads picks up the bottles and places them in the open case. Alternatively, a drop packer gently drops the bottles into the case with as little mechanical agitation as possible. Another unit glues and closes the case. Sometimes bottles are packed in cardboard multipacks, usually six packs. The familiar cardboard bottle carriers are open at the top. They can be set up in the case before the bottles are inserted. When multipacks are not used, cardboard separators may be placed into the case to prevent the bottles from rubbing against one another. Another way to prepare the case is to first stage the cans or bottles and then to wrap the case blank around them. This method provides tighter packaging, allowing bottles to be packaged without multipacks or separators.

Cans do not come in carrier packs, and cardboard separators are not used. Often cans are restrained in a plastic web called a six- or four-pack ring. The rings are stretched onto the cans over cones that allow the rings to clear the seam and grip the neck of the can. The cans, with or without rings, are packed into cardboard boxes. Alternatively, the cans can be placed on open cardboard trays.

Tertiary packaging refers to assembling consumer-sized cases into larger units for shipping. This is also referred to as palletizing. The cases are usually stacked on pallets to a maximum of about 1000 kg (2200 lb). The cases must be secured to one another and to the pallet with strapping tape or shrink wrap. This prevents dangerous and expensive accidents resulting from inevitable shifts in the load. Beer packaged in kegs or casks also needs secondary packaging. Kegs and casks should be stacked on the pallet no higher than two layers. As with cases, they must be secured to the pallet and to one another.

Packaging Line Design

Bottles and cans are carried by a conveyor system through each of the key steps of the packing process. Stainless tabletop conveyors are used in many industries but are not commonly used for beer, because they require lubrication to reduce bottle-to-conveyor friction and drag. Plastic tabletop or mattop conveyors are more common because they do not require lubrication. Belted or roller conveyors may be used for transportation of secondary packaging. Package conveyor systems should include accumulation tables. Accumulation before or after a certain process is needed to allow continued operation of other machines. To stop the entire bottling line in the event of one malfunction would be very costly and hard on the machines.

Each packaging operation should be capable of slight increases or decreases in speed through a variable frequency drive. The primary machine in any packing operation is typically the filler. The speed of the filler is the metric upon which the rest of the line is designed. Typically, the further from the filler, the faster the equipment should be running. This "v-curve" design allows the line to continue operation in the event of a small stoppage. For example, the machines upstream of the filler can speed up after a stop in order to catch up to the filler and to ensure the filler always has bottles to fill. The machine downstream of the filler can speed up to ensure the filler always has space for bottle discharge. In a hypothetical scenario to further illustrate the v-curve, suppose the depalletizer runs at 400 bpm (bottles per minute), the rinser at 300 bpm, the filler at 100 bpm, the labeler at 200 bpm, and the packer at 300 bpm. If one operation fails, slight adjustment to speed of the others will ensure the filler maintains operation. Here the use of accumulation tables in between each operation is critical. Efficient operation of a packaging line will minimize downtime and maximize efficiency.

11.7 BEER SERVICE AND PACKAGING QUALITY

Beer Dispense Quality

Hygiene   The key issue in beer dispense quality is hygiene. All parts of the system that contact beer must be cleaned and sanitized on a regular schedule, generally every two weeks. Long, elaborate draft systems are more difficult and expensive to clean. The system should be designed for minimum length. This will save time, chemicals, and beer. Cleaning solutions are necessarily corrosive. Worker training and equipment must be in place to minimize the frequency and severity of accidents. Perhaps the most important safety precautions are procedures that eliminate any possibility of inadvertently delivering cleaning chemicals to a customer's glass. Many bars use outside vendors to clean their beer dispense systems.

Microbial contamination can jeopardize the quality of beer at the point of dispense. Beer dispense lines are a common source of off-flavors, as certain bacteria can take up residence in the line by forming biofilms. Routine cleaning every two weeks as described above should prevent unwanted bacterial growth; otherwise the lines will need to be replaced. **ATP luminometer** testing, as discussed in Chapter 14, can also reveal the presence of microbes at various points in the system. Another issue in draft lines is transfer of beer flavors especially after changing from one style to another. For example, a cinnamon-spiced winter warmer is followed by a pilsner. The pilsner will likely suffer from cinnamon contamination. If flavors from strongly flavored beer are turning up in mildly flavored beer, the beer lines should be replaced. The *Draught Beer Quality Manual*, cited below, has detailed recommendations for beer line cleaning and maintenance.

Gas Leaks

WARNING: A carbon dioxide or nitrogen gas leak can be a deadly hazard in a poorly ventilated space, like a keg cold room. When a leak in an enclosed space is suspected, the atmosphere should be checked for breathability with a gas monitor. The maximum allowable carbon dioxide concentration is 1% by volume; the minimum required oxygen concentration is 19.5%.

Gas leaks can injure workers, waste gas, cause unexpected interruptions in service (always at busy times), and cause the pressure in parts of the dispense system to be lower than intended. Low pressure can lead to slow, foamy dispense, beer loss, and a bad customer experience. The main challenge in repairing leaks is to find them. This is done in a step-by-step methodical procedure using the regulator gauges.

Close all keg couplers (pick their handles up). Open the valves from the regulator to the manifold. If the system has secondary regulators, set them to deliver their normal pressure. Open the gas supply valve. Set the main regulator to deliver its normal pressure. Close the gas supply valve. The entire gas system is now pressurized, but there is no new source of gas. If gas is leaking, the reading on the upstream pressure gauge will decrease slowly as gas escapes through the leak. If the upstream pressure is steady for at least five minutes, there is no leak in the line. This does not account for leaks in the couplers or kegs, once they are engaged.

If a leak is discovered, the exact source can be determined by applying soapy water to fittings that may be leaking. Bubbling will reveal the source of the leak. The usual problem is the high-pressure fitting connecting the regulator to the supply. Check the washer (for a CO_2 regulator) or the gland (for a nitrogen regulator). This check should be performed whenever a tank is changed or other actions are performed on the high-pressure side of the regulator.

Measurement of Beer Packaging Quality

Packaging affords many opportunities to ruin good beer. The filling equipment must be designed to be easily cleaned and sanitized. The staff must be well trained and committed to quality. There are several tests that can help maintain and check consistency in packaging quality. Not included here is microbiological testing, which is covered in Chapter 14.

Measuring Carbon Dioxide   Carbon dioxide is a primary flavor component. Many brewers measure CO_2 in every batch, either at the bright beer tank, in the package, or both. In either case, a sample must be obtained with no loss of pressure, and the CO_2 levels must be consistent for the style. Samples for measurement of carbonation in packaged beer are obtained by piercing the package. Bottles are pierced through the cap; cans are pierced through the bottom. The usual method for CO_2 analysis is to fill a pressure cylinder with the beer, shake it to establish gas–liquid equilibrium, and measure the pressure and temperature. The carbonation level, in volumes, can be read from a table like Table 10.1. Oxygen at 200 ppb will give 0.16 mL of gas per liter of beer at room temperature. Nitrogenated beer at 30 ppm will give 27 mL of nitrogen gas. The normal carbonation level for beer is over 2000 mL gas per liter.

Several proprietary methods are available that claim to be more selective for carbon dioxide in the presence of other gases. One method is to increase the volume of a pressure/vacuum vessel containing a fixed sample of beer (volume expansion method), and another is to collect gas through a membrane and measure the thermal conductivity.

Measuring Package Oxygen   The usual limit for total package oxygen (TPO) is 200 µg/L (200 ppb w/v). Less is better. A major problem with measuring package oxygen is that inadvertent exposure to the atmosphere will cause erroneous results. Air contains about 270 000 µg/L of oxygen. Sampling for oxygen should be conducted with a piercing device, as discussed above, to avoid introduction of oxygen during sampling.

Another sample collection problem is that oxygen in packaged beer has two phases: a solution phase (in liquid beer) called *dissolved oxygen* and a gas phase (in the headspace) called *headspace oxygen*. If the sample is mixed well by shaking, the concentrations of oxygen in the two phases will be related by Henry's law. In a newly filled bottle, the phases have not yet come to equilibrium, so the concentrations of the gaseous headspace and the dissolved gas in beer are independent. It is important to take these measurements immediately after filling; otherwise the concentrations will approach equilibrium. The sum of the headspace oxygen and the dissolved oxygen is the *total package oxygen* (TPO). A comparison of TPO and the individual measurements for the headspace and dissolved oxygen at packaging can be helpful to isolate sources of oxygen pickup. Excessive headspace oxygen suggests issues in the filling operation. Dissolved oxygen implicates processing after the fermenter up to the bright beer tank, including filtration and centrifugation.

Oxygen meters usually measure dissolved oxygen; they operate on one of two principles. The electrochemical oxygen sensor has a membrane that is selective to oxygen. Electrical pulses are sent to contacts that can interact with any oxygen. The oxygen drives the electrochemical reaction $O_2 + 2H_2O + 4e^- \rightarrow 4OH^-$, where e^- represents an electron. The electric current is a measure of the oxygen present. A major advantage of the electrochemical sensor is that a single sensor can measure dissolved oxygen in the range from 0 to 200 ppm, easily covering the dissolved oxygen expected in pitching wort (8–15 ppm) and in packaged beer (20–200 ppb). A disadvantage of the electrochemical sensor is that it consumes the oxygen, so the measurement drifts down with time. The electrochemical sensor requires a good deal of maintenance.

The fluorescence quenching **detector** has a fluorescent polymer, one side of which is in contact with the fluid being measured. Blue light shines on the other side of polymer. When no oxygen is present, the polymer absorbs the blue light and releases red light. When oxygen is present, it suppresses the production of red light, a process called *quenching*. Less red light means more oxygen. The optical sensor is fast and resistant to interference and drift; it requires minimal maintenance. The major disadvantage is that different probes are needed for the different ranges of oxygen concentration in wort and packaged beer.

CALCULATING DISSOLVED AND HEADSPACE OXYGEN

To calculate dissolved oxygen (DO), headspace oxygen (HSO), and total package oxygen (TPO), the dissolved oxygen must be measured before shaking (DO_{still}) and after shaking for five minutes (DO_{shake}). The sample temperature must also be measured. The mass of dissolved oxygen in mg, before or after shaking, is the dissolved oxygen concentration in mg/L multiplied by the liquid volume in

(continued)

L. DO after shaking is related to the headspace oxygen by Henry's law: $DO_{shake} = K_H \times$ pressure (atm), so the partial pressure of oxygen in the headspace will be pressure $= DO_{shake}/K_H$. The mass of oxygen in the headspace is determined by combining this with the ideal gas law:

$$n = \frac{pV}{RT}$$

The moles are multiplied by 32000 mg/mol, the molar mass of O_2; the volume is divided by 1000 to change milliliters to liters; and $R = 0.08206\,L\,atm\,K^{-1}\,mol^{-1}$, the gas constant, leading to

$$mass_{mg} = \frac{390 \cdot DO_{shake} \cdot V_{mL}}{K_H\left(t_C + 273.15\right)},$$

where DO_{shake} is the measured oxygen concentration in milligrams per liter after shaking, V_{ml} is the headspace volume in milliliters, K_H is Henry's law constant from Figure 11.4, and t_C is the sample temperature in °C. The total package oxygen mass is the sum of the masses (not concentrations) of dissolved and headspace oxygen. The total package oxygen mass is the same before and after shaking, so the before-shaking headspace oxygen is the difference between the total package oxygen mass determined after shaking and mass of dissolved oxygen before shaking. TPO concentration is the total package oxygen mass divided by the volume of beer.

Example: A can with 355 mL (=0.355 L) of beer and 20 mL of headspace has 185 ppb dissolved oxygen at 5 °C before shaking and 155 ppb after shaking. Calculate the total package oxygen and the headspace oxygen before shaking.

Solution:

Set up a table to organize the calculation.

	Before Shaking	After Shaking
Dissolved	A measured conc × volume	B measured conc × volume
Headspace	C total – dissolved	D equation
Total	E same as after shaking	F sum dissolved + headspace

A: Dissolved oxygen before shaking is the concentration times the volume (185 ppb = 0.185 mg/L): 0.355 L × 0.185 mg = 0.066 mg.

B: Dissolved oxygen after shaking is the concentration times the volume: 0.355 L × 0.155 mg/L = 0.055 mg.

D: Headspace oxygen after shaking is determined from the equation above, where K_H at 5 °C is 64 mg/L/atm, from Figure 11.4. The equation gives 0.068 mg oxygen in the headspace.

F: TPO is the sum of B and D: 0.055 mg + 0.068 mg = 0.123 mg. On a per liter basis, this comes to 0.123 mg/0.355 L = 0.346 mg/L = 346 ppb.

E: Same as F.

C: E − A: 0.123 mg − 0.066 mg = 0.057 mg.

The results are summarized in the table below.

	Before Shaking	After Shaking
Dissolved	A 0.066 mg	B 0.055 mg
Headspace	C 0.057 mg	D 0.068 mg
Total	E 0.123 mg	F 0.123 mg
TPO conc	346 ppb	346 ppb

Closure Testing   For crown caps, the diameter of the crimped cap is measured as an indication that the correct crowning force was applied. The crimp tester is simply a plate with several holes that differ in diameter by 0.1–0.2 mm. Some gauges are go/no-go gauges. These have a series of holes that span the crimp size specification range. If the small no-go holes will accept the cap (too tight), or the large no-go holes will not accept the cap (too loose), the crimp is out of specification. Small crimps may be excessively tight, leading to damage to the glass upon opening. Large crimps are too loose and may leak or admit oxygen.

Can seams are made in two operations. The accuracy of each of the seaming operations should be regularly evaluated. The seam dimensions may drift because of wear on the parts or adjustment issues. If not corrected, seam failure may result. The thickness of the seam after the first and second operations is specified by the can manufacturer. To measure a first operation seam thickness, the seaming machine must be operated manually, and a can must be taken out after the first operation only. Any can from production may be used to measure the second operation seam thickness. Figure 11.13 shows the seam thickness dimension on a filled can. It is possible to measure seam

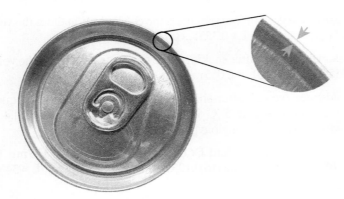

Figure 11.13 Seam thickness.

thickness with an ordinary micrometer or Vernier caliper, but the measurement is faster and more accurate with a special type of micrometer caliper with a narrow gap and a short measuring screw that can easily grip the seam.

The dimensions of internal parts of the seam can be measured by disassembling the seam. This is manually performed with a metal snipper. The lid is cut off and torn away, leaving the lid hook still engaged to the body flange. A cut is made through the seam, so the lid hook can be removed. Manual teardown is tedious and potentially dangerous because of the sharp metal edges. Care should be taken to avoid distorting the seam. There are devices that attach to a drill press to tear down a can in seconds. Another method is to use a very sharp saw to make slots in the wall of the can through the seam. This provides access to a side view of the seam. Measurements can be made with image analysis optics and software.

CHECK FOR UNDERSTANDING

1. What are kegs and casks? How are they used?

2. What are the distinctions among the various Sankey coupling types?

3. Calculate the mass of oxygen and nitrogen in a 355 mL sample of air (21% oxygen, 79% nitrogen) at 1.01 bar and 22 °C.

4. What is the function of FOB valves? How can they save money?

5. What is the role of nitrogen in dispensing beer that is not nitrogenated?

6. Describe the various temporary draft systems. What sort of events are each type most suitable?

7. What are primary, secondary, and tertiary packaging?

8. Describe tests to evaluate the quality of the seals on bottles and cans.

9. How is a sample obtained for package oxygen measurement?

10. A partially used keg is returned to the brewery. Describe how the ullage is removed and the keg is refilled.

11. List in correct order and describe the function of each step in a small pack packaging line.

12. Regarding packaging line design, what is meant by the "v-curve?" Why is this design optimal?

13. What is the typical TPO threshold for packaged beer?

14. Dissolved oxygen concentration is measured at 9.0 °C after shaking in bottles with 355 mL of liquid and a headspace volume of 18 mL. DO after shaking is 125 ppb. Henry's law constant is 59 mg/L/atm. Calculate TPO in ppb.

CASE STUDY

Case Study 1

You work in the quality assurance lab for a packaging brewery with an annual production of 50 000 bbl. During a bottling run on Monday, you sampled several bottles off the line, from start to finish. The TPO was around 800 ppb for all of them. On Tuesday, you repeated the tests on a second bottling run, and again, all beers measured a TPO of 800 ppb. Explain how you will evaluate, identify, and solve the problem.

Case Study 2

You work in the quality assurance lab for a packaging brewery with an annual production of 50 000 bbl. During a bottling run on Monday, you sampled several bottles off the line, from start to finish. The TPO increased slowly during the run, starting out below 100 ppb at the start but approaching 1 ppm by the end. On Tuesday, the same trend was observed. Explain how you will evaluate, identify, and solve the problem.

Case Study 3

You work in the quality assurance lab for a packaging brewery with an annual production of 50 000 bbl. During a bottling run on Monday, you sampled several bottles off the line, from start to finish. The TPO in the first bottles was 1 ppm, but shortly after, the remaining bottles were 200 ppb. Explain how you will evaluate, identify, and solve the problem.

BIBLIOGRAPHY

Barth R. 2013. The science of carbonation. *Zymurgy* 36(6):65–69.

Bates S. 2006. Bottle Shop Operations. In Ockert K (editor). *Fermentation, Cellaring, and Packaging Operations*. Master Brewers Association. ISBN 0-9770519-2-7.

Brewers Association Technical Committee. 2017. *Draught Beer Quality Manual, 3rd ed.* http://www.draughtquality.org.

Bückle J. 2009. Labeling. In Eßlinger HM (editor). *Handbook of Brewing*. Wiley-VCH. ISBN 978-3-527-31674-8.

Hornsey IS. 1999. *Brewing*. Royal Society of Chemistry. ISBN 0-85404-568-6.

Seaman R. 2015. How Can We Improve Our Packing Line Efficiency? Presentation at the Master Brewers Association of the Americas Eastern Technical Conference, October 23. Philadelphia, PA.

CHAPTER 12

FLAVOR

Beauty is in the eye of the **beer** holder. This phrase captures the subjective nature of enjoying a beer. **Brewing** is largely about **flavor**, getting the desired flavor, protecting the flavor, and verifying the flavor. But flavor can vary between individuals. Different people may have different **thresholds** for perception of certain **molecules**. Flavor sensations are extremely prone to bias or external influence.

There are three components to flavor, **taste**, **aroma**, and **mouth feel**. These components are often processed together in the brain to generate a single composite flavor sensation. There are hundreds of distinct molecules that contribute to beer flavor. The purpose of this chapter is to understand the perception of flavor and to identify the most prominent molecules responsible. Though sensory training is necessary to fully master the ability to isolate and detect flavor flaws in a product, this chapter explains the derivation of the most common beer flavors.

Mastering Brewing Science: Quality and Production, First Edition.
Matthew Farber and Roger Barth.
© 2019 John Wiley & Sons, Inc. Published 2019 by John Wiley & Sons, Inc.

12.1 FLAVOR ANATOMY AND CHEMISTRY

Taste

Taste is a chemical sense that responds to dissolved substances in the mouth. **Gases** can only be tasted if they dissolve. Taste is sensed by taste receptor cells in structures called **taste buds** (Figure 12.1). There is a pore in each taste bud to give the receptor cells access to the liquid in the mouth. The receptor cells are connected to neurons (nerve cells) that are connected indirectly (via cranial nerves) to the brain. Each type of taste has its own receptor cells. There are five currently recognized tastes: sweet, sour, salty, bitter, and umami (meaty or brothy taste like monosodium glutamate).

The sweet taste is evoked by **sugars**. Most sweet sugars normally present in beer wort are removed by **fermentation**. The most concentrated sugar in finished beer is maltotetraose (four-glucose **units**) followed by maltotriose (three-glucose units) typically present at the level of about 5 g/L (ppm) in **lager beer** and 10 g/L in **ale**. These sugars are not very sweet. The concentration of sucrose (table sugar) that would give the same sweetness would be 1–1.5 ppm. Evidently the sugars present in beer do not account for significant perceived sweetness. It is likely that most beer characterized as sweet has aromas associated with sweetness, rather than a true sugary taste. Lactose, which is 16% as sweet as sucrose, has found favor as an adjunct to sweeten the beer. Lactose is unfermented by **yeast** so it is retained in the finished beer. Because some people are lactose intolerant, any lactose additions should be clearly noted on the beer label.

The sour taste is evoked by **acids**, which are **compounds** that release the hydrogen **ion** (H^+). The situation is complicated; the rank ordering of various

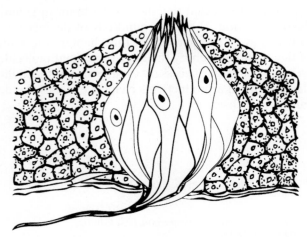

Figure 12.1   Taste bud. *Source*: From Barth (2013). Used with permission.

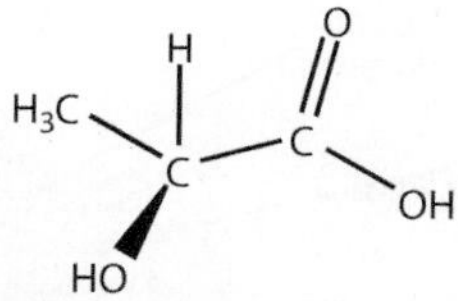

Figure 12.2   Lactic acid.

acids does not follow **pH** or amount of acid. Standard finished beer typically has a pH in the range of 3.9–4.3. Most consumers would not consider it to be sour. Sour ("tart") beer styles have pH in the range of 3.3 (very sour: new **lambic**) to 3.6 (Berliner Weisse). Sourness in these styles comes mostly from lactic acid (Figure 12.2), traditionally introduced by **bacterial** fermentation. For safety, beer must have a pH of 4.6 or lower.

The bitter taste is evoked by many natural compounds with no evident chemical similarity. There are at least 40 different bitter **receptors**. For over a thousand years, bitterness has been a primary flavor in standard beer. The most important bitter compounds in beer are the iso-**alpha acids** derived from **hops**, shown in Figure 8.11.

The salty taste is stimulated by sodium ions. Other alkali metal ions may taste salty, but the further away on the periodic table they are from sodium, the less they are perceived as salty. The concentrations of salt in brewing liquor should not be high enough to elicit a flavor, except that pronounced salt additions are common in *gose*-style beer. Human perception of sodium occurs around 250 ppm. Levels approaching this salinity can create a slick minerality perception on the palette. Levels at or exceeding 250 ppm will taste salty. Excessive sodium has a negative effect on yeast growth and viability.

Umami (glutamate) tastes are seldom important in beer.

Aroma

Aroma, or the **olfactory** sense, is a chemical sense that responds to molecules in the **gas phase**. The olfactory receptors in humans are located at the top of the nasal airway, just below the brain case. Aroma compounds ("odorants") can arrive through the nostrils (orthonasal), or they can drift up from the mouth and throat (retronasal). Retronasal aroma is a key component of flavor. Consider what happens with a cold and severe nasal congestion. The ability to taste food is lost. The next time you sip a beer, pinch your nose closed to illustrate the effect. Blocking the nasal cavity blocks flavor perception from aroma. However, it does not block the taste receptors, the perception of salty, sweet, sour, bitter, or umami. To take advantage of retronasal aroma, it is highly recommended that after a first swallow of beer, the taster exhales through the nose while forcing air into the nasal cavity with the tongue.

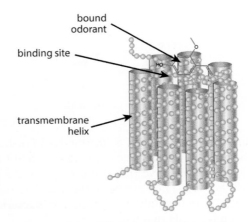

bound
odorant

binding site

transmembrane
helix

Figure 12.3   Olfactory receptor protein.

The olfactory receptors are **neurons**; they are directly stimulated by molecules, unlike the taste receptors, in which neurons are activated by specific receptor cells in the taste buds. Olfactory receptor neurons (ORNs) for different aromas are very similar to one another and have the same **mechanism** of action. The ORNs have cilia (hairlike projections) that extend into the nasal cavity. The other end of the ORN is a fiber called the axon that connects to a part of the brain called the **olfactory bulb**. On the ORN cells are olfactory receptor proteins. These proteins belong to a family of receptors called G protein-coupled receptors (GPCRs). All GCPRs are transmembrane proteins with seven alpha helices (refer to Section "Proteins," Chapter 3). The coils cluster in a circle forming the structure of the receptor. Figure 12.3 shows the protein as a string of beads, with the transmembrane helices wrapped around imaginary cylinders for illustration. On the extracellular side of the receptor is a pocket called the **binding site**, shown binding a molecule of 4-vinylguaiacol, which gives a phenolic flavor to some styles of wheat beer. Olfactory receptor proteins for different families of compounds differ at the binding site. Just like the **active site** on an **enzyme**, the active site for a receptor is shaped by the protein structure and the identity of the **amino acids**. When these amino acids in the active site bind to the target molecule, the receptor changes its shape, causing the release of a signal molecule inside the ORN. The signal molecule switches on an enzyme that produces many molecules of a second messenger, amplifying the response. This is an example of outside-in signaling first discussed in Section "Regulation of Gene Expression" in Chapter 3 with **transcription factors**.

In the case of ORNs, the signal molecule turns on an enzyme that hydrolyzes ATP to **cyclic adenosine monophosphate (cAMP)**, the second messenger. The second messenger opens the gates to a calcium **channel**, allowing calcium to flow into the cell, depolarizing it, that is, making it less negatively charged. When the cell **voltage** reaches a threshold, **voltage-gated** sodium channels open, bringing a flood of positive charge that causes more sodium channels to open. This sends a wave of depolarization throughout the neuron, along the axon, and to the brain where it is interpreted as a flavor.

GPCRs are found in cells of all **eukaryotes**, including humans, **barley**, hops, and yeast. In humans they are involved in vision; in the sensation of sweet, bitter, and umami tastes; and in many other functions. The mechanisms of sour and salty taste processing are less understood, but there are indications that they may involve the binding of ions to receptors. Humans have about 800 genes for olfactory receptor GPCRs, much fewer than the number of different aromas that can be sensed. This suggests that multiple olfactory receptors participate in producing an aroma. Each receptor responds with a different intensity to a variety of different odorants. The brain interprets the pattern of responses of different ONR giving a sensation characteristic of the odorant.

Mouth Feel

The mouth is well supplied with nerve fibers that sense touch, **pressure**, vibration, **temperature**, and pain, all derived from the trigeminal nerve. These sensations are processed to give sensations of **carbonation**, **astringency**, hot or cold, **viscosity**, texture, fullness, and thinness. The importance of these sensations is highlighted by the typical customer's reaction to warm, flat beer. Mouth feel is largely, but not entirely, a set of physical sensations. The physical properties of beer are, of course, related in complex ways to its chemical composition. Some of the key players in mouth feel are carbon dioxide, which gives beer its lively **carbonated** feel, and **tannins**, which provide **astringency**, a drying, puckering sensation. Sulfate ion enhances astringency. Astringency is believed to result from non-specific binding of compounds to proteins in the sensory system. **Dextrins**, which are dissolved unfermentable **carbohydrates**, are said to provide a sensation of fullness. The role of dextrins in mouth feel is not fully understood. Metallic sensations evoked by iron ions are another chemically related mouth feel. Brewer's yeast can also produce **glycerol** during fermentation, with concentration varying by strain. Glycerol is an odorless, sweet-tasting molecule that enhances fullness on the palette.

12.2   FLAVOR COMPOUNDS

There are hundreds of flavor-active compounds in beer. For many of them, there is no consensus on their flavor contribution. The situation is further complicated by changes that occur in beer composition and flavor as it ages.

More complex still is the ability of yeast to **biotransform** flavors of other ingredients like hops. Some compounds that are present in very small traces are highly flavor active and make an outsize contribution, for better or worse, to the quality of the beer. Isolating these compounds and associating them with the flavor of the beer can be very challenging. Here we focus on the most pronounced flavors.

Flavor Threshold and Flavor Units

The intensity of the flavors evoked by the flavor compounds that sometimes show up in beer varies by huge factors. The smallest concentration of ethanol, a mildly flavored compound, that can be detected reliably by the average taster is 14 g/L (1.4% w/v). This concentration is described as the **flavor threshold** for ethanol. The corresponding flavor threshold for 3-methylbut-2-ene-1-**thiol** (MBT), the compound responsible for the skunky aroma of lightstruck beer, is 7 ng/L (7×10^{-9} g/L). Based on these thresholds, MBT is 2 billion times (14 g/L $\div 7 \times 10^{-9}$ g/L) as strongly flavored as ethanol. Most flavor thresholds are reported in mg/L or ppm.

Flavor scientists may also use the **flavor unit**. The number of flavor units for a compound is the concentration of the compound divided by its threshold. For example, beer with 40 g/L ethanol (5.1% ABV) provides 40 g/L $\div$ 14 g/L = 2.9 flavor units. The concentration of MBT required to also yield 2.9 flavor units is 2.9×7 ng/L = 20 ng/L (20 ppt). For compounds that have not been isolated or identified, it is possible to estimate their flavor units by dilution. Suppose we take 100 mL of beer and add 200 mL of water. This is a 3 : 1 dilution. If the flavor can still be discerned, it is present at 3 flavor units or more in the undiluted beer. If we cannot discern it in a 4 : 1 dilution, then it is present at less than 4 flavor units.

Flavor compounds in beer can be characterized as primary, secondary, tertiary, or background. Primary flavor compounds are those present at 2 flavor units or more. They are easily discerned; their absence is instantly noticed. Secondary flavor compounds are present at 0.5–2 flavor units. They distinguish different styles and brands from one another. Tertiary flavor compounds are those present at 0.1–0.5 flavor units. They are less obvious, but they can work together to influence the perception of beer. Compounds present at less than 0.1 flavor units are part of the background. Table 12.1 lists compounds often present in beer and their contributions to the flavor.

Flavor Categories

Beer styles can be placed into general categories of ale or lager and of bitter, hoppy, or malty. An additional category is needed for styles with unusual flavors (smoked, sour, chili peppers, chocolate, etc.). Most of the commercial

TABLE 12.1   **Beer Flavor Compounds**

Compound	Official Name	Flavor Units	Character
Carbon dioxide	**Carbon dioxide**	**3.0–5.5**	**Carbonation**
Hop acids	**Beer bitter substances**	**2–12**	**Bitter**
Ethyl alcohol	**Ethanol**	**1.4–6.4**	**Alcohol**
Myrcene	3-Methylene-1,6-octadiene	0–2.2	Resinous
Ethyl acetate	Ethyl ethanoate	0.3–1.0	Fruit/solvent
Ethyl caproate	Ethyl hexanoate	0.2–1.3	Apple
Ethyl caprylate	Ethyl octanoate	0.1–0.6	Apple
Isoamyl acetate	3-Methylbutyl ethanoate	0.1–0.8	Banana
Ethyl isobutyrate	Ethyl 2-methylpropanoate		Fruit/aromatic

Primary flavor compounds highlighted.

beer in the world can be classified in a bitter lager style sometimes called international lager. There are substantial sources on specific beer style guidelines from the Brewers Association and the Beer Judge Certification Program, among others. Depending on the viewpoint of the brewer, styles may be embraced or rejected as constraining. Either way, when submitting a beer for a competition based on styles, make sure the beer aligns with the guidelines.

Flavors are derived from **malt**, **hops**, yeast, and water – with type, quality, and process variables controlling intensity. For an illustration of which flavor categories are produced by yeast over time, refer to Figure 9.4. To re-examine the influence of water on flavor, consult Table 4.5. Here we discuss bitter, malty, esters, and hop aroma.

Bitter   Bitterness in beer is derived from **isomerized** alpha acids from hops, shown in Figure 8.10. The bitterness is often reported as **international bitterness units (IBU)**. One IBU is roughly equivalent to 1 mg of **isohumulone** per **liter**. Four IBU is about 1 flavor unit. The IBU method is challenged by significant **dry hopping**. Humulinones, hulupones, and other structurally similar molecules are detected with this method, but they are less bitter. This often leads to higher IBU measurements but lower perception of bitterness. Perceived bitterness is also influenced by pH.

Standard international pilsner typically has a bitterness level of 8–12 IBU (2–3 flavor units). Many European pilsners are a good deal more bitter, up to about 30 IBU. Some craft ales are said to have 100 or more IBU, but these measurements are based on calculations only. Scientific evidence argues that the limited solubility of iso-alpha acids in beer restricts beer to no more than 70 IBU.

Astringent bitterness is described as harsh, unpleasant, and mouth coating. This type of bitterness may be derived from inappropriate **milling**, **lautering**, or **sparging** techniques due to extraction of tannins from the barley **hull**.

It can also be indicative of **autolyzed** yeast. Yeast will self-destruct, or lyse, given prolonged or extreme stress. The most common sources of autolyzed yeast are keeping the beer on the yeast too long, excessive temperatures, and extreme mechanical agitation.

Malty   Aptly named, malty flavor is derived from malt. Several manufacturers of malt have developed malt flavor wheels to provide better descriptive analysis of this characteristic. Such descriptions include roasted flavors like coffee, cacao, dark chocolate, and toast; nutty flavors like almond, hazelnut, raisin, and vanilla; malty flavors like honey, biscuit, cracker, and marmalade; and variations of caramel.

Maltiness comes from Schiff base products as shown in Figures 4.11 and 4.12. Malty compounds usually arise from heat treatment, especially during drying and roasting of malt. Bock is a familiar malty lager style. Its maltiness arises mostly from Munich malt.

Esters   Fruity flavors in beer arise from **esters**, compounds that arise from condensation of an **alcohol** and a carboxyl group, commonly produced by yeast. Figure 12.4 shows the structures of some esters often found in beer. Their flavors are given in Table 12.1. Most of the esters in beer are produced by the yeast during fermentation. Ester concentrations are usually higher with ale strains than lager strains. Esters can be increased by raising the fermentation temperature, increasing the aeration of wort, decreasing the pitch count, brewing at **high gravity**, and fermenter geometry (increases with more hydrostatic pressure). Esters have also been shown to increase based on the carbohydrate content of the wort, with an increase in glucose percentage as compared with maltose. Hops are a secondary, but still a potentially significant, source of esters. Ethyl isobutyrate (Figure 12.4E) is often provided by hops.

Hop Aroma   In addition to bitter taste, hops can serve as a source of aroma compounds that can have a big influence on beer character. Many of these are **terpenes** or their oxidation products. Terpenes are biological **hydrocarbons**, the most important of which are structurally related to isoprene, shown in Figure 12.5. They are members of a larger family called **terpenoids**, which include terpenes modified by addition of atoms of oxygen, phosphorus, sulfur, or others. Terpenes are classified by the number of isoprene units that they include. A monoterpene has two isoprene units, sesquiterpene has three, and diterpene has four.

Figure 12.6 shows four of the hundreds of terpenes found in hop lupulin. The drawing of myrcene is highlighted to show the constituent isoprene units. Terpenes are **volatile**; they escape from wort easily during boiling. When a beer style calls for hop aroma, it is often necessary to add hops at the end of

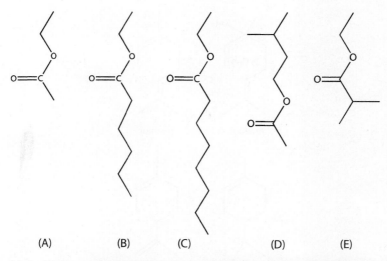

(A)   (B)   (C)   (D)   (E)

Figure 12.4   Flavor esters. (A) Ethyl acetate. (B) Ethyl caproate. (C) Ethyl caprylate. (D) isoamyl acetate. (E) Ethyl isobutyrate.

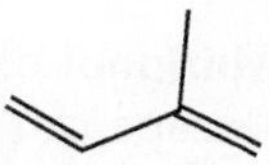

Figure 12.5   Isoprene.

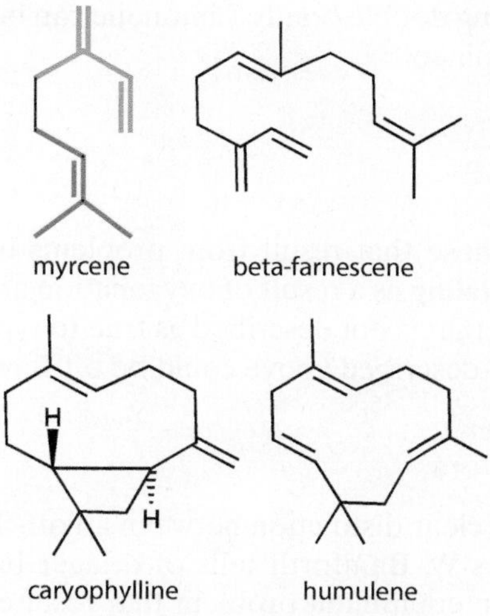

myrcene   beta-farnescene

caryophylline   humulene

Figure 12.6   Hop terpenes.

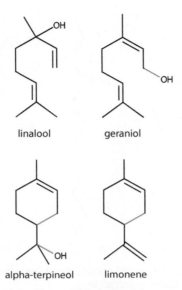

Figure 12.7   Terpenoids derived from myrcene.

boiling, or after boiling in the **whirlpool** or **hop back**, or after **chilling** (dry hopping). Terpenes that do not escape can easily react, especially with oxygen, to give a variety of flavored products. Figure 12.7 shows a few of the many terpenoids that can occur in beer. These terpenoids are derived from myrcene. The three alcohols, linalool, geraniol, and alpha-terpineol, can all be made by addition of water across one of the double **bonds** in myrcene, in some cases also shifting double bonds. Limonene can be made by eliminating water from alpha-terpineol.

12.3   OFF-FLAVORS

Off-flavors include those that result from problems in **brewery** processing, poor ingredients, or **staling** as a result of oxygenation and time. In practice, an **off-flavor** is anything that is not described as true to type for a specific brand. The flavor categories described above could be off-flavors in certain styles.

Processing Off-Flavors

There is not always a clear distinction between an off-flavor and a style characteristic. Dr. Charles W. Bamforth tells of a lager brewery in the eastern United States with an equipment problem that resulted in condensate from the kettle running back into the wort. This issue resulted in an undesired cooked vegetable flavor in the beer. The problem persisted for years until wear and tear made it necessary to replace the boiler. The opportunity to correct the technical problem was seized, and the brewery started producing beer without the off-flavor. Customer reaction was intensely negative; they liked the beer the way it was. Of course, the brewery immediately modified the kettle plumbing to give the original flavor. There are two lessons to be learned from this. A flavor that customers like, irrespective of what the beer judges say, is not an off-flavor; it is a style characteristic. Your customers are the people who like your beer. If you change the flavor, you may find your brewery looking for new customers. In this section we will cover what are conventionally regarded as off-flavors in roughly the order of the processes that introduce (or fail to remove) them. Their true designation as an off-flavor depends on the intention of inclusion.

Medicinal   Medicinal or adhesive tape aromas are often the result of chlorophenols. There are many phenolic compounds in beer wort. If the beer is made with chlorinated water or if chlorine-containing sanitizers are left in the vessels, phenolic compounds from **grain** or hops can react to give highly flavor-active compounds. Figure 12.8 shows 2-chlorophenol, whose aroma threshold is 0.1 µg/L (ppb).

Astringency   Astringency is a mouth feel described as dry or puckering. It can come from sulfate ion (SO_4^{2-}) in the brewing water or from polyphenols extracted from grain hulls during **mashing** and wort separation. High to moderate astringency is characteristic of pale ale and related styles but is an off-flavor in most other styles. Polyphenols are slightly **acidic**; they are more soluble at high pH and high temperature. Sparging should be carried out at moderate temperature (<80 °C, 176 °F) and low pH (<6). Small hull particles make better contact with the liquid phase, so compounds are more easily extracted from them. Grain handling and milling processes should avoid abrasion of the hull. Excessive agitation during mashing should also be avoided. When **hammer-milled** grain is used (usually in conjunction with **mash filtration**), the small hull particles should be compensated by lowering the sparging temperature and maintaining a low sparge pH. Some brewers acidify the sparge water to minimize polyphenol pickup.

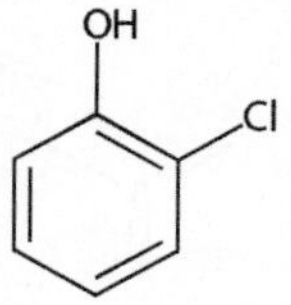

Figure 12.8   2-Chlorophenol.

Dimethyl Sulfide Dimethyl sulfide (DMS), shown in Figure 12.9, is a volatile compound (boiling point 38 °C) that gives an undesired aroma of boiled vegetable. DMS has a flavor threshold of about 30 µg/L; when present at levels above 50 µg/L, it is considered objectionable. DMS is produced during various stages of malting and brewing from precursor compounds *S*-methyl methionine (SMM) (Figure 12.9) and dimethyl sulfoxide (DMSO) (Figure 12.10), which itself originates from SMM. Sometimes SMM and DMSO are collectively called DMS precursors (DMS-P). The DMSO route to DMS occurs during fermentation.

S-methyl methionine (SMM) is not present in unmalted barley, but it forms during germination. Heat treatment including kilning causes some of the SMM to be hydrolyzed to DMS and homoserine as shown in Figure 12.9. DMS formed during malt kilning is mostly released as **vapor**. Some of it is oxidized to DMSO, which is very soluble and not volatile (boiling point 189 °C) and is likely to end up in the fermenter. Malt that is cured at low temperature like pale lager-type malt can retain as much as half of its SMM. Higher cured malt, like ale malt, loses much of its SMM but can retain a significant amount of DMSO. SMM in the malt readily dissolves in the wort during mashing.

Under typical infusion mashing conditions of 65–70 °C (149–158 °F), DMS synthesis from SMM is barely significant, but the rate doubles for every 6 °C temperature increase. Higher temperatures during sparging and decoction can greatly increase the rate of conversion. During boiling, about one-third of the SMM in the kettle is converted to DMS for each hour at 100 °C. If the boil is vigorous or if other measures are taken to enhance removal of volatile substances, DMS is eliminated. The process of DMS synthesis continues until the wort is chilled. This includes the time it takes to **pump** the wort into the whirlpool or hop back, the time spent in the wort clarification process, and

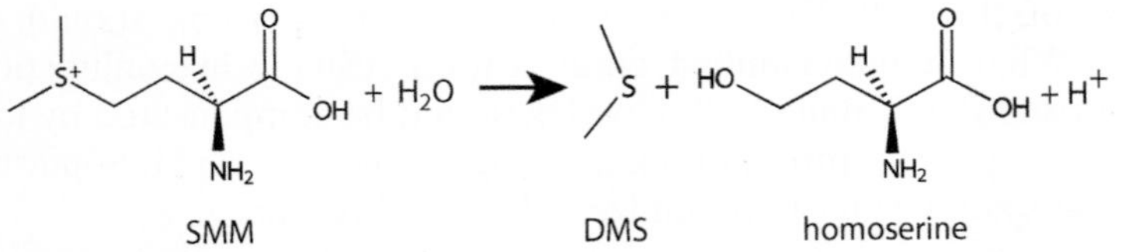

Figure 12.9   DMS synthesis. SMM, *S*-methyl methionine. DMS, dimethyl sulfide.

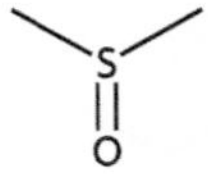

Figure 12.10   DMSO.

the time it takes to pump the wort through the chiller. Once the wort is chilled, additional conversion of SMM to DMS becomes insignificant.

Brewhouse practices that limit DMS in beer include:

- Use of more highly kilned malt.
- Prolonged and/or vigorous boiling.
- Wort stripping after clarification.
- Less time between the end of boiling and chilling.
- Rapid chilling.

DMSO is a nearly flavorless nonvolatile liquid that is formed from DMS during malting. DMSO carried to the fermenter will be reduced to DMS by yeast enzymes and by some bacteria. This activity is greater when FAN is insufficient, when the initial gravity is high, and when the fermentation temperature is low. Yeast converts only a fraction of the DMSO to DMS, but **spoilage** bacteria make a greater contribution to DMS.

Sulfur Compounds   Sulfur compounds are typically produced by yeast in their use of sulfur-containing amino acids and dissolved sulfate ions. Hydrogen sulfide (H_2S) has the aroma of rotten eggs. Sulfur dioxide (SO_2) has the aroma of burnt matches. These flavors may be appropriate in small amounts for lagers. Sulfur aromas may be removed during vigorous fermentation. Their formation is promoted by excessive FAN, carryover of trub into the fermenter, insufficient aeration, poor yeast health, and bacterial contamination.

Solvent   Solvent aromas arise from alcohols with three or more carbon atoms (fusel alcohols) and from some esters, especially ethyl **acetate** (Figure 12.4A). Figure 12.11 shows the biosynthesis of isoamyl alcohol (3-methyl-1-butanol), a common fusel alcohol. The process recovers the nitrogen from a leucine molecule by transamination. The nitrogen-free product is decarboxylated, and the resulting **aldehyde** is hydrogenated to the alcohol. **Fusel alcohols** are also called higher alcohols and are derived from yeast during fermentation. Higher alcohols are acceptable in high-gravity beers but are generally considered off-flavors, providing a hot alcohol sensation. Higher alcohols must be controlled during primary fermentation as **conditioning** has little effect on their removal. Higher alcohols are increased by high fermentation temperatures, physical forces on brewing yeast, **high gravity** brewing, and excessive aeration. Higher alcohols should not exceed 100 ppm. Fusel alcohols are removed by **condensation** reactions that produce flavorful esters.

Phenolic   The phenolic flavor is a characteristic of certain styles of ale, notably Bavarian weizen and Belgian wheat ales. It is considered an off-flavor

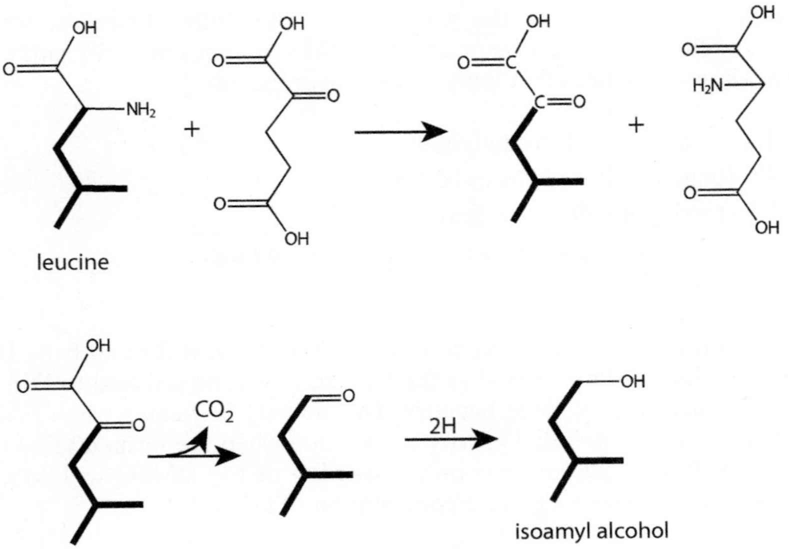

leucine

isoamyl alcohol

Figure 12.11   Isoamyl alcohol synthesis.

in other styles. The flavor has been described as the aroma of cloves, but not everyone discerns cloves in phenolic beer. The phenolic flavor comes from 4-vinylguaiacol (4VG), which certain strains of yeast produce by enzymatic decarboxylation of ferulic acid (FA), shown in Figure 12.12. The flavor threshold for 4VG is 0.3 mg/L (ppm). 4VG is the precursor to 4-ethylguaiacol (4EG), produced by certain **Brettanomyces** strains, creating notes of spiciness or smokiness.

Ferulic acid (FA), the precursor to 4VG, is found in **bound** form in plant cell walls, especially in those of the **aleurone layer** of wheat and barley. It is released during mashing, the highest rate of release occurring at 45 °C (113 °F) and in the first 30 minutes of mashing. FA is not converted to 4VG during mashing, but a small amount of conversion can occur during wort boiling. The phenolic flavor mostly comes about by conversion of FA by enzymes in yeast during fermentation. Even if the FA concentration is high, unless the yeast strain makes the enzyme that converts it to 4VG, the phenolic flavor will be barely detectable. Some yeast strains make enough 4VG for it to be a primary flavor compound, which is one present at more than 2 flavor units.

Cidery   The cidery off-flavor comes from **acetaldehyde** (Figure 12.13), an intermediate in fermentation. It is secreted by yeast typically in the first three days of fermentation and has a strong taste of green apple. Acetaldehyde will be reduced by yeast over time. Initial amounts of acetaldehyde in beer are

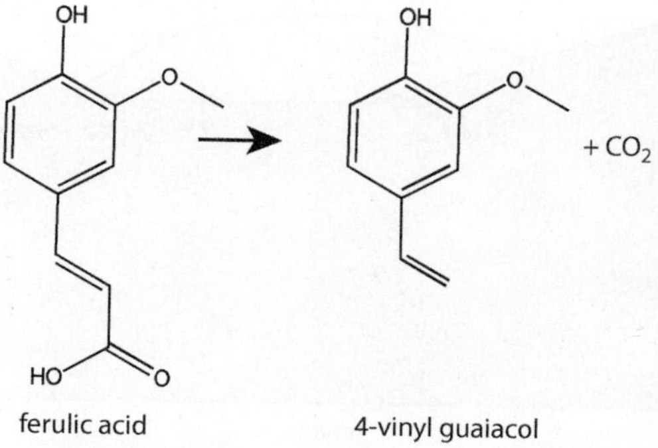

Figure 12.12   Production of 4-vinyl guaiacol.

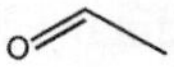

Figure 12.13   Acetaldehyde.

increased by rapid fermentation, warm temperature, high pitch rates, too little aeration, and pressure. Another potential source of acetaldehyde is oxidation of ethanol.

Buttery   Buttery flavors are a significant concern, especially in lager beer brewing. These flavors come from **diacetyl** (officially called butanedione) and 2,3-pentanedione, collectively known as **vicinal diketones (VDK)**, meaning compounds with $C=O$ groups on each of two adjacent carbon atoms. The VDK compounds are shown in Figure 10.1. VDK is easily the costliest off-flavor issue that brewers address in terms of time and money. The VDK is produced by yeast during fermentation from intermediates in the biological synthesis of amino acids. Diacetyl comes from the synthesis of valine and leucine; 2,3-pentanedione comes from the synthesis of isoleucine. The origins and removal of VDK are discussed in Section 10.1.

Stale Off-Flavors

The flavor of beer changes, mostly for the worse, from the time it leaves the brewery. Figure 12.14 gives a graphic representation of a typical **staling** profile. There is considerable variation in the staling off-flavors and their development depending on the characteristic of the beer and the storage and packaging conditions. For example, the **ribes** (catty) flavor does not develop at all if the total package oxygen is very low. Some stale flavors are listed in Table 12.2 along with the molecules with which they are associated.

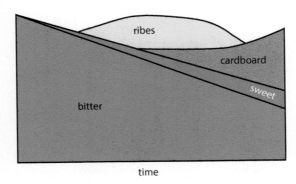

Figure 12.14   Flavor changes in beer with time. *Source*: After Dalgliesh (1977).

Lightstruck   Beer made with ordinary hop products is susceptible, upon exposure to light, to an off-flavor reminiscent of the skunk (*Mephitis* spp.). The beer is said to be **lightstruck** or, informally, skunked. Light in the **wavelength** range of 350–500nm is most efficient at producing the lightstruck flavor. The compound responsible for the lightstruck flavor is 3-methylbut-1-thiol (**MBT**). Figure 12.15 shows the sequence leading to MBT. In the first step, energy from light breaks the C—C bond between the **carbonyl** group (C=O) and the ring. The energy can be transferred directly if the light is of low wavelength, but for ordinary visible light, the energy goes first to riboflavin, which deposits the energy into the C—C bond. When a bond is broken directly or indirectly by light, the electron pair is divided between the two fragments, each of which gets one unpaired electron. An atom or molecule with an unpaired electron is called a free radical. Free radicals are usually very reactive. The second step shows a **free radical** ejecting carbon monoxide to make a smaller free radical. In the third step, the new free radical abstracts (grabs) a sulfur atom, probably from an amino acid. This gives MBT, the skunky thiol.

The most straightforward way to control the lightstruck flavor is to prevent light in the 350–500nm range from reaching the beer. Kegs, **casks**, and cans admit no light. Brown bottles filter out nearly all the light in the problem wavelength range but will still be affected over time. Green bottles and colorless bottles are ineffective. Isomerized hop acids that have been treated with hydrogen to eliminate some double bonds (Figure 4.31) do not form MBT. These special hop products can be added, to the exclusion of natural hops, to prevent lightstruck flavor.

Certain brands of beer have lightstruck as a flavor characteristic. One brand actually adds a measured amount of a lightstruck-flavored compound to introduce a consistent level of skunkiness. These are beers best drunk with a lime wedge. It bears repeating that it is the customer who decides what is an off-flavor and what is a style characteristic.

TABLE 12.2 Stale Off-Flavors in Beer

Flavor	Description	Source(s)
3-Methyl-2-butene-1-thiol (MBT)	Skunky, lightstruck	Iso-alpha acids exposed to light
4-Vinyl syringol	Burnt, smokey	
Acetic acid	Vinegar, sour	Oxygen. Contamination by bacteria
Astringent bitterness	Sharp, lingering bitterness	Autolysis of yeast
Caproic acid	Goaty, cheesy	Breakdown of fatty acids

(Continued)

TABLE 12.2 (Continued)

Flavor	Description	Source(s)	
Catty, ribes 3-sulfanyl-3-methylbutyl formate	Tomcat urine, ribes, or blackcurrant	Oxidation. Hops	
Dimethyl trisulfide	Onion	Hops	
Ethyl phenylacetate	Honey sweetness, sherry, meat	Oxidation	
Isovaleric acid	Cheesy, sweaty	Stale hops	
Methional	Onion, meaty, cooked potato	Strecker degradation of methionine	
(E)-2-Nonenal	Cardboard, paper	Oxidation of malt lipids	

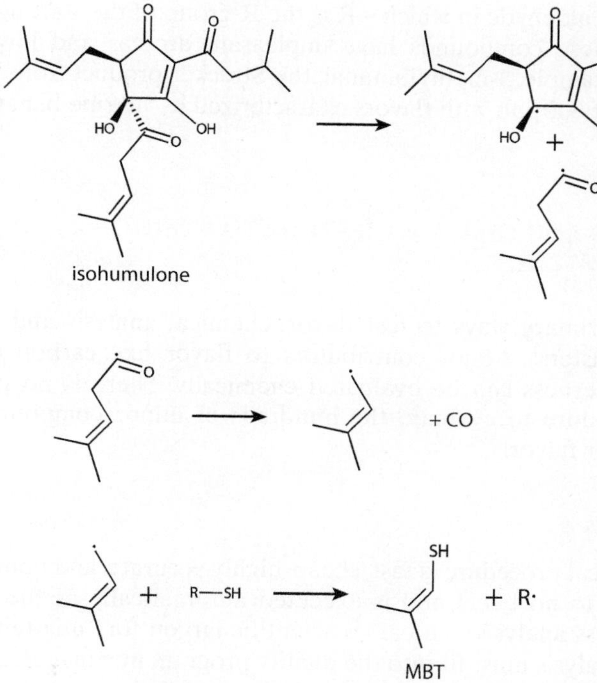

isohumulone

Figure 12.15   Lightstruck sequence.

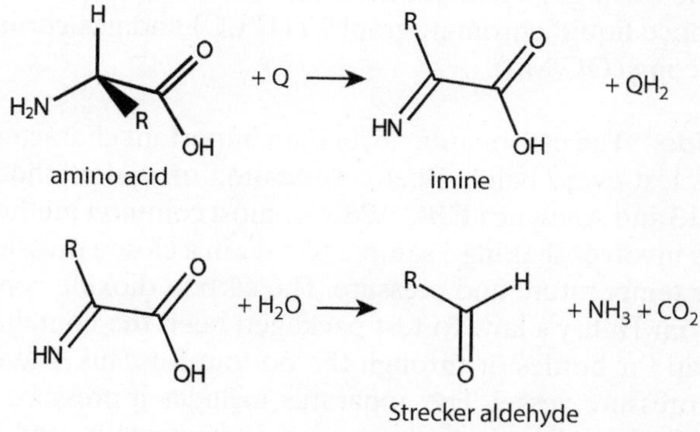

amino acid

imine

Strecker aldehyde

Figure 12.16   Strecker degradation.

Oxidized   Oxidation can produce compounds that give stale flavors to beer. One example of an oxidation process is the Strecker degradation of amino acids, shown in Figure 12.16. The first step is an oxidation, where Q is a generic oxidizing agent. The oxidation number of the left carbon in the amino acid is 0, and that in the imine is +2. The second step is not an oxidation–reduction.

The product is an aldehyde in which —R is the R-group of the starting amino acid. Many of these compounds have unpleasant aromas and low flavor thresholds. For example, 3-methylbutanal, the Strecker product from leucine, has a threshold of 600 ppb with flavors characterized as "unripe banana" and "cheese."

12.4 ANALYSIS AND QUALITY CONTROL TO MONITOR FLAVOR CONSISTENCY

There are two primary ways to test flavor, chemical analysis and sensory panel (human tasters). Major contributors to flavor like carbon dioxide, ethanol, and bitterness can be evaluated chemically. There is no practical laboratory procedure to evaluate the hundreds of minor compounds that contribute to beer flavor.

Chemical Analysis

The ideal analytical procedure is fast, cheap, highly accurate, and nondestructive (ruins little to no beer) and is executed automatically as the process continues (process analysis). "Ideal" is scientific jargon for "unattainable in practice." The analysis must fit into the quality program in a meaningful way, that is, it tells you something you need to know, and there are **control limits** on the measurement and procedures for dealing with deviations. Advanced methods for measuring the compounds that contribute flavor in beer include high performance liquid chromatography (HPLC) and **gas chromatography/mass spectroscopy** (GC/MS).

Carbon Dioxide   The carbonation level is an important characteristic of beer. Many brewers test every batch. Beer carbonation official methods are ASBC Method Beer 13 and Analytica **EBC** 9.28. The most common method for carbon dioxide testing involves shaking a sample of beer in a closed pressure vessel and measuring the temperature and pressure. The carbon dioxide concentration is determined from Henry's law. To test packaged beer, the container is pierced through the cap for bottles or through the bottom for cans. The bottle or can serves as the pressure vessel. The apparatus includes a pressure gauge and a **thermometer**. Carbon dioxide concentration in volumes is read from a table. Table 10.1 contains this information. There are more accurate tables that take the effects of **final gravity** and ethanol content into account, which can be significant in high-alcohol beer. A variation is to use a syringe to maintain a constant pressure and measure the volume. The pressure–volume–temperature methods all have the inherent limitation that they do not distinguish between carbon dioxide and other dissolved gases potentially found in beer.

A tedious but potentially accurate method is to make the gas from the beer bubble through a **solution** of potassium hydroxide. Carbon dioxide is acidic; it reacts according to $2KOH + CO_2 \rightarrow K_2CO_3 + H_2O$, neutralizing some of the KOH. The amount of acid needed to neutralize the unreacted KOH is determined. The amount of KOH determined in this way is compared with the amount before reacting with carbon dioxide. The amount of carbon dioxide on the beer sample is then calculated.

There are proprietary methods used by dedicated instruments that exploit properties of gases and gas **mixtures** containing carbon dioxide. For example, one instrument uses the big difference in thermal conductivity between CO_2 and other gases found in beer. Some of these methods claim to have good selectivity for CO_2 in the presence of other dissolved gases.

Ethanol   Analysis of ethanol is covered in Section 10.4.

Hop Bitter Acids (IAA or IBU)   The analysis of hop bitter acids is covered in ASBC Methods of Analysis Beer 23 and in Analytica EBC 9.8. These methods do not correlate exactly with the perceived bitterness, but they are nonetheless useful. Many breweries measure international bitterness units (IBU) of every batch as a quality check on the brewing process. In this method, a sample of beer is acidified to neutralize the charge on the phenolic —OH groups. The sample is then extracted 2,2,4-trimethypentane (isooctane), a nonpolar solvent that absorbs the iso-alpha acids. The absorbance of the **organic** layer is measured at 275 nm. The bitterness unit value is calculated from $IBU = 50 \times A_{275}$. A similar extraction procedure followed by dilutions with methanol and measurement of the absorbance at 255 nm can be used to calculate the iso-alpha acid (IAA) concentration.

WARNING: Isooctane is highly flammable. Its storage, handling, and disposal must be under the supervision of qualified chemists.

There are also some methods that use HPLC, which involves injecting samples into a solvent that flows through a tube (the column) packed with coated silicon dioxide particles. As the solvent emerges from the column, it enters a **detector** that measures a property that varies with the concentration of the stuff being analyzed, ultraviolet light absorbance in this case and refractive index for other HPLC analyses. An advantage of HPLC is that it can determine the individual concentrations of the different iso-alpha acids. A disadvantage is the high cost of the equipment and supplies.

Sensory Analysis

The most versatile piece of analytical equipment for beer is the human flavor palate. All raw materials, wort, green beer, and finished beer should be taste tested by the brewer for a quick, on-the-spot go/no-go assessment.

As operations scale larger, so does the need for a more formal sensory training program, especially for sensory analysis of beer.

There are many resources available to assist in the development and execution of a sensory program. This is only an introduction that describes common biases in brewery sensory programs, design of sensory labs, and useful tests for a sensory panel to conduct.

Sensory Bias Sensory results usually have a high degree of variability (high **standard deviation**) and are prone to bias. Bias is nonrandom error caused by a personal feeling or inclination toward an outcome. For example, if the color of the sample is dark, scores for maltiness flavor will be increased compared with the same samples for which the color was concealed. Variability can be factored out by training tasters, with repetitive testing, and through application of statistical methods. Bias can only be reduced by designing the experiments to minimize the factors that introduce bias. The following sources of bias should be avoided.

Adaptation is a change in sensitivity as a result of continuous exposure, otherwise known as palate fatigue. An example of adaptation bias occurs in the sampling order of beers from high bitterness to low bitterness. The analyst becomes desensitized to the bitterness over time. To avoid this bias, the tasters should test a limited the number of samples, and they should take regular breaks from tasting.

Enhancement and suppression describes the phenomena where the presence of one substance will increase or decrease the taster's sensitivity to another. For example, sulfur off-flavors are particularly pungent and may cause perception of other attributes to be less. To avoid this bias, palate cleansing and regular breaks should be included.

Expectation is when preconceived ideas influence responses. No sensory participant should know the purpose of the sensory test. Panelists should be given a blind tasting with clear, concise questions.

Habituation is the result of repeated exposure to the same product over several days. Because of habituation, minor or gradual change may not be perceived. Keep sensory panelists alert by occasionally spiking samples with off-flavors.

Stimulus error is a major source of bias where irrelevant criteria influence the response. The example above with dark beer color influencing perception of malt flavors is an example. To avoid expectation, sampling should be blind. The cups should be opaque.

The *halo effect* describes the association of two or more characteristics in the minds of the taster. For example, hoppy beers may be assumed to be more bitter. Overcoming the halo effects requires the use of a trained sensory panel or carefully designed questions.

The *order of presentation* can also be a major source of bias. Patterns, positions, and contrasts can influence perception. Samples should always be presented in a randomized order with randomized three-digit codes such as 326, 298, and 172.

Mutual suggestion is common when verbal or facial expressions of others influence a response. If one person in the room states that a beer is horrible, others may follow the opinion. If someone grimaces, that person may inadvertently influence someone else. Sensory analysis should not be conducted at a round table. At minimum it should be conducted with everyone sitting in the same direction. At best, panelists should use isolated booths.

Design of the Sensory Laboratory   For analytical tasting, the selection of tasters is important. The tasters will need to go through a recruitment/ selection process and undergo training to discern and describe flavors according to a standardized vocabulary. Standardized sample sets are available for flavor training. Training tools like the beer flavor wheel and flavor maps can be helpful in explaining flavor terminology.

The testing area should be located away from areas that have odors, like lunch rooms, loading docks, and production areas. Tasters should be seated in booths, so they do not see one another. The furniture should be comfortable and the atmosphere pleasant. If color is potential distraction, the lighting and the sample glasses can be adjusted to conceal the color. Sessions should not be too long and should not include too many samples. Trained tasters are a valuable resource that should be protected from burnout.

Sensory Tests   There are many testing methods. In all cases, the order of presentation and the coding must be randomized, and the tasters must be given no indication of the expected outcome. In the *paired comparison test*, samples are presented in pairs; the tasters are asked to compare them according to a certain characteristic. In the *triangular test*, three samples are be presented, two of which are identical. Tasters are asked to pick the one that is different. In the *duo–trio test*, the tasters are first presented with a standard that is identified to them as such. Then they taste two unknown samples, one of which is identical to the standard. They must choose the one that is different from the standard. In the *ranking test*, three to six samples are simultaneously presented. The taster is asked to arrange them in order of a specified flavor characteristic. Simple *trueness-to-type* tests may be designed to survey consistency of a brand.

CHECK FOR UNDERSTANDING

1. What is the difference between taste and aroma?

2. What is retronasal sensory? How does one use the retronasal sensation when tasting a beer?

3. What is the most prominent mouth feel in beer?

4. What is a flavor threshold? How is it determined?

5. A certain compound has a flavor threshold of 50 ppm. A sample has 75 ppm of this compound. How may flavor units is this?

6. List the three compounds that are primary flavors in nearly all forms of beer.

7. What would you change in your process to promote an increase in fruity esters in a beer?

8. What are fusel alcohols and where do they come from?

9. Which off-flavors will be removed as a result of conditioning?

10. What are some flavors that come from hops?

11. What is an off-flavor? Are there compounds that are sometimes off-flavors and sometimes desirable flavors?

12. Describe the sensory changes in beer as it ages?

13. What are common sources of bias in a sensory program?

14. Money is not an issue; design a sensory lab.

15. Describe how you would design a sensory panel experiment to evaluate potential differences in beer flavor after the addition of a new piece of brewing equipment.

16. Describe two methods to determine alcohol by volume (ABV).

CASE STUDY

Case Study 1

You are the brewer at a small 7 bbl brewery with only three employees, the owner/taproom manager, the assistant brewer, and you. You do not have access to a trained sensory panel, so much of the quality control is done by you and your assistant. In the last two brews, you noticed the presence of astringency, a harsh, lingering bitterness that had not been there before. Explain how you will identify, evaluate, and solve the problem.

Case Study 2

You are the brewer at a 10 bbl brewery. The brewery recently purchased a small canning line to widen distribution of its best brands. Unfortunately, the shelf life of the IPAs is poor, losing substantial hop bitterness and flavor after only a few weeks in storage and developing flavor notes of onion and garlic. Explain how you will identify, evaluate, and solve the problem.

Case Study 3

Congratulations! You were hired as the new Director of Quality at Brewery X, a 40 bbl brewery with about 20 employees. One of your first orders of business is to develop a sensory program for the brewery that incorporates a panel of trained panelists. Describe your equipment and space needs to build a sensory laboratory. Explain how and with what you will train panelists. Explain how you will manage the sensory program to ensure consistency across all brands.

BIBLIOGRAPHY

ASBC Methods of Analysis Beer 4, 13, 23.

Bamforth CW. 2014. *Flavor*. American Society of Brewing Chemists. ISBN 978-1-881696-23-0.

Barth R. 2013. *The Chemistry of Beer: The Science in the Suds*. Wiley.

Barth R. 2018. Beer Conditioning, Aging, and Spoilage. In Bordiga M (editor). *Post-Fermentation and -Distillation Technology*. CRC. ISBN 978-1-4987-7869-5. p. 186–190.

Briggs DE, Boulton CA, Brookes PA, Stevens R. 2004. *Brewing: Science and Practice*. CRC. ISBN 0-8493-2547-1. Chap. 8.

Dalgliesh CE. 1977. Flavour Stability. *Proc. Eur. Brew. Conv. Congr. Amsterdam*. DSW. p. 623–659.

Pelosi P. 2016. *On the Scent*. Oxford University Press. ISBN 978-0-19-871905-2.

Spedding G, Aiken T. 2015. Sensory Analysis as a Tool for Beer Quality Assessment with an Emphasis on Its Use for Microbial Control in the Brewery. In Hill AE (editor). *Brewing Microbiology*. Woodhead. ISBN 978-1-78242-331-7.

Thompson S. 2010. Sensory analysis: a matter of taste. *ASBC Brewmedia Modules*. https://www.asbcnet.org/lab/webinars/webinars/BrewMedia/ThompsonSue/player.aspx

CHAPTER 13

COLOR, FOAM, AND HAZE

The consumer's first impression of **beer** is by sight. Beer is meant to be seen and appreciated in appropriate glassware. Bottles and cans are simply the means to get it there. From the subtleties of color to the consistency of clarity and to the appealing qualities of stable **foam**, beer is beautiful. While consistency in **flavor** is essential to reproducing **quality** beer, so too are the physical attributes of color, foam, and **haze**. The appearance of the beer depends on the interaction of light with the beer, its foam, and the **glass**. Everything starts with light.

13.1 LIGHT

Light, for our purposes, is a traveling wave, like a wave on the ocean, as opposed to a standing wave, like that on the string of a musical instrument. Traveling waves have four characteristics, two of which are shown in Figure 13.1. The distance between peaks (or other equivalent points) on two successive waves is the **wavelength**, λ (lambda). The peak displacement of the wave is the amplitude (A). The number of waves that pass a fixed point in one second is the frequency, ν (nu). Frequency is measured in reciprocal time (s^{-1}),

Mastering Brewing Science: Quality and Production, First Edition.
Matthew Farber and Roger Barth.
© 2019 John Wiley & Sons, Inc. Published 2019 by John Wiley & Sons, Inc.

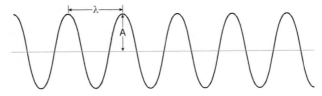

Figure 13.1   Wave.

TABLE 13.1   **Types of Light**

Type	Wavelength Range	Frequency	Effect
Radio	$>1\,m$	$<3\times10^8$ Hz	Induce signal in antenna
Microwave	$1\,m\text{–}0.001$	$3\times10^8\text{–}3\times10^{11}$	Molecular rotation
Infrared	$0.001\text{–}7\times10^{-7}$	$3\times10^{11}\text{–}4.2\times10^{14}$	Molecular vibration
Visible	$7\times10^{-7}\text{–}4\times10^{-7}$	$4.2\times10^{14}\text{–}7.5\times10^{14}$	Color/chemical bonds
Ultraviolet	$4\times10^{-7}\text{–}1\times10^{-8}$	$7.5\times10^{14}\text{–}1\times10^{16}$	Valence electrons
X-Ray/gamma ray	$<1\times10^{-8}$	$>1\times10^{16}$	Core electrons/nuclear

also called hertz (Hz). The speed of the wave is c. The speed, wavelength, and frequency are related by the equation

$$c = v\lambda$$

Based on this equation, as the frequency (v) becomes higher, the wavelength (λ) becomes shorter. The speed of light in a vacuum is a constant of nature with a value close to $3.00\times10^8\,m/s$, so if the wavelength is known, the frequency can be determined.

Ranges of wavelengths are named by detection methods and effects on matter, as given in Table 13.1. Different wavelengths of visible light give rise to the sensation of color (Table 13.2). The human eye has three types of color-sensing organs, called cones, each responsible for an overlapping set of wavelengths. The color perceived is determined by the relative response of the three types of cones. Many colors that we see are combinations of the single-wavelength colors. Natural white light has a maximum intensity at a wavelength of around 480 nm, dropping off rapidly at lower wavelengths and more slowly at higher wavelengths. Light delivers **energy** in packets that depend on the frequency or the reciprocal of the wavelength ($1/\lambda$). Green light at 500 nm (6×10^{14} Hz) delivers energy equivalent to 240 kJ/mol, enough to break a weak chemical **bond**.

When light hits a glass of beer, three things can happen. The light can be transmitted, reflected, or absorbed. *Transmitted* light goes through a material and comes out the other side. *Reflected* light strikes a **surface** and bounces off.

TABLE 13.2   **Colors of Visible Light**

Color	Wavelength Range (nm)
Red	700–620
Orange	620–590
Yellow	590–570
Green	570–495
Blue	495–450
Violet	450–400

Absorbed light does not emerge on the other side; it is converted to another form of energy by the material. Light that is strongly absorbed by a material does not reach the eye; thus the perceived color of an object results from transmitted or reflected light.

Beer strongly absorbs wavelengths corresponding to blue and green, so its color results from yellow and red light that it transmits or reflects. The amount of light that a material transmits depends on the distance through which the light travels in the object (the path length), the inherent intensity of color for the **molecules** absorbing the light (the absorptivity), and the concentration of these molecules. The **transmittance** of a sample is the ratio of light power emerging from the sample (I) over that entering (I_o). The transmittance is defined by

$$T = \frac{I}{I_o}$$

where I is the light power emerging from the sample and I_o is the light power that enters the sample. The percent transmittance is the transmittance expressed as a percentage: $\%T = T \times 100\%$. The light power is usually measured at a specific wavelength. The **absorbance** is a measure of the amount of light absorbed by the sample. It is calculated from

$$A = -\log(T) \quad \text{or} \quad A = 2 - \log(\%T)$$

The relationship among the absorbance, the path length (distance the light travels in the sample), and the **concentration** of the light-absorbing material is called **Beer's law**:

$$A = \varepsilon bc$$

where A is the absorbance, ε is called the absorptivity (a characteristic of the light-absorbing material and of the wavelength), b is the path length, and c is the concentration of the colored material. Many analytical procedures use Beer's law to determine the concentration of a **compound**.

The principle of Beer's law is manipulated by the bartender to influence the appearance of beer. For example, when paleness is a desired style characteristic, as with Pilsner beer, the beer is often served in a tall, narrow glass. The narrow glass provides a short path length for light, providing less perceived color.

When light passes from one material to another, it may be transmitted or reflected at the **interface**, usually both. The details of this behavior depend on the speed of light moving through the material and the angle between the direction of the light ray and the direction of the interface between the materials. The speed of light in a material is usually expressed as the **refractive index**, which is the ratio of the speed of light in a vacuum (close to that in air) to that in the material. The more a material retards light, the greater its refractive index. The refractive indices of some common materials are shown in Table 13.3.

Figure 13.2 shows a ray of light, the incident ray, moving from a high index material, like water or beer, to a low index material, like air. At the interface where the materials meet, some of the light is transmitted – it continues to travel into the other material – while some is reflected. The imaginary line at a right angle to the interface is the normal (N). The angle between the normal and the incident ray (θ_i) is equal to that of the reflected ray (θ_r). The transmitted ray changes direction, a process called **refraction**. When the ray is transmitted from a high to a low index material, the light is bent away from the normal, that is, down toward the interface. This behavior is captured in the mnemonic "high to low, dive below." When the incident angle is small, the bending is small. The amount of bending increases as the ray moves closer to parallel with the interface (higher θ_i). The fraction of the light that is transmitted from a high to a low index material decreases, and the fraction that is reflected increases.

The application of this principle is that every time the light ray encounters an interface, the light is scattered; the interface might be a haze particle or a **gas** bubble, resulting in light that is both reflected and transmitted. When there are many such interactions, the light follows a complicated path, breaking up any image. In this case, the material appears cloudy or opaque. This is why we cannot see very far through foam or a New England IPA.

TABLE 13.3   **Refractive Indices of Some Materials**

Material	RI
Vacuum	1
Air	1.000
Carbon dioxide	1.001
Water	1.333
5% ethanol in water	1.336
10% sucrose in water	1.348
Ethanol	1.361
Glass	1.518

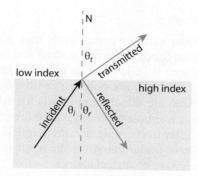

Figure 13.2 Transmission and reflection.

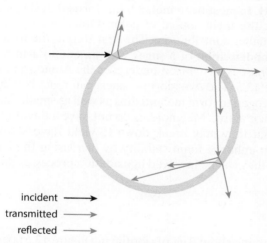

incident ⟶
transmitted ⟶
reflected ⟶

Figure 13.3 Reflection and refraction in a bubble.

Some of the paths that a ray might take in a single bubble are shown in Figure 13.3. The liquid film around the bubble has two interfaces, one with air on the outside and one with gas on the inside. Each time a ray encounters one of these interfaces, it is subject to reflection and refraction, only a few instances of which are shown in the figure.

13.2 COLOR

The most significant source of color in beer is a family of complex network **polymers** called **melanoidins**. Riboflavin, a yellow compound also called vitamin B_2, is another source whose color is mostly evident in very pale styles of beer. Using the properties of light and Beer's law, the color of **wort** or beer is measured to ensure consistency and trueness to type.

Melanoidins

Melanoidins are the same class of compounds that give a familiar brown color to roasted and baked foods, including **malted** and roasted **grain**. Despite a good deal of effort, there is little agreement about their molecular structures. One suggestion is that they are polymers with a backbone as shown in Figure 13.4. The wavy lines are attachment points for additional **units** of the structure. This suggestion is simpler, although no more plausible than several other possibilities that differ in every significant detail. There is general agreement that melanoidins have **molar masses** in the range of 10000–60000 g/mol and that they carry a negative electrical charge.

The sequence leading to the formation of melanoidins is called the **Maillard** (MAY-yard) reaction. It is favored by high temperature, low water concentration, and high **pH**. In most beer, melanoidins formed during malting or roasting of grain are the main source of color. Other hot processes, especially boiling, make smaller contributions. The first step in the formation of melanoidins is the **condensation** of a **sugar** with an **amino acid** to form a Schiff base. The Schiff base is unstable; it undergoes the Amadori rearrangement as shown in Figure 13.5. The Amadori products can react by additional loss of water and of nitrogen to form melanoidins as well as smaller molecules, many of which are flavor active. Melanoidins do not have a flavor of their own, but it is believed that they may break down to yield flavor-active compounds. Melanoidins can enhance foam stability by increasing the local **viscosity** at the bubble, and they can be involved in oxidation processes that contribute to beer **staling**.

Riboflavin

Barley malt contains about 3 µg of riboflavin (Figure 13.6) per gram, most of which ends up in the beer. The intense yellow color of riboflavin is most

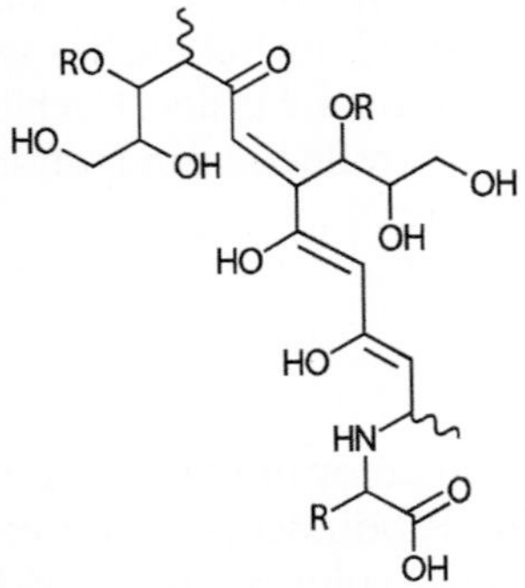

Figure 13.4   Melanoidin backbone (one possibility).

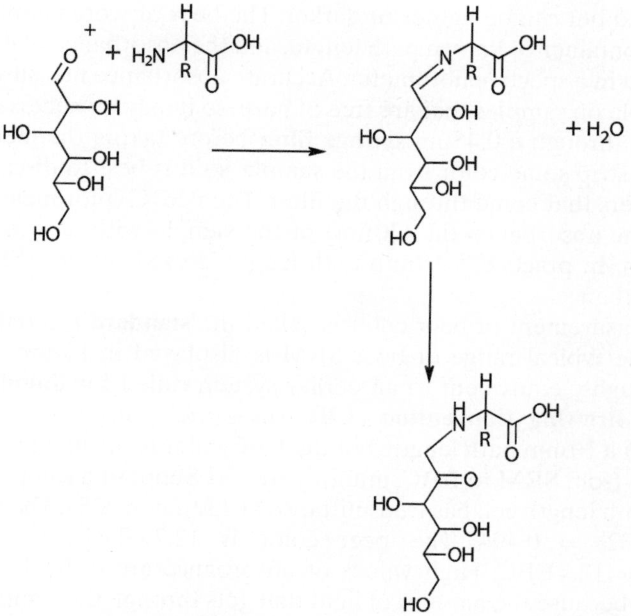

Figure 13.5   Schiff base formation and Amadori rearrangement.

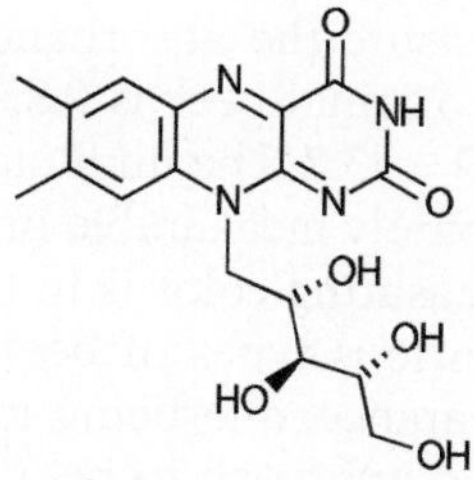

Figure 13.6   Riboflavin.

evident in beer styles made from very pale malt, such as Mexican pale **lager**. The color comes from the alternating single and double bonds in the three-ring portion of the molecule. Riboflavin can capture light energy and transfer it to **isomerized hop** acids. Ultimately this gives rise to the skunky aroma of **lightstruck** beer as discussed in Chapter 12.

Analysis and Quality Control of Color

There are two distinct approaches to measuring beer color (also appropriate for **wort** color). The simplified approach is to assume that beer is essentially beer colored but can be lighter or darker. The beer or wort sample is placed in a glass container of known path length, and the absorbance of 430 nm light is measured in a spectrophotometer. Accurate absorbance measurements are only possible on samples that are free of **haze**, so it may be necessary to drive the sample through a 0.45 μm syringe **filter** before taking the measurement. Filters can strip some color from the sample, so it is best to discard the first few milliliters that come through the filter. The ASBC color measurement is 10 times the absorbance (at 430 nm) of the sample with a ½ in. (12.7 mm) path length. In practice, a 10 mm path length is used so the ASBC color is 12.7 × absorbance.

This measurement of beer color is called the **standard reference method** (SRM). The typical range of beer SRM is displayed in Figure 13.7. ASBC color is roughly equivalent to an earlier system called **Lovibond** color. The **European Brewing Convention (EBC)** uses the same wavelength as the ASBC with a 10 mm path length, but the EBC color is calculated from $25 \times A$. To convert from SRM to EBC, multiply by 1.97. Suppose a sample of beer in a 10 mm path length cell has transmittance at 430 nm of 32%. The absorbance is $2 - \log(32) = 0.494$. The beer color is $12.7 \times 0.494 = 6.3$ SRM or $25 \times 0.494 = 12.4$ EBC. High values of absorbance are difficult to measure accurately because the amount of light that gets through the sample becomes too small. For dark samples, the beer can be diluted to bring the absorbance into a measurable range; the absorbance of the undiluted sample is then calculated. For example, suppose we take 25 mL of beer and add enough water to give 100 mL total. We measure the absorbance of the diluted sample as 0.85. The absorbance of the original beer is $0.85 \times (100 \text{ mL})/(25 \text{ mL}) = 3.40$. The ASBC color is $12.7 \times 3.40 = 43.2$. The undiluted beer would have a transmittance of 0.04%, which is barely measurable on many instruments.

The other approach to measuring color is to take into account the actual differences in the hues of various types of beer. Beers can have the same SRM but still differ in appearance, one being more red and another being more yellow. A more detailed approach to beer color is to report the color using the method of the *Commission Internationale d'Eclairage*, called CIE LAB color. The CIE LAB method yields three numbers, L*, a*, and b*. L*

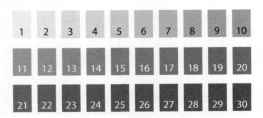

Figure 13.7   Beer color SRM chart.

corresponds to the lightness, a* to position on the red-green axis, and b* to position on the yellow-blue axis. To measure CIE LAB color, the transmittance must be measured at many points across the entire visible range. Software is available in many spectrophotometers to calculate and report CIE LAB color from these measurements. CIE LAB color is a more comprehensive description of color, but most color specifications, for example, those in style guidelines, go by ASBC or EBC color.

13.3 FOAM

Foam makes a major contribution to the visual impact of beer. The **head** of foam on a glass of beer is instantly recognized as a characteristic distinguishing beer from other sparkling beverages. Foam enhances the **organoleptic** quality of the beer, that is, it improves character perceived by the senses. Foam is visually appealing. It changes the texture of the beverage on the palate. It provides effervescence and aromatic scents because **volatile** flavors are released as the foam bubbles break. In beer, a long-lasting head of foam is desirable, but the foam must not be so copious that the beer becomes difficult to serve or drink.

By contrast, any foam during wort boiling, in most **cleaning** systems or in utilities like boilers or **chillers**, interferes with their functions and must be suppressed. Certain cleaning chemicals are specifically designed as foams to achieve effective coverage of equipment exterior surfaces.

Surfaces

A **surface** is the place where a liquid or solid meets a **gas**. The general term for the place where phases meet is an **interface**. Energy is needed to create or extend a surface (but not necessarily an interface). This **surface energy** is released when the surface is destroyed. The energy needed to make a unit area of surface is the **surface tension**, symbolized γ (gamma).

Surface energy results in a spontaneous tendency for particles, including bubbles of foam, to take forms that minimize their surface areas. Because of this minimizing tendency, bubbles act in some ways like elastic balloons; they squeeze the gas inside, raising its **pressure**. The smaller the bubble and the higher the surface tension, the greater the excess pressure inside. A very small bubble would have unrealistically high inside pressure. For example, a 1 µm (0.001 mm) gas bubble in water ($\gamma = 72$ **dyne**/cm $= 0.072$ N/m) would have an inside pressure nearly 3 bar (44 **psi**) higher than the outside pressure. This makes it difficult for a new bubble to form in the middle of the liquid.

Under most circumstances, bubbles form at **hydrophobic** locations, called **nucleation sites** that **attract** gas molecules (Figure 13.8). Once a bubble forms,

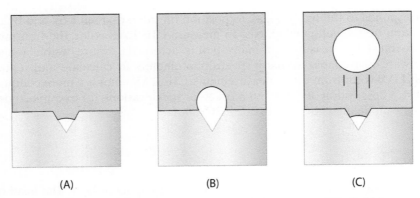

(A)   (B)   (C)

Figure 13.8   Foam formation. (A) Nucleation site in glass. (B) Bubble growth. (C) Bubble breaks off.

it can easily absorb dissolved gas from the liquid, thus growing larger until big enough to break free and float to the surface. Some beer glasses are etched on the bottom to provide nucleation sites. One surprising source of nucleation sites is tiny **cellulose** fibers that may adhere to the insides of glasses. In this manner, a dirty glass is often revealed by excessive bubbles on the walls of a recently filled glass of beer.

Foam Formation, Persistence, and Deterioration

A bubble on a nucleation site experiences a force attracting it to the site and an opposing buoyant force pulling it upward. As the bubble grows, the buoyant force increases until it tears most of the bubble off the site and into the bulk liquid. A remnant remains to nucleate the next bubble (Figure 13.8). The free bubble is pulled toward the liquid surface by **buoyancy**, growing as it rises because of additional gas absorbed and because the pressure exerted by the **weight** of the liquid becomes less near the surface. When the bubble reaches the liquid surface, it is pushed up and out of the liquid by other bubbles arriving from below. It then joins other bubbles to form a foam layer. Up to this point, the behavior of a beer bubble is the same as that of a bubble of any other **carbonated** liquid, like soft drinks and sparkling wine. The difference is that beer foam persists for several minutes, in contrast to the foam of other beverages, which collapses in seconds. To understand why beer foam persists, we first need to evaluate what makes foam collapse.

The bubbles in foam are typically about 0.2 mm in diameter but can be as large as 0.5–2 mm. The liquid content of fresh foam is about 0.2 mL/mL of foam. There are about 170 000 bubbles in 1 mL of foam. The surface area, that is, the area of contact with the gas inside the bubbles, comes to about 220 cm^2/mL of foam. The surface area of the same amount of liquid if it were in a single spherical drop would be 1.65 cm². Thus, in the form of foam, the surface of the liquid increases by a factor of 133! This also means that a certain volume of foam has 133 times the surface energy of the liquid it contains. When the foam collapses, energy is released as heat, a highly spontaneous process. There are two approaches that can control a spontaneous process. It can be made less spontaneous, or it can be slowed. Both approaches contribute to foam stability, known in **brewing** jargon as **head retention**.

To make foam collapse less spontaneous, we might lower the surface energy. The surface energy is the surface area multiplied by surface tension. Lowering the surface area is counterproductive, because it would mean less foam. On the other hand, the surface tension can be lowered by substances called **surfactants**. A surfactant molecule has a **polar (hydrophilic)** end that is attracted to the water in the beer and a nonpolar **(hydrophobic)** end that can extend into the gas in a bubble. Very small quantities of surfactant lower the surface tension a good deal, which results in less energy release when the foam collapses and thus a more stable foam. Surfactants in beer are mainly **protein** fragments derived from the malt but may also be hop-derived **polyphenols** or modified **alpha acids**.

If we turn our attention back to natural foam collapse and a consideration of how to slow the process, the four **mechanisms** for foam deterioration over time are draining, **disproportionation, diffusion**, and **coalescence**.

Draining   When first formed, foam can be as much as 40% liquid by volume. The liquid drains down, through the foam and into the beer, under the influence of gravity. After a few minutes, the foam typically drains to about 20% liquid. This amounts to a decrease in foam volume by 20% without any change in gas content, bubble size, or number of bubbles. The foam continues to drain more slowly, but other processes become more dominant. The local **viscosity** of the liquid in the space between the bubbles has a big effect on drainage. Higher viscosity yields slower drainage. Bubble size also affects drainage. Smaller bubbles make the drainage path longer and more convoluted, slowing drainage. Drainage shortens the distances between the bubbles and from the inside to the outside of a bubble. These distances affect the speed of other foam deterioration processes.

Disproportionation   Disproportionation is the gain of material by larger particles at the expense of smaller particles, giving a net increase in particle size and loss of surface area. When the particles are bubbles, disproportion occurs because the smaller bubbles have a higher internal pressure than the larger bubbles. Higher pressure gives higher solubility, according to Henry's law. In the presence of small and large bubbles, the gas from the small bubbles will dissolve and be absorbed by the large bubbles. Disproportionation makes the foam coarser with larger but fewer bubbles; the foam volume is not directly affected. Foam coarsening is slowed by lowering the surface tension. This lowers the pressure difference between large and small bubbles. Wetter foam yields thicker bubble walls, slowing down the movement of the dissolved gas through the liquid. The inclusion of nitrogen, which has very low solubility in beer, inhibits disproportionation. Lower temperature slows down the diffusion of dissolved gas through the liquid. Materials that stick to the bubble surface can physically block the migration of gas from a bubble. Disproportionation is regarded as the most significant mechanism of foam deterioration.

Diffusion   Diffusion is the movement of material from a region of high concentration to a region of low concentration. The top layer of bubbles is in contact with the atmosphere. The atmosphere consists of about 80% nitrogen and 20% oxygen by volume. The interior of a bubble consists mostly of carbon dioxide. Carbon dioxide from a bubble in contact with the atmosphere diffuses across the bubble wall to the atmosphere. Air gases diffuse in, but because oxygen and nitrogen are hardly soluble in the bubble wall, the exit of carbon dioxide is much faster than the entrance of air. The bubble shrinks to a small fraction of its original size. The shrinkage of the bubbles on top exposes the next layer of bubbles to the atmosphere. Eventually a thin layer of very small bubbles filled with air remains on top of the beer. Diffusion can be slowed down by covering the beer, increasing the concentration of carbon dioxide in the atmosphere in contact with the foam.

Coalescence   Coalescence is the merging of smaller particles to form larger particles. Coalescence of bubbles is manifested by the breaking of the wall between two bubbles, so they merge to form a single bubble. Breaking a wall between bubbles makes the foam coarser, that is, it has larger bubbles, but the total volume of foam is not directly affected. If the walls between the atmosphere and bubbles in the top layer break, those bubbles are lost and the foam volume decreases.

Lacing

Lacing, shown in Figure 13.9, is foam that clings to the sides of a glass after the beer has been consumed. Customers vary in their perception of lacing. Some regard lacing as highly desirable; some dislike it; and some are ambivalent. The authors consider lacing to be a beautiful quality that enhances enjoyment of a beer and thus deserving of attention.

Lacing requires that the materials at the surface of bubbles be very stable, virtually solid. The glass must be free of **fats** and **detergents**. The glass is said to be "beer clean." It is ironic that dislike of lacing is associated with a perception that a laced glass looks dirty.

Figure 13.9   Lacing. *Source*: Photo: Naomi Hampson.

Foam-Active Materials

There are three related aspects to foam activity: foamability, foam stabiliza-tion, and foam quality enhancement. *Foamability* is the ability to form foam without regard to its characteristics. *Foam stabilization* is the ability to make the foam last. We can highlight the distinction between foamability and foam stabilization by considering a beer with excessive foamability and insufficient foam stabilization. The beer will be difficult to serve because foam will fill the glass as it emerges from the **tap**. But the foam quickly collapses. The bar-tender will have to fill the glass in several stages, allowing the foam to subside each time. Because of the poor stabilization, the foam will subside shortly after the customer starts drinking. *Foam quality* is the visual appeal of the foam to the drinker and includes aspects like color, bubble size, and lacing (foam clinging to the sides of the empty glass). Consumer perception of foam quality varies geographically and culturally.

For a material to be foamable, it must travel rapidly to the bubble inter-face, stick to it, and deform to match the contours of the bubble. Foamable materials have small or flexible molecules with both hydrophobic and hydro-phobic regions. They lower the surface energy, allowing the foam to form. Short **polypeptides** and molecules like hop bitter compounds tend to be foamable as they stabilize the surface tension of the bubbles. Foam stabilizers need to form a **tough** network around the bubbles to inhibit migration of the gas from the bubble. Their molecules stick together and pack well, covering the bubble surface. Because of their ability to lower surface energy, foamable molecules tend to drive foam stabilizers from the bubble surface. An illustra-tion of the foam stabilizing ability of proteins and hop polyphenols is demon-strated in Figure 13.10.

Proteins Hop bitter compounds and polyphenols are thought to form foam-positive associations with water-soluble proteins and protein fragments (polypeptides) derived from malt. The identity of these proteins is an area of debate. On the one hand, it is believed that specific barley proteins contribute to foam. LTP1 (lipid transfer protein) and protein Z have been shown to contribute to foam stability and represent the most abundant proteins in beer. On the other hand, it is argued that all proteins may contribute to beer foam given the right biochemical properties. To support this idea, one group demonstrated that certain foam-active proteins are actually derived from the surface of yeast cells.

Irrespective of the identity of the foam stabilizing proteins, their biochemi-cal quality is key. To be soluble in beer, a polypeptide must be at least partly hydrophilic, that is, it must have charged or polar groups or regions that are attracted to water. To stabilize foam, the polypeptide must also have

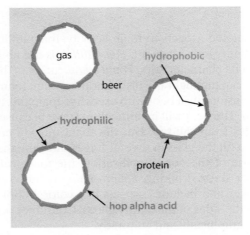

Figure 13.10 Foam-stabilizing proteins and hop acids. Hydrophobic regions on proteins stick to gas. Hydrophilic regions stick to beer.

hydrophobic (nonpolar) regions that are attracted to the gas in the bubble. A model of this concept is demonstrated in Figure 13.10.

The most effective foam stabilizing polypeptides are those with the greatest hydrophobicity, up to the limit that extremely hydrophobic polypeptides will not dissolve in beer. When the polypeptide is in **solution**, the hydrophobic regions tend to fold together to avoid the water. In contact with the bubble, a foam stabilizing polypeptide must reorient itself to bring the hydrophobic regions into contact with the bubble. The polypeptides make a coating on the bubble that lowers surface tension and serves as a barrier to migration of the gas from the bubble. This reorientation is not always reversible, so the polypeptides are good for only one use. If foam is formed during the brewing process, subsequently formed foam in finished beer is usually of inferior quality.

Hop Bitter Acids   Hop bitter acids, shown in Figure 8.10, help stabilize foam and are particularly important for lacing. The iso-alpha acids are believed to bind to foam-active polypeptide molecules and make links between them, as shown in Figure 13.10. This cross-linking stabilizes and stiffens the coating. Iso-alpha acids are **amphiphilic**, they have polar and **hydrogen bonding** groups, and they have nonpolar side chains. Hydrogenated iso-alpha acids (Figure 4.30) tetrahydroisohumulone and especially hexahydroisohumulone are more foam active than the natural hop acids. Hydrogenation of the double bonds on the iso-acid side chains increases their effective volume by an unexpectedly large amount, increasing their hydrophobicity.

Melanoidins   Melanoidins from dark malt can stabilize beer foam and give rise to smaller bubbles. This may result from attraction between negative charges on the melanoidins and positive charges on the foam stabilizing polypeptides.

Ethanol   "Beer" with no **ethanol** usually has very deficient foam. At low concentration (<1% by volume), ethanol enhances foam stability, perhaps by lowering the surface tension. As the ethanol concentration increases above 1%, foam stability decreases.

Foam Inhibitors   Common foam inhibitors in beer are **lipids** from barley, lipids from yeast, **detergents**, and other contamination from brewing equipment or processes, like lubrication. Lipids from barley can be minimized by reducing **trub** carryover. Lipids from yeast can be minimized by promoting yeast **viability** and preventing **autolysis**. Lipids, especially fats and **esters** of fatty acids, are strong foam inhibitors. They also have a negative effect on lacing. Detergents, although foam enhancers on their own, inhibit foam and lacing in beer. It has been suggested that the detergent molecules insert themselves into the bubble surface and interfere with the linking between iso-alpha acids and foam polypeptides. Ethanol in high concentration inhibits foam.

Additives   Some foam-enhancing additives are believed to increase the local viscosity of the beer in the bubble walls. The most commonly used additive is **propylene glycol alginate** (PGA), a copolymer of mannuronic acid and guluronic acid, with some of the **carboxylic acid** groups condensed with propylene **glycol** ($CH_3-CHOH-CH_2OH$). The uronic acids are sugars whose end carbon is a carboxyl (COOH) group instead of a hydroxyl (OH) group. A fragment containing one mannuronic acid ester (left) and one guluronic acid (right) is shown in Figure 13.11. In practice, there are usually blocks of many of the same uronic acid, rather than a simple alternation. PGA is made by reacting alginic acid, which is derived from seaweed, with propylene glycol. It is usually dissolved in water and dosed at about 50 ppm before the final filtration. PGA is incompatible with some **finings**, forming visible fluffy or slimy **precipitates** in packaged beer after several months. It has been suggested that the problem may be alleviated if the concentration of alginic acid, a common impurity in PGA, is very low.

Metal ions with a +2 charge enhance foam stability. These ions may bind to **carbonyl** (C=O) groups on the ring of iso-alpha acids helping to link the acids to one another, perhaps extending the range of linking of the acids to foam-active proteins. Cobalt sulfate was used as a foam-enhancing additive in some beers until the 1960s when its use was discontinued because of concerns about the toxicity of cobalt.

Influence by Process Steps

Grist   More protein in the **grist** gives more foam, possibly at the expense of other aspects of beer quality, like haze formation. The use of less **modified** malt enhances foam. Wheat malt, and even wheat flour, can increase foam.

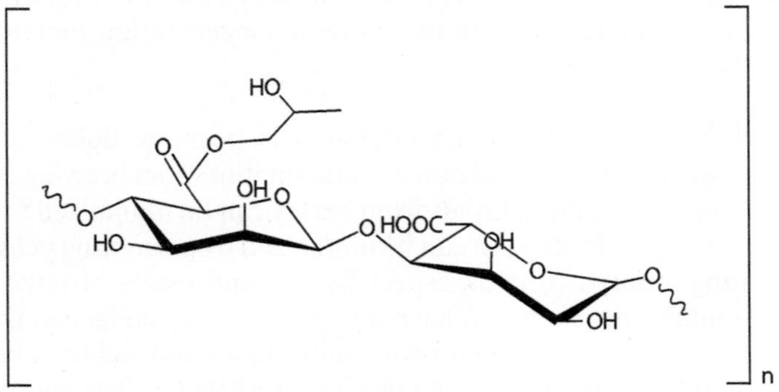

Figure 13.11   Propylene glycol alginate.

Dark malt adds foam-enhancing melanoidins. Low-protein adjuncts, including sugars, rice, and maize, provide little or no protein and decrease foam.

Mashing/Wort Separation   High temperature **mash**-in is said to increase foam. The use of a protein rest around 44–59 °C (113–138 °F) is thought to reduce foam stability due to increased activity of protein-**hydrolyzing** enzymes at the lower temperature. Lipid pickup during mashing and wort separation is also detrimental to foam.

Boiling/Hops   Prolonged boiling slightly inhibits foam. Boiling has the beneficial effect of extracting foam-enhancing iso-alpha acids from hops. The use of hops products with reduced tannins enhances foam. Interestingly, **dry hopping** can influence foam depending on the variety. When dry hopping at 0.4 kg/hL (1 lb/US bbl), some hop varieties were shown to enhance foam stability, while others resulted in a reduction of foam.

Fermentation   **High-gravity fermentation** yields beer with reduced concentrations of foam-active polypeptides, significantly decreasing beer foam. The formation of foam during fermentation depletes the activity of the foam-enhancing materials. Yeast releases a proteinase that can hydrolyze foam-active polypeptides. If this enzyme is not destroyed by **pasteurization**, it will slowly degrade foam quality.

Conditioning/Packaging   Any procedure that generates foam during processing depletes foam stabilizing proteins in finished beer. Dissolved gas, normally carbon dioxide, is essential for the formation and persistence of a head. The inclusion of some nitrogen can greatly increase foam stability quality. Details of this phenomenon are provided in the box below.

NITROGENATED BEER

Nitrogenated beer, also called nitrogenized beer or "beer on nitro," is beer that has been equilibrated with a relatively low pressure of carbon dioxide and a high pressure of nitrogen (typically 25% CO_2/75% N_2). Nitrogenated beer can be served with a creamy head with very fine bubbles that last for a long time. Beer served in this way has a smoother mouth feel than that served at ordinary levels of carbon dioxide and no nitrogen. The unusual properties of the head derive from differences in the chemistry of the two gases and from the physics of bubbles. At 4 °C, carbon dioxide is about 70 times more soluble in water (on a volume basis) than nitrogen is under the same conditions of pressure.

(continued)

This property of nitrogen results in slower gas diffusion as compared with CO_2, leading to smaller bubbles and more stable foam. Nitrogen bubbles release hop aroma less effectively than those of carbon dioxide; for this reason nitrogenation is often considered less suitable for hoppy, aromatic beer styles.

When a nitrogenated beer is poured into a glass, it has a homogenous, creamy foam. Because the bubbles are smaller, the buoyant force pulling them to the top is less, and they rise slowly. At the wall of the glass, downward flow of beer can carry the bubbles with it, resulting in a beautiful cascading effect.

Analysis and Quality Control of Foam

The major technical challenge in foam analysis and quality control is the difficulty of generating foam from beer in a reproducible, yet realistic, way. Perhaps the easiest method is to simply dispense the beer into a clean, clear glass, recording impressions on the stability and attractiveness of the head. The problem here is reproducibility and quantification. The following are general descriptions of several methods used to assess head retention. They are organized in increasing order of trouble and expense, going from free to over $50000 (US).

The Shake Method   The shake method, described by Kapp and Bamforth (2002), is a simple test that involves physical agitation in a tube to generate foam. In brief, 5 mL of beer is added to a 15 mL tube. The tube is capped tightly and shaken in a 40 cm arc with 10 oscillations to be completed in 10 seconds. The initial foam height is measured by calculating the distance (mm) from the top of the foam to the foam/liquid interface. The cap is removed and final foam height is measured after 30 minutes. Foam height or percent change of foam height is reported.

Constant Method   In the Constant method, a bottle of beer at the measuring temperature is mounted in a clamp on a ring stand over a 1 L Berzelius (tall-form) glass beaker. The pouring apparatus is arranged so that the pouring angle is determined by a mechanical stop. During the pour, the container opening is 4 cm above the beaker, and the beer is directed to the bottom of the beaker, rather than the sides. When all the beer has been poured, a timer is started. The height of the liquid–foam interface and the top of the foam are read at intervals. The height of the liquid after all foam has collapsed is also measured. The logarithm of the foam height against time is fit to a straight line, and various parameters are derived. This method is slightly more costly than the shake method, but it yields numerical results that can be useful for monitoring foam quality. This method is most applicable to beer in bottles or cans.

Sigma Value Method (ASBC Beer 22a)   The beer is poured into a cylindrical glass funnel of specific dimensions (pictured in Figure 13.12), with a stopcock at the bottom, until the foam reaches a set height. Thirty seconds later, the stopcock is opened and the liquid drained at a moderate rate. The clock is started. After the clock runs 200 seconds, the liquid is drained into a graduated cylinder over a period of 30 seconds, after which the clock is stopped and the liquid volume (volume b) is measured. The foam remaining in the vessel is discharged with a measured volume of a defoaming agent, and the liquid is collected in a graduated cylinder. The volume is corrected by subtracting the volume of the agent giving volume c. A calculation involving the clock time (about 230 seconds) and volumes b and c yields a metric termed the sigma value. Sigma values in the range of 95–105 seconds indicate average quality. For best results, beer must be dispensed into the funnel in a consistent and repeatable manner.

Rudin Method   A glass tube with a porous disk on the bottom is marked at intervals above the disk. Beer at the measurement temperature is added up to the 10 cm mark. Carbon dioxide is blown through the disk, converting all the liquid beer to foam until the foam reaches the 32.5 cm mark. The time it takes for the liquid level from the collapsing foam to go from the 5 cm mark to the 7.5 cm mark is the Rudin head retention value (HRV).

Flash Foam Method (ASBC Beer 22b)   Beer is driven through a 0.8 mm orifice, converting it entirely into foam, 200 mL of which is collected in a graduated cylinder. After 90 seconds, the volume of liquid is measured. The remaining foam is collapsed with isopropanol. The ratio of the volume of beer released in 90 seconds to the entire volume of beer in 200 mL of foam is multiplied by 200 to give the foam value units (FVU) of the beer.

NIBEM Method (ASBC Beer 50)   A NIBEM apparatus generates a specific amount of foam and has **electrodes** that detect the top of the beer head. As the head recedes, the electrodes are automatically lowered to maintain contact with the foam. The rate of lowering is recorded and used to generate a head retention time called the NIBEM value.

Lacing   One method for lacing analysis is to generate a 2 cm head on a glass of beer. Beer is withdrawn from the bottom of the glass at intervals, as though someone were sipping at it. When all the liquid has been removed, the lacing is washed down to give a known volume of solution. The concentration of

Figure 13.12   A foam funnel used to measure the foam sigma value.

beer in the solution is analyzed by light absorption at 230 nm and used to generate a lacing index.

13.4   HAZE

Clear beer is generally considered an important quality factor. Considerable effort has been placed on ensuring clarity and its stability in beer over time. Here we discuss what constitutes beer haze, how it is reduced, and how it is measured. On the other hand, hazy or New England-style IPAs promote and embrace haze. Just as consistent clarity is a quality characteristic of **bright beer**, consistent haziness is a quality characteristic of hazy beer. The promotion of haze will not be discussed here.

Cloudiness, termed haze, is regarded as a **defect** in most beer styles. Beer that is adequately clear is said to be bright. Haze in beer is caused by the presence of particles, especially those whose size is on the order of the wave-length of visible light. Particles that are large enough to be seen individually are called **bits**.

Bits

Bits, also called floaters or snowflakes, are visible particles in the beer often associated with age. Bits are easy to remove by filtration, so any bits in the consumer's beer will have been formed during handling and storage of the beer after packaging or from prolonged age. One source of bits is handling of the beer in ways that raise foam. The bubble films are sometimes stable enough that they remain as bits after the foam has collapsed. Finings and other additives can interact in the package to produce bits. PGA (Figure 13.11), and to an even greater extent its common impurity, alginic acid, can interact with finings like **papain** and **isinglass** to produce a blizzard of bits. Brewers should be cautious about introducing new additives. It may take months for the bits to appear. The revised product should be subjected to several weeks of accelerated aging as a minimum precaution. Other potential sources of bits include impurities from processing such as dust from container closures, lubricants, and **filtration media**.

Protein–Polyphenol Haze

Certain polypeptides can be can **bound** together by polyphenol molecules to generate particles large enough to scatter light but small enough to remain in suspension. Grain storage proteins that contain a high fraction of proline and glutamine (Figure 13.13) yield fragments that are most active for haze formation. The polyphenols most implicated in haze formation are proanthocyanidins, which are composed of two connected polyphenols, each of which has a pair of

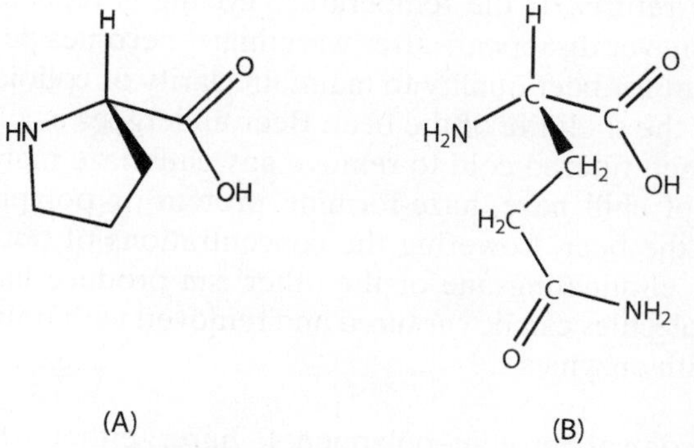

(A)   (B)

Figure 13.13   (A) Proline. (B) Glutamine.

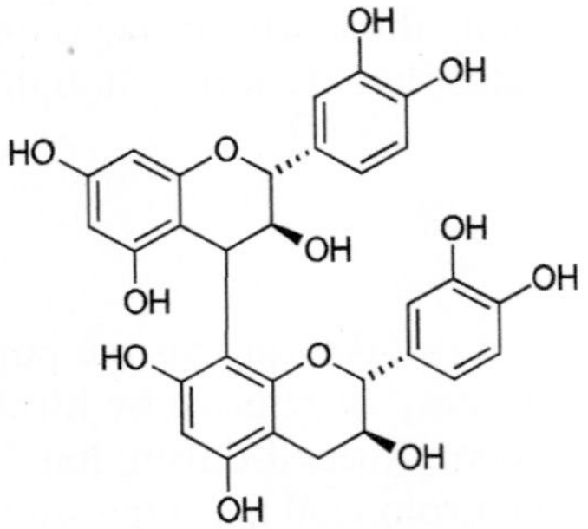

Figure 13.14   Procyanidin B3.

fused six-member rings connected to a third six-member ring. Figure 13.14 shows procyanidin B3, a haze-active proanthocyanidin found in beer.

Protein–polyphenol particles with only a few polypeptide molecules are loosely held together. They break up at room temperature and only display visible haze at low temperature. This type of haze is called **chill haze**. Polyphenols can polymerize, especially in the presence of oxygen, which gives them a tendency to bind more polypeptide molecules into larger particles. The larger particles do not break up at room temperature. They give rise to permanent haze that is visible at any temperature. Permanent haze is promoted by age, heat, oxidation, heavy metals, and physical forces. Haze may also be formed by excessive β-**glucan**, excessive polyphenols, yeast in suspension, and microbial contamination.

Promoting Colloidal Stability of Beer

A cold bottle of beer is poured into a glass. The beer is slightly hazy. As the beer warms up, the haze disappears. This is chill haze. When the beer is chilled again, the haze returns. If the temperature cycling is repeated a few more times, the haze never disappears after warming; it becomes permanent haze.

It is important for beer quality to maintain clarity or colloidal stability for the duration of the shelf life of the beer. Beer undergoes a cold conditioning period and is then filtered cold to remove any chill haze material. But even after removal of chill haze, haze-forming protein or polyphenols are still likely to be in the beer. Lowering the concentrations of both proteins and polyphenols or eliminating one or the other can produce haze-stable beer. Haze-active molecules can be captured and removed with finings, or they can be degraded with enzymes.

Finings To control protein–polyphenol haze, either the haze-active polypeptides or the polyphenols are removed, usually by the addition of materials called finings. Finings clarify the beer by binding to haze-active molecules forming aggregates that can be removed by **sedimentation** or filtration. Any treatment that removes or destroys proteins can be expected to have some effect on head retention, because some haze-active polypeptides help promote foam. Some of the more common finings are listed below.

Silica gel is prepared by dissolving sand in sodium hydroxide to give sodium silicate (Na_2SiO_3), acidifying the solution to **precipitate** silicon hydroxide (silicic acid, $Si(OH)_4$), and then drying in a **kiln** to give an **amorphous** (noncrystalline), highly porous solid. Certain grades of silica gel selectively bind haze-active polypeptides and have a lower (but not zero) affinity for larger foam-active polypeptides. Silica gel is usually mixed with deaerated water and added to the beer on the way to the final filter.

Gelatin is a **mixture** of polypeptides made by hydrolyzing **collagen**. It is usually prepared from animal skins. Gelatin selectively binds and **coagulates** haze-active polypeptides, which can then be allowed to settle. Gelatin molecules are positively charged at the pH of beer. They can precipitate yeast, which has negatively charged cells. Gelatin forms **hydrogen bonds** with haze-forming polyphenols allowing them to sediment out. An issue with gelatin is that, as an animal product and potentially containing pork, it is unacceptable to some customers.

Isinglass is a refined form of collagen derived from the swim bladders of fish. In modern times, certain fish from between the tropic of cancer and the equator, mostly in the region of Southeast Asia, are used. Like gelatin, isinglass has a positive charge at beer pH. It is attracted to yeast cells, which have a negative charge, allowing them to form large particles that can be removed by sedimentation. Usually isinglass is used to precipitate yeast from top fermenting beer. It is typically added as a **slurry** after primary fermentation. Like gelatin, isinglass is not acceptable to all customers.

Tannic acid is a polyphenol of plant origin. It binds and precipitates proteins. Its use as a beer fining is not widespread, but advocates claim that, in purified form, it selectivity precipitates haze-active proteins.

Polyvinylpolypyrrolidone (**PVPP**), an insoluble, artificial polymer shown in Figure 13.15, selectively removes polyphenols from beer. PVPP can accept hydrogen bonds from the —OH groups of polyphenols. The bonding is supplemented by **stacking forces**. The PVPP particles with bound polyphenols are filtered out. PVPP is classified either as one-use or as regenerable. The one-use grade has smaller particles giving a larger surface for interaction. The regenerable grade requires a dedicated filter to capture the PVPP, which is then washed with a sodium hydroxide solution and **neutralized**. One-use PVPP is often introduced together with silica gel, both of which are removed with the same filter. This practice lowers the required dosage of both finings. Because of the lower dose of silica gel, the risk of head retention loss is decreased.

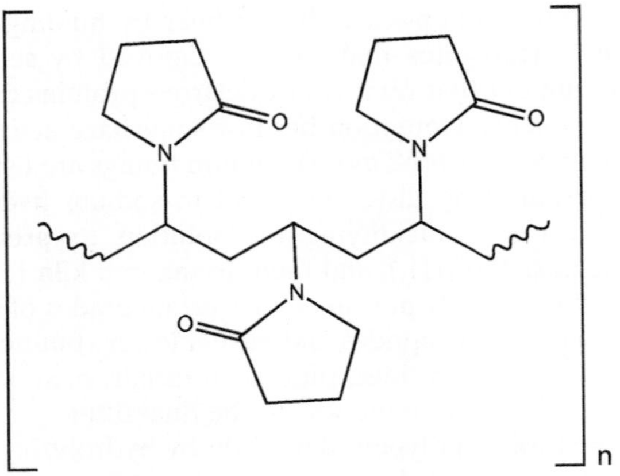

Figure 13.15   PVPP.

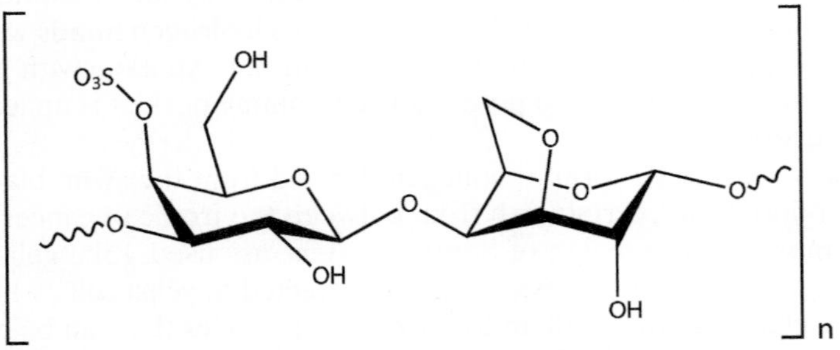

Figure 13.16   κ-Carrageenan repeating unit.

Carrageenan, also called **Irish moss**, is a preparation derived from seaweed. The active fining is kappa-carrageenan (κ-carrageenan), a polymer of **galactose** modified with sulfate shown in Figure 13.16. Unlike the other finings mentioned in this section, carrageenan is added to the **kettle** at a rate of 4–8 g/hL near the end of the boil. The molecules are negatively charged; they attract and coagulate positively charged (**basic**) polypeptides that can give rise to haze. The coagulated material comes out in the kettle trub.

Enzymes   Papain is a protein hydrolysis enzyme derived from papaya. It is effective for removing haze-active polypeptides. It is not highly **selective**; it can degrade foam-active as well as haze-active polypeptides. The problem is exacerbated because papain is a very hardy enzyme; it can survive pasteurization and be carried into the finished beer, hydrolyzing polypeptides all the while. Often PGA is added to make up for the loss of foam-active polypeptides. The polypeptides hydrolyzed by papain yield free amino acids, which can encourage growth of **spoilage** organisms.

An enzyme, called Brewers Clarex®, specifically hydrolyzes proline-rich polypeptides, which are regarded as haze active. The enzyme in liquid form is added to the fermenter at the start of fermentation. An additional advantage of Clarex is that the polypeptides that it hydrolyzes are also responsible for celiac disease and **gluten** sensitivity. Beer made with barley, wheat, or rye treated with Clarex may qualify as a "reduced gluten" product, but it cannot be labeled "gluten free" under current US regulation. It has been reported that Brewers Clarex has no negative effects on beer foam.

Another enzyme used in the brewing industry is not a hydrolase but rather a microbial transglutaminase (mTG). It functions by **covalently** cross-linking glutamine residues. The formation of **colloids** is promoted by the addition of mTG after beer maturation but before filtration. In this manner, the protein-rich colloids created by mTG are then removed with filtration, improving colloidal stability of the beer.

Other Hazes

Invisible Haze
Invisible haze is detected by light scattering at 90° but is not visible as cloudiness in the beer. It is caused by very small particles (~100 nm). Invisible haze is an analytical issue rather than a quality issue.

Starch/Carbohydrates **Starch** particles resulting from incomplete hydrolysis during mashing can make the beer hazy. Other carbohydrate materials from partial degradation of **cell walls** include beta glucan gums and partially hydrolyzed **hemicellulose**.

Oxalate Oxalic acid (HOOC—COOH) is present in grain, including barley. It can form an insoluble precipitate with calcium and other metal ions. If the oxalate is not removed by precipitation during mashing, it is likely to form a fine precipitate of calcium oxalate in the package. In addition to causing haze, the precipitate particles can serve as nucleation sites for carbon dioxide release, causing **gushing**. Oxalate precipitation in beer can be avoided by using mash-in water with sufficient calcium to remove the oxalate as insoluble CaC_2O_4 during mashing.

Filter Aids Filter aids, like **diatomaceous earth** and perlite, can become suspended in beer because of errors in procedure or equipment malfunctions. Like oxalate, these particles can also cause gushing.

Packaging Materials Linings of bottle caps or can heads, processing lubricants, dust in the empty package, fibers from packaging, and the like can leave haze or bits in the beer. Identifying the source of the problem is often more difficult than solving it. The ASBC *Beer Inclusions* publication, cited below, has micrographs of many common problem materials.

Microbes   Yeast that is not removed by filtration can be a significant cause of haze. Some styles of beer are intended to have yeast haze. Yeast that is in poor condition can break up and release fragments into the beer that are difficult to remove. **Bacteria** can give rise to haze and other issues. Dead bacteria can come with the malt and persist into the beer. Live bacteria growing in the beer can give haze as well as films or rope (strings of gelatinous **polysaccharides**) that customers will not tolerate.

Analysis and Quality Control of Haze

Haze is known in analytical chemistry as **turbidity**. Turbid samples scatter light. The details of the scattering depend on the direction of light in and out, the distribution of particle sizes, and the shapes of the particles. Figure 13.17 shows red laser light entering from the right passing through two samples. The sample on the right is pure water. No scattering is evident. The sample on the left is highly diluted milk, which provides small droplets of fat that scatter the light.

Turbidity Standards   The biggest issue in measuring haze is getting standard samples for calibration. The official primary scattering standard for beer analysis is formazin, a polymer suspension prepared from hydrazine sulfate ($N_2H_6SO_4$) and hexamethylenetetramine (formin, $C_6H_{12}N_4$). Formazin has a wide range of particle sizes and shapes. Working standards prepared from these reagents have a short lifetime. The reagents, especially hydrazine sulfate,

Figure 13.17   Light scattering by clear and hazy samples.

are toxic. Many laboratories use purchased standards prepared from suspensions of polymer particles. These too have a limited lifetime. There are permanent secondary turbidity standards (Gelex® from Hach, Inc.) that are suspensions in a block of plastic. These last indefinitely.

Turbidity Units   The major units for turbidity measurement are the formazin turbidity unit (FTU); nephelometric turbidity unit (NTU), which is a version of the FTU specified by the US Environmental Protection Agency (EPA); and the formazin nephelometric unit (FNU), which is a version of the FTU specified by the International Standards Organization (ISO). All involve measurement of light scattered at 90° from the incident (incoming) light. Instruments that measure NTU use a white light source. Those that measure FNU use **infrared** light. Many instruments use angles other than 90°; some measure multiple angles and make calculations to factor out light absorption by the sample. NTU, FTU, FNU, and many other units of turbidity are defined by the same standard suspensions. Solutions of 1 NTU (FNU, FTU) contain 1.25 mg/L of hydrazine sulfate and 12.5 mg/L of hexamethylenetetramine. The ASBC standard solution prepared according to ASBC Method Beer 26 yields a solution of 580 NTU. It is reported in the Method as a 10 000 FTU turbidity standard, so an ASBC FTU is completely different from a standard FTU. We will call this unit an ASBC turbidity unit. The European Brewing Congress uses a unit called the EBC turbidity unit, which is equal to 0.25 NTU. Table 13.4 summarizes the differences.

For example, to convert 20 EBC turbidity units to ASBC units, we look at the intersection of the EBC row and the ASBC column. The entry is 69, ASBC = 69 × EBC = 20 × 69 = 138, so 20 EBC units is equal to 138 ASBC units.

Secchi Disk   A Secchi disk is a circle with a cross. Two opposite segments are black and the others are white. The disk is mounted at the bottom of a cylinder. The sample is added while viewing the disk from above until the distinction between the light and dark areas of the disk are obscured by the turbidity. The height of the sample is noted and compared with standards of known turbidity. The turbidity can be calculated from the known turbidity of

TABLE 13.4   **To Convert the Unit in the First Column to That in the First Row, Multiply by the Table Entry**

	To Convert To		
From	NTU	ASBC	EBC
NTU	1	17.24	0.25
ASBC	0.058	1	0.0145
EBC	4	69	1

the standard and the heights of the sample and the standard. This method depends on the lighting in the room and on the judgment of the analyst.

Visual Method (ASBC Beer 27A)   Visually compare a beer sample to series of standards, viewed in uniform red light. The turbidity is reported as equal to that of the standard that seems closest in haziness to that of the sample.

Nephelometer   Light scattered by the sample is compared with that scattered by a standard and by turbidity-free water. The sample scattering should be between that of the water and of the standard. The **nephelometer** light source can be a white incandescent bulb, an infrared or colored LED, or white light passed through a wavelength filter. The scattering angle can be 90°, less than 90° (backscatter), or more than 90° (forward scatter). Many units measure at several angles, including 0° (straight across from the source), to correct for color in the sample.

Forcing Tests   One major issue with haze is that it can take a long time to form. It is routine for beer to have an expected shelf life of six months. Beer that was brilliant when it left the **brewery** may be unappetizingly hazy after a few months. The brewer needs to know within a few days of packaging if the beer meets quality standards. There are two general methods to estimate the lifetime of a batch of beer. The first is to subject the sample to conditions that accelerate the aging process. These always involve high temperature and sometimes other stresses. The second is to measure the concentrations of haze-active proteins and polyphenols.

ASBC Method Beer 27 II subjects the packaged beer to an elevated temperature (40, 50, or 60 °C) for one week, followed by cooling to 0 °C for 24 hours. *WARNING: Sealed packages heated to elevated temperatures can develop dangerous internal pressure that could cause the package to burst. Shielding must be provided to protect against can or bottle fragments.* The chill haze resulting from this process is compared with that in samples that are kept at 22 °C. If a sample of a certain beer takes nine weeks at 22 °C to throw the same haze as it does after one week at 50 °C, then one week at 50 °C is equivalent to nine weeks at room temperature. This result can then be used for other batches of the same beer.

Another accelerated aging process involves decarbonating the beer and putting a small sample (10 mL) into a big (30 mL) test tube. The tube is held for a certain time at 60 °C and then for a certain time at 0 °C. Periods of 24 hours at each temperature give results that correlate well with in-package methods. The availability of oxygen picked up during decarbonating and from the headspace of the tube increases the intensity of the test. Subjecting a small sample of decarbonated beer to 60 °C is a good deal less risky than to do so with a sealed, carbonated package.

Haze-active proteins can be measured by treating the sample with tannic acid and measuring the amount of precipitation that forms. Haze-active polyphenols can be precipitated with polyvinylpyrrolidone (PVP), a non-cross-linked, soluble version of PVPP. A refined variation of these methods is nephelometric titration. A solution of tannic acid (for polypeptides) or PVP (for polyphenols) is added while measuring the turbidity with a nephelometer. Light scattering increases as the target material precipitates. When the material is depleted, precipitation stops, and the light scattering decreases because additional solution dilutes the haze already present.

CHECK FOR UNDERSTANDING

1. Calculate the frequency of light whose wavelength is 730 nm (7.30×10^{-7} m).

2. Why is the head on beer opaque?

3. What are two systems for reporting beer color?

4. What is CIE LAB color and how can it be used?

5. Describe the four mechanisms for foam collapse.

6. What is a surfactant?

7. What are the distinctions among foamability, foam stabilization, and foam quality?

8. What are the sources of haze?

9. Discuss various finings and enzymes for haze-proofing beer.

10. How is haze measured?

11. There has been increasing interest in making hazy or "juicy" beers. Using what you know about making clear beer, how would you change your process to produce hazy beer? How would you ensure that the beer is consistently hazy?

CASE STUDY

Case Study 1

You work for a brewery with an excellent quality control lab. You have noticed a recent change in color for most of the brands, all increasing by about 2 SRM. Explain how you will identify, evaluate, and resolve the problem.

Case Study 2

You were recently hired as a brewer at a 40 bbl brewery with annual production around 50000 bbls. Your Director of Operations has expressed concern over the substantial lack of foam stability in most of the brands. Develop a plan to review and evaluate all points of production that could be the root cause of the poor foam stability. Explain how you will identify, evaluate, and resolve the problem.

BIBLIOGRAPHY

American Society of Brewing Chemists Identification Guide. 2008. *Beer Inclusions: Common Causes of Elevated Turbidity*.

Aron PM, Shellhammer TH. 2012. A discussion of polyphenols in beer physical and flavour stability. *J. Inst. Brew. 116*(4):369–380.

ASBC Methods of Analysis: Beer 10 Color. 2015. doi:10.1094/ASBCMOA-Beer-10.

ASBC Methods of Analysis: Beer 22 Foam Collapse Rate. 2018. doi:10.1094/ASBCMOA-Beer-22.

ASBC Methods of Analysis: Beer 27 Physical Stability. 1975. doi:10.1094/ASBCMOA-Beer-27.

Bamforth CW. 1985. The foaming properties of beer. *J. Inst. Brew. 91*:370–383.

Bamforth CW. 1999. Beer haze. *J. Am. Soc. Brew. Chem. 57*(3):81–90.

Bamforth CW. 2004. The relative significance of physics and chemistry for beer foam excellence: theory and practice. *J. Inst. Brew. 110*(4):259–266.

Bamforth CW. 2006. *Scientific Principles of Malting and Brewing*. American Society of Brewing Chemists. ISBN 978-1-881696-08-7.

Bamforth CW. 2012. *Foam*. American Society of Brewing Chemists. ISBN 978-1-938119-00-2.

Barth R. 2018. Beer Conditioning, Aging, and Spoilage. In Bordiga M (editor). *Post-Fermentation and -Distillation Technology Stabilization Aging, and Spoilage*. CRC. ISBN 978-1-4987-7869-5.

Blasco L, Veiga-Crespo P, Sanchez-Perez A, Villa TG. 2012. Cloning and characterization of the beer foaming gene *CFG1* from *Saccharomyces pastorianus*. *J. Agric. Food Chem. 60*(43):10796–10807.

Constant M. 1992. A practical method for characterizing poured beer foam quality. *J. Am. Soc. Brew. Chem. 50*:37–47.

Gabriel P, Sladký P, Sigler K. 2016. A new rapid high-throughput method for prediction of beer colloidal stability. *J. Inst. Brew. 122*:304–309.

Hughes PS, Baxter ED. 2001. *Beer: Quality, Safety, and Nutritional Aspects*. Royal Society of Chemistry. ISBN 0-85404-588-0.

Kapp GR, Bamforth CW. 2002. The foaming properties of proteins isolated from barley. *J. Sci. Food. Agric. 82*:1276–1281.

Maye JP, Smith R, Leker J. 2016. Dry-hopping's effect on beer foam. *Proceedings of the World Brewing Congress,* Dever, Colorado, USA.

Ronteltap AD, Hollemans M, Bisperink CGJ, Prins A. 1991. Beer foam physics. *MBAA Tech. Q. 28*:25–32.

Taylor JP, Jacob F, Arendt EK. 2015. Fundamental study on the impact of transglutaminase on hordein levels in beer. *J. Am. Soc. Brew. Chem. 73*(3):253–260.

Wang H-Y, Qian H, Yao W-R. 2011. Melanoidins produced by the Maillard reaction: structure and biological activity. *Food Chem. 128*:573–584.

CHAPTER 14

BIOLOGICAL STABILITY

The advantage of a small brewpub with only **draft** sales over a production **brewery** with a distribution network lies in product turnover. If the beer is quickly sold with sales only in the **pub**, then stability is less of a concern. However, when beer is packaged and distributed, with product potentially sitting on warm shelves for two to six months, then the importance of stability becomes critical. With competition for shelf space high, successful brewers will be those who put forward the best and most stable product. Inadequate attention to stability could cost the brewery customers, or worse yet, recalled product.

There are three main areas of beer stability that require attention: microbial **spoilage**, **flavor** staling, and **colloidal** stability. This chapter focuses on **microbial** stability. **Flavor** stability was discussed in Chapter 12; **colloidal** stability was covered in Chapter 13.

Although **yeast** is a microbial organism at the core of beer production, all steps of production, from raw materials through packaging and on to shelf-life stability, can be threatened by unwanted microorganisms commonly found in the environment. Microbes are everywhere, so the brewer must always

Mastering Brewing Science: Quality and Production, First Edition.
Matthew Farber and Roger Barth.
© 2019 John Wiley & Sons, Inc. Published 2019 by John Wiley & Sons, Inc.

be careful to detect and avoid contamination. Common sources of microbial spoilage in the brewery include the following:

- Raw materials, especially **malt**.
- Untreated water.
- Airborne dust and debris.
- Insects.
- Dirty equipment and inefficient **clean-in-place** (**CIP**).
- Unclean floor drains
- Boots worn outside of the brewery
- Brewers.

After boiling, beer is free of live microbes. The brewery must focus attention on sources of potential spoilage at all steps after kettle **knock out**. Meticulous and thorough cleaning is vital to maintaining microbial stability. Chemical **sanitization** and **sterilization** is covered in Chapter 16. Cleanliness extends beyond equipment and should also encompass floors, walls, and drains. It also includes employee cleanliness. Proper hygiene, work apparel, and boots for brewery use only are good manufacturing practices to help avoid contamination.

Microbial stabilization is paramount to beer **quality** and consistency. Contamination can lead to the following:

- Changes in **alcohol** content and **attenuation** after packaging.
- Excessive CO_2 production, overcarbonation, and ruptured bottles or cans.
- Beer **off-flavors**.
- Gushing.
- Premature yeast **flocculation**.
- Excessive **acidity**.
- **Turbidity** (**haze**).

14.1   IDENTIFYING SPOILAGE MICROORGANISMS

Identification of **bacteria** through traditional microbiological approaches aids the identification of potential spoilage bacteria. By knowing the identity of the contaminating microorganisms, the brewer may better understand the source of contamination and potential risk of spoilage. The Gram **stain**, oxidase test, and catalase test are simple methods that assist in identification and characterization of spoilage microorganisms. To run these tests, a bacterial isolate must first be obtained through the methods described in Section 14.3.

Chemical Tests

Gram Stain   The Gram stain is a method for the differential labeling of the bacterial **cell wall**. Gram-positive cells have a thick cell wall outside of the **cell membrane**. The cell wall retains crystal violet stain. Gram negative cells have a thinner cell wall that is enclosed in a second membrane. In Gram-negative cells, the crystal violet stain is washed away. The application of a red counter-stain, safranin, colors the cells pink (Figure 14.1). Using this method, Gram-positive cells stain purple, obscuring the pink stain; Gram-negative cells stain pink. A mixed culture of Gram-positive and Gram-negative bacteria is shown in Figure 14.1. The procedure for the Gram stain is as follows:

- Flame the top side of a clean slide.
- Draw a 1 cm circle with a wax pencil on the slide.
- Inside the circle, add a drop of **sterile** water with a sterile **inoculation loop**.
- Add the bacteria sample, smear it out.
- Heat-fix the sample by passing the bottom of the slide through a flame quickly three times.
- Turn off all flames.
- Add a few drops of crystal violet **solution** to the sample.
- Wait 30 seconds, then rinse by dipping the slide in pure water three times.
- Add a few drops of Gram's iodine solution.
- Wait 30 seconds, then rinse by dipping the slide three times in pure water, then three more times in fresh pure water.

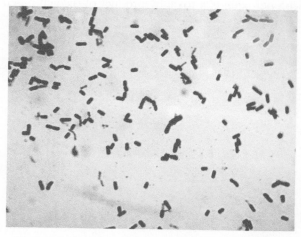

Figure 14.1   Gram stain. *Source*: Photo: Kent Pham.

- Hold the slide at a 45° angle and allow drops of decolorizer solution to wash across the sample until the flow is clear (~15 seconds).
- Add a few drops of Safranin stain to the sample. Let stand 60 seconds.
- Rinse in pure water. Allow to dry.
- Inspect the sample under a **microscope** with an oil-immersion **objective** to visualize the bacteria.

Oxidase Test   The oxidase test detects the presence of cytochrome C oxidase. Most bacteria that are capable of **aerobic respiration** use the **enzyme** cytochrome C oxidase to accept **electrons**, transferring them to oxygen. In brief, an oxidation indicator, tetramethyl-*p*-phenylenediamine (TMPD) is mixed with a bacterial sample. Cytochrome C oxidase, if present, accepts electrons from the indicator, causing it to turn purple, a positive oxidase test (Figure 14.2).

The test can be applied in several ways. A drop of indicator solution can be added to a sample of bacteria on a slide. A sample of bacteria can be added to a piece of filter paper impregnated with the **indicator**. A drop of indicator solution can be added directly to a colony on a plate. A **swab** impregnated with indicator solution can be wiped across a **surface** with the bacteria. In each case, if a purple color develops within 10 seconds, the sample is oxidase-positive. If the color develops between 10 and 20 seconds, the result is recorded as a delayed positive. If no color develops within 20 seconds, the sample is oxidase-negative.

Catalase Test   The catalase test identifies the presence of catalase, an enzyme that catalyzes the decomposition of hydrogen **peroxide** (H_2O_2) to oxygen and water. Hydrogen peroxide is a toxic by-product of oxygen-based respiration.

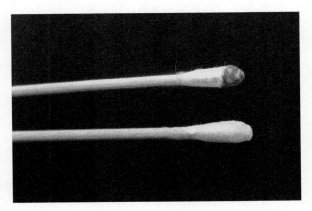

Figure 14.2   Oxidase test. *Source*: Photo: Kent Pham.

Figure 14.3   Catalase test. *Source*: Photo: Kent Pham.

Most organisms that can use oxygen make catalase to protect themselves from hydrogen peroxide toxicity. To run the test on a **glass** slide, a drop of 3% hydrogen peroxide solution is placed on the slide, and an inoculum of bacteria is added. Alternatively, the test can be run in a test tube with several milliliter of hydrogen peroxide. The transfer of the reagent and the bacteria should be done with a wooden stick or a glass rod, not any wire containing iron. If catalase is present, the catalase-dependent cleavage of hydrogen peroxide into water and oxygen will produce copious bubbling; a positive catalase test (Figure 14.3).

Cell Morphology

When preparing cells for Gram staining, cell shape can also be noted. Identification of bacterial **morphology** is essential for proper identification. Most species of bacteria have a characteristic shape that does not change. However, because some bacteria genera may share similar shapes and sizes, differentiation by morphology alone is not enough. Figure 14.4 illustrates different cell shapes and sizes of **prokaryotes**. Typical bacteria are 0.5–2 μm in size. **Bacilli** are rod-shaped. **Cocci** are spherical. Spirochetes are spiral-shaped. The prefix strepto- means long chains. The prefix staphylo- means clusters. In this manner, streptobacilli are chains of rod-shaped cells and staphylococcus are clusters of cocci. Further differentiation is warranted for pairs of cells, called Diplo-, or for clustering of exactly four cells, called tetrads. **Pleomorphic** bacteria demonstrate variability in the size and shape.

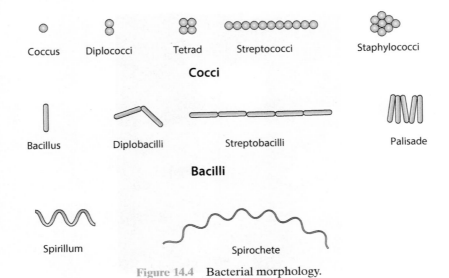

Figure 14.4   Bacterial morphology.

Common Spoilage Bacteria

A combination of the tests described above along with cellular morphology and point of contamination records are essential for the identification of bacteria. The following is a discussion of the most common spoilage bacteria in the brewery broadly grouped into categories that make them unique among beer spoilage microorganisms. Table 14.1 summarizes general characteristics of the most common beer spoilers.

Lactic Acid Bacteria (LAB)   *Lactobacillus* spoilage results in acidification through lactic acid production and turbidity. All *Lactobacillus* are Gram-positive rods (Figure 14.5). They are commonly aerotolerant **anaerobes** with typical risk in **fermentation** vessels, **bright tanks**, pitching yeast, and packaged beer. While there are many *Lactobacillus* species, *Lactobacillus brevis* is the most common spoiler, isolated in over 50% of contaminated samples. *L. brevis* is **heterofermentative** and grows best at 30 °C and **pH** 4–6. It is generally resistant to **hop** compounds and can utilize **dextrins** and **starch**, making it a severe risk for over-attenuation.

In addition to *L. brevis*, several other *Lactobacillus* species are known spoilers of beer. These include *L. amylovorus, L. backii, L. brevisimilis, L. buchneri, L. casei, L. collinoides, L. coryniformis, L. delbrueckii, L. fermentum, L. frigidus, L. harbinensis, L. hilgardii, L. homohiochi, L. lindneri, L. malefermentans, L. parabuchneri, L. paracasei, L. paracollinoides, L. pentosus, L. plantarum, L. perolens, L. reuteri, L. rhamnosus,* and *L. sakei. L. casei* is known to produce **diacetyl**, yielding an unacceptable buttery off-flavor.

TABLE 14.1 Characteristics of Spoilage Bacteria

Bacteria		Gram	Shape	Acid	O$_2$ Growth	No O$_2$	Catalase	Oxidase
Lactic acid bacteria (LAB)	Lactobacillus	(+)	Rods, chains	(+)	(+)	(+)	(−)	(−)
	Pediococcus	(+)	Cocci, tetrads	(+)	(+)	(+)	(−)	(−)
	Leuconostoc	(+)	Pleomorphic	(+)	(+)	(−)	(−)	(−)
Acetic acid bacteria (AAB)	Acetobacter	(−)	Pleomorphic	(+)	(+)	(−)	(+)	(−)
	Gluconobacter	(−)	Pleomorphic	(+)	(+)	(−)	(+)	(−)
Coliforms	Enterobacter	(−)	Short rods	(−)	(+)	(+)	(+)	(−)
	Klebsiella	(−)	Short rods	(−)	(+)	(+)	(+)	(−)
Bacillus	Bacillus	(+)	Streptobacilli	(+/−)	(+)	(−)	(+)	(+/−)
Anaerobes	Megasphaera	(−)	Curved rods	(−)	(−)	(+)	(−)	(−)
	Pectinatus	(−)	Curved rods	(−)	(−)	(+)	(−)	(−)
	Zymomonas	(−)	Short rods	(−)	(+)	(+)	(+)	(−)

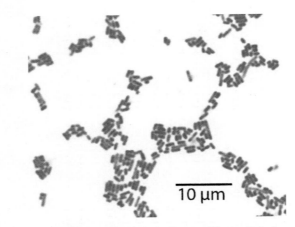

Figure 14.5   *Lactobacillus brevis. Source*: Photo: Eric Jorgenson.

Mechanisms of lateral **gene** transfer or evolution of novel resistance mechanisms may cause expansion of this list.

Pediococcus   *Pediococcus* spoilage may result in acidification, turbidity, and/or diacetyl production. All *Pediococcus* species are Gram-positive cocci that grow in pairs or tetrads (Figure 14.6). They are aerotolerant anaerobes that typically infect **fermentation vessels, bright beer tanks,** pitching yeast, and packaged beer. They are **homofermentative** bacteria that commonly produce excessive diacetyl. In addition, some species can produce exopolysaccharides (EPS). EPS causes a thickening of the beer known as rope. In spontaneous fermentations like **lambics,** ropiness is resolved over time by wild yeast such as ***Brettanomyces.*** *Pediococcus* species can be resistant to hops and are very ethanol tolerant. Identified spoilage species in beer include *P. acidilactici, P. claussenii, P. damnosus, P. inopinatus, P. pentosaceus,* and *P. parvulus.*

Leuconostoc   *Leuconostoc mesenteroides* is the only known species to contaminate beer. It requires a small amount of oxygen for growth. It is a heterofermentative lactic acid-producing bacterium and is Gram-positive with a pleomorphic morphology, forming either short rods or cocci. It is ethanol and hop tolerant with very high acid tolerance.

Acetic Acid Bacteria (AAB)   *Acetobacter* and other AAB are differentiated from LAB in that they produce acetic acid through fermentation, they are obligate aerobes, and they are Gram-negative (Figure 14.7). Because of their dependence on oxygen, *Acetobacter* can cause spoilage at beer-air interfaces. Common problem areas are pitching yeast, balance lines, draft lines, **cask**

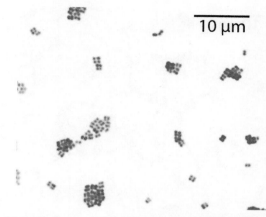

Figure 14.6   *Pediococcus damnosus*. *Source*: Photo: Eric Jorgenson.

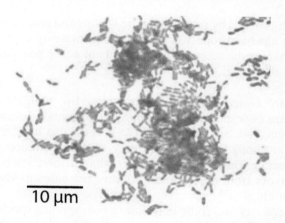

Figure 14.7   *Acetobacter aceti*. *Source*: Photo: Eric Jorgenson.

beers, and **barrels**. With modern **brewing** practices that severely reduce dissolved oxygen in finished beer, AAB rarely spoils packaged product.

A. aceti and *A. pasteurianus* are known beer spoilers. *Acetobacter* oxidizes ethanol into acetic acid, otherwise known as vinegar, the primary off-flavor caused by infection. *Acetobacter* can also oxidize **acetate** to carbon dioxide and water. AAB are hop and ethanol (6–8% **alcohol by volume** [ABV]) tolerant and very acid tolerant. *Acetobacter* are pleomorphic, shaped as rods or ellipsoids, curved or straight. They may grow singly, in pairs, or as chains.

Because *Acetobacter* are obligate aerobes, they represent a major infection risk in beer conditioning in wooden barrels due to openings in the wood and between the barrel staves. For this reason, effort should be made to limit

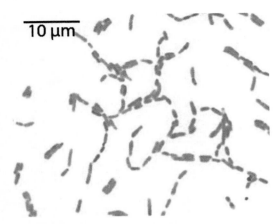

10 μm

Figure 14.8   *Gluconobacter oxydans*. *Source*: Photo: Eric Jorgenson.

oxygen entering barrels by preswelling the barrels and minimizing **headspace** after filling. It has been reported that barrels previously contaminated with *Acetobacter* that produced excessive amounts of acetic acid can be saved by CO_2 purges of the barrel.

Gluconobacter   *Gluconobacter* are Gram-negative obligate aerobes, with risk areas where beer meets air including pitching yeast, balance lines, draft lines, cask beers, and barrels (Figure 14.8). They also oxidize ethanol into acetic acid, but unlike *Acetobacter*, they do not produce CO_2 through oxidation of acetate. They utilize sugar over ethanol. Infection is characterized by vinegar flavors from acetic acid, turbidity, increases in **viscosity**, and/or **pellicle** formation.

  Gluconobacter oxydans is the only known beer spoiling species, often forming single or short chains of bacilli-shaped cells. *Gluconobacter* may be motile or nonmotile, with motility enabled by multiple flagella. Cells may produce pink or brown pigments that may assist differentiation of colonies on **agar media**. *Gluconobacter* are hop, ethanol (12–13% ABV), and acid resistant, with growth occurring in beer as low as pH 3.6. Because they are obligate aerobes, minimizing oxygen at every step of production will prevent spoilage.

Enterobacter/Coliforms   *Enterobacteriaceae* are a large family of Gram-negative, often rod-shaped, facultative **anaerobic** bacteria including the genera *Citrobacter, Enterobacter, Hafnia, Klebsiella, Rahnella, Salmonella, Serratia*, and *Shigella*. Some species are encapsulated, which aids identification through Gram staining (Figure 14.9). *Hafnia protea* and *Rahnella aquatilis* (formerly *Enterobacter agglomerans*) are known contaminants of pitching yeast, negatively affecting fermentation rates and leading to downstream

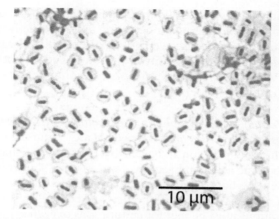

Figure 14.9   *Klebsiella pneumoniae. Source*: Photo: Eric Jorgenson.

contamination issues when yeast is repitched from cone to cone. Off-flavors of *H. protea* include parsnip-like flavor and **aroma**. Off-flavors from *R. aquatilis* include diacetyl and **DMS**. Both species are hop and ethanol resistant. Species of *Klebsiella, Citrobacter, Enterobacter, Obesumbacterium,* and *Escherichia* are known **wort** spoilers, producing dimethyl sulfide (DMS), **organic** acids, and diacetyl. These species of *Enterobacter* are hop resistant but are very sensitive to ethanol and do not thrive under anaerobic conditions, so they are not found in finished beer as active spoilage organisms, but typically originate from contamination of wort or pitching yeast.

Zymomonas   *Zymomonas* is a genus of anaerobes. *Zymomonas mobilis* subsp. *mobilis* and *Z. mobilis* subsp. *pomacea* have been identified as packaged beer spoilers, particularly finished beer that has been primed for secondary fermentation in the package such as cask beer or bottle-conditioned beer. *Zymomonas* contamination is characterized by strong turbidity, acetaldehyde, hydrogen sulfide, and overcarbonation. The reason *Zymomonas* contaminates packaged beer for secondary fermentation is that it can only utilize **glucose**, fructose, and sucrose. Secondary fruit additions may also promote contamination. Strains are very alcohol tolerant (>16% ABV), hop resistant, and acid tolerant (<pH 3.5). *Zymomonas* are reportedly aerotolerant anaerobes, so culture conditions for detection should be anaerobic as they do not thrive in the presence of oxygen. *Zymomonas* are Gram-negative with a rod-shaped morphology. Cells grow singly, in pairs, or in a rosette pattern.

Megasphaera   *Megasphaera cerevisiae, Megasphaera paucivorans,* and *Megasphaera sueciensis* have been identified packaged beer spoilers. Infection is characterized by turbidity with high levels of hydrogen sulfide, butyric acid, and caproic acid. *Megasphaera* cells are Gram-negative cocci that are obligate anaerobes. Sampling and **plating** for detection must be strictly anaerobic with minimal exposure to air, because even brief oxygen exposure can compromise culturability. *Megasphaera* are sensitive to low pH and do not thrive in beer above 3% ABV. For these reasons, *Megasphaera* often infect early or stalled fermentations or low-alcohol beer.

Pectinatus   *Pectinatus cerevisiiphilus* and *Pectinatus frisingensis* are known packaged beer spoilers. *Pectinatus* species are Gram-negative, obligate anaerobes with contamination indicated by turbidity and high amounts of acetic acid, propionic acids, and acetoin, resulting in a sour rotten egg aroma and flavor. They are hop tolerant, acid tolerant, and ethanol tolerant up to approximately 5% ABV. Cell morphology are curved or spiral rods. Strict anaerobic conditions must be met for culturing, as oxygen exposure is lethal.

Bacillus   ***Bacillus*** *coagulans* and *Bacillus stearothermophilus* are known spoilers of wort and malt, producing lactic acid in hot, sweet wort. Contaminating *Bacillus* species can grow singly, in pairs, or in chains and are Gram-variable, appearing Gram-positive during stationary growth (Figure 14.10). They grow best in aerobic environments but can survive through anaerobic conditions. *Bacillus* are very hop sensitive and thus do not actively contaminate finished beer. However, they are very well known for their ability to form endospores, which are highly heat tolerant. They can persist through wort boiling, though subsequent germination requires ideal nutrient and growth conditions that are not present in finished beer. Because of the hardiness of *Bacillus* endospores, *Bacillus* is a common contaminant of dirty **zwickels**. If identified on nutrient-rich agar media, one should check the

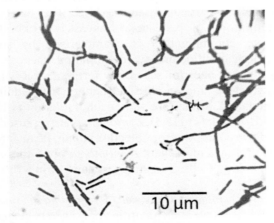

Figure 14.10   *Bacillus* spp. *Source*: Photo: Michelle McHugh.

possibility that the problem comes from the environment as a result of poor **aseptic** conditions during sampling, rather than from the beer.

Wild Yeasts

Any yeast species that is unintentionally present in beer is considered a wild yeast, including certain strains of *Saccharomyces cerevisiae*. Wild yeast spoil beer through secondary fermentation, which leads to overcarbonation, thin body, noncompliance with **ABV**, and potential off-flavors. Yeasts that produce phenolic off-flavors, like 4-vinylphenol and 4-vinylguaiacol, are considered *POF+*.

Brettanomyces (also called *Dekkera*) was first discovered as a beer spoiler in 1904. Five species have been identified: *Brettanomyces anomalus* (also called *Brettanomyces clausenii*), *Brettanomyces bruxellensis* (also called *Brettanomyces intermedius* or *Brettanomyces labbicus*), *Brettanomyces custerianus, Brettanomyces naardenensis, and Brettanomyces nanus*. *Brettanomyces* are common beer spoilers because, like *S. cerevisiae*, they are **Crabtree** positive. They typically require more oxygen than *S. cerevisiae* to reach critical biomass, and they grow more slowly. *Brettanomyces* species present greater spoilage risk due to their ability to utilize **cellobiose** and dextrins. For this reason, they commonly take up residence in the wood of barrels and may referment the dextrins in beer. They will survive in very poor, nutrient-limited conditions.

Brettanomyces species can provide sour off-flavors through a process known as the *Custer effect*. The Custer effect is the ability to produce acetic acid (vinegar) from acetaldehyde in the presence of oxygen. This process is very slow, requiring up to six months and exposure to oxygen. Other flavors vary greatly by strain and by brewery. Predominant flavors include vinyl and ester phenols, **esters**, and isovaleric acid, which contribute to flavors described as clove, spicy, horsey, barnyard, smoky, medical, band-aid, goaty, floral, and tropical fruit.

There are a variety of non-*Saccharomyces* yeasts that can grow in beer. Their spoilage potential is variable and, if isolated by the brewery lab, should be further evaluated for spoilage risk by testing over-attenuation and production of phenolic off-flavor. Common genera found in beer include *Pichia, Candida, Saccharomyces, Torulaspora, Zygosaccharomyces, Kluyveromyces, Rhodotorula, Aureobasidium, Schizosaccharomyces, Hanseniaspora, Wickerhamomyces*, and many others.

A major beer spoilage yeast in unfiltered, unpasteurized beer is the yeast *Saccharomyces cerevisiae* var. *diastaticus*, informally called *diastaticus*. While some *diastaticus* strains are true wild yeast contaminants, other strains are commercial ale strains selected for their high attenuation and production of phenolic flavor. Excessive attenuation by *diastaticus* is caused by the secretion of a **glucoamylase**, which catalyzes breakdown of dextrins into fermentable **sugars**. Contamination and refermentation in packaged products can lead to overcarbonation, changes in flavor, and increases in ABV above that which may be legally permissible. *Diastaticus* strains represent great heterogeneity in their ability to overattenuate and create off-flavors, so their spoilage potentials vary.

14.2   MICROBIAL STABILIZATION METHODS

There is a difference between a *contaminating* microorganism and a *spoilage* organism. All bacteria or yeast present in processing or finished beer where they do not belong are contaminating microorganisms. Contaminating microorganisms that damage the product for example by lowering the pH, producing excess carbon dioxide, or changing the flavor profile are spoilage microorganisms. Beer is susceptible to contamination, especially in small operations without filtration or **pasteurization**, but there are contaminants that may represent low to no risk in product quality. Determining the spoilage potential of contaminating microorganisms is an important role for a quality control scientist in a brewery.

Spoilage Potential

From a food safety perspective, beer is inherently free of **pathogenic** bacteria. Pathogenic bacteria are those that cause sickness in humans, most commonly including species of *Salmonella, Clostridium, Campylobacter, Staphylococcus, Listeria, Yersinia*, and a specific *Escherichia coli* strain O157:H7. The US Food and Drug Administration defines acid foods as those that have a natural pH of 4.6 or below. The reason for this distinction is that no pathogenic bacteria can grow from **spores** at a pH lower than 4.6. Because beer is boiled and naturally has a pH lower than 4.2, it is not susceptible to dangerous contamination. Some of the factors that account for the safety of beer include boiling, low pH, hops, alcohol, and low sugar. A beer with low alcohol and hops, perhaps mixed with fruit juice, could move out of the safety zone. Brewers who create innovative beers with atypical ingredients should not take food safety for granted.

Although beer microbes are not normally a safety issue, beer is not free from all spoilage microorganisms that threaten beer quality. Certain process points may be more prone to infection than others. For example, consider the properties of wort and beer as outlined in Table 14.2. In addition to pH, the lack of simple carbohydrates and oxygen, and the presence of ethanol, carbon dioxide, and hop bitter compounds in beer all provide a **bacteriostatic** environment; there are few microorganisms that can tolerate such conditions. Those that may survive and may even thrive are discussed in Section 14.1.

TABLE 14.2   **Spoilage Potential of Wort Versus Beer**

Wort	Beer
Rich in carbohydrates	Unfermentable dextrins
pH 5.2–5.8	pH < 4.2
Aerobic	Anaerobic
Variable temperature	Presence of CO_2
	>4% Ethanol
	Hops

Unlike beer, wort is a microbial haven, rich in simple sugars for food and oxygen that supports aerobic bacteria growth. With a pH of 5.2–5.8, it is not food safe; it can also promote the growth of aerobic beer spoilers. Therefore, wort should not be stored for a long time in the brewery without sufficient sterilization. The major use of stored wort is for yeast propagation. For small-scale preparations, if wort cannot be fermented quickly, it can be **autoclaved** and stored at 4 °C before use. For larger-scale preparations, an in-line pasteurization device during wort transfer from the brewhouse to a wort storage tank in the yeast propagation room is required.

Pasteurization

Pasteurization is a method of enhancing microbial stability of food by the controlled application of heat. First applied to **wine** in 1864 by Louis Pasteur, the pasteurization process is practical and effective. To pasteurize beer, the temperature of the beer is increased to a high enough temperature for a long enough time to kill most spoilage microbes. The key consideration here is that pasteurization kills *most* of the spoilage microorganisms; it does not sterilize.

The severity of a pasteurization treatment is measured in **pasteurization units** (PU). A pasteurization unit is a degree of heating (time and temperature) that kills that same fraction of the target microbes as does the exposure of the sample to 60 °C for one minute. For the spoilage organisms normally present in beer, the number of pasteurization units accomplished can be estimated from the equation

$$PU = t \times 1.389^{(T-60\,°C)},$$

where PU is the number of pasteurization units, t is the time in minutes, and T is the temperature in degrees Celsius. For example, if a brewer wished to pasteurize beer to 25 PU by holding at 60 °C, then solving for t yields a hold time of 25 minutes. To shorten the time required to reach 25 PU, the brewer can raise the temperature. Increasing the pasteurization temperature to 75 °C with a goal of 25 PU yields a hold time of 11 seconds. This is the principle of **flash**

pasteurization. Because pasteurization can change the flavor of beer, the lowest effective severity is applied. Beer is most effectively pasteurized at 15–30 PU with 15 PU enough for low-risk beers (high alcohol, low **final gravity**...) and 30 PU for higher-risk beers.

The usual pasteurization regimes are effective for organisms *normally* found in beer. The use of unconventional ingredients and processes can radically alter the picture. As one example of this, a brewery experimenting with post-boil additions added cacao nibs after fermentation. It was expected that any potential spoilers would be neutralized by pasteurization. But months later, bottles picked up excessive **carbonation** and off-flavors. The cause was determined to be an unconventional spoiler of the genus *Sporolactobacillus*. As the name suggests, these bacteria form temperature-tolerant spores. They were introduced by the post-fermentation addition and resisted pasteurization. When considering PU target values, it is also important to consider that different microorganisms respond differently to PU values. For example, *Lactobacillus* are more resistant than *S. cerevisiae* and wild yeast are more resistant than *Lactobacillus*.

Tunnel Pasteurization   Tunnel pasteurization is performed on bottled or canned beer only. The packages travel on a conveyor and are sprayed with water at a controlled temperature, as shown in Figure 14.11. In the example shown, there are two preheat and two cooldown zones. In the first preheat zone, the beer is heated with a water spray at around 35 °C. In the second preheat zone, the spray is about 50 °C. The heating zone uses water several degrees higher than the pasteurization temperature to bring the internal temperature of the bottle to the pasteurization temperature, typically about 60 °C. The hold zone keeps the pasteurization zone constant. Two cooldown zones follow with temperatures a few degrees lower than the corresponding

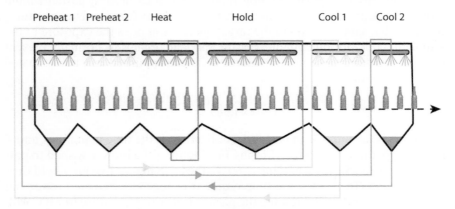

Figure 14.11   Tunnel pasteurization.

preheat zones. The preheat water exits cooler than it entered, it is then used in a cooldown zone. The cooldown water picks up heat from the hot bottles. It then is used as preheat water. The heat zone is hotter than the desired pasteurization temperature. When the beer reaches the pasteurization temperature, the packages enter a zone in which the water spray is at the pasteurization temperature to prevent the beer from cooling. Because heat exchange between the spray and the beer is slow, it can take 40–60 minutes for a package to go from start to finish through a tunnel pasteurizer.

A tunnel pasteurizer takes up a lot of floor space because the system must be designed to hold beer at the appropriate time and temperature to achieve the target number of PUs. In a conveyor system, treatment time is the length of the treatment zone divided by the speed. The tunnel length can be shortened if the speed is slower, but then the throughput will be limited. Tunnel pasteurization is not practical for large packages like kegs because it takes too long for them to heat up and cool down. Other considerations are pressure and volume in the package. The volume of beer increases by 2% when the temperature increases to 65 °C. **Gas** pressure in the headspace (empty space above the beer) increases as temperature increases and as volume decreases. Evaporation of water and ethanol adds to this. For glass bottles, a minimum of 4% headspace is recommended to keep the pressure at a safe level; for a 355 mL (12 fl. oz) serving, that comes to 14 mL (0.5 oz). Cans are much less stiff than bottles; they deform under pressure, so the internal temperature should be limited to 62 °C.

At 60 °C, the equilibrium pressure of carbon dioxide of beer with 2.8 volumes of carbonation can be 7 bar (85 psig) or more. Tunnel pasteurization is only practical because equilibrium is reached slowly, and the actual pressure is temporarily lower than the equilibrium pressure; the beer is supersaturated with carbon dioxide. Any jostling or bumping of the packages can cause the internal pressure to reach dangerous levels almost instantly. The major advantage of tunnel pasteurization is that it is applied after the package is sealed. There is no opportunity for microbes to enter after pasteurization. Some disadvantages are high cost of operation, high space requirement, and a lot of ruined beer if the unit stops during a run.

Flash Pasteurization   In flash pasteurization, the pasteurization temperature is high, and the hold time is low. The typical maximum temperature is in the range of 71–79 °C for a period of 15–60 seconds. The volume under treatment at a given time is small to allow the temperature to rapidly rise and fall. The rapid, turbulent flow ensures that all beer experiences the designed time-temperature program, but it makes it impossible to maintain supersaturation of carbon dioxide. The unit must be run under a high pressure of carbon dioxide to maintain carbonation. The beer is first **pumped** through a plate **heat exchanger** called the regeneration section. Here, it is preheated by hot beer

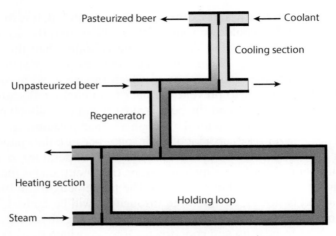

Figure 14.12   Flash pasteurization.

exiting from the holding loop. The warm beer now enters another plate exchanger, the heating section, where it is heated to the pasteurization temperature by steam or hot water. Next, the holding loop keeps the beer hot for the correct length of time. Beer from the holding loop enters the other side of the regeneration section, where it is cooled by the entering cold beer. Finally, the pasteurized beer is chilled to packaging temperature by refrigerated coolant in the cooling section. The process is illustrated in Figure 14.12. Because of the shorter residence times at high temperatures, there is less of a negative impact on beer color and flavor. The flash pasteurization unit feeds directly into the packaging unit to minimize the chance of contamination of the pasteurized beer. The major drawback of flash pasteurization is that keg and package filling must be accomplished under near-sterile conditions to avoid reintroduction of spoilage organisms.

Sterile Filtration

Sterile filtration is a method of microbial stabilization that does not require heat. Sterile filtration requires the beer to be passed through a **membrane filter** with a pore size of 0.45 μm or smaller, which effectively removes all yeast and bacteria. These filters can provide greater than 99.9% sanitation, because the pore sizes are smaller than all brewing microorganisms (Table 14.3). The term "sterile" filtration is a misnomer because a pore size of 0.45 μm may not exclude the smallest biological units such as spores and viruses, thus they are not truly sterile. Most filter systems used for beer clarification do not provide effective microbial filtration.

TABLE 14.3 **Cell Sizes of Common Brewery Microorganisms**

Microorganism	Size
Saccharomyces cerevisiae	9.5 × 10 μm
Saccharomyces pastorianus	7.0 × 8.5 μm
Brettanomyces bruxellensis	3.0 × 5.0 μm
Escherichia coli	0.5 × 2.0 μm
Lactobacillus brevis	0.5 × 3.0 μm
Pediococcus claussenii	1.0 × 2.5 μm
Acetobacter aceti	0.6 × 2.0 mm
Megasphaera paucivorans	1.5 × 1.2 μm

Commonly used sterile filters consist of cartridge/canister filters and lenticular filters. Their operating principles and designs are discussed in Chapter 10. Because of the small pore sizes employed in sterile filtration, the filters can be easily clogged as material collects on the membrane. Therefore, it is important to reduce yeast and large particles before sterile filtration. Particle reduction can be accomplished by prolonged cold maturation times, the use of **fining** agents such as gelatin or **isinglass**, and incorporation of upstream rough filtration or **centrifugation**. Like flash pasteurization, any processes downstream of the microbial filter, including packaging, must be done under near-sterile conditions to prevent reintroduction of contamination.

Bacteriostatic Properties of Hops

Hop bitter compounds are toxic to most Gram-positive beer-spoilage bacteria. This property may, in part, account for the prominence of hopped beer in international trade under the auspices of the Hanseatic league after 1200. Hanseatic trade was a key factor in popularizing the use of hops in European beer. Hops preserved the beer, so it could be shipped for long distances. Before hops were used, long-lasting beer needed a high alcohol content to keep spoilage bacteria in check. High alcohol means extra fermentable material, hence higher cost. Today, the use of hops in beer is nearly universal; in some cases, it is incorporated into the legal definition of beer.

Hop **iso-alpha acids** are not as antimicrobial as alpha acids, but they are present in beer at much higher concentration because of their higher solubility. They are most active at low pH. The mechanism of their toxicity is still under study, but it is believed that they assist the transport of hydrogen ions across the cell membrane, lowering the internal pH of bacteria. They also bind manganese ions, which play a role in the bacterial defense against oxidative stress. There are strains of bacteria that are hop tolerant. These can cause significant spoilage, especially those that are also alcohol tolerant and that can survive at low pH and under anaerobic conditions, like those in packaged beer.

14.3   ANALYTICS AND QUALITY CONTROL OF MICROORGANISMS

No brewery is free from contaminating microbes. Therefore, it is essential to monitor process points to detect and correct the presence of spoilage yeast or bacteria. Contamination may result in off-flavor development, turbidity, and in extreme cases, over-attenuation and package rupture resulting in costly product recalls. Microbe detection in the brewery is essential for any quality program.

Aseptic Technique

Sampling for microbiological stability requires strict attention to aseptic technique. Aseptic technique describes procedures to minimize contamination by foreign microorganisms. Failure to prevent outside contamination from entering sample points and microbiological media may lead to false positives. In general, aseptic technique requires the following:

- Wearing nitrile gloves and spraying them with 70% ethanol or other disinfectant.
- Disinfecting the sample container and work surfaces before and after use.
- Working in a draft-free, clean environment.
- Working with sterile goods (glassware, tubes, and transfer pipettes).

It is essential to prevent aerial contamination by dust-borne bacteria and yeast while sampling in the brewery to prevent false positives. There are several simple practices to maintain aseptic conditions. All sterile tubes or glassware for sampling should be kept covered unless in active use. While filling or transferring from the container, the lid or cover should be kept above the opening to avoid drop-in contamination. Never completely remove the lid nor set it aside on a table as the lid itself could become contaminated. Figure 14.13 demonstrates this technique.

All processing of beer samples for microbiological analysis must avoid additional contamination from the environment. This is best achieved by testing in a separate, enclosed room away from the brewhouse and cellar. While working at a lab bench or microbiology station, the surface should be disinfected with 70% ethanol. Drafts should be avoided by either working near a flame or in a **laminar** flow hood. As illustrated in Figure 14.13, when working near a flame, hot air rises, thus preventing drop-in from dust and airborne microbes. As a more expensive but more aseptic option, laminar flow cabinets use a high efficiency particulate air (**HEPA**) filter to remove microbes from incoming air, which is then gently pushed out through the working area, creating an aseptic environment as demonstrated in Figure 14.14. The laminar flow cabinet protects the sample, but provides no protection for the operator, who works in the air flow from the sample.

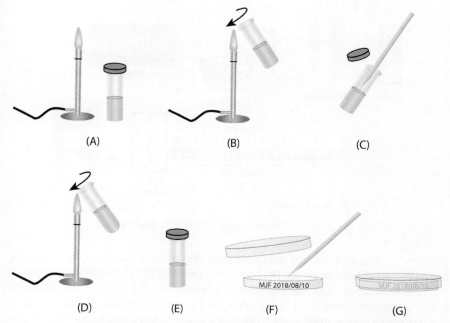

Figure 14.13 Aseptic technique. (A) Work by flame. (B) Flame glass container before use. (C) Use sterile pipet; protect interior of sample. (D) Flame container after use. (E) Keep container closed. (F) Transfer sample to sterile, labeled container minimizing the time the container is open. (G) Cover container.

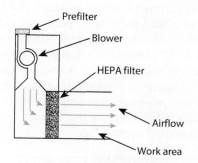

Figure 14.14   Laminar flow cabinet.

Sampling Points

Beer should be sampled at critical sample points throughout the brewing process. Boiling sterilizes the wort, so most microbiological analysis in the brewery focuses on steps downstream of the **wort chiller**. Critical sample points are identified as any point through which the beer passes to a new piece of process equipment. During brewhouse design, this requires conscious

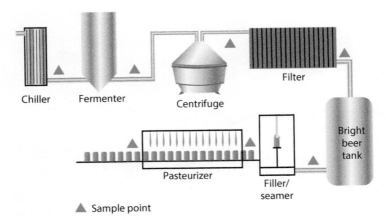

▲ Sample point

Figure 14.15   Critical sample points in the brewery.

placement of sample points upstream and downstream of all equipment. For example, consider a brewery that has only two sample points downstream of the boil kettle, one after the chiller and a second on the bright tank. Spoilage bacteria were isolated from the bright tank but not the chiller. In this scenario, the brewer cannot pinpoint the source of the contamination. It may have been in dry-hopping equipment, the fermenter, the centrifuge, the filter, the carbonation unit, or the apparatus to introduce finings. Ideal design includes sample points before and after each critical process. This permits a much more accurate determination of where contamination originates (Figure 14.15).

Beer is often sampled through sanitary zwickels. The zwickel must be properly cleaned before and after sampling to prevent false positives identified during analysis. The brewer should ensure that zwickels are sanitary in design, can be integrated into a CIP procedure, are regulatory compliant, and can be removed for sterile autoclaving or further cleaning. It can be advantageous to place a sanitary valve between the equipment and the zwickel in case removal is necessary while the equipment is in process or full of beer. The frequency of beer sampling should be defined by the brewer and is dependent upon frequency of use and potential for spoilage.

A PROTOCOL FOR SAMPLING FROM A ZWICKEL

Dirty zwickels are the most common source of false positives while sampling for microbiological analysis. Proper sampling technique, illustrated in Figure 14.16 is essential for maintaining aseptic technique while sampling.

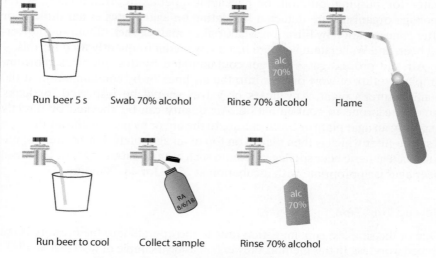

| Run beer 5 s | Swab 70% alcohol | Rinse 70% alcohol | Flame |
| Run beer to cool | Collect sample | Rinse 70% alcohol | |

Figure 14.16   Aseptic sampling from a zwickel.

1. Spray the inside and outside of the sample port with 70% ethanol. You may wish to use a sterile cotton swab dipped in ethanol to scrub the inside of the port.
2. Briefly flame the zwickel to ignite the ethanol, and then let it burn off completely. Do not continue heating the zwickel as this can lead to charring of the port and burning of the beer.
3. Let the zwickel cool for approximately 15 seconds.
4. Open the zwickel and collect beer in a waste container for approximately 10 seconds. Discard this beer.
5. Collect the sample for analysis in a sterile container.
6. Spray the inside and outside of the sample port again with 70% ethanol to remove any residual beer. Ignite the ethanol and allow it to burn off.
7. Transfer sample to the laboratory.

Water provided by the town or city should arrive at the brewery free from contamination according to specifications set by public health regulations. In addition, typical water spoilers such as *E. coli* and other *Enterobacteriaceae* do not survive in finished beer. Contamination is even further limited because water is usually boiled as part of the brewing process. However, any water used for cold-side processing should be tested for contamination. This includes water for pushing, dilution, or packaging operations. Traditionally, water spoilage organisms are detected by plating on agar plates either directly or after concentration by filtration. MacConkey, **malt extract**, sabouraud, universal beer, and Wallerstein agar are just a few media frequently used for this.

Air and process gases can get contaminated by dust particles, moisture droplets, reflux of wort or beer into the air lines, or by contamination at the manufacturer's plant. Any hoses or valves should be inspected for leaks, which are prone to contamination. Air quality can be checked by directly exposing an agar plate or broth culture to the air, or by passing the air through a sterile filter which is then placed on top of an agar plate. For brewery applications, a generic beer spoilage medium such as malt extract agar or universal beer agar is appropriate with incubation at 25 °C for 48–72 hours.

Forced Wort Test

One of the simplest contamination tests to incorporate into the brewery is the forced wort test. In this method, brewers set aside a sample of wort in a sanitized container and place it in an incubator or warm place in the brewery ranging from 25 to 30 °C for a set period. The sample lid may be left loose or sealed to promote aerobes or anaerobes, respectively. Contamination is monitored by **sediment** formation, CO_2 evolution, and/or turbidity. The method can be expanded to yeast **slurry** and beer by using a nutrient-rich broth medium containing a pH indicator. After incubation with a sample, any acid producers will turn the medium from red to yellow. These methods are quick and simple tests to check for potential spoilers, however they do not measure the number of contaminating cells nor do they identify the microorganisms, thus they are only useful to identify potential problems but do not work well to quantify potential risk.

Plating Techniques

To identify spoilage risk, collected beer samples are traditionally evaluated through plating techniques on nutrient-rich agar plates. Proper detection of contaminating microorganisms requires correct selection of the medium type, the right plating methods for screening, and the right incubation and culture conditions for growth.

Plate Spreading   Proper spreading and cell concentration will isolate individual cells on the agar medium. With appropriate incubation time and temperatures, ranging from 25 to 30 °C and one to seven days depending on the medium and organism, individual cells will multiply, each cell forming a single **colony** (Figure 14.17). The colony represents 10–100 million cells that divided over time on the plate, but which originated from a single cell in the beer sample. Proper sample volume, concentration, and spreading technique

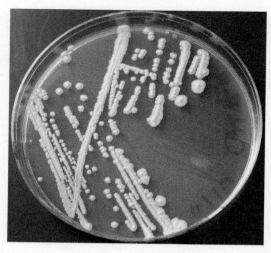

Figure 14.17   Streaked plate demonstrating distinct yeast colonies.

are required to obtain individual colonies for subsequent analysis. A properly spread agar plate resulting in distinct colonies is shown in Figure 14.17.

A colony that grows is considered one **colony forming unit** (CFU). A CFU represents a viable cell in the beer sample that was able to grow on the medium. Thus if 10 colonies are on a plate, there are 10 CFUs. It is important to note the limitation of this approach. Just because there are no colonies on a plate, it does not mean there are zero CFUs. First, the type of medium and culture conditions may limit the growth of certain bacteria. Second, the number of CFUs in the sample may be too low to be captured on the plate. For example, suppose a beer contains 1 CFU per **liter** but only 1 mL was plated, then the bacterium may not have been captured on the plate. For this reason, *if no growth is observed on a plate, this is described as <1 CFU.*

CFUs are typically reported as CFUs per mL. Thus, it is essential that the original volume plated is known. For example, if 100 μL (0.1 mL) is plated and 10 CFUs were measured, then 100 CFUs/mL are present in the sample.

Depending on the sample source, the sample may be plated directly, diluted before plating, or concentrated through membrane filtration. If plated directly, volumes should not exceed 200 μL on a 100 mm plate. If dilution is necessary, a 1 : 10 or 1 : 100 dilution in sterile media or **phosphate** buffered saline (PBS) is appropriate. If CFUs are likely low, as in finished beer, then membrane filtration is required. *Unfiltered beer cannot be membrane-filtered as the yeast and colloids will quickly clog the filter.* Two techniques for spreading the sample on a medium are shown in Figure 14.18. A sample is pipetted to the center of a plate. Then a sterile, "hockey-stick"-shaped glass rod is used

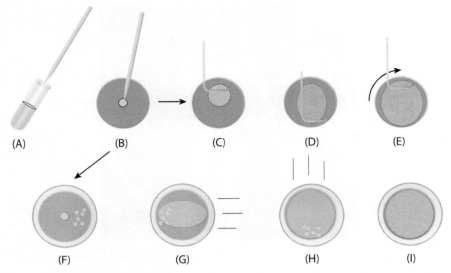

(A)   (B)   (C)   (D)   (E)

(F)   (G)   (H)   (I)

Figure 14.18   Sample spreading techniques. (A) and (B) Sample pipetted to plate. (C)–(E) Hockey stick spreader method. (F)–(H) Copacabana method. (F) Add glass balls, cover. (G) and (H) Shake left and right and up and down (not in a circle). (I) Remove beads, cover and incubate upside down.

to spread the liquid. Alternatively, sterile glass beads are added and the sample is spread by shaking the plate, a technique called the Copacabana method.

Streaking   **Streaking** is a technique to isolate cells by serial dilution on an agar plate. It is not used for colony counting, rather it is used to prepare pure cultures derived from individual cells. In brewing, streaking is often used to prepare pure yeast cultures (Figure 14.17). Using aseptic technique (Figure 14.19A), a sample is picked up on a sterile **inoculating loop** or toothpick (Figure 14.19B). The loop is streaked across the agar surface near one edge of the dish (Figure 14.19C). The dish is turned 90°. The loop is flamed again (Figure 14.19D), or a new sterile stick is used to streak through the edge of the previous streaks into the second quadrant (Figure 14.19E). This takes a small amount of sample from the first streak and puts it on the second streak; a dilution. The loop is flamed again, and the sample is streaked from the edge of the second quadrant into the third quadrant (Figure 14.19F), a dilution of the dilution. The last streak is pulled into the middle of the plate (Figure 14.19G). If all goes well, this region should produce individual colonies, each originating from a single cell. The plate is covered and incubated upside down to prevent condensate from dripping into the culture. In between each step, it is essential to allow the loop to cool after flaming to prevent

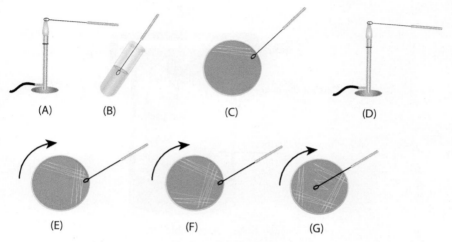

(A)   (B)   (C)   (D)

(E)   (F)   (G)

Figure 14.19   Streaking a plate.

scorching and killing the microorganisms. Plates should always be incubated in an upside-down position, agar on top, lid on bottom.

Anaerobic Methods

Depending on the medium and the type of bacteria being detected, establishing anaerobic conditions may be warranted. Typical incubators in ambient air support the growth of aerobic bacteria or aerotolerant anaerobes but may prevent the growth of strict anaerobes. Certain lactic acid bacteria grow more quickly in an incubator with 5% CO_2 and do not require strict anaerobic growth. Anaerobic cultures require attention to the availability of oxygen from the environment.

The simplest method to create anaerobic conditions is to place plates in an airtight container, inserting a small, lit candle in the jar before sealing. The oxygen is depleted as the flame burns, extinguishing once the oxygen is gone. A better method is to use an anaerobic chamber or jar with a gas generating insert (Figure 14.20). The material in the insert releases carbon dioxide and hydrogen. A catalyst in the jar causes the hydrogen and oxygen to react, producing water and depleting the air of oxygen. Typically, a card impregnated with **methylene blue** is placed in the jar as an indicator to show that oxygen has been successfully removed.

Individual colonies can then be screened with Gram stain, catalase, and oxidase tests, or other advanced methods to facilitate identification and spoilage potential.

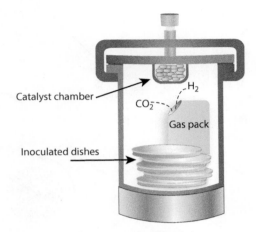

Figure 14.20   Anaerobic jar.

Microbiological Media Used in the Brewing Industry

There are two concepts in selecting the right microbiological media to test for brewery contamination: find everything or find only the high-risk, beer spoilage microorganisms. Generally, in the brewery lab, multiple microbiological media (Table 14.4) are employed to cover a wide spectrum of contamination but also to create redundancies and support further differential identification. The media discussed in this section are not exhaustive but rather represent a good spectrum for the detection of most brewery microorganisms.

TERMINOLOGY FOR DIFFERENT TYPES OF MEDIA USED IN MICROBIOLOGICAL EVALUATION

Nutrient media: Synthetic, complete media that contain all nutrients required for growth.

Selective media: Encourage growth of some microorganisms while preventing others often through addition of chemicals. Common selective media additions include the following:

    Cycloheximide – inhibits yeast

    Chloramphenicol – inhibits bacteria

    Penicillin/Streptomycin – inhibits bacteria

Differential media: An observable change in the media occurs after a biochemical reaction from the presence of a microorganism.

Enriched media: Encourage growth of a particular microorganism over others through addition of special nutrients.

TABLE 14.4   **Commonly Used Media in the Brewing Industry**

Name	Used for
Yeast Extract Peptone Dextrose (YPD)	General yeast media
Yeast Extract Peptone Maltose (YPM)	General yeast media
Wort Agar	Wort spoilers
Universal Beer Agar (UBA)	Beer spoilers
Wallerstein Labs Nutrient (WLN)	LAB, AAB, *Enterobacter*
Lee's Multi-Differential Agar (LMDA)	LAB, AAB, *Enterobacter*
Schwarz Differential Agar (SDA)	LAB, AAB, *Enterobacter*
Raka Ray	LAB
De Man, Rogosa, and Sharpe (MRS)	General bacteria media
Hsu Lacto Pedio (HLP)	LAB, AAB
Nachweismedium für bierschädliche Bakteriën (NBB)	LAB, AAB, *Enterobacter*
Selective Media Megasphaera and Pectinatus (SMMP)	Strict anaerobes
Lin's Wild Yeast Medium (LWYM)	Wild yeast (*Saccharomyces* spp.)
Lin's Cupric Sulfate Medium (LCSM)	Wild yeast (non-*Saccharomyces* spp.)
Lysine	Wild yeast (non-*Saccharomyces* spp.)
Xylose, Mannitol, Adonitol, Cellobiose, Sorbitol (XMACS)	Wild yeast

Preparation of Microbiological Agar

The brewer can save money by preparing his or her own microbiological media. Many recipes are published or are sold as pre-weighed and mixed reagents. In most cases, the brewer combines the solid nutrient chemicals with distilled water. The solution is gently heated to dissolve the contents, and then it is sterilized in an autoclave. In the case of agar media, the sterilized media is cooled to approximately 50 °C and then poured into sterile plates using aseptic technique. Typical plates are 100 mm in diameter, but some brewers prefer smaller 50 mm plates to conserve reagents and space. Before stacking and storage, plates should remain covered and allowed to **polymerize** overnight at room temperature to prevent condensation on the lids. Plates are then stored inverted at 4 °C. Careful attention to manufacturer's recommendations for shelf-life is important as selectivity may decline over time.

Sterilization of microbiological media is essential and routinely requires the use of an autoclave, the most effective method of sterilization, killing all living microorganisms and destroying any spores or viruses. Autoclaves provide wet heat under pressure. By increasing pressure in the device, the boiling point temperature can be increased. Common autoclave parameters are 121 °C at 1 bar gauge (15 psig) for 15 minutes. A large autoclave requires a steam generator and significant floor space. Smaller tabletop autoclaves are available with their own steam generators.

TIPS FOR AUTOCLAVING MICROBIOLOGICAL MEDIA

- NEVER place a sealed vessel in an autoclave or dangerous pressure could build up. If the container is open, cover the opening with foil. If the container has a screw cap, leave the cap on loosely.
- Use autoclave tape to indicate sterilization temperatures were reached. The tape can be left on in the laboratory to show that glassware or reagents were autoclaved.
- For agar media, only fill a container about 25–30% full, as **boil over** can occur in the autoclave. Containers with broth media can be 75–80% full.

PCR and Advanced Detection Methods

Traditional microbiology plating methods require one to seven days of culture time for results, depending on the media type and microorganisms present. If analysis for contamination in beer is required before release for sale, this may pose an issue. Several biotechnology companies have utilized the polymerase chain reaction (**PCR**) for rapid contamination detection. PCR-based methods of microorganism detection can be completed in as little as 4–12 hours.

PCR is a technique that makes millions of copies of a specific piece of **DNA** in a tube. The method uses the fundamentals of DNA **replication** in which a **primer** pair specifies the area of the DNA to be copied, typically a gene or gene sequence within the genome. After primer binding, DNA polymerase makes the copies. The process is repeated many times, amplifying the specific region of DNA as defined by the primers. Each cycle consists of three steps. First, a high-temperature step denatures the DNA, dissociating the double strand. Second, the temperature is reduced to a target that promotes primer binding to the DNA template. Then the temperature is further reduced promoting the activity of the DNA polymerase and copying the DNA sequence. This temperature cycle is repeated 30–40 times. The number of target DNA molecules doubles with each cycle.

In applying the PCR method to the identification of spoilage bacteria and yeast in beer, companies have identified key genes that either convey spoilage potential or that are specific to a problematic species. For example, current PCR methods for *diastaticus* detection use primers that amplify the *STA1* gene. The *STA1* gene encodes the glucoamylase enzyme responsible for hydrolysis of dextrins to glucose. Another example of a PCR target is the *HorA* gene, which renders certain *Lactobacillus* species resistant to hops. The detection platform typically includes a method for semiquantitative analysis of the amplified products.

Many platforms that use PCR technology differentiate themselves with proprietary means of quantifying the PCR results in addition to keeping the primers and conditions a trade secret. Sensitivity, time to results, number of species detected, and cost per test should all be carefully considered when evaluating a PCR platform. While rapid results are a major advantage of PCR-based methods, limitations involve sensitivity and focus on only a few genes.

Typical PCR-based methods have a minimum sensitivity of 10 CFUs/mL, otherwise requiring an enrichment step in a culture medium before the PCR test. To overcome the limitations in sensitivity, several platforms utilize quantitative PCR (qPCR). qPCR employs sensitive fluorescent dyes that enable the measurement of PCR products in real time, greatly increasing the sensitivity of the method. The cost of qPCR machines has limited their use in breweries, but as new platforms develop cheaper methods, qPCR use is on the rise.

PCR methods are further limited by the complexities of gene expression (see Section "Regulation of Gene Expression" in Chapter 3). There is no single gene that enables spoilage by all species or strains. For example, there are *Lactobacillus* strains that lack the *HorA* gene but that still spoil beer. PCR platforms with multiple gene targets aim to overcome this limitation. Nonetheless, there will always be exceptions and evolving mechanisms of contamination. PCR-based methods represent the means for rapid spoilage detection for the most common beer spoilers, but a sound traditional microbiology program with plating on selective media is recommended to further support the detection of all possible microorganisms in beer.

CHECK FOR UNDERSTANDING

1. Describe the purpose of aseptic technique. Explain best practices, equipment, and space necessary for practicing aseptic technique.

2. What are some characteristics of wort and beer that encourage or discourage microbial growth?

3. One cannot practically use every possible microbiological medium in the laboratory. Propose a reasonable set of media to use in a quality control lab that covers the detection of typical beer spoilers. Explain your rationale for their inclusion.

4. How do the methods for sample spreading and streaking differ? What is the goal of each?

5. Discuss the advantages and disadvantages of pasteurization and sterile filtration.

6. What is flash pasteurization?

7. A sample is brought to 64 °C for 4.0 minutes. How many pasteurization units is this?

8. What is an anaerobic jar or chamber? For what purpose is it used in a microbiology program?

9. A sample derived from beer wort gave a positive Gram test, a negative catalase test, and looks like circles in the microscope. What could it be?

10. Describe some hazards and precautions for using an autoclave.

CASE STUDY

Your brewery recently installed a canning line with a four-head filler. After placing some cans in a warm area of the brewery, you noticed three out of six of the cans are bulging, developing excessive pressure due to overcarbonation. Assuming you have access to a lab with microbiology capabilities, describe your approach to determine the source of contamination including the point of infection and the identity of the responsible microorganism. Once you have isolated colonies, how will you evaluate spoilage risk?

BIBLIOGRAPHY

Bamforth CW. 2017. *Freshness*. American Society of Brewing Chemists. ISBN 978-1-881696-27-8.

Bokulich NA, Bamforth CW. 2013. The microbiology of malting and brewing. *Microbiol. Mol. Biol. Rev.* 77(2):157.

Boulton C, Quain D. 2006. *Brewing Yeast and Fermentation*. Blackwell Science. ISBN 978-1-4051-5268-6.

Haakensen MC, Butt L, Chaban H, Deneer HG, Ziola B, Dowgiert T. 2007. *horA*-Specific real-time PCR for detection of beer-spoilage lactic acid bacteria. *J. Am. Soc. Brew. Chem.* 65(3):157–165.

Jorgenson E. 2017. An overview of bacteria found in brewing exosystems. *Tech. Q. Master Brew. Assoc. Am.* 54(2):95–102.

Lewis MJ, Young TW. 2002. *Brewing*, 2th ed. Springer. ISBN 978-1-4615-0729-1.

Menz G, Andrighetto C, Lombardi A, Corich V, Aldred P, Vriesekoop F. 2010. Isolation, identification, and characterisation of beer-spoilage lactic acid bacteria from microbrewed beer from Victoria, Australia. *J. Inst. Brew.* 116(1):14–22.

Sakamoto K, Konigs WN. 2003. Beer spoilage bacteria and hop resistance. *Int. J. Food Microbiol.* 89:105–124.

Sanders ER. 2012. Aseptic laboratory techniques: plating methods. *J. Vis. Exp.* 2013(63):3064.

Vriesekoop F, Krahl M, Hucker B, Mens G. 2012. 125th Anniversary review: bacteria in brewing: the good, the bad and the ugly. *J. Inst. Brew.* 118:335–345.

CHAPTER 15

MATHEMATICS OF QUALITY

Any manufacturing process has a strong quantitative component. In this chapter, we will introduce some aspects of **quality** management that provide valuable tools to improve the quality of beer, the efficiency with which it is produced, and the quantity of profits realized.

15.1 STATISTICS FOR QUALITY

Normal Distribution

Beer **brewing** is subject to variation in outcomes. If the target for **wort strength** is 12.00 °P, and only 11.8 °P was reached, that might be acceptable by some standards. If only 10.75 °P was attained, this would be a problem. Small variations in outcomes may be acceptable, but large deviations represent process problems or mistakes.

A **histogram** plots the number of observations against the range of measurement. Figure 15.1 is a histogram showing strengths of 30 representative batches of wort. The histogram is a bar graph showing the number of measurements whose outcomes fall in the indicated range. Superimposed on the histogram is

Mastering Brewing Science: Quality and Production, First Edition.
Matthew Farber and Roger Barth.
© 2019 John Wiley & Sons, Inc. Published 2019 by John Wiley & Sons, Inc.

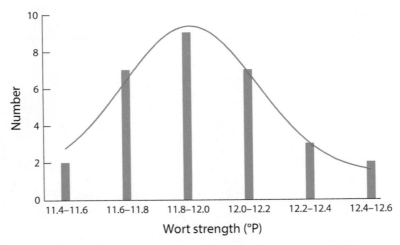

Figure 15.1   Histogram. Red, number of measurements in each range. Blue, normal distribution.

a curve showing the best fit of the results to the **normal distribution**. The normal distribution is a mathematical function, also called the Gaussian, that describes the probability of manufacturing and measurement outcomes in many, but not all, situations. The familiar bell-shaped curve of the normal distribution is a widely recognized result of probability. The normal distribution for any specific case or measurement in the **brewery** depends on two parameters. The **mean**, or average, locates the center of the distribution, and the **standard deviation** (σ, Greek sigma) relates to the width of the curve. For the 30 measurements in Figure 15.1, the average is 11.96 °P and the standard deviation is 0.24 °P. It is easier to work with the normal distribution in a standard form, called the Z-score. The conversion to Z-scores is given by $Z = (x - \text{avg})/\sigma$. The average of the Z-scores is always zero and the standard deviation is one. The idealized normal distribution plotted against Z-scores is shown in Figure 15.2.

The mean and standard deviation can be calculated from a list of numbers. The relevant equations are given below. The symbol Σ (sigma) means the sum of all items with subscripts; x_i is one of the measured values, so Σx_i means to add all the measured values. To calculate the mean (average), the sum of the numbers is divided by the number of values (N) in the analysis. Means and standard deviations are most accurately calculated on a computer spreadsheet, like Excel®. The Excel format for the mean is =average(*list*), where *list* is the list of numbers to be averaged. The format for the standard deviation is =stdev(*list*):

$$\text{Average} = \bar{x} = \frac{\Sigma x_i}{N} \qquad \sigma = \sqrt{\frac{\Sigma x_i^2 - \left(\Sigma x_i\right)^2 / N}{N-1}}$$

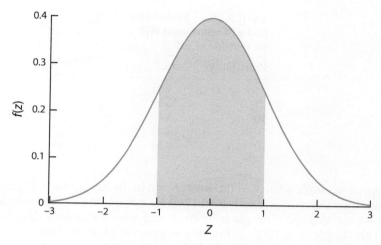

Figure 15.2   Normal distribution.

If we take the samples in groups, the average remains the same, but the standard deviation of the means of groups of *n* items is the individual standard deviation divided by the square root of the number of items in the group. For example, suppose the standard deviation for **carbonation** for individual bottles is 0.2 volumes. If we use averages of the carbonation of groups of four bottles, the standard deviation of these averages will be 0.2 volumes divided by the square root of 4, which comes to 0.1 volumes.

The normal distribution is an important tool for monitoring the quality and consistency of brewing processes. The significance is that the probability of any measurement of the variable being between two values of *Z* is the area under the curve between those values. The colored area shown on Figure 15.2 represents the probability of a measurement of *Z* coming in between −1 and +1. The probability is 0.683 or 68.3%. Thus, 31.7% of the measurements will be further than one standard deviation from the average. The area between −2 and +2 is 95.45%, and that between −3 and +3 is 99.73%, so about 5% of measurements lie outside of two standard deviations from the mean, and about three of every thousand lie outside of three standard deviations. For measured numbers, the range of values whose area contains a specified probability, often 95%, is called the **confidence interval** for the measured number. Tables and spreadsheet functions with cumulative areas of the normal distribution from −∞ to a specified value of *Z* are available. The Excel® function for the cumulative area of the normal distribution is =norm.dist(*x,mean,std dev,1*), where *x* is the measurement. If the measurement is given in *Z*-scores, the mean is 0 and the standard deviation is 1. The difference between the cumulative areas for two measurements is the area between them. Table 15.1 gives the *Z*-score limits for various levels of probability.

TABLE 15.1 Probability that a Measurement Will Be Outside of ±Z

Probability	±Z
0.05	1.96
0.001	3.29
10^{-4}	3.89
10^{-5}	4.42
10^{-6}	4.89

Suppose we made a batch of the wort shown in the histogram in Figure 15.2 and measured the wort strength as 11.17 °P for a batch of the wort mentioned above whose average strength is 11.96 °P and whose standard deviation is 0.24. This batch is 0.79 °P below the mean of 11.96 °P. The Z-score is $Z = -0.79\,°P/0.24\,°P = -3.29$. The probability of a wort strength that far from the average is 0.001. Either we witnessed a one-in-a-thousand event, or this measurement is not actually part of the original distribution, meaning something went wrong. Perhaps the measuring device is not working properly, or maybe the **mash** temperature (or its measuring device) was not correct. When a measurement is very improbable, we suspect that there is an issue with the process or the measurement, and it should be investigated.

T-test

Measurements themselves are subject to random variations. Procedures like weighing and pipetting can introduce variations. Measuring devices are subject to electronic effects, like noise and drift. As a result, multiple measurements of the same sample may yield a normal distribution of results. The standard deviation is seldom independently known; it must be estimated from the measurements that are taken. To correct for an imperfect knowledge of the standard deviation, the **T-test**, also called the Student's T-test, is used. The T-test has a distinguished history in brewing. "Student" was a pseudonym of William Sealy Gosset (1876–1937), a chemist/mathematician who worked for the Guinness brewery.

The T-test provides information about the probability that a series of measurements is different from a target value or from a second series of measurements. The **null hypothesis** is that the observed result does not differ significantly from the target or that the two series of measurements do not differ significantly from one another. The null hypothesis is rejected if its probability is below a certain level, typically 0.05 = 5%. If the null hypothesis is rejected, the conclusion would be that the sample is off target or that the two series of measurements are different.

Suppose a certain beer has a color target of 30 **SRM**. We make three independent measurements of the color with results 27.4, 26.5, and 28.2 SRM. The mean of these measurements is 27.36, and the standard deviation is 0.85 SRM. We calculate the t-statistic from

$$t = \frac{|\bar{y} - y_0|}{\left(s / \sqrt{n}\right)}$$

where $|\bar{y} - \bar{y}_0|$ is the absolute value (always positive) of the difference between the mean and the target value, n is the sample size, and s is the sample standard deviation. For the three colors measured,

$$t = |27.4 - 30| \times \frac{\sqrt{3}}{0.85} = 5.36$$

The *degrees of freedom* in this case are the number of values minus one, in this case, $3 - 1 = 2$. The critical value of t for a specific probability and number of degrees of freedom is returned by the Excel spreadsheet function =t.inv. 2t(*prob,df*). Tables with the critical t values are also available in books and on the web. In this case, the Excel entry would be =t.inv.2t(0.05,2), which returns 4.302. If the t value exceeds the critical value, the null hypothesis can be rejected. In this case, the t value is 5.36 and the critical value is 4.302. The measured color is statistically different from the target. Statistically different is not the same as discernably different. Customers may not notice a difference of a few degrees SRM, even though it can be confidently detected analytically.

There is another type of T-test that is used to compare two sets of measurements. Suppose the brewery is considering changing the boiling routine from 15% evaporation to 10%. Six batches are prepared alternating between the old and new routines. The resulting beers are taste tested with the following scores rated out of 5:

Old	New
4.65	4.22
4.55	4.5
4.28	4.45

The t-statistic for comparison of two means is

$$t = \frac{|\bar{y}_1 - \bar{y}_2|}{\sqrt{\left(s_1^2 / n_1\right) + \left(s_2^2 / n_2\right)}}$$

The number of degrees of freedom is $n_1 + n_2 - 2$. For the example above $\bar{y}_1 - \bar{y}_2 = 0.103$, $s_1 = 0.1914$, $s_2 = 0.1493$, and $t = 0.7373$. The critical value of t =t.inv.2t(0.05,4) is 2.776. The null hypothesis cannot be rejected; these measurements do not provide evidence that the new boiling routine affects the **flavor** score.

Poisson Distribution

If we take a small sample of beer and count the number of microbes in it, the count is exact. But if we take a duplicate sample and count the microbes in it, the result is likely to be different. The variation arises from the random distribution of microbes throughout the sample, not from any error in the count itself. Counted items in samples follow the Poisson [pwa-SONE] distribution. Because a count is never negative, the Poisson distribution is not symmetrical about the average. If the count is 10 or more, the asymmetry becomes less significant, and the Poisson distribution becomes reasonably close to a normal distribution with the same mean and standard deviation.

One very useful feature of the Poisson distribution is that the standard deviation is equal to the square root of the expected count. Suppose a 10 mL sample of beer wort is diluted to 100 mL (dilution factor of 10) in a **sterile phosphate**-buffered saline (PBS) **solution**. Then 100 μL (0.1 mL) of the diluted sample is **plated** on growth medium supported by agar in a **Petri dish**. After an appropriate period of incubation, 10 bacterial **colonies** are observed. Ten **colony-forming units** (CFU) in 0.1 mL comes to 100 CFU in 1 mL. Multiplication by the dilution factor by 10 yields 1000 CFU/mL in the original wort. The estimated standard deviation for the *original count* is $\sqrt{10} = 3.16$. The 95% confidence interval is $\pm 2 \times 3.16 = \pm 6$. This gives 1000 ± 600 CFU/mL, a very wide range of uncertainty. Suppose the dilution step had been omitted and 100 μL of undiluted wort was applied directly to the plate. If the count is 100 CFU, the standard deviation would be $\sqrt{100} = 10$ for a confidence interval of ±20. The final result would be (100 ± 20 CFU/100 μL) × 1000 μL/mL = 1000 ± 200 CFU/mL, a much more precise count. This is the reason that the sample should be diluted to give a count of no less than 50 CFU on the plate. Note that the standard deviation must be calculated from the *actual count*, before any adjustments for sample size and dilution are made.

15.2   CONTROL CHARTS

When a process gives an average and standard deviation that, over time, remain the same, the process is said to be *in control*. This is not to say that the process cannot be improved, but rather, based on current methods, a range of data is expected. If the mean drifts up or down, or the degree of scatter from batch to batch changes (especially upward), the process is no longer in control. The beer may still be salable, but trouble is on the horizon. Something has changed. A **control chart** is a graphic device to monitor whether the process is in control.

X-bar Chart

The most basic type of control chart, sometimes called an X-bar chart, is a graph of some process variable, like wort strength, against time or batch number. A different control chart is necessary for different recipes, because the performance of the process is compared with its own history. A common refinement of the X-bar chart is to mark the historical mean across the center of the chart and "control lines" at three standard deviations above and below the mean. Sometimes "warning lines" are added customarily at two standard deviations above and below the mean (Figure 15.3). If the samples are taken in groups, the group standard deviation should be used to locate the lines.

A control line is not the same as a specification. Specifications are set by what is acceptable to the customer or some other brewing necessity. For example, customers may not be able to tell the difference between 26 and 30 **IBU** (International Bittering Units), so beer that is 26 IBU should not be discarded. By contrast, if the standard deviation on **hop** bitterness is 1 IBU, the controls lines would be at 27 and 33 IBU. Statistically speaking, the process should remain in that region if it is under control. If specifications are too close to the control lines, the process is not sufficiently refined to meet the specification reliably. The process should be adjusted to lower the standard deviation.

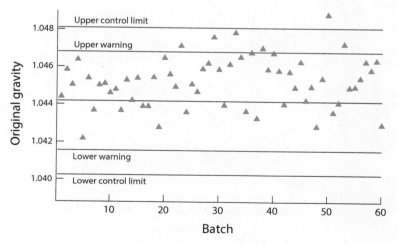

Figure 15.3   Control (X-bar) chart.

There are several ways for a process to be out of control. First, a single measurement outside of the **control limits** has a probability of 0.003 and indicates that the process is out of control. Second, nine measurements in a row on the same side of the mean has a probability of $(0.5)^9$, which is 0.002, indicating loss of control. Third, two of three points outside the 2σ line or six points in a row all trending up or down indicate loss of control. And finally, if multiple batches measure outside of the warning lines, all on an upper or lower limit, there may be loss of control.

Figure 15.3 shows a control chart for wort **original gravity**. Until the out-of-control point at batch 50, the chart appeared in control. If scrutinized, there is a feature that could have been cause for concern if it were noticed. Of the six batches outside of the warning lines, all were above the upper warning; none fell below the lower warning line.

CUSUM Chart

Another type of control chart is the cumulative sum (CUSUM) chart. In the **CUSUM chart**, the deviations from the mean or the target are calculated and converted to Z-scores by dividing by the standard deviation. The sum of all Z-scores up to the current batch is plotted for each batch. Figure 15.4 shows the CUSUM chart of the same data as those plotted in Figure 15.3. In the CUSUM chart, if the target is lower than the average measurement, the points will drift upward at a constant slope (steepness of rise or fall). If the average changes, the slope changes. In Figure 15.4, the slope of the graph increases at about batch 19. This suggests that the average original gravity

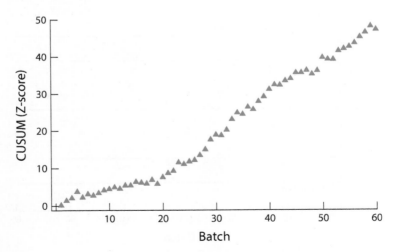

Figure 15.4   Cumulative sum (CUSUM) chart.

increased at about this time. It could have been a batch or two earlier or later, but batch 19 is the place to start looking. CUSUM charts are more sensitive to small changes in the process average than X-bar charts.

15.3 DIMENSIONS, UNITS, AND CONVERSIONS

A **dimension** is a quantity that can be measured, like mass or power; a **unit** is a standard quantity in which the dimension is measured, like kilograms or watts. Volume is a dimension; it can be measured in units of **liters**, gallons, or **barrels**. Every dimension can be expressed in terms of some combination of seven basic dimensions: length, time, mass, electric current, temperature, amount of substance, and luminous intensity. Combined dimensions are called derived dimensions. Volume is a derived dimension based on length. If something fills a box of a certain length, width, and height (each of which has the dimension of length), the volume is given by the product of these three lengths, so the dimension of volume is length cubed.

There is a system of units called the International System (**SI**) that is accepted internationally and that is coherent. A coherent system of units can be used with the equations of physics to give the correct results. Suppose we would like to know the kinetic energy of a 10 kilogram mass moving at 6 meters per second. The formula for kinetic energy is $E = \frac{1}{2}mv^2$, so the energy is $(0.5)(10\,kg)(6\,m/s)^2 = 180\,kg{\cdot}m^2{\cdot}s^2$. It turns out that the dimension of mass·length²/time² comes up often enough that it has its own name, energy, and a derived unit, the **joule** (J). The kinetic energy comes to 180 J. In noncoherent units, like American customary units, this calculation does not work well. Table 15.2 gives the six of the seven basic dimensions (we will not use luminous intensity), their base units, and the symbols for the units.

The base units can be combined using the equations of physics to produce derived units. For example, force is described by Newton's second law as mass times acceleration. Acceleration has dimensions of speed divided by time. Speed has dimensions of length divided by time. So force = mass × length/time². The SI unit of force is $kg{\cdot}m/s^2 = kg{\cdot}m{\cdot}s^{-2}$. This

TABLE 15.2   SI Dimensions and Units

Dimension	Base Unit	Symbol
Length	Meter	m
Mass	Kilogram	kg
Time	Second	s
Electric current	Ampere	A
Temperature	Kelvin	K
Amount of substance	Mole	mol

TABLE 15.3 Derived Units

Dimension	Unit	Symbol	Base units
Energy	Joule	J	$kg \cdot m^2 \cdot s^{-2}$
Force	Newton	N	$kg \cdot m \cdot s^{-2}$
Pressure	Pascal	Pa	$kg \cdot m^{-1} \cdot s^{-2}$
Power	Watt	W	$kg \cdot m^2 \cdot s^{-3}$

TABLE 15.4 Unit Prefixes

Multiplier	Prefix	Symbol
10^{-9}	Nano	n
10^{-6}	Micro	µ
10^{-3}	Milli	m
0.01	Centi	c
100	Hecto	h
10^3	Kilo	k
10^6	Mega	M
10^9	Giga	G

derived unit is called a **newton** (N). Table 15.3 shows some important named derived units and their symbols.

Base units are not always a convenient size. A **yeast cell** may be 0.000004 m in diameter; the pressure of the atmosphere is about 110000 Pa. To simplify the measurement, prefixes are combined with unit symbols to describe smaller or larger SI units. Table 15.4 shows useful unit prefixes.

For example, 0.000004 m = 4 µm; 100000 Pa = 100 kPa. Prefixes that are not powers of 1000, like hecto and centi, are only used in certain cases. There are a few units with special names: 100 kPa = 1 bar; $0.001 \, m^3$ = 1 liter (symbol L). These units can take the standard prefixes: mL, Mbar. When writing a number with a unit symbol, there should be a space between the number and the unit symbol. Unit symbols do not get a period unless they end a sentence.

Despite the elegance and universality of the SI system, there are many measurements reported in other traditional unit systems, especially in English-speaking countries. In some cases, the same name applies to different units. For example, a US gallon is 3.785 L but an Imperial (UK) gallon is 4.546 L, a difference of more than 20%. The volume unit that is called a barrel differs from one place to another and by what is being measured, like oil or beer. Even countries that follow the SI system use some nonstandard units, like hours and days.

Some English-speaking countries retain an eclectic system of customary units, which can vary from place to place and by what is being measured. A **pump** may be rated in gallons (US) per minute and a tank volume in US barrels (bbl US). The brewer is faced with problems like the following: how long will it take a 55 gal/min pump to fill a 30 bbl tank? To solve such problems, the same quantity must be expressed in different units, a procedure called unit conversion. Unit conversion starts with an equivalence, an equation that relates the units. In this example a suitable equivalence would be 1 bbl = 31 gal. From the equivalence, conversion factors are derived: (1 bbl)/ (31 gal) and (31 gal)/(1 bbl). Because the numerator (top) and denominator (bottom) of the conversion factors are "equal" in terms of the quantity they measure, they are equivalent to 1; a quantity can be multiplied or divided by the conversion factor without changing its value. The objective is to get the volume units for the pump rate to match the units for the tank capacity. To do so, the pump rate can be converted to barrels per minute, or the tank volume can be converted to gallons. We will change the tank volume to gallons. The units we want to get rid of are barrels, and the units we want to bring in are gallons. We multiply the tank volume in barrels by the conversion with the gallons on top and the barrels on the bottom, as shown in the equation below:

$$30 \, \cancel{bbl} \left(\frac{31 \, gal}{1 \, \cancel{bbl}} \right) = 930 \, gal$$

The barrels on top and bottom cancel out, leaving gallons. The pumping time can be calculated: (930 gal)/(55 gal/min) = 16.9 min. Table 15.5 has some equivalences useful in brewing.

15.4 BREWING CALCULATIONS

Dilution

One common calculation in the brewhouse is **composition** before or after dilution. Suppose 50 L of clean in place (**CIP**) solution containing 2% (2 g/100 mL) sodium hydroxide (NaOH) is needed. The NaOH product, also called the stock solution, is 50% w/v (50 g/100 mL). How much of the stock solution is needed? In this case, the dimensions of the composition are mass of NaOH divided by volume of solution: c = mass/vol. The key insight for concentration calculations is that dilution does not change the quantity of the material being diluted, in this case NaOH. The mass of NaOH is given by mass = conc·vol = cV. The mass is the same before and after dilution, so $c_1 V_1 = c_2 V_2$, where the ones refer to the stock solution and the twos refer to the CIP solution to be prepared. c_1 is 50 g/100 mL, c_2 is 2 g/100 mL, and V_2 is 50 L. The result is $V_1 = c_2 V_2 / c_1 = (2 \, \text{g/mL})(50 \, \text{L})/(50 \, \text{g/mL}) = 2 \, \text{L}$.

The calculation can also be performed in the other direction; the concentration of the stock sample can be calculated from the concentration in the

TABLE 15.5   **Unit Equivalences**

Mass

1 Pound (lb)	=453.5 grams (g)
1 Avoirdupois ounce (oz avoir.)	=28.35 grams (g)
1 Grain (gr)	=64.799 milligrams (mg)

Length

1 Foot (foot)	=0.3048 meter (m)
	=12 inches (in.)
1 Mile	=5280 ft
	=1609.34 m

Volume

1 Cubic meter (m³)	=1000 liters (L)
	=10 hectoliters
1 Cubic centimeter (cc or cm³)	=1 milliliter (mL)
1 US gallon (gal US)	=3.785 liters (L)
1 Fluid ounce (fl oz)	=29.57 mL
1 US quart (qt)	=0.9464 liter (L)
	=57.75 cubic inches (in.³)
1 Imperial gallon (gal UK)	=4.5466 liters (L)
1 US beer barrel (bbl US)	=31 gal US
	=117.35 L
1 UK beer barrel (bbl UK)	=163.66 L
	=36 UK gal
1 Firkin	=0.25 bbl UK
	=40.91 L
1 Cubic foot	=28.32 L
	=1728 cubic inches (in.³)

Pressure

1 Pound per square inch (psi)	=6895 pascal (Pa)
1 Atmosphere	=101325 Pa
	=14.696 psi
	=1.01325 bar
1 Bar	=100000 Pa
	=14.504 psi
	=0.98692 atm

Concentration

1 Degree Plato (°P)	=1% by weight
	$\approx (SG - 1)/0.004$

Gases

1 Volume of carbonation	=1.963 g CO_2/L

Color

Standard Reference Method (SRM)	=1.97 × EBC
	$\approx$ Lovibond (°L)

diluted sample. Suppose a 10.00 mL sample of yeast **slurry** is diluted to 1000 mL with saline solution to count the cells in a **hemocytometer** (Figure 9.5). The cells are counted in five of the squares in the central region. Each of these squares is 0.02 cm (0.2 mm) on an edge and 0.01 cm deep. The volume of each square is $0.02\,cm \times 0.02\,cm \times 0.01\,cm = 4 \times 10^{-6}\,cm^3 = 4 \times 10^{-6}\,mL$. There are 225 cells in five squares. The total volume of the squares is $5 \times 4 \times 10^{-6}\,mL = 2.0 \times 10^{-5}\,mL$. The cell concentration after dilution is 225 cells/$2.0 \times 10^{-5}\,mL = 1.13 \times 10^{7}$ cells/mL.

To get the concentration in the original slurry, the dilution equation is used: $c_1 V_1 = c_2 V_2$, where the ones refer to the original slurry and the twos refer to the diluted slurry. The volumes apply to the dilution (not the part we looked at in the hemocytometer), so $V_1 = 10\,mL$ and $V_2 = 1000\,mL$. The slurry concentration is $c_1 = c_2 V_2/V_1 = (1.13 \times 10^{7}\,cells/mL) \times (1000\,mL/10\,mL) = 1.13 \times 10^{9}$ cells/mL. The original count was 225; the standard deviation is $\sqrt{225} = 15$, so the 95% confidence interval for the count is ±30 cells. Dividing by the volume and factoring in dilution gives $30/(2.0 \times 10^{-5}\,mL) \times (1000\,mL/10\,mL) = \pm 1.5 \times 10^{8}$ cells/mL, so the concentration is $1.1 \pm 0.2 \times 10^{9}$ cells/mL. (The confidence interval was rounded to one digit, and the cell count was rounded to match the confidence interval.)

Alcohol

Alcohol by volume (ABV) is the volume of pure ethanol divided by the total volume times 100%, as measured at a reference temperature (usually 20 °C). **Alcohol by weight (ABW)** is the mass of ethanol divided by the total mass, measured at any temperature. ABV is always higher than ABW, because a gram of alcohol (density = 0.7893 g/mL) occupies about 25% more volume than a gram of water (density = 0.9982 g/mL). The equations below relate ABW and ABV:

$$ABW = \frac{0.7907 \times ABV}{SG_{beer}}$$

$$ABV = \frac{ABW \times SG_{beer}}{0.7907}$$

Extract and Degree of Fermentation

In the lexicon of brewing, **extract** means solids content of wort or beer. The **original extract (OE)** is the percent solids (**degrees Plato**) in the wort before fermentation. The **real extract** (RE) is the percent solids in the beer after fermentation. The **real degree of fermentation** (RDF) is given by

$$RDF = \frac{\text{Mass wort solids} - \text{Mass beer solids}}{\text{Mass wort solids}} \times 100\%$$

One complication is that beer has ethanol and solids dissolved in water. Simple analytical methods, like **specific gravity** and **refractive index**, are affected by both. Another complication is that the mass of the beer is not the same as that of the wort, because material is lost as carbon dioxide and as yeast biomass. To account for this, CJN Balling created a formula in 1865: 2.0665 g extract gives 1.0000 g ethanol + 0.9565 g carbon dioxide + 0.011 g yeast (dry weight). This means that for every gram of extract consumed in fermentation, $(0.9565 + 0.11)/2.0556 = 0.5161$ g is lost as carbon dioxide and yeast, and 0.4839 g is converted to alcohol.

There are issues with this formula. The mass of yeast varies with wort oxygen content, yeast species/strain, temperature, and even the shape of the fermenter. Another issue is that the formula is based on fermentation of a six-carbon **monosaccharide**; a **disaccharide**, like **maltose**, will yield about 10% more product than a monosaccharide per gram consumed. Nonetheless, after 150 years, the **Balling** formula still holds relevance in the industry. Its errors tend to be less than the measurement errors, so it gives a reasonable estimate. Note that the four decimal places in the coefficients can give an exaggerated illusion of accuracy. The best-known equation derived from the Balling formula is the Balling equation

$$OE = \frac{100\left(2.0556 \times ABW + RE\right)}{100 + 1.0665 \times ABW}$$

Both parameters of the Balling equation, the ABW and the RE, can be difficult to measure directly. There are online tables that provide reasonable estimates for these parameters from specific gravities before and after fermentation.

Many brewers dodge the complications of calculating RE and RDF by using apparent extract and apparent attenuation. The apparent extract, before, after, or during fermentation, is the extract that one reads from specific gravity tables, such as that in the **American Society of Brewing Chemists** (ASBC) Methods, not correcting for alcohol content. The apparent attenuation is given by

$$AA = \frac{SG_{\text{init}} - SG_{\text{final}}}{SG_{\text{init}} - 1} \times 100\%$$

The apparent attenuation is always a good deal higher than the RDF because the alcohol content after fermentation lowers the final specific gravity.

The apparent extract can be used to estimate the RE:

$$RE = 0.1948 \times OE + 0.8052 \times AE$$

Combining this equation with the Balling equation gives this estimate for ABW:

$$ABW = \frac{0.8052 \times OE + 0.8056 \times AE}{2.0556 - 0.010665 \times OE}$$

CHECK FOR UNDERSTANDING

1. The target carbonation for a certain beer is 2.6 volumes. Three bottles are sampled with the following results: 2.5, 2.3, and 2.6. Perform a T-test to determine if there is evidence at the 95% level that the carbonation is off target.

2. To secure vegan certification, a brewery is considering switching from gelatin to a different fining. Turbidity measurements give the following readings: for gelatin, 18.2, 16.6, and 19.4 and for the new fining, 20.1, 18.6, and 17.5. Use the T-test to determine if these results provide evidence at the 95% level that the finings give different turbidities.

3. A brewery with a 50 hL mash tun would like to buy a hot liquor tank with a capacity of 2.5 times that of the mash tun. What does this come to in US gallons?

4. Iodophor stock solution has 16 g/L iodine. What volume in milliliter would be needed to make 10 US gallons of a 25 ppm (mg/L) solution?

5. A batch of beer has an original extract of 14.3 °P and an apparent extract of 2.8 °P. Calculate the percent of alcohol by weight (ABW).

6. For the beer mentioned in problem 5, calculate the percent of alcohol by volume (ABV).

7. For the beer mentioned in problem 5, calculate the real degree of fermentation (RDF).

CASE STUDY

You are the Director of Quality at a brewery. You are interested in tracking the data collected at critical **control points** in the brewery, and you feel like your process has been fairly controlled to date. You decide to start with mash pH.

Prepare an X-bar chart to visualize the data below. Graph the mean, the upper and lower warning lines, and the upper and lower control limits.

Mash pH data points (in order by production lot): 5.2, 5.23, 5.14, 5.31, 5.18, 5.23, 5.25, 5.24, 5.17, 5.29, 5.19, 5.32, 5.18, 5.24, 5.24, 5.17, 5.28, 5.31, 5.29, 5.15, 5.21, 5.29, 5.19, 5.31, 5.15, 5.25, 5.22, 5.2, 5.18, 5.29.

After preparing the X-bar chart above, you continue monitoring mash pH over 30 more production lots with results below. Using the warning lines and control limits established above, analyze the following data. Is this data within the established limits? Identify the points at which the pH is outside of control. Were there any data trends that indicated the specification was out of control?

Mash pH data points (in order by production lot): 5.18, 5.25, 5.13, 5.27, 5.22, 5.23, 5.21, 5.14, 5.32, 5.31, 5.26, 5.31, 5.33, 5.33, 5.36, 5.37, 5.38, 5.33, 5.43, 5.24, 5.41, 5.34, 5.29, 5.32, 5.27, 5.45, 5.34, 5.29, 5.45, 5.38

If the process is out of control, how will you will identify, evaluate, and resolve the problem with mash pH?

BIBLIOGRAPHY

American Society of Brewing Chemists. 1940. *Tables Related to Determinations on Wort, Beer, and Brewing Sugars and Syrups*.

American Society of Brewing Chemists. 2018. *Methods of Analysis*. Beer–6.

Bamforth CW. 2002. *Standards of Brewing*. Brewers Publications. ISBN 978-0-937381-79-3. Chap. 3.

Holle SR. 2010. *A Handbook of Basic Brewing Calculations*. Master Brewers Association. ISBN 0-9718255-1-3.

Pellettieri M. 2015. *Quality Management Essential Planning for Breweries*. Brewers Publications. ISBN 978-1-938469-15-2.

Spedding G. 2016. Alcohol and Its Measurement. In Bamforth CE (editor). *Brewing Materials and Processes*. Academic Press. ISBN 978-0-12-799954-8. Chap. 7.

CHAPTER 16

CLEANING, CIP, AND SANITIZATION

Beer has been **brewed** for thousands of years. Before the development of our scientific understanding of microorganisms and their impact on beer, people still made **alcohol**. The beer was probably terrible by modern standards, but they liked it. Today, beer competes on **quality**; strict attention is now placed on maintaining a **clean** facility, not only for government regulation and food safety, but to ensure high-quality product.

Do not let the allure of the brewing industry fool you. To brew beer is to clean equipment. The brewer spends substantial and significant effort cleaning in any **brewery**. From **kettles** to **kegs**, all areas of the brewery are susceptible to soils that compromise beer quality or production efficiency. Here we discuss the most essential definitions of cleaning, some simple concepts for effective cleaning, and the most common chemicals used by the brewing industry. Your chemical suppliers are excellent resources and should always be consulted for best use and application.

Mastering Brewing Science: Quality and Production, First Edition.
Matthew Farber and Roger Barth.
© 2019 John Wiley & Sons, Inc. Published 2019 by John Wiley & Sons, Inc.

16.1 CLEANING A BREWERY

It is not possible to make good beer with dirty equipment. Deposits of soil contribute off-**flavors**, provide refuge to **spoilage** organisms, and interfere with the proper operation of **heat** transfer surfaces, **valves**, **pumps**, and sensors. Cleaning refers to the removal of unwanted materials from **surfaces**. **Sanitizing** refers the processes that *reduce* **microbes** to acceptable levels. Sanitization should not be confused with **sterilization**. *Sterilization* effectively destroys all living cells, **spores**, and viruses. *Sanitization* only eliminates the living cells.

Sinner Circle

The **Sinner circle** is a concept put forward in 1959 by Dr. Herbert Sinner of Henkel, AG. There are four essential elements to cleaning: temperature, time, mechanical action, and chemicals. These are represented in a divided circle shown in Figure 16.1. The concept of the Sinner circle is that all four elements are required for effective cleaning. If there is less of one element, the other elements need to be increased to maintain cleaning quality.

Regarding temperature, higher temperature makes molecules move faster. Reactions go faster at high temperature, so many cleaning chemicals become more effective. In addition, many constituents of soils become more soluble as the temperature is raised, improving their removal.

Chemicals, such as **acids** and **bases**, serve as complexing agents that bind constituents of soil and help them dissolve. **Surfactant**s enhance contact between the water and the soil. Oxidants react with soils to improve their solubility. Chemical suppliers innovate and often protect their specific formulations as trade secrets. The best designed chemical cleaners help reduce the other aspects of the Sinner circle.

Mechanical action can include scrubbing, stirring, and even ultrasonic treatment. In most brewery cleaning situations, mechanical action is provided

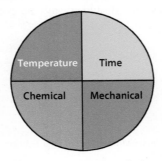

Figure 16.1   Sinner circle.

by jets of cleaning **solution** under pressure. Mechanical action is increased by turbulent flow, correct fluid flow velocities, and pulse cleaning, which creates spikes of **turbulence**. **Foaming** chemicals and sprayers are effective for cleaning large, complex exterior surfaces like packing lines. In these foam applications, mechanical action is limited to the popping of bubbles, and it is not usually possible to increase the temperature, so cleaning effectiveness depends on long contact time and strong chemicals.

Sanitary Design

Brewery design, installation, and planning for future expansion must provide for effective cleaning. Valves and **fittings** that carry beer, **wort**, brewing **liquor**, or anything that goes into the beer must be free of cracks and crevices that are difficult to clean. **Sanitary fittings** are used to connect **pipes** and hoses. The most commonly used fittings in the brewing industry are **stainless steel** tri-clamps (Figure 16.2). Tri-clamp fittings involve two flat flanges or ferrules that come together around a gasket. The flanges are pulled together by a clamp.

Threaded **pipe fittings** are not suitable, because the spaces between the threads prevent effective sanitation. All contact surfaces, including the insides of pipes and tanks, should be smooth. Abrupt bends or size changes should be avoided. Elbows and corners should have a radius at least 1.5 times the **tube** inside diameter. Transitions from one tube diameter to another should be made with smoothly changing transition fittings.

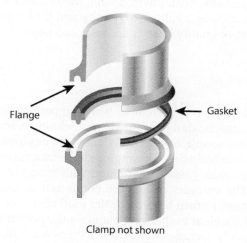

Flange ← → Gasket

Clamp not shown

Figure 16.2   Sanitary fitting.

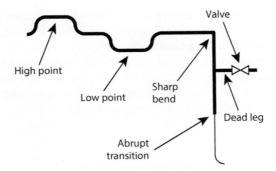

Figure 16.3   Piping problems.

Pipe layout is important for sanitary design (Figure 16.3). Branches in the plumbing can lead to "dead legs" in which flow stagnates. High points without vents can collect air, leading to stagnant zones. Low points without drains can collect fluid. This applies to sagging pipes as well. Abrupt bends or changes in pipe diameter interfere with the flow pattern, which can give rise to zones of incomplete mixing.

Exterior Cleaning

The floors, walls, ceilings, storage areas, and exterior surfaces of equipment must be kept clean. A regular program of sweeping, mopping, vacuum cleaning, application of cleaning foam, and hosing down should be implemented. Areas under and behind things need special attention. Exterior cleaning helps maintain a healthy work environment, holds down insect and rodent infestation, and reinforces a culture of quality and professionalism that is as much an ingredient of quality beer as are **malt** and **hops**.

16.2 CLEAN IN PLACE

Clean in place, or **CIP**, is a system in which apparatus and plumbing for cleaning are built into the design of the plant. Cleaning is accomplished without the need to disassemble or physically enter the equipment. CIP is nearly universal in the food (including brewing) and pharmaceutical industries. In a CIP-enabled plant, each vessel is equipped with sprayers or jets to project cleaning agents. The sprayers or jets are connected to tanks with cleaning agents or rinse water. Return lines carry the used agents or rinse water back to tanks for recovery. **Heat exchangers** maintain the proper temperature. The cleaning cycle usually includes a prerinse to remove bulk soil and soluble materials, a cleaning step; intermediate rinse to remove cleaning agents,

TABLE 16.1   **The Steps of a Typical Clean in Place (CIP) Cycle**

Step	CIP Program	Time (min)
1	Purge out CO_2	5
2	Prerinse with water	5
3	Wash with caustic (recirculation)	30–50
4	Rinse with water	5
5	Wash with acid wash (recirculation)	10–15
6	Rinse with water	5
7	Wash with sanitizer (no recirculation)	15–20

sanitation to kill microbes; and final rinse to remove sanitation chemicals unless no-rinse sanitizers are used. These steps are often subdivided, for example, an **alkaline** cleaning step may be followed by an acid cleaning step. The cycle is often automated, allowing the reproducible application of material to standardize cleaning and minimize waste. At minimum, a pump is required to push chemical into tanks and a second pump for draining.

In steps where mineral deposits are unlikely, the acid wash may be employed less frequently. Table 16.1 lists the typical steps in a CIP process using a no-rinse sanitizer. Draining steps are omitted. Temperature and time are highly variable depending on the chemicals used and the CIP system. To conserve water, the initial rinse for removing spent **grain**, wort, or beer may be recycled rinse water from previous CIP cycles.

Cleaning of Vessels

Vessels, in contrast to pipes, tubes, hoses, and the like, have a large cavity. Sprayers installed in the cavity can spray cleaning agents onto the walls. The two classes of sprayers are **spray balls** and jets. Spray balls are hollow fixtures with multiple holes to spray the cleaning or rinsing solutions. They distribute cleaning agent widely throughout the vessel but without much force. Figure 16.4 shows a spray ball and its action in a vessel. The pattern and position of the ball is critical to ensure full coverage of the vessel surfaces. Sometimes the spray is pulsed to increase mechanical scouring.

Jets have a few (typically 2–4) moving nozzles that drive cleaning agent against the surface under high pressure. The impact of the fluid on the surface provides mechanical scouring. Jets can project fluid a longer distance than spray balls; they are often used in large vessels or those that are tall compared to their width, or vice versa. The nozzles usually rotate in two planes to give thorough coverage. Figure 16.5 shows a four-nozzle jet that rotates in the horizontal and vertical planes. The black line in Figure 16.5B shows the spray pattern of one nozzle.

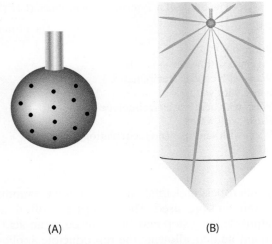

(A) (B)

Figure 16.4   (A) Spray ball. (B) Spray ball action.

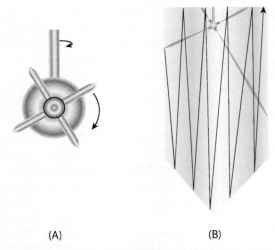

(A) (B)

Figure 16.5   (A) Clean in place (CIP) jet. (B) Jet action.

For both spray balls and jets, the sprayers and other internal equipment must be located so that the spray is not blocked in parts of the vessel. Such "shadow" spots may include **manways**, temperature probes, sample lines, racking arms, and spaces around gaskets. All equipment should be visually inspected on occasion for CIP effectiveness with additional scrutiny of potential shadow spots.

Cleaning Pipes and Hoses

Pipes and hoses are cleaned by circulating cleaning solutions through them with a pump. In permanently installed lines, a device called a *pipeline pig* may be inserted to promote mechanical action. The cleaning cycle is the same as in vessels. Mechanical scouring is provided to a limited extent by turbulent flow. Because the flow is parallel to the walls, it cannot scour the pipe walls directly. The key issue is to make sure every surface inside the pipe or hose contacts the cleaning agent. Flow upward is more effective than flow downward, as shown in Figure 16.6. It may be helpful to run the cleaning cycle in both directions.

Hoses have the additional problem of deterioration. They are made of polymer, their interiors are difficult to inspect, and the inside usually takes more wear than the outside. The result is that hoses can look fine but be riddled with cracks inside making them impossible to clean effectively.

Cleaning Out of Place

Some parts of a system are not compatible with the CIP regimen. They can be damaged by the heat or chemicals, or they have interior channels that require special treatment. These must be removed and cleaned out of place (COP) before the CIP cycle is initiated. Procedures for reinstalling COP parts must ensure that the parts are not forgotten and that they are not contaminated during storage or installation. Tri-clamps, gaskets, and other fittings can be soaked in cleaner according to manufacturer specifications.

Figure 16.6   Cleaning pipe. (A) Flow down. (B) Flow up.

16.3 CLEANING CHEMICALS

There is no purpose in cleaning an extensively soiled surface. It will stay soiled. For example, you would never add caustic to a **mash tun** still coated with spent grain. A system should look significantly clean before attempting the CIP cycle. This is accomplished by prerinsing with water.

Water

Water is an outstanding cleaning agent. It can be highly effective, especially with **polar** or **ionic** soils. It is cheap, readily available, nontoxic, noncorrosive, free of off-flavors, and environmentally friendly. All cleaning is done with solutions in which water is the major **component**. Water must be supplemented with other agents, all of which lack one or more of water's desirable properties.

Sodium Hydroxide

Sodium hydroxide, NaOH, is also called **caustic** soda, lye, **alkali**, or just caustic. It is an inexpensive agent whose action results from its strong **basicity**. Sodium hydroxide is particularly effective on protein soils and biofilms. It is often used at concentrations of 2–5% and temperatures of 70–90 °C. The main mode of action of sodium hydroxide is **hydrolysis**. It hydrolyzes **fats** into more soluble fatty acid salts, called soaps, and **glycerol** as shown in Figure 16.7. It hydrolyzes **proteins** into smaller fragments and **amino** acids.

WARNING: Sodium hydroxide solutions, especially when hot, are very corrosive to the skin and eyes. Careful attention to training, equipment, operating and maintenance procedures, **personal protective equipment** (PPE), and procedures for handling spills is necessary.

Some disadvantages of sodium hydroxide are that it can form insoluble salts with some metal ions; it is not a surfactant; it is not active at binding (sequestering) ions like calcium; it can foam; it sticks to surfaces, it can be difficult to rinse off, and it reacts with carbon dioxide to produce sodium bicarbonate: $NaOH + CO_2 \rightarrow NaHCO_3$. A **hectoliter** of carbon dioxide can neutralize the sodium hydroxide in 3 L of cleaning solution. The loss of sodium hydroxide diminishes the effectiveness of the cleaning program. The loss of carbon dioxide can produce an unexpected vacuum in a vessel, potentially crushing it. For this reason, CO_2 must always be purged from tanks before cleaning. Effective **pressure relief valves** must be installed as a backup. Please review Section "Cylindroconical Fermenters" (Chapter 9), which covers the sources of implosion risk in brewing vessels. Not only will implosion ruin an expensive piece of equipment, but it is also a severe safety risk associated with CIP processes.

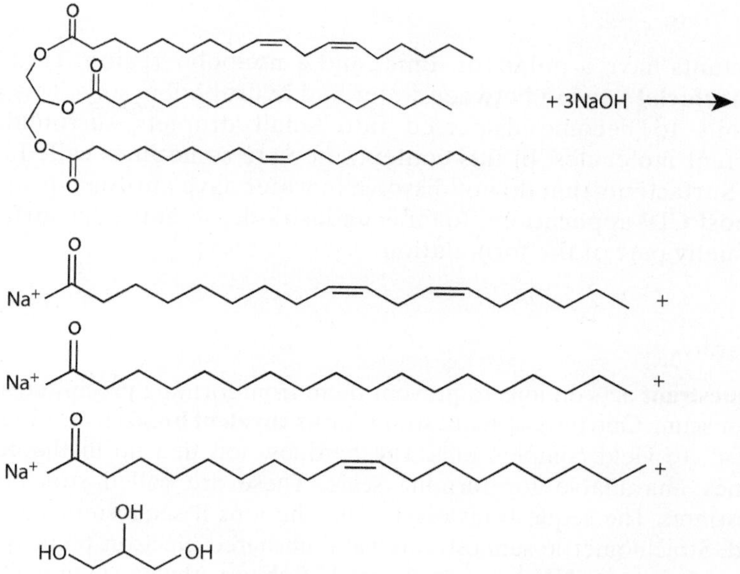

Figure 16.7   Hydrolysis of fat.

Sodium hydroxide is incompatible with aluminum. Despite the disadvantages, sodium hydroxide is widely used, usually with additives to make up for its deficiencies.

Sodium Metasilicate

Sodium metasilicate, Na_2SiO_3, provides alkalinity and helps disperse soils. It is often used as a component of cleaning preparations, especially for beer dispense systems.

Phosphoric Acid

Phosphoric acid (H_3PO_4) is the most commonly used acid. Acid cleaners do not react with carbon dioxide. They rinse freely. Sometimes they are used to remove sodium hydroxide. Acid cleaners are used to dissolve mineral deposits. They are particularly effective at removing calcium oxalate, also known as **beer stone**. They are less effective than alkaline cleaners for fat and protein removal.

Nitric acid is a strong oxidizing acid that may be blended with phosphoric acid to promote cleaning and protection of stainless steel. Nitric acid is effective at removing rust and producing a thin, corrosion-resistant oxide coating on stainless steel, a process called *passivating*.

Surfactants

Surfactants have a polar (or ionic) and a nonpolar region. They lower the **interfacial tension** between water and hydrophobic soils. This allows the soils to become dispersed into small droplets surrounded by surfactant molecules. In this context, they are sometimes called *dispersants*. Surfactants that do not dissolve in water have antifoam properties. For most CIP applications, foam is undesirable, so antifoam surfactants are usually part of the formulation.

Sequestrants

A **sequestrant** acts on ions to prevent them from forming **precipitates** giving scale or scum. One type of sequestrant forms **covalent bonds** to **hardness** ions, like Ca^{2+} to yield complex ions. The hardness ion tied up in the complex becomes unavailable for forming scale. These are called stoichiometric sequestrants. The sequestrant reacts with the ions it sequesters and is consumed. Stoichiometric sequestrants have unshared electron pairs, often on oxygen or amino ($-NH_2$) groups. Figure 16.8 shows ethylene diamine tetraacetate (EDTA) ion, a powerful sequestrant of calcium ions frequently used in cleaning formulations.

The other type of sequestrant is the **threshold sequestrant**. A threshold sequestrant does not react directly with the hardness ions. It modifies the surface of the solid so that it does not stick to form scale. Most threshold sequestrants contain several **phosphate** groups. An example is 1-hydroxyethane diphosphonic acid (HEDP) shown in Figure 16.9.

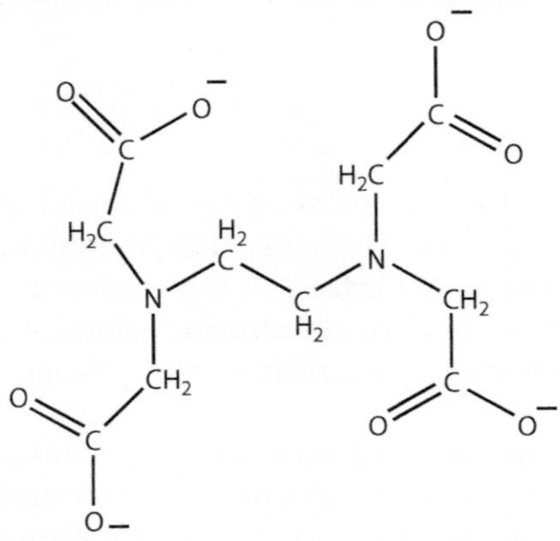

Figure 16.8   Ethylene diamine tetraacetate (EDTA).

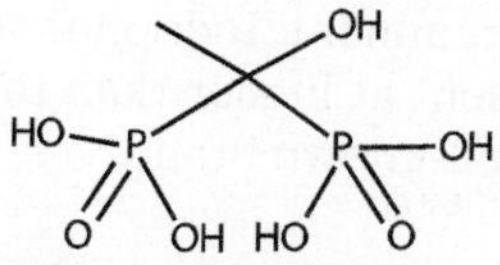

Figure 16.9   1-Hydroxyethane diphosphonic acid (HEDP).

16.4   SANITIZERS

Sanitizers are compounds that kill microbes on surfaces. Ordinary sanitizers will not make a surface **sterile**. They will, if properly applied, lower the concentration of microbes to an acceptable level. Chlorine and **bleach** are powerful sanitizers that should never be used with brewing equipment due to their corrosive action on stainless steel.

Acid Sanitizer

Acid sanitizers contain about 300 mg/L phosphoric acid and an equal amount of a surfactant. Acid sanitizers are used at room temperature. They sanitize in one to two minutes and do not require rinsing. The concentrate, as delivered, is irritating, but the diluted solution is harmless. The major disadvantage for CIP use is foaming. They are quite suitable in soak baths for COP. Star San® is a well-known brand.

Peracetic Acid

Peracetic acid (PAA), $CH_3C(=O)OOH$, also called peroxyacetic acid, is an **organic peroxide**. Its antimicrobial activity comes from oxidation. It is used at or below room temperature at 75–300 mg/L. PAA is smelly and decomposes over time to **acetic acid** and hydrogen peroxide. The hydrogen peroxide decomposes to water and oxygen **gas**, so the storage containers should allow gas to escape. It is suitable for use in a final CIP rinse; no further rinsing is required. Perasan® is a well-known brand.

Iodophor

Iodophor is iodine stabilized with a surfactant. It is an oxidizer. It is used at room temperature at 10–25 mg/L iodine. Iodophor is not suitable for use at high pH. If the equipment had been treated with an alkaline agent, it should be neutralized with acid prior to the iodophor rinse. Iodophor can be corrosive to stainless steel at high concentration. The undiluted stock solution should not be allowed to make contact with the vessel. Iodophor can be used for CIP final rinse and in COP soak baths. The solution should be allowed to drain for 15 minutes after treatment. Iodophor stains plastic, and it can lend an off-flavor to beer if present at higher than the recommended concentration. BTF Iodophor® is a well-known brand.

Chlorine Dioxide

Chlorine dioxide, ClO_2, is a powerful sanitizing agent used at room temperature at concentrations of 0.5–1 mg/L. It is an oxidizer. Chlorine dioxide has many desirable properties. It introduces no flavor taints; it is not corrosive; it can be used as a final CIP rinse. Its major disadvantage is that it is not chemically stable. Chlorine dioxide is usually generated by a chemical reaction at the point of use. One common method is the reaction of sodium chlorite ($NaClO_2$) with an acid:

$$5NaClO_2 + 4H^+ \rightarrow 4ClO_2 + 4Na^+ + NaCl + 2H_2O$$

The solution must be used at once. Sodium chlorite is not to be mistaken for sodium hypochlorite ($NaClO$, bleach).

Ozone

Ozone dissolved in water at a rate of 1–2 mg/L is an oxidizing sanitizer similar in some respects to chlorine dioxide. Ozone is generated at the point of use by treating air or oxygen with an electric discharge (arc). The resulting ozone decomposes to oxygen, so the solution must be used at once. It leaves no flavor taint and no harmful residue. It is even approved for direct application to food surfaces. At normal concentration it is not harmful to brewing equipment.

Hydrogen Peroxide

Hydrogen peroxide, H_2O_2, is a neutral oxidizing sanitizer. It eventually decomposes to water and oxygen. Nonfood grades can contain harmful stabilizers. Hydrogen peroxide can be used as a final rinse in CIP systems, but its high effective concentration of 3% can be inconvenient.

Bleach

Sodium hypochlorite ($NaClO$), called bleach, is a very popular disinfectant in water systems and swimming pools. It is still used for food and beverage sanitation at a concentration of 50–300 mg/L. Bleach has several serious disadvantages. It can react with beer constituents, especially polyphenols, to produce off-flavors in the beer. It can produce pitting **corrosion** in stainless steel. Its effectiveness is strongly dependent on pH. WARNING: Bleach must not be mixed with acid; toxic chlorine gas will be released. The use of bleach in breweries should be avoided.

16.5   ANALYSIS AND QUALITY CONTROL OF CLEANING EFFECTIVENESS

The easiest quality control test for cleaning is to visually inspect surfaces. With attention to problems areas like **brandhefe** rings, potential shadow points, and beer stone deposits, if a surface looks dirty, then it is dirty. No further tests are needed. Once a surface looks clean, a sensitive method to confirm cleaning effectiveness is to use an **adenosine triphosphate (ATP)** luminometer. Rinsing effectiveness should be checked with a **pH meter**, and chemical concentration should be validated with **titration**. Many chemical suppliers provide kits or instructions on tests for validation of the recommended working concentrations of their products

Surface ATP Tests

The ATP surface test is deemed essential because of its easy use and rapid results. It is most commonly employed to check for general CIP cleaning effectiveness of brewery surfaces. Any surface with routine exposure to wort, beer, or yeast should be regularly checked. Though a surface may look clean, the presence of bacteria is nearly impossible to see with the naked eye unless there is a severe problem, like a biofilm. Biofilms that harbor bacteria can be very difficult to remove completely. The ATP test can also be adapted to check for air and water quality. There are many suppliers of ATP systems, including Hygiena® (Camarillo, CA), VWR® (Radnor, PA), and 3M® (Maplewood, MN).

The ATP test measures the presence of bacteria or yeast. The reagents contain an **enzyme** called luciferase and a substrate called luciferin. The luciferase **catalyzes** a reaction of luciferase with ATP resulting in emission of light (luminescence), the very same reaction that occurs in fireflies, which is how these reagents were discovered. The amount of light emitted depends upon the availability of ATP from the environment. ATP is found in all bacteria and yeast, living or dead; the more bacteria that are present, the more luminescence is detected by the ATP test.

To run the test, a sterile **swab** is brushed in a pattern against a 10×10 cm (4×4 in.) swatch of the surface in question to collect any surface bacteria or yeast (Figure 16.10). The swab is exposed to the reagents and placed back into the collection tube. The collection tube is then placed into a

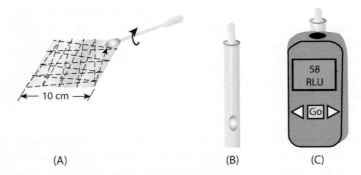

(A)   (B)   (C)

Figure 16.10   Adenosine triphosphate (ATP) swab testing procedure. (A) Swab surface. (B) Activate reagents. (C) Results.

luminometer to record relative light **units** (RLUs). Swabs with samples can be stored, before exposure to reagents, for up to four hours. Once exposed, they should be quickly analyzed. The acceptable RLUs for a brewery depend upon specifications set by each brewery. Typically, a reading of 0–10 RLUs is acceptable; 10–40 RLUs warrants caution; greater than 40 RLUs is unacceptable.

Liquid ATP Test

The same ATP test, with a revised sample collection procedure, can be used to detect contamination in liquids. Liquid samples from equipment with inaccessible surfaces, like pipes and hoses, can be evaluated for cleanliness. Liquid testing can also be used on the final rinse for brewing vessels. Liquid testing gathers a sample from the entire surface, so it is good for detecting localized contamination that could be missed by swabbing, but it gives no indication of exactly where the contamination is located. In liquid testing, the equipment is rinsed, and ATP testing is conducted on a sample of the rinse water. The sample is collected with a probe that picks up a measured volume. The sample in the probe is processed in the same way as a swabbed sample.

pH Check of Rinse Water

It is important to remove all soils from equipment, but it is equally important to remove all traces of the cleaning reagents. Because most cleaning reagents are caustic or **acidic**, checking the rinse water pH is a quick and effective method to monitor rinsing effectiveness. Checking the pH of the rinse water may also help to prevent excess rinsing, thereby saving water.

CHECK FOR UNDERSTANDING

1. What is the Sinner circle?

2. A brewer cannot mechanically scrub all surfaces of every piece of equipment that contacts beer. Using the Sinner circle with practical examples, explain how a reduction in mechanical scrubbing is overcome in the brewery.

3. What are some features of sanitary fittings?

4. Identify potential sanitary problem areas in pipe layouts.

5. Define clean in place.

6. What are some common design elements in brewing vessels to promote CIP effectiveness?

7. How does sodium hydroxide work as a cleaning agent?

8. What are three ways to avoid collapsing a tank during CIP cleaning?

9. What are suitable applications for acid cleaners?

10. What are the roles of surfactants and sequestrants in cleaning?

11. Discuss the application of ATP testing in validating cleaning effectiveness.

CASE STUDY

You were recently hired to run quality control at a brewery with a 10 bbl brewhouse and 20 bbl fermenters. The brewery currently makes their own caustic. They clean with 2% (w/v) sodium hydroxide solution at 70 °C. The brewery does not use an acid wash step, and they use iodophor for sanitation. You are eager to start working at the brewery but cannot relocate for another few weeks. You decide to get started by creating an auditing checklist, developing a plan to evaluate effectiveness, and doing some research on new chemicals to purchase.

As one of your first initiatives, you will audit the design of the CIP system and its effectiveness. In anticipation of auditing the brewery, create an auditing checklist of essential CIP design elements for each processing step that you deem essential (brewhouse, chiller, fermenter, and bright tank).

Describe how you will use the auditing checklist to validate efficacy of the current CIP system.

Finally, recommend a caustic cleaner, acid cleaner, and sanitizer for implementation in the brewery. You are encouraged to reach out to local chemical suppliers or brewers in your area to inquire about their recommended products. In your recommendation consider the cost of implementing new chemicals on the bottom line of the business. Explain the rationale for the new expense.

BIBLIOGRAPHY

Briggs DE, Boulton CA, Brookes PA, Stevens R. 2004. *Brewing Science and Practice.* CRC. ISBN 0-8493-2547-1. pp. 637–647. Covers sanitary design, cleaning agents, and sanitizers.

Lewis A. 2006. Plant Sanitation. In Ockert K (editor). *Brewing Engineering and Plant Operations.* Master Brewers Association. ISBN 0-9770519-3-5. Chap. 2.

Praeckel U. 2009. Cleaning and Disinfecting. In Eßlinger HM (editor). *Handbook of Brewing.* Wiley VCH. ISBN 978-3-527-31674-8. Chap. 25.

GLOSSARY

1000 kernel weight *n* Mass of 1000 seeds estimated from a representative clean sample of grain. Also called the *1000 corn weight* or **thousand corn weight**.

α-acid *n* **Alpha acid.**

α-amylase *n* **Alpha-amylase.**

β-acid *n* **Beta acid.**

β-Amylase. Beta-amylase.

absorbance *n* Measure of light absorbed by a sample calculated from $2 - \log(\%T)$, where $\%T$ is the transmittance of a sample, expressed as a percentage.

ABV *n* Alcohol by volume.

ABW *n* Alcohol by weight.

acetaldehyde *n* **Organic compound** CH_3CHO. Precursor to **ethanol** in **fermentation.**

acetate *n* Ionized form of **acetic acid**. Final product of **ethanol** metabolism.

acetic acid *n* **Organic compound**: CH_3CO_2H. Gives vinegar flavor.

acetobacter *n* **Bacteria** that can gain **energy** by **oxidizing ethanol** to **acetic acid** in the presence of oxygen.

Mastering Brewing Science: Quality and Production, First Edition.
Matthew Farber and Roger Barth.
© 2019 John Wiley & Sons, Inc. Published 2019 by John Wiley & Sons, Inc.

acetyl coenzyme A *n* Biological thioester used to transport **acetic acid**.

acid 1. *n* Any chemical **compound** that provides a hydrogen ion. 2. *adj* Having a **pH** of less than 7. 3. *adj* Exhibiting sourness characteristic of **acidity**.

acidic *adj* Dominated by **acid**.

acidity 1. *n* Quantity or concentration of **acid** in a sample. Also called titratable acidity. 2. *n* Sour flavor or other property characteristic of acid.

acid rest *n* A period during which the **mash temperature** is held at about 35–40 °C (95–105 °F) to allow phytase **enzyme** to release phosphoric acid lowering the **mash pH**.

acidulated malt *n* **Malt** that has been subjected to a bacterial process to generate **acid**, usually lactic acid. Used to lower **mash pH** without contravening the **Reinheitsgebot**. Also called *Sauermalz*.

acrospire *n* Shoot issuing from the **embryo** of a **germinating** seed.

activated carbon *n* Highly porous carbon made by treatment of **coke** or **charcoal** at high **temperature** with a material that reacts with it to give a gas. Used to purify water and other **fluids**.

activation energy *n* Minimum **energy** needed for a **chemical reaction** event.

active site *n* Location on a **catalyst molecule** or surface responsible for the catalytic **reaction**.

actuator *n* Device that automatically operates a mechanical system, like a **valve**, in response to a control signal.

addition *n* **Chemical reaction** in which the second or third **bond** in a multiple bond breaks and groups bond to each end of the broken bond.

adenosine diphosphate *n* ADP. Adenosine ester with pyrophosphoric acid ($H_4P_2O_7$) attached to the **ribose** portion, the low-energy form of ATP. Chemical name: adenosine pyrophosphate.

adenosine triphosphate *n* ATP. Adenosine ester with tripolyphosphoric acid ($H_5P_3O_8$) attached to the **ribose** portion, the molecule that carries **energy** in **cells**. Chemical name: adenosine tripolyphosphate.

adjunct *n* Source of **fermentable** material other than **malted** grain.

administrative control *n* Policy or procedure that reduces the risk or severity of a hazard. For example, a policy that sets a schedule for testing **relief valves**.

admix *n* **Body feed**.

ADP Adenosine diphosphate.

advanced hop product *n* Flavorings for **beer** derived from **hops** by liquid extraction, hydrogenation, etc.

aerobe *n* Organism that uses oxygen.

aerobic *adj* Using oxygen.

agar *n* Jellylike material derived from seaweed. Used to support **growth media** for **microbial cultures**.

aging *n* Deliberate process of storing an **alcoholic beverage** or other food for an extended time to develop characteristic **flavors**.

air-actuated valve *n* **Valve** that is opened or closed by the application or removal of air **pressure**.

air knife *n* Device that blows compressed air in a **laminar** flow across bottles or cans to facilitate drying. Used in packaging operations.

alcohol *n* 1. **Organic** compound with the —OH group as its principal **functional group**. 2. **Ethanol**.

alcohol by volume *n* Volume fraction of **ethanol** in a solution measured at 20 °C, usually expressed as a percent. Abbrev: ABV.

alcohol by weight *n* Mass fraction of **ethanol** in a **solution**, usually expressed as a percent.

alcohol dehydrogenase *n* **Enzyme** that catalyzes the dehydrogenation of ethanol to acetaldehyde or the reverse. Abbrev: ADH.

alcoholic beverage *n* Liquid containing **ethanol** for human consumption.

alcoholic fermentation *n* **Fermentation** in which the major **product**s are **ethanol** and carbon dioxide.

alcoholism *n* Addiction to **ethanol**.

alcoholometer 1. *n* **Hydrometer** designed for measuring **ethanol concentration** in water **solution**. 2. *n* Any instrument for measuring ethanol concentration.

aldehyde *n* Compound with a C=O group bound to one or two hydrogen atoms.

aldehyde dehydrogenase *n* **Enzyme** that **catalyzes** the dehydrogenation of an **aldehyde** to a **carboxylic acid**. Abbrev: ALDH.

aldose *n* **Monosaccharide** originating as an **aldehyde**.

ale 1. *n* Formerly, **beer** made with **gruit** instead of **hops**. 2. *n* Beer fermented at a temperature above 15 °C (60 °F), typically with *Saccharomyces cerevisiae* ale strains known at top cropping **yeast**.

aleurone layer *n* Thin layer of living cells beneath the bran or **hull**. A source of **proteins** and **lipids**. Produces **enzymes** when the plant **embryo** **germinates**.

alkali 1. *n* **Base**. 2. *n* Sodium hydroxide, potassium hydroxide, or a **mixture**.

alkaline *adj* **Basic**.

alkalinity *n* Measure of the amount of **base** in water.

alkaloid *n* Naturally occurring **amine**.

alkane *n* **Organic compound** with only carbon and hydrogen and only single **bonds**.

alkene *n* **Hydrocarbon** with one or more C=C **double bonds**.

alkyl group *n* Part of a **molecule** including only C and H atoms with only single **bonds**.

alloy *n* **Mixture** displaying metallic character comprising one or more **metals** and possibly **nonmetals**.

alpha *n* First letter of the Greek alphabet: α.

alpha-acetolactate *n* **Compound: anion** of $C_5H_8O_4$. Precursor for **diacetyl**.

alpha acid (α-acid) *n* One of three **compounds** in **hops**, humulone, cohumulone, or adhumulone that, after **isomerization**, gives hop bitterness.

alpha-amylase (α-amylase) *n* **Enzyme** that **hydrolyzes** alpha($1\to4$) **glycosidic links** on a **starch** chain at random locations.

American Society of Brewing Chemists *n* Learned society that devises analytical procedures and sets standards for **malting**, **brewing**, and packaging.

amide *n* Carbonyl compound with a nitrogen atom bound to the carbonyl carbon.

amine *n* Compound with an amino or substituted **amino** group as its principal functionality.

amino *n* $-NH_2$ group.

amino acid *n* Molecule with the **carboxyl group** ($-COOH$) and the **amino** group ($-NH_2$), usually connected to the same carbon.

amino nitrogen *n* **Free amino nitrogen**.

aminopeptidase *n* **Enzyme** that **hydrolyses** **peptide molecules** at the **amino**-terminus.

Amorphous *adj* Arrangement of **atoms** or **molecules** in which there is no regular pattern or order.

amphipathic *adj* **Amphiphilic**.

amphiphilic *adj* Partly **polar** and partly nonpolar. Said of a **molecule**. Also called amphipathic.

amylase *n* **Enzyme** that **hydrolyzes** alpha($1\to4$) glucose links on **starch** chains.

amyloglucosidase *n* **Enzyme** that **hydrolyzes** all **glycosidic links** in **starch**. Also called **glucoamylase**.

amylolytic *adj* Having the ability to **hydrolyze starch**. Said of an **enzyme**.

amylopectin *n* **Polymer** of **glucose** with mainly alpha($1\to4$) links with alpha($1\to6$) branches every 30–50 units. A type of **starch**.

amylose *n* Unbranched **polymer** of **glucose** with alpha($1\to4$) links. A type of **starch**.

anabolism *n* Biological process in which simple **molecules** are assembled to give complex molecules.

anaerobe *n* Organism that does not use oxygen.

anaerobic *adj* Without air or oxygen.

analyte *n* Substance whose quantity is being measured.

analytical balance *n* Laboratory **balance** with a sensitivity of 0.1 mg or smaller.

anion *n* **Atom** or group of atoms with a negative electric charge.

anomeric carbon *n* Hemiacetal or hemiketal carbon atom in a cyclic **carbohy-drate** molecule. Has **bonds** to two oxygen **atoms**.

anticodon *n* Sequence of three **RNA nucleotides complementary** to a **codon**. Found on **tRNA**.

antifreeze *n* Substance, such as ethylene glycol, added to a **solution** to lower the freezing point.

antioxidant *n* Substance that inhibits **oxidation** either as a reducing agent or a **free radical** trap.

antiparallel *adj* Oriented with the major axes parallel but with opposite direc-tionality or spin. Said of **polymers** that have inherent directionality, like **proteins** or **nucleic acids**, and of bodies with angular momentum, like atomic **nuclei**.

apparent attenuation 1. *n* Difference between **specific gravity** before (initial gravity: OG) and after (final gravity: FG) **fermentation**. 2. *n* **Apparent degree of fermentation**.

apparent bulk density *n* **Mass** divided by the bulk volume of a sample, with-out taking void space into account. Sometimes called *unit weight*.

apparent degree of fermentation *n* Difference between **specific gravity** before (initial gravity: OG) and after (final gravity: FG) **fermentation** reported as a percentage of OG. The measure is "apparent" because after fermenta-tion, **ethanol**, which is less dense than water, makes the specific gravity an inaccurate estimate of the solids content. Abbrev: ADF.

apparent density *n* **Density** calculated from a **mass** determined without taking **buoyancy** in air into account.

apparent extract *n* Solids content equal to that of a sucrose–water solu-tion of the same **specific gravity**. Usually said of **beer** during or after **fermentation**.

apparent gravity *n* **Final gravity**. This term is not recommended; the **specific gravity** is real. The solids content estimated from it is apparent.

aqueous *adj* In water **solution**.

arabinose *n* Specific five-carbon **aldose** often found in **cell walls**.

Archimedes principle *n* Law of physics: a body immersed in a **fluid** experi-ences an upward force equal to the **weight** of the fluid whose volume the body displaces.

aroma *n* Sensation caused a material in the **gas phase** detected by smell.

aroma hops *n* **Hops** added shortly before or after the end of boiling to impart a characteristic **aroma** to the **beer**.

aromatic *adj* Having a ring containing alternating single and **double bonds**. Said of an **organic compound**.

aseptic *adj* Free from contamination of unwanted microorganisms.

ash *n* Solid remaining after a sample has been completely burned. Often expressed as the **mass fraction** of ash to the original dry mass of the sample.

as-is *adj* Based on the mass of a sample without correction for moisture. Said of an analytical result. *See* **dry basis**.

assortment *n* Size distribution. Said of **grain** and **malt**.

astringent 1. *adj* Dry or puckering mouth feel. 2. *n* Astringent (1) substance.

asymmetric carbon *n* Carbon **atom** with four different types of atoms or **groups** attached.

atom *n* Tiny particle that is the smallest single unit of a particular **element**.

atomic number *n* Number of **protons** in an **atom**. The atomic number determines each unique **element**.

atomic weight *n* **Molar mass** of atoms of a particular **element**.

ATP *n* **Adenosine triphosphate**.

ATP bioluminescence *n* Test for the presence of **microbes**. Microbes, living or dead, contain adenosine triphosphate (**ATP**). The ATP is made to generate light. The intensity of light generated is an indication of the number of microbes in the sample. Used to test the effectiveness of **sanitation**.

attenuate *v* Lower the solids concentration by **fermentation**.

attenuation *n* Difference between initial (before **fermentation**) and final (after fermentation) solids content of **wort** or **must**, often expressed as a fraction of the initial solids.

attenuation limit *n* Maximum **attenuation** possible with a particular strain of yeast and a particular **wort composition**. Typically determined by **forced fermentation**.

attract *v* Exert a force that pulls something closer.

auger *n* **Screw conveyor**.

autoclave *n* Container or device that enables high **pressure** and **temperature** for **sterilization**, often through steam.

autolysis *n* Rupture of a **cell membrane** by action of the cell's own **enzymes**.

awn *n* Long spiky extension of the **hull** of a seed of grain. Also called a *beard*.

axial 1. *adj* In a direction perpendicular to the plane of a ring. 2. *adj* In the direction of the **optical axis**.

bacillus *n* Rod-shaped **bacterial cell**. *pl* bacilli.

back-titrate *v* Analyze by mixing the **analyte** with a known amount of **reactant** that is enough to be an excess. Then analyze the excess reactant by **titration**.

bacteria *n* Organisms belonging to a large domain of single-cell **prokaryotes**. Singular: bacterium.

bacteriostatic *adj* Biological or chemical agent that prevents **bacteria** from reproducing but does not necessarily kill them.

balance 1. *n* Instrument to measure the **mass** of an object. 2. *v* Adjust the **coefficients** of a **chemical equation** to get the same number of each type of **atom** and the same total **charge** on the reactant and product sides of the equation.

balanced 1. *adj* Having equal numbers of each type of atom on each side; said of a chemical equation. 2. *adj* Having a pleasing ratio of flavor elements, especially maltiness to bitterness in beer.

Balling *n* Scale relating specific gravity to sugar content for **aqueous solutions** of sucrose. Considered obsolete. *See* **Plato**.

ball valve *n* **Valve** with a rotating ball with a hole bored completely through it. The valve is operated by positioning the hole parallel to the flow path to open it or perpendicular to the flow path to close it. Can be operated partially open to provide coarse control of flow rate. Is not capable of **CIP**.

bar 1. *n* A counter for serving **alcoholic beverages**. 2. *n* An establishment that sells alcoholic beverages for consumption on the premises. 3. *n* A unit of **pressure** equal to 100000 pascals (0.987 atm or 14.50 **psi**).

barley *n* Grassy plant of the genus *Hordeum*.

barm beer *n* The liquid **beer** present within and/or on top of a **yeast slurry**.

barrel 1. *n* Cylindrical container often made of wood staves. Usually refers to sizes small enough to be moved. 2. *n* Any of various measures of volume, for example, the US beer barrel is 31 US gallons (117.35 L). Abbrev: bbl. *See* **cask** (2), **foeder**.

barrier tubing *n* **Tubing** lined with an oxygen-impermeable coating, usually polyethylene terephthalate (PET). Used in **beer lines**.

base 1. *n* Substance whose **molecules** accept hydrogen **ions**. 2. *n* **Nucleobase**.

base malt *n* **Malt** with at least sufficient **diastatic power** to **saccharify** its own starch.

base pairing *n* Selective **hydrogen bonding** interaction among **nucleobases**.

basic *adj* Dominated by **base** (1).

batch sparge *v* **Sparge** with a technique where the **wort** is completely drained during the **lauter**. The mash/lauter is then refilled with water; the grain bed is re-established, and the second runnings are drained.

bauxite *n* Ore of aluminum with about 50% aluminum oxides and hydroxides and the rest mostly iron oxides.

beer 1. *n* **Fermented alcoholic beverage** made from a source of **starch** without concentrating the **alcohol**. 2. *n* Beer made with **hops** in contrast to **ale** (1).

beer hall *n* A large establishment that sells **beer** for consumption on the premises.

beer line *n* **Tubing** to carry **beer** in a dispensing system.

Beer's law *n* Also known as the Lambert–Beer law. Describes, in spectroscopy, the relationship between light absorption by a material and the **concentration** of the material.

beer stone *n* Solid whose major component is calcium oxalate. Beer stone can stick to **beer brewing** and dispensing equipment.

beta *n* Second letter of the Greek alphabet: β.

beta acid (β-acid) *n* One of three **compounds** in **hops**, lupulone, colupulone, or adlupulone. These compounds do not **isomerize** to give bitterness, but their **oxidation** products can provide bitterness. Also called **lupulone**.

beta-amylase (β-amylase) *n* **Enzyme** that **hydrolyses** the second alpha$(1{\rightarrow}4)$ **glycosidic bond** from the nonreducing end of a **starch** chain to yield **maltose**.

beta-glucan (β-glucan) *n* **Gum** comprising **glucose** mostly with beta$(1{\rightarrow}4)$ linkages but with beta$(1{\rightarrow}3)$ linkages every four to six glucose units. Increases the **viscosity** of water **solutions**.

bill of lading *n* Document issued by a carrier to acknowledge receipt of cargo. When signed by the receiver, serves as proof that the cargo arrived.

binding site *n* Location on one **molecule** that has a specific **attraction** for another molecule or part of a molecule.

biopolymer *n* **Polymer** synthesized by a cell, such as a **protein**, **DNA**, or **polysaccharide**.

biotransformation *n* Chemical change of a substance resulting from the action of a **cell**.

bits *n* Particles large enough to be individually distinguished by eye rather than as a homogenous **haze**.

bittering hops *n* **Boiling hops**.

bitterness unit *n* Measure of **isomerized alpha acids** in **beer** according to ASBC Method Beer 23A: bitterness units (BU) or the equivalent EBC method. Abbrev: BU. Roughly equivalent to one milligram of isohumulone per liter. Also called international bitterness unit. Abbrev: IBU.

bleach 1. *n* **Solution** of about 5% sodium hypochlorite (NaOCl) in water. Used to remove color and as a disinfectant. 2. *v* Cause color to fade.

body feed *n* **Filtration medium** suspended in the material to be clarified. *See* **precoat**.

boiling hops *n* **Hops** added when boiling starts to add bitterness to beer. Also called **bittering hops**.

boiling point *n* **Temperature** at which the **vapor pressure** of a liquid is equal to the surrounding pressure.

boil over *v* Rise above the opening and flow out of a container of boiling liquid because of excessive buildup of **foam**.

boilover *n* Instance of **boiling over**. Also spelled *boil-over*.

bond *n* Force of **attraction** holding **atoms** together.

bond energy *n* Amount of **energy** needed to break a particular **bond**. Also called *bond strength*.

bonding pair *n* Pair of **valence electrons** shared between two **atoms**.

bound *adj* Held to something by forces of **attraction**.

boundary layer *n* Thin, stationary, non-flowing region of a moving **fluid** (or of a fluid through which a solid moves) against the immersed solid surface. Matter and **energy** are exchanged between the bulk (moving) fluid and the boundary layer largely by **diffusion**.

bouza *n* North African style of **beer** made from bread and **malt**.

brandhefe *n* Cold break **lipids**, **protein**, and **yeast** left behind as a ring near the top of a fermenter after the **krausen** subsides [*Ger. burnt yeast*].

brass *n* **Alloy** of copper and zinc.

breathalyzer *n* Portable device to measure **alcohol concentration** in exhaled breath.

Brettanomyces *n* Genus of **yeast** used to make certain **beer** styles.

brew *v* Prepare **beer**.

brewer's grain *n* Spent **grain** after **mashing**.

brewery *n* Facility for making **beer**.

brewhouse *n* Portion of a **brewery** with equipment for **mashing, wort separation**, boiling, **wort chilling**, and associated processes.

brewhouse efficiency *n* Ratio of actual carbohydrate extracted to potential **extract** of **carbohydrate** sources in the **mash**.

brewing liquor *n* Water that is to be made into **beer**.

bright *adj* Clear; free of **haze**.

bright beer *n* **Beer** from which **yeast** has been removed.

bright beer tank *n* Vessel for **bright beer**. Also called *bright tank* or *brite tank*.

brink *n* Vessel for storing used **yeast** prior to **pitching** into a new **fermentation**.

brittle *adj* Susceptible to cracking or shattering.

Brix 1. *n* Scale relating **specific gravity** to content for **aqueous solutions** of **sucrose**. 2. *n* A **degree Brix**, which closely approximates 1% **sugar** by weight. Abbrev: °Bx.

BTU *n* British thermal unit. Unit of **energy** equal to 1054.35 **Joules**. In heating and cooling systems, often erroneously used for BTU/h, a measure of **power**. 1 BTU/h = 0.2929 W.

bud 1. *n* Immature **yeast cell** still attached to the parent cell. 2. *v* Produce a **bud** (1).

budding index *n* Fraction or percent of a sample of cells that are producing **buds** (1).

buffer 1. *n* **Solution** containing a partially neutralized weak **acid** or weak **base**, allowing it to maintain a relatively constant **pH** upon the introduction of small amounts of acid or base. 2. *v* Resist change in pH because of the presence of a **buffer** (1).

bung 1. *n* Stopper for the side opening of a **cask** or **barrel**, also called **shive**. 2. *v* To close with a **bung** (1). 3. *v* **Spond**.

buoyancy *n* Force pushing a submerged or floating object upward.

buoyancy correction *n* Correction applied to the measured **mass** of an object taking into account the effect of **buoyancy** of air. For materials whose density is near that of water, the correction is about 0.11%.

burr *n* Female **hop** flower.

butterfly valve *n* **Valve** with a plate that, when it is perpendicular to the flow direction, blocks flow. The valve is opened by rotating the plate to be parallel to the flow.

calandria *n* Tubular **heat exchanger** that provides **heat** to a liquid to boil it.

calorie *n* Unit of **energy** equal to 4.184 joules. Symbol: cal.

Calorie *n* Unit of **energy** equal to 1000 **calorie**s or 4184 **Joules**; a kilocalorie.

cAMP *n* **Cyclic adenosine monophosphate**.

Campaign for Real Ale *n* Consumer movement to maintain the traditional forms of serving **ale** to preserve a **cask-conditioned** flavor.

CAMRA *n* **Campaign for Real Ale**.

candi sugar *n* **Brewing adjunct** prepared by **heat** treatment of unrefined beet **sugar**.

capacitance *n* Ability to store electrical charge, defined by $C = Q/V$, where C is the capacitance (farads), Q is the electric charge (coulombs), and V is the electrical potential (volts).

caramel color *n* Material prepared by **heat** treatment of **sugar** alone or with the addition of **acid** or **base**. Caramel color has an intense brown color that can be used to adjust the color of food and beverages. Caramel color for beer is usually prepared by heating sugar with ammonium compounds.

caramel malt *n* **Crystal malt**.

carbohydrate *n* Substance with only carbon, hydrogen, and oxygen, having two hydrogen **atom**s for each oxygen atom, a **carbonyl** group and one —OH group on each of the other carbon atoms, or a substance made by condensing two or more of these with the elimination of water.

carbonate 1. *v* Dissolve carbon dioxide in a **solution**. 2. *n* **Compound** containing the **carbonate ion**.

carbonate ion *n* CO_3^{2-} ion.

carbonyl *n* C=O group.

carboxylic acid *n* **Organic compound** with the —COOH **group**.

carboxypeptidase *n* **Enzyme** that **hydrolyzes peptide molecules** at the carboxyl-terminus.

carboy *n* Very large bottle made of thick glass or plastic.

carrageenan *n* Polymer of sulfated **galactose** units prepared from seaweed. Used as a **fining** during the **wort** boil. Also called **Irish moss**.

cask 1. *n* Cylindrical container with two or more openings used to transport and serve **beer**. 2. *n* Container of any size made of wood **staves** for storage or transportation of materials. *See* **barrel** (1), **foeder**.

cask breather *n* **Valve** to admit carbon dioxide to a **cask** during **beer** dispense to keep air out.

cast 1. *v* Transfer liquid or **slurry** from a vessel, especially to transfer **wort** from the **kettle**. 2. *adj* Having been cast.

catabolism *n* Biological process in which complex **molecules** are broken down to simpler molecules.

catabolite repression *n* **Sugars** other than **glucose** are not **fermented** when the glucose **concentration** is high. Also called glucose repression. *See* **Crabtree effect**.

catalase *n* **Enzyme** that **catalyz**es the decomposition of hydrogen peroxide to oxygen and water, used to characterize **bacteria**.

catalyst *n* Material that speeds up a chemical **reaction** but is not a **reactant** or a **product**.

catalyze *v* Speed up a chemical **reaction** without becoming a **reactant** or a **product**.

cation *n* [CAT-eye-on] An **atom** or group of atoms with a positive charge.

caustic 1. *adj* Strongly **basic**. 2. *n* Sodium hydroxide.

celiac disease *n* Autoimmune response to **gluten** that damages the small intestine. Also spelled coeliac disease.

cell *n* Basic unit of life containing structures and substances surrounded by a **membrane**.

cellar 1. *n* Part of a **brewery** in which **fermentation** takes place. 2. *n* Room for storing **beer** before it is served.

cell membrane *n* Thin film surrounding a cell, also called plasma membrane. Sometimes mistakenly called the **cell wall**.

cellobiose *n* **Disaccharide** comprising two **glucose** units with a beta(1→4) link.

cellulose *n* Long chain **carbohydrate** comprising **glucose** units with beta(1→4) links.

cell wall *n* Rigid layer outside the **cell membrane** in plants, **fungi**, and **bacteria**.

centiliter *n* Non-recommended unit equal to 10 milliliters. Symbol: cL.

centrifuge *n* Device to spin a sample rapidly causing heavier particles to separate from lighter ones by **sedimentation**.

cereal *n* Grassy plant that yields edible **grains**. These include many of the grains used to make **beer**.

cereal cooker *n* Vessel to boil unmalted **grain** used as an **adjunct**.

chalk *n* Calcium carbonate.

channel *n* **Protein** structure extending through a **membrane** that can open to allow certain **ions** or **molecules** to flow through the membrane.

chaperone protein *n* **Protein** that folds another protein into a specific **tertiary structure**.

charcoal *n* Carbon made by **heat** treatment of plant or animal material.

check valve *n* **Valve** that allows a **fluid** to flow in one direction only.

chelating agent *n* **Compound** that can provide two or more **electron** pairs that form **covalent bonds** with **metal ions**. This lowers the effective **concentration** of the bound ions. Used in **cleaning solutions**.

chemical bond *n* **Bond**.

chemical equation *n* Symbolic statement of a chemical **reaction** showing how much of each item is consumed and how much is produced.

chemical formula *n* Symbolic representation of a **compound** or **molecule** in which **element** symbols are given subscripts showing how many atoms of each element are in a unit of the compound or molecule (subscripts of one are omitted).

chicha *n* South American **beer** style made from **maize**.

chiller 1. *n* **Heat exchanger** used to lower the **temperature** of **wort** after boiling. 2. *n* Refrigeration unit.

chill haze *n* Aggregation of **hop polyphenols** and **barley proteins** in **beer** that results in cloudiness that appears at low **temperature** and disappears at room temperature.

chime 1. *n* Portion of the **staves** of a **barrel** that extend beyond the head. 2. *n* Protective rail around the ends of a **keg**. Also spelled chimb.

chiral *adj* Having a mirror image that is different from the original **molecule**.

chit 1. *n* Small protrusion of root from a seed during sprouting. 2. *v* (Of seeds) to display a chit (1).

chitin [KITE'n] *n* **Polymer** made of modified **glucose units** connected by beta(1→4) linkages. Structural material in **fungus cell walls**.

chit malt *n* Seeds that are **kilned** as soon as they **chit** (2), so they are minimally modified.

chromatography *n* Method of separation and analysis that uses difference in affinities of the **molecules** under study for a stationary and a mobile **phase**.

chromosome *n* Long strand of tightly coiled **DNA** in living **cells** that encodes genetic information or **genes**.

CIP *adj* **Clean in place**.

citric acid cycle *n* Biochemical pathway in which **acetate** ion reacts to give carbon dioxide and several hydrogenated molecules that can be combusted to yield **energy** in the form of **ATP**. Also called Krebs cycle.

clean 1. *adj* Free of undesired material. 2. *v* Remove undesired material from a surface.

clean in place *adj* Relating to a method of **cleaning** inside surfaces of equipment and plumbing while it is still assembled. Abbrev: **CIP**.

coagulate *v* Form solid particles in a liquid.

coalescence *n* Particle growth by smaller particles colliding and sticking together to form larger particles. *See* **disproportionation**.

coccus *n* Spherical **bacterial** cell. *pl* cocci.

codon *n* Sequence of three **RNA nucleotides** that specifies a particular **amino acid** in biological **protein** synthesis.

coefficient *n* Number in a **chemical equation** indicating the number of **molecules** or **ions** consumed or produced.

cohesive *adj* Having **attractive** forces that hold parts of a substance together.

coke *n* Carbon made by **heat** treatment of coal.

cold break *n* Solid material precipitated upon chilling of **wort**. *See* **trub**.

collagen *n* **Protein** in animal connective tissue.

colloid *n* Heterogenous material that consists of a continuous **phase** and a dispersed phase, especially one in which the phases do not separate quickly. Examples include **beer haze** and **foam**.

colloidal stability *n* Resistance to the appearance of **haze**.

colony *n* Cluster of **microbes** originating from a single **cell**.

colony forming unit *n* Unit estimating the number of live microbes in a sample based on formation of visible **colonies**. Abbrev: CFU.

column *n* Tube or cylinder.

complementary *adj* Engaging in a specific binding interaction. Said of **nucleobases**: guanine is complementary to cytosine; adenine is complementary to thymine in **DNA** and to uracil in **RNA**.

component *n* One of the substances in a **mixture**.

composition 1. *n* Relative amount of each **component** of a **mixture**. 2. *n* Relative amount of each **element** in a **compound**, often called *elemental composition*.

compound *n* Single substance with more than one type of **atom**.

compression *n* Force tending to decrease the length or volume of an object; push. *See* **tension**.

compression fitting *n* **Fitting** for **tubing** comprising a body with a cup-shaped sealing surface, a deformable ring called a *ferrule*, and a cap called a *nut* that can be tightened to threads on the body. To make the joint, the ferrule is place in the body, and the nut screwed on lightly. The tube is inserted into a hole in the nut and through the ferrule. The nut is tightened, deforming the ferrule and making it grip the tube and press against the sealing surface in the body. *See* **pipe fitting**.

concentration *n* Ratio of a measure of **mass**, **moles**, volume, or other units of a **component** to a measure of the amount of the mixture.

condense 1. *v* Change from the **vapor** to the liquid state. 2. *v* React to connect two **molecules** and release a water molecule.

condenser 1. *n* Part of a refrigerator in which coolant **vapor** is **condensed** under **pressure**, releasing **heat** into the warm side of the cooling cycle. 2. *n* Part of a light **microscope** that illuminates the sample. 3. *n* Laboratory **heat exchanger** used to **condense** (1) vapors during **distillation**.

condition *v* Process **fermented beer** to give improved clarity and **flavor**.

cone hops *n* **Leaf hops**.

confidence interval *n* Range of values for a measurement within which there is a specified probablity that the average of all possible measurments would lie.

confined space *n* Place that is possible for a person to enter but with limited access making it unsuitable for habitation. Special precautions must be observed when it is necessary for a person to enter a confined space.

Congress mashing *v* Small-scale, standardized laboratory method of **mashing** to assess **malt** quality.

conjugated *adj* Having alternating single and double **bonds** or **unshared pairs**.

continuous sparge *v* The rinsing of **grains** after **lautering** whereby water is sprayed over the grain bed at similar rate to **wort** runoff, maintaining a thin level of water above the grain.

control chart *n* Graph showing a time sequence of measurements (or statistics derived from the measurements) taken at a specific point in a process. The chart is usually marked with the **mean** and with upper and lower **control limits**, usually three **standard deviations** above and below the mean. Used to estimate the inherent variability in the process and to diagnose loss of control.

control limit *n* Boundary of the region outside of which a process is regarded as out of control.

control point *n* Step or process at which control can be applied to reduce risk of a **defect**.

conveyor *n* Device for continuously moving objects or material.

coolship *n* Wide shallow open vessel in which hot **wort** is allowed to cool by contact with ambient air.

copper 1. *n* Highly conductive red **metal**, element 29, symbol Cu. 2. *n* (chiefly British) **Kettle**.

core 1. *n* Part of an **atom** including the **nucleus** plus **electrons** of lower **energy** than the outmost electrons. 2. *adj* Belonging to the core.

corn 1. *n* Seed of **grain**. 2. *n* **Maize**.

corrosion *n* Loss of **metal** due to **oxidation**.

countercurrent *adj* **Fluid** being processed flows in the opposite direction of the fluid moving **heat** in or out.

coupler *n* **Fitting** that connects a **keg** to a dispensing system. Has provision to provide **gas** to the headspace of the keg and to accept beer from the **spear** and deliver it to the **beer line**.

covalent *adj* Sharing **electrons**.

covalent bond *n* Chemical **bond** resulting from sharing **electrons**.

Crabtree effect *n* Alcoholic **fermentation** despite the presence of oxygen.

crash *v* Quickly lower the temperature of **beer** after **fermentation** to cause the **yeast** to **sediment**.

crimp 1. *v* Join two parts by bending one around the other. Said of **crown caps** and bottles. 2. *n* Seal between a crown cap and bottle.

critical point *n* **Temperature** and **pressure** for a particular substance at which the distinction between **vapor** and liquid disappears. Not to be confused with **triple point**.

crop *v* Collect **yeast** from **fermentation** for reuse.

crown cap *n* Cap with puckers along its rim. Used for sealing bottles.

croze *n* Groove in the **staves** of a **barrel** to accommodate the head.

crush *v* **Mill** (2).

crystal *n* Regular three-dimensional pattern of **atoms**, **ions**, or **molecules** to form a solid.

crystalline structure *n* Arrangement of **atoms** or **molecules** into a regular pattern in three dimensions.

crystal malt *n* Class of **malt** prepared by **stewing** to convert starch in the kernel to **sugar** or **dextrins**. Also called *caramel malt*.

culinary steam *n* Hot water **vapor** of a quality suitable for direct contact with food. *See* **process steam**.

culm *n* Root from sprouted **grain** seeds.

culture 1. *n* **Cells** growing under cultivation in an artificial **medium** (which could be beer wort). 2. *v* Prepare or perpetuate a **culture** (1).

curing *n* Final stage of **malt** drying.

current *n* Rate of transfer of electrical charge. Sometimes called "amperage."

CUSUM chart *n* **Control chart** in which the cumulative sum of deviations from the **mean** or from a target value is shown on the *y*-axis.

cyclic adenosine monophosphate *n* **Nucleotide** comprising adenosine, **ribose**, and a phosphate group attached at two points to the ribose to make a ring. Serves as a second messenger in some **flavor transduction** pathways, including those for **aroma**. Abbrev: cAMP.

cylindroconical *adj* Shaped as a cylinder mounted upon a cone with the apex pointing down. Said of **brewing** vessels.

cytoplasm *n* Liquid material inside a **cell** excluding the nucleus and other organelles.

Darcy's law *n* Flow of a **fluid** through a permeable **medium** is directly proportional to the flow area and the **pressure** drop and inversely proportional to the **viscosity** of the fluid and the length of the medium.

deciliter *n* Non-recommended **unit** equal to 100 milliliters. Symbol: dL.

decline phase *n* Final stage in a **microbial** growth curve in which overall **cell** death leads to a loss in cell number.

decoction *n* Raising the **temperature** of **mash** by boiling a portion of it and returning the hot portion to the **mash tun**.

defect 1. *n* Property that fails to meet a **quality** standard. 2. *n* Deviation from perfect regularity in the structure of a **crystal**.

degree of fermentation *n* Fraction or percentage of the mass of **wort** solids that is converted during **fermentation**. Slightly lower than the fractional **real attenuation** because of the loss of carbon dioxide and yeast. Also called *real degree of fermentation* (RDF).

degrees Balling 1. *n* Mass percent solids as estimated from the obsolete **Balling** scale. 2. **Degree Plato**. Abbrev: °B.

degrees Brix *n* Mass percent solids as estimated from the **Brix** scale, often adapted to **refractive index** measurements. Equivalent in practice to **degrees Plato**. Abbrev: °Bx.

degrees Oeschle *n* **Points**. Abbrev: °Oe.

degrees Plato *n* Mass percent total dissolved solids as estimated from the **Plato** scale. Abbrev: °P.

dehydration 1. *n* Chemical **reaction** in which water is removed and a **bond** is formed between two **molecules** to make a larger molecule or between two parts of a molecule to make a ring. 2. *n* Loss or removal of water, drying.

deionize *v* **Demineralize**.

demineralize *v* Subject to **ion exchange** treatment in which undesired **ions** are replaced with hydrogen ions and hydroxide ions, which combine to yield water.

denature 1. *v* Cause **protein** or **DNA molecules** to lose their shape and function as a result of high **temperature**, extreme **pH**, mechanical stress, etc. 2. *v* Add a toxic or repellant substance to **ethanol** to make it undrinkable.

densitometer *n See* **oscillating density meter**.

density *n* Ratio of **mass** to volume.

deoxyribonucleic acid *n* **DNA**.

detector *n* Sensor for an analytical instrument.

detergent *n* **Surfactant** used for **cleaning**.

dextrin *n* Soluble but unfermentable **carbohydrate** with several **sugar** units resulting from partial **hydrolysis** of **starch**. *See* **maltodextrin, unfermentable dextrin**, and **dextrin malt**.

dextrinase *n See* **limit dextrinase**.

dextrinizing unit *n* Measure of **alpha-amylase** content in terms of **enzymatic** activity. Abbrev: DU.

dextrin malt *n* A type of **crystal malt** that is roasted at a low **pH** and high **temperature** to promote formation of 1,6 **glycosidic bonds** without development of color. Contributes body and **foam** stability to **beer**.

dextrose *n* **Glucose** (2).

diacetyl 1. *n* **Ketone** $CH_3C(=O)-C(=O)-CH_3$ that has a buttery flavor. Official name: butanedione. 2. *Informally* **vicinal diketone (VDK)**.

dial thermometer *n* **Thermometer** in which a bimetal strip arranged in a spiral moves a pointer across a graduated dial face.

diastase *n* **Amylase**.

diastatic power *n* Amount of **starch hydrolysis** that a certain amount of **malt** can produce. Abbrev: DP.

diastereomer 1. *n* One of a set of **isomers** distinguished by different configurations about **asymmetric carbons**, but that are not mirror images of one another. 2. *n* One of a set of isomers with the **atoms** connected in the same way but oriented differently in space that are not mirror images of one another.

diatom *n* Algae with a silicon dioxide-containing **cell wall**.

diatomaceous earth *n* Porous claylike material composed of **diatom cell walls**, used as a **filtration medium**. Also called *kielseguhr*. Abbrev: DE.

differential media *n* Microbiological **medium** that provides an observable change after a biochemical **reaction** induced by the presence of a specific microorganism.

differential pressure *n* The difference in pressure between two parts of a system than can drive flow, such as the **pressure** above and below the **grain** bed in a **lauter tun**.

diffusion *n*. Spontaneous transport of matter from higher to lower **concentration**.

dimension 1. *n* Type of measurable quantity, like length or speed. 2. *n* Length measured in a particular direction.

dimer *n* **Molecule** consisting of two identical parts.

dimethyl sulfide *n* $(CH_3)_2S$, gives an off-flavor to **beer**. Abbrev: DMS.

dioecious *adj* Having the two sexes on different individuals. **Hops** are dioecious; there are male plants and female plants.

diplo– *adj* Prefix: occurring in pairs. Said of **bacteria**.

dipole *n* Particle with a positive end and a negative end.

dipole–dipole interaction *n* Force of **attraction** between **polar molecules**.

dipole moment *n* Vector quantity representing the extent and direction of **polarity**.

direct-fired *adj* Heated by the application of flame, said of a **kettle**.

direct steam injection *n* Method of heating **wort** by adding **superheated culinary steam**.

disaccharide *n* **Carbohydrate** with two simple **sugar** units.

dispersion *n* Forces resulting from temporary **polarity**.

disproportionation *n* Particle growth by the dissolving or evaporating of material from small particles and the deposition of material onto larger particles. Also called *Ostwald ripening*. *See* **coalescence**.

distill 1. *v* Purify by evaporating and **condensing**. 2. *v* Loosely, to concentrate **ethanol** by any means.

dm³ *n* Cubic decimeter. Properly called a **liter**.

DNA *n* Deoxyribonucleic acid. **Polymer** of **nucleotides** whose sequence codes for the **amino acids** in **proteins**.

DNA polymerase *n* **Enzyme** that adds deoxyribonucleotides to the 3′ end of a strand of DNA.

dockage *n* Plant material that contaminates **grain** including seeds of other plants, parts of plants other than seeds, or shriveled or broken seeds.

DON *n* Deoxynivalenol, mycotoxin that frequently contaminates **grain**. Also called *vomitoxin*.

dormancy 1. *n* State of being **dormant**. 2. *n* Fraction of seeds that are **dormant**.

dormant *adj* Temporarily unable to **germinate**; said of **grain** seeds.

double bond *n* **Covalent bond** formed by sharing two pairs of **electrons**.

draff *n* **Brewer's grains**.

draft *adj* Drawn from a **keg** or **cask**, said of **beer**. Also spelled *draught*.

drauflassen *v* To top up a **fermentation** with fresh **wort**, commonly used to **propagate yeast** due to lack of a propagation system or larger **fermenter** volume as compared the **brewhouse** volume.

drunk *adj* Incapacitated to some extent as a result of consumption of **ethanol**.

dry basis *adj* Corrected to reflect only the non-water portion of the sample, usually by dividing by $(1 - wf)$, where wf is the **mass fraction** of water. Said of an analytical result.

dry hopping *n* Process of adding **hops** to **wort** or **beer** after chilling.

dyne *n* Unit of force equivalent to $g \cdot cm \cdot s^{-2} = 10^{-5}$ **newtons** (exactly) = 2.248×10^{-6} pound-force (approximately).

EBC *n* **European Brewing Convention**.

electrical potential *n* **Energy** required to move a unit charge between two points. Also called **voltage**.

electrode *n* Solid electrical conductor that makes contact with a **nonmetallic** part of a circuit.

electron *n* Elementary negatively charged subatomic particle.

electronegativity *n* Tendency to **attract** shared electrons in a **covalent bond**.

electron group *n* Unshared pair of **electron**s or a **bond** (single, double, or triple).

element *n* Substance composed of **atoms** with a particular **atomic number**.

elementary charge *n* Charge on a proton: 1.602×10^{-19} C.

elongation *n* Step of **transcription** or **translation** in which the chain length is increased.

Embden–Meyerhof pathway *n* Version of **glycolysis** used by all organisms higher than **bacteria**. Makes two **molecules** of **pyruvic acid**, two of **ATP**, and two of **NADH** for each **glucose** molecule consumed.

embryo *n* Baby plant or animal in a seed or egg.

emmer *n* Early form of wheat: *Triticum dicoccon*.

enantiomer *n* One of a pair of **compounds** that are mirror images of one another.

endopeptidase *n* **Enzyme** that **hydrolyzes peptides** in the interior of the chain.

endoplasmic reticulum *n* Network of tubes attached to the **nucleus** (3) of a **cell**. Tags **proteins** and is involved in **protein** transport and secretion. Abbrev: ER.

endoprotease *n* **Enzyme** that **hydrolyses** large **proteins** in the interior of the protein chain.

endosperm *n* Part of a seed containing **starch** and some **protein**.

energy *n* Quantity measuring the capacity to make something move or do work.

energy level *n* Value of the **energy** permitted by physics that a particle can have. Particles cannot have energies between the energy levels.

engineering control *n* Design, built-in device, or modification to reduce the risk of injury or harmful exposure. For example, ventilation to capture and remove dust from a **mill**. *Contrast* **personal protective equipment**.

enol *n* **Compound** with an —OH group attached to a carbon **atom** that also has a **double bond** to another carbon atom.

enriched media *n* Microbiological **medium** that encourages the growth of some microorganisms while preventing the growth of others through the addition of special nutrients.

enzyme *n* **Protein** that acts as a **catalyst** and speeds up specific chemical **reactions**.

equatorial *adj* Directed in the plane of a ring pointing from the center to the outside. *See* **axial**.

equilibrium *n* State of a system in which the rates of forward and reverse progress are equal, so no change is evident.

equilibrium constant *n* Numerical value of the **reaction quotient** of a chemical **reaction** at **equilibrium**.

equivalent 1. *n* Amount of a material that can provide or accept one **mole** of hydrogen **ions**. 2. *n* Amount of a material that can provide or accept one **mole** of **elementary charge**. Abbrev: Eq.

ester 1. *n* **Organic compound** that can be formed by **condensation** of a **carboxylic acid** and an **alcohol**, releasing water. 2. *n* Compound that can be formed by condensation of an alcohol with an inorganic **oxyacid**.

ethanol *n* Two carbon **alcohol**, CH_3CH_2OH, also called ethyl alcohol.

ether *n* Compound with C—O—C as the principal **functional group**.

eukaryote *n* Organism whose cells have a **nucleus** and possibly other **organelles**.

European Brewing Convention *n* **Beer** standard setter for Europe. Also applied to any of their analytical methods, especially color. Abbrev: EBC.

exopeptidase *n* **Enzyme** that **hydrolyzes protein molecules** at points near the ends of the protein chains. *See* **aminopeptidase** and **carboxypeptidase**.

exponential 1. *n* Mathematical function characterized by increase or decrease by a fixed fraction or percentage in each time period. 2. *adj* Increasing or decreasing according to the **exponential** (1) function.

exponential phase *n* Second stage of a **microbial** growth curve where **cell** division occurs rapidly, resulting in **exponential** increases in cell number.

extensive property *n* Property of a system that scales with the mass. An example is the volume. *See* **intensive property**.

extract 1. *n* Soluble **carbohydrates** recovered from **mash**. 2. *n* Fully or partly dehydrated **beer wort**. 3. *n* Solids content of beer or beer wort. 4. *v* Retrieve or pull out.

eyepiece *n* Lens of a light **microscope** nearest the eye of the viewer. Also called ocular.

facultative *adj* Optional. Said of organisms that can survive with or without oxygen.

false bottom *n* Perforated plate to hold up **grain** during **wort separation** or **hops** in a **hop back**.

FAN *n* **Free amino nitrogen.**

Farbebier *n* Very dark **malt fermentation** product used to adjust **beer** color.

fass *v* Transfer **beer** from a **fermenter** to a vessel or container.

fassing *adj* Used for or associated with transfer of **beer** from a **fermenter**. Like a fassing **centrifuge**.

fat *n* **Ester** of **glycerol** with three **fatty acids**.

fatty acid *n* **Carboxylic acid** with un unbranched carbon chain, usually of four or more carbon **atoms**. Natural fatty acids have even numbers of carbon atoms.

ferment 1. *v* Be consumed by **fermentation**. 2. *v* Cause a material to ferment (1).

fermentability *n* Extent to which **carbohydrates** are usable by **yeast**.

fermentation *n* **Energy** production by **microbes** from a source of food without the consumption of oxygen. *See* **alcoholic fermentation**.

fermentation, degree of *n* **Degree of fermentation**.

fermentation lock *n* Tube filled with liquid that allows **gas** out of the **fermenter** and prevents the entry of air.

fermentation vessel *n* **Fermenter** (1). Abbrev: **FV**.

fermented beverage *n* Liquid containing **ethanol** prepared for human consumption by **fermentation** of **sugars**.

fermenter 1. *n* Vessel in which **fermentation** is carried out. Also called **fermentation vessel**. 2. *n* Any agent or substance that causes fermentation. *See* **fermentive organism**. Also spelled *fermentor*.

fermentive organism *n* **Microbe** that performs **fermentation**, especially one that produces desired products.

FG *n* **Final gravity.**

FIBC *n* **Super sack**. For flexible intermediate bulk container.

film boiling *n* Mode of boiling in which the bubbles form on the surface and then coalesce to form a film that interferes with the transmission of **heat** to the liquid.

filter 1. *n* Device or material used to **filter** (2). 2. *v* Remove particles from a **fluid** by passing it through a **medium** with small holes or channels.

filtration medium *n* Material through which a **fluid** is passed to remove particles.

final extract *n* Solids content after **fermentation**. Also called **real extract**.

final gravity *n* **Specific gravity** after **fermentation**.

fining *n* Material added to **beer** or **wort** to bind and **precipitate haze**-causing materials and to clarify the beer.

finish *n* Part of a bottle that makes a **gas**-tight seal with the cap.

firkin *n* **Unit** of volume equal to 40.91 L = 9 imp gal. Used for **casks**.

fitting *n* Device to make a connection between one or more pipes, tubes, valves, or other components of a **fluid** handling system. *See* **pipe**, **tubing**, **sanitary fitting**, **pipe fitting**, **compression fitting**.

flash pasteurize *v* **Pasteurize** a liquid by subjecting it to a high **temperature** for a short time in a flow system.

flavor 1. *n* Composite sensation resulting from **aroma**, **taste**, and **mouth feel**. 2. *v* Provide with flavor (1). Also spelled *flavour*.

flavor threshold *n* Lowest concentration at which a **flavor compound** can be detected by **aroma**, **taste**, or **mouth feel**.

flavor unit *n* Concentration of a **flavor**-active substance given in multiples of the **flavor threshold**.

flocculant 1. *adj* Tending to **flocculate**. 2. *n* Additive that causes particles to flocculate.

flocculate *v* Stick together in clumps.

flower hops *n* **Leaf hops**.

fluid *n* Material that flows. Liquid, **gas**, or **slurry**.

flux *n* Material added to lower a melting point.

fly sparge *v* **Continuous sparge**.

foam *n* Particles of **gas** in a liquid or solid matrix.

fob 1. *n* **Foam**. 2. *v* Emit foam. 3. *v* Cause to emit foam, as during packaging to purge air from the package.

FOB detector *n* *Foam-on-beer* detector. **Valve** to prevent the flow of **foam** into a **beer line**.

foeder [FOO-der] *n* Large vessel made of wood **staves**. Used to **ferment** and **age beer**. *See* **barrel** (1), **cask** (2).

forced carbonation *n* Adding carbon dioxide by introducing it under **pressure**.

forced fermentation *n* Test to determine **attenuation limit**. After **yeast** is **pitched**, a sample of the mixture is subjected to constant stirring at an elevated **fermentation temperature** (typically 27 °C). The dissolved solids content after activity stops is the unfermentable portion of the **wort**.

fork lift *n* Motor vehicle equipped with a hydraulic lift that can raise a frame called a *mask*. Long narrow protrusions, called *tines*, extending parallel to the floor, are often attached to the mask. The tines can engage a **pallet** or **skid** and raise it off the floor or shelf to transport it. Also called a *fork truck*.

formal charge *n* Fictive electric charge calculated by allocating shared **electron**s equally between the two bound **atom**s.

formula *n* Representation of a **molecule** or **compound** using the symbols for the **elements** in it with subscripts to show how many **atoms** of each element are in a **unit** of the compound.

fouling *n* Accumulation of solid deposits on a surface, interfering with the function of the device.

foundation water *n* Water added to cover the **false bottom** of a **lauter tun** or **mash tun** (2) before **mash** is added.

free amino nitrogen *n* Concentration of nitrogen from **amino acids** not including proline and di- and tripeptides that cannot be assimilated by **yeast**. Some testing methods also include ammonia and inorganic ammonium compounds, in which case the result is properly called **yeast assimilable nitrogen**. Abbrev: FAN.

free radical *n* **Atom**, **molecule**, or **ion** with an unpaired **electron**.

freezing point depression *n* Lowering of the melting point of a substance by the presence of a dissolved material.

fresh hops *n* Whole **hop strobiles** that have not been dried.

friabilimeter *n* Instrument to measure **malt friability**.

friability *n* Tendency to crumble into small pieces under **pressure**.

full mash *adj* Following a process to make **beer** starting with preparation of **wort** by **mashing malt** in hot water.

functional group *n* Group of **atoms** that give a **molecule** characteristic behavior or properties.

fungus *n* Member of a eukaryotic kingdom characterized by the presence of **chitin** in the **cell wall**. *pl* fungi.

fusarium head blight *n* Fungal disease of **grain**. Produces **DON**, a mycotoxin. Infested **malt** may cause **gushing** in the **beer**. Also called *head scab*.

fused quartz *n* Glass made from pure silicon dioxide. Used in high **temperature** applications.

fusel alcohol *n* Any **alcohol** with three or more carbon **atoms** found in an **alcoholic beverage**. Also called **higher alcohol**.

FV *n* **Fermentation vessel**.

galactose *n* Specific aldohexose.

gas 1. *n* Fluid phase of matter that expands to fill its container. 2. *n* Substance above its critical pressure but below its critical temperature. *See* **vapor**.

gas chromatograph *n* Instrument to analyze **mixtures** of substances that can evaporate by allowing them to separate on a packed or coated **column**.

gas law *n* Any of several equations that relate volume, **pressure**, **temperature**, and/or amount of a sample of **gas**.

gate valve *n* **Valve** with a plate or wedge that slides in and out of the flow path at right angles to the flow direction. Usually used only in the full open or closed positions.

gelatin *n* **Hydrolyzed** form of **collagen** often derived from animal skin or bones. Used as a **fining**.

gelatinize *v* Cause the absorption of water between the chains of **molecules** of substances made up of long chains, like **starches** or **proteins**.

gene *n* Part of a **DNA** molecule that carries the code for a single **protein**.

gene expression *n* Process of **transcription** of a **gene** and **translation** into a **protein**.

gene regulation *n* Inhibition or enhancement of **gene expression**.

geometric isomer *n* One of a set of **isomers** distinguished by the arrangement of groups on the same side or across from one another with reference to some feature of the **molecule**. An example would be the cis- and trans- forms of some **compounds** with **double bonds**.

geometry *n* **Underlying geometry**.

germinate *v* Begin to grow.

germinative capacity *n* Ability of seeds to **germinate** eventually. Live seeds have germinative capacity even if they are **dormant**.

germinative energy *n* Ability of seeds to **germinate** vigorously. Seeds that are dead or **dormant** lack germinative energy.

gibberellin *n* Plant growth **hormone**.

glass 1. *n* Brittle, transparent, noncrystalline mixture of silicon dioxide and other oxides. 2. *n* Any noncrystalline solid that softens without a definite melting point as the **temperature** is increased.

glass electrode *n* **Electrode** with a special **glass** envelope making it selective for the hydrogen ion. Used in sensors of **pH** meters.

glassy *adj* **Steely**.

globe valve *n* **Valve** in which **fluid** flows through a port that can be covered with a plug that can be moved into or out of the port by a rod called the **stem**. *See* **needle valve**.

glucan 1. *n* Any **compound** made by connecting **glucose molecules** with **glycosidic links**. 2. *n* (informal) **Beta-glucan gum**.

glucoamylase *n* *See* **amyloglucosidase**.

glucose 1. *n* A **aldose** sugar, $C_6H_{12}O_6$ in a specific form. 2. *n* One of the **enantiomers** of glucose (1), D-glucose, which is found in living **cells**.

gluten 1. *n* Mixture of storage **proteins** in seeds of grassy plants such as wheat, barley, oats, and rye. 2. *n* Protein or fragment that causes an adverse reaction in persons with **celiac disease**.

glycerol *n* **Organic compound** $C_3H_8O_3$, an **alcohol** with an —OH group bound to each carbon **atom**. Used to prevent damage to cells during freezing of **cultures**.

glycogen *n* **Starch** chain with branches every 10–12 **glucose** units. Used to store **energy** by animals and **fungi**.

glycol 1. *n* **Organic compound** with two **hydroxyl groups** on adjacent carbon **atoms**. 2. *n* Ethylene glycol, officially named 1,2-ethanediol. 3. *n* (informal) Heat transfer **fluid** consisting of ethylene glycol or propylene glycol dissolved in water.

glycolysis *n* Pathway to convert **glucose** to **pyruvic acid**. **ATP** and **NADH** are produced.

glycoprotein *n* A class of protein to which an **oligosaccharide** chain is **covalently** attached.

glycosidic link *n* **Bond** between an **anomeric carbon atom** on a **carbohydrate molecule** and an —OR **group**.

Golgi apparatus *n* **Organelle** consisting of stacks of **membrane**-bound compartments. Facilitates **posttranslational modification** of newly formed **protein molecules**.

grain 1. *n* Dry edible seed. Also called **corn**. 2. *n* Plant grown for grain (1).

gram stain *n* **Staining** technique for the characterization of **bacteria**. Gram positive bacteria have a thick, external peptidoglycan layer in the **cell wall** and stain purple. Gram negative bacteria have a thin internal peptidoglycan layer and stain pink.

grant *n* Vessel that holds the **fluid** output of another vessel at atmospheric **pressure** prior to the transfer to another vessel or a **pump**. Prevents the application of excessive suction to the first vessel. Often used to receive the output of a **lauter tun**. Also called *underback*.

gravity 1. *n* (informal) **Specific gravity**. 2. *n* (very informal) Wort solids content.

green beer *n* **Beer** before **conditioning**, also called **ruh beer**.

green hops *n* **Fresh hops**.

green malt *n* **Malt** that has been **steeped** and **germinated**, but not dried.

grist *n* Crushed **grain**.

grist case *n* Vessel for holding dry **grist** from the **mill** and delivering it to the **mash tun**.

grits *n* **Milled grain** with particle size between 0.25 and 1.25 mm.

group 1. *n* Column of the **periodic table**. 2. *n* **Functional group**.

growth medium *n* Substance prepared to support the growth of certain types of **microorganisms**. *pl growth media.*

gruit *n* Mixture of herbs formerly used to flavor **beer**.

gum *n* Non-**starch**, water-soluble **polysaccharide** that can cause a large increase in **viscosity**.

gushing *n* Premature or excessive release of dissolved **gas** from **beer**.

HACCP *n* Hazard analysis and critical control points. Set of procedures designed to minimize the risk of unsafe **defects** in food and beverages. Also used to minimize risk of nonhazardous defects.

half-reaction *n* **Oxidation** portion or **reduction** portion of an oxidation–reduction **reaction**.

Hall–Heroult process *n* Electrochemical method to make aluminum metal from aluminum oxide.

hammer mill *n* Device to break up a material to small particles using rotating metal bars called hammers. Most have a device, such as a screen, to allow the release of sufficiently small particles while retaining particles that require additional milling.

hard 1. *adj* Resistant to scratching and piercing. 2. *adj* Having a high **concentration** of calcium, magnesium, or other **ions** with an electrical charge of +2 or more, said of water. 3. *adj* Containing **ethanol**, said of a beverage.

haze *n* Cloudiness.

head 1. *n* **Foam** on a serving of beer. 2. *n* Seed-bearing part of a **grain** plant. 3. *v* Develop a head (2).

head retention *n* Stability of **foam** on **beer**.

headspace 1. *n* Space in a filled container that is not actually occupied by liquid. 2. *adj* Regarding the **gas** or **vapor** in the headspace (1). *See* **ullage**.

heat 1. *n* **Energy** that flows from a region of high **temperature** to a region of lower temperature. Not to be mistaken for temperature. 2. *v* Provide or apply heat (1).

heat capacity *n* Amount of **energy** needed to increase the **temperature** of an object by $1\,°C$. *See* **molar heat capacity**, **specific heat capacity**.

heat exchanger *n* Device to transfer **heat** from one **fluid** to another.

heat of fusion *n* **Energy** needed to melt a specific quantity of a substance. Also called *latent heat of fusion.*

heat of vaporization *n* **Energy** needed to convert a specific quantity of a substance from liquid to **gas** or **vapor**. Also called *latent heat of vaporization.*

hectoliter *n* **Unit** of volume equal to 0.1 cubic meter (100 L). Used in **brewing**, but otherwise not recommended.

hedonistic 1. *adj* Judgment based on liking or disliking. 2. *adj* Concerned with pursuit of pleasure.

helical *adj* Having the shape of a **helix**.

helicase *n* **Enzyme** that separates paired strands of **DNA**.

helix *n* Twisted, corkscrew shape. *pl* helices.

hematite *n* Geological term for iron(III) oxide (Fe_2O_3).

hemiacetal *n* Product of the **reaction** of an **aldehyde** with an **alcohol**.

hemicellulose *n* A polysaccharide that can contain many different monomers including **glucose, xylose, arabinose, galactose**, and **mannose**. A main component of the **endosperm cell wall** in **barley**.

hemiketal *n* Product of the **reaction** of a **ketone** with an **alcohol**.

hemocytometer *n* Thin transparent chamber into which a liquid can be injected to a precise depth. The bottom of the device has lines defining a grid pattern with precise dimensions. Used in a **microscope** to count **cells** in a known volume of liquid. Also spelled *haemocytometer*.

Henry's law *n* **Gas** dissolves in proportion to its **pressure** at the surface of a liquid.

HEPA *adj* Capable of removing 99.97% of particles larger than 300 nm. Said of a filter. For high efficiency particulate air.

heteroatom *n* **Atom** other than carbon and hydrogen **covalently** bound in an **organic molecule**.

heterofermentive *adj* Producing more than one fermentation product. *See* **homofermentive**.

heterogenous *adj* Having regions, called **phases**, with different **compositions** or properties.

hexose *n* **Sugar** with six carbon **atoms**.

higher alcohol *n* Any **alcohol** with three or more carbon **atoms** found in an **alcoholic beverage**.

high gravity *adj* Regarding a **brewing** technique in which **wort** is made and processed at substantially higher **original extract** than indicated by the desired strength of the **beer**. The resulting strong beer is diluted with deaerated water to bring it to the desired **ethanol concentration**. This method can make more beer without increasing the volume or number of vessels.

histogram *n* Bar chart showing a frequency distribution. The horizontal axis shows ranges of outcomes, and the vertical axis shows number of occurrences.

homofermentive *adj* Producing a single **fermentation product**. *See* **heterofermentive**.

homogenous *adj* Displaying the same **composition** and properties in all regions down to the **molecular** level.

hop 1. *n* Plant, *Humulus lupulus*, whose **strobiles** (cone-like fruits) are used to **flavor beer**. 2. *v* Add hops to, as "beer hopped with Citra." *See* **hops**.

hop back *n* Straining device using **hops** as a **filtering medium**.

hop cannon *n* Device for delivery and dispersal of **hops** into **beer** via pressurized ejection.

hop creep *n* **Secondary fermentation** of **beer** observed after **dry hopping** as a result of the introduction of **glucoamylase** from **hops**.

hop doser *n* Any of several types of devices to introduce and disperse **hops** or hop products into **wort** or **beer** at some stage of **brewing**. Many are named after weapons.

hop jack *n* **Hop back**.

hopped wort *n* **Wort** that has been treated with **hops**.

hops *n* **Strobiles** (cone-like fruits) of the **hop** plant. Used to **flavor beer**.

hop utilization *n* Fraction of a desired component or derivative of **hops**, usually **iso-alpha acids**, that survives in the **beer**.

hordein *n* Storage **protein** in **barley**. Also the specific name of a **gluten** protein in the barley plant.

Hordeum vulgare *n* Common **barley**.

hormone *n* **Compound** used by one part of an organism to signal other parts of the organism.

hot break *n* **Protein** and other solids that **precipitate** from **wort** during boiling. *See* **trub**.

hot liquor tank *n* Vessel for holding or heating the hot water that will be used in the **brewing** process, especially the water for **mashing** and **sparging**.

hot water extract *n* Measure of malt **extract** potential reported in **points** × liters per kilogram of **malt**.

huangjiu *n* Chinese **beer** style made from rice.

hull *n* Woody outer covering of a seed. Also called *glume*.

humulone *n* Principal **alpha acid** produced by **hop strobiles**.

Humulus lupulus *n* Botanical name for the **hop**.

hydrate 1. *n* **Compound** whose **formula** includes one or more water **molecules**. 2. *v* React by the addition of water to a **compound**.

hydrated *adj* Surrounded by water **molecules**.

hydrocarbon *n* **Compound** with only hydrogen and carbon.

hydrogen bond *n* Force of **attraction** between a hydrogen **atom** bound to an N, O, or F atom and an unshared pair of **electrons** on another **molecule** (or on a remote place on the same molecule).

hydrolyze *v* Add a water **molecule** across a **bond** in a larger molecule resulting in two smaller molecules.

hydrometer *n* Instrument to measure **density** or **specific gravity** of a liquid by how high it floats in the liquid.

hydronium ion *n* H_3O^+ **ion**. Signature of an **acid** in water.

hydrophilic *adj* **Polar**, charged, or **hydrogen bonding**, said of a **molecule** or a region of a molecule.

hydrophobic *adj* Nonpolar, said of a **molecule** or a region of a molecule.

hydrophobic force *n* Tendency for **hydrophobic molecules** or regions of molecules to repel water, other highly **polar** compounds, or **hydrophilic** regions of molecules.

hydrophobin *n* Highly **hydrophobic polypeptide** produced by certain **fungi**.

hydroxyl group *n* —OH group.

hypothesis *n* Proposed explanation for an observation. *pl* hypotheses.

IBU *n* **Bitterness unit**.

ideal gas law *n* Equation relating **pressure**, volume, **temperature**, and amount **(moles)** of a **gas** under moderate conditions of temperature and pressure. Abbrev: IGL.

IKE *n* **Isomerized kettle extract**.

Imhoff cone *n* A long, transparent conical cup with volumetric gradations used to quantify **sedimentation**.

immersion lens *n* Light **microscope objective lens** designed for use with **immersion oil**.

immersion oil *n* Liquid with a high **refractive index** (typically >1.5) that occupies the space between the **objective lens** and the sample to enhance the **numerical aperture**.

in² *n* Square inch ($6.452\,cm^2$).

indicator *n* **Compound** whose color depends on **pH**.

induced fit *n* Mechanism of **enzyme** function in which binding of the **reactant molecule** causes a change in the shape of the **enzyme** molecule.

inebriated *adj* **Drunk**.

infrared *adj* Having a **wavelength** in the range of 700 nm to 1 mm. Said of light.

infusion mashing *n* **Mashing** without raising the mash **temperature**.

initiation *n* First step of **transcription** or **translation** process.

in-line *adj* Located and installed to allow continuous measurement of some property of the contents of a vessel, **pipe**, or other conduit. Said of an analytical instrument or procedure.

inoculate *v* Transfer a sample of **microbes** to **medium** in a container.

inoculation loop *n* Tool with a loop to transfer samples of **microbes**. The loop can be wire, in which case it is flame **sterilized** before each use, or single-use sterile plastic.

intensive property *n* Property of the system that is unchanged by the size of the system. An example is **temperature**. If two 1 kg samples of water each at 40 °C are combined, the combined system still has a temperature of 40 °C. *See* **extensive property**.

interface *n* Location where one **phase** of material meets another phase or vacuum. *See* **surface**.

interfacial tension *n* **Energy** needed to make a unit area of **interface**. Usually applied to interfaces not involving **gas** or vacuum.

intermolecular force *n* Force attracting **ions** or electrically neutral **molecules** toward one another.

international bitterness unit *n* **Bitterness unit**. Abbrev: IBU.

intoxicated *adj* **Drunk**.

iodine test *n* Test for **starch** by treating a sample with iodine dissolved in potassium iodide **solution**. **Amylose** shows a blue-black color; **amylopectin** shows blue with a slight red cast; large **dextrin molecules** show a red color. Small dextrins and **polypeptides** give no **reaction**.

ion *n* Electrically charged particle.

ion–dipole force *n* **Intermolecular force** between **polar** molecules and dissolved **ions**.

ion exchange *n* Method to soften or purify water by replacement of undesirable **ions** with acceptable ions.

ionic *adj* Having or caused by charged particles.

ionic bond *n* Chemical **bond** resulting from force of **attraction** between a positive **ion** and a negative ion.

Irish moss *n* Seaweed: *Chondrus crispus*. Natural source of **carrageenan**.

isinglass *n* **Collagen** derived from fish swim bladders. Used as **finings**.

iso-alpha acid *n* Bitter **isomerization** product of an **alpha acid**.

isoelectric point *n* **pH** at which the electric charge on a **molecule** is, on average, neutral.

isohumulone *n* Bitter, soluble **isomer** of **humulone** formed during boiling.

isomer *n* One of a set of **compound**s with the same number of each type of **atom** and distinguished by differences in structure or geometry.

isomerize *v* Change the form or structure of a **molecule** without adding or removing any **atom**s.

isomerized kettle extract *n* **Hop** extract that has been pre-**isomerize**d intended for addition to the boiling **kettle** to adjust bitterness. Abbrev: IKE.

isotope *n* Version of an element with a particular number of **neutrons**.

jockey box *n* Temporary **beer** dispense system in which beer is driven from a **keg** by **gas pressure** and through a **heat exchanger** in a box of ice water.

joule *n* **SI unit** of **energy** equal to $kg \cdot m^2 \cdot s^{-2}$.

keg *n* **Metal** cylinder for transporting and serving **beer**. Has one opening and a dip tube.

keg coupling *n* **Valve** to connect a **keg** to **beer** dispensing lines and a source of **gas**.

ketone *n* **Compound** with a **carbonyl** group attached to two carbon **atoms** (and no hydrogen).

ketose *n* **Monosaccharide** originating as a **ketone**.

kettle *n* Vessel for boiling **wort**, also called **copper**.

keystone *n* Fitting for the hole on the end of a **cask** to accept a **valve**.

kieselguhr *n* **Diatomaceous earth**.

kilderkin *n* **Unit** of volume equal to 81.83 L = 18 imp gal. Used for **casks**.

kiln 1. *n* Furnace or oven. 2. *v* **Heat** a material in a kiln (1).

knock out 1. *v* Transfer **wort** from the **kettle** at the end of boiling. 2. *v* Eliminate or deactivate a specific **gene**. 3. *n* (loosely) Procedures after boiling and before **fermentation**.

koji *n* Mold: *Aspergillus oryzae*. Used to provide **amylases** to **hydrolyze** rice **starch** for *sake* and similar beverages.

Kolbach index *n* **S/T ratio**.

kosher *adj* Compatible with Jewish dietary laws.

krausen *n* The rocky heads of **foam** produced during **fermentation**. 2. *v* Add actively fermenting wort, rich in active **yeast**, to another fermentation such as before bottling to aid in **carbonation** [*Ger.* ruffle].

krausen ring *n* **Brandhefe**.

kvass *n* Eastern European **beer** made from bread.

lacing *n* Fragments of **beer foam** adhering to the sides of a glass after the beer is consumed.

lager 1. *v* To **condition beer** at low **temperature** for several weeks or months. 2. *n* **Lager beer**.

lag phase *n* First stage of a **microbial** growth curve where **yeast** or **bacteria** assimilate nutrients and prepare for **cell** division. There is no change in cell number during this stage.

lager beer *n* **Beer fermented** at **temperatures** below 60 °F (15 °C) typically with *Saccharomyces pastorianus*.

lambic *n* Belgian style of **ale** made with naturally occurring **bacteria** and **yeast**.

lamella *n* A thin layer or **membrane** that is part of the **endosperm cell wall**. In **barley**, the outer layer is rich in **hemicellulose, beta-glucan**, and ferulic acid with a middle layer rich in **proteins**.

laminar *adj* Occurring in parallel layers. Said of **fluid** flow. *See* **turbulent**.

Laplace law *n* Law of physics that the **pressure** inside a bubble or droplet is higher than that on the outside by a factor proportional to the reciprocal of the radius of the bubble or droplet. Because of the very high pressure inside a very small bubble, bubbles form only with difficulty without a **nucleation site**. Also called Young–Laplace equation.

latent heat *n* Heat absorbed or released at constant **temperature** as a result of a **phase** change.

lauter 1. *adj* For the purpose of or having been subjected to the **lauter process**. 2. *n* (informal) **Lauter tun**. 3. *v* Subject to the **lauter process**.

lauter process *n* Method of **wort separation** using the **grain** as a **filter medium**.

lauter tun *n* Vessel for the **lauter process**.

leaf hops *n* Dried whole **hop strobiles**. Also called *cone hops* and *flower hops*.

Le Chatelier's principle *n* When a system at **equilibrium** is subjected to a change, like a change in **pressure**, **temperature**, or **concentration**, the system will shift to a new equilibrium that partly offsets the change.

Lewis structure *n* **Structural formula** for a **compound** showing **atoms** with their symbols and shared pairs of **valence electrons** as lines. Unshared **valence electrons** may be shown as dots.

ligand 1. *n* **Ion** or **molecule** that provides an **unshared pair** of **electrons** forming a **covalent bond** to a **metal atom** forming a complex ion or coordination compound. 2. *n* A small molecule that binds to a **protein**, usually by intermolecular (noncovalent) interaction.

light beer *n* **Beer** brewed to have a low **concentration** of **carbohydrate** after **fermentation** but with the same or slightly lower **ethanol** content than standard **beer**.

lightstruck *adj* **Off-flavor** in **beer** resulting from exposure to light. Also called *skunked*.

limit dextrinase *n* **Enzyme** that **catalyzes** the **hydrolysis** of alpha($1\rightarrow6$) **glycosidic bonds**. Also called **pullulanase**.

linear 1. *adj* Forming a straight line. 2. *adj* Providing an output or response that can be related to the input by the equation of a line.

lipid *n* Family of biological **compounds** whose **molecules** are at least partly **hydrophobic**.

lipid bilayer *n* Filmlike structure comprising two layers of **molecules** with **polar** heads and nonpolar tails. The molecules in each layer are organized with nonpolar regions pointed to the other layer and polar regions directed away from the other layer. Lipid bilayers are the main structural element in **membranes**.

lipid raft *n* Region of a **membrane** with a high **concentration** of **sterols**. Serves as an anchor point for **proteins**.

lipoxygenase *n* Type of **enzyme** that causes an **unsaturated fat** to react with oxygen.

liquefaction *n* **Hydrolysis** of **starch** into water-soluble **molecules**. Also called *liquefication*.

liquor *n* Water for **mashing** and **sparging**.

liter *n* **Unit** of volume equal to 0.001 cubic meters. Symbol: L. Sometimes called "cubic decimeter."

load cell *n* **Transducer** for measuring force. Often used to measure the mass of the contents of a vessel.

lock out *v* Secure a device to prevent unwanted actuation or release of stored **energy** to allow safe maintenance or repair.

lockout 1. *n* Mechanism to deny access to the controls of a device that has been **locked out**. Normally the key to the lockout is held by the person performing the service for which locking out was occasioned. 2. *n* Status of having been locked out, as "the mill is on lockout."

log phase *n* **Exponential phase**.

lone pair *n* **Unshared pair**.

Lovibond *n* Color scale used for **beer** and other foods.

luminometer *n* Instrument for measuring **ATP** bioluminescence.

lupulin *n* Sticky yellow powder produced by **hop strobiles**. Source of **beer flavor compounds**.

lupulone 1. *n* A specific **beta acid**, $C_{26}H_{38}O_4$. 2. *n* Any beta acid.

lysosome *n* **Organelle** containing **hydrolysis enzymes** that degrade a variety of **biopolymers**.

macromolecule 1. *n* **Molecule** with many **atoms**. 2. *n* (informal) Substance or material composed of macromolecules (1).

Maillard reaction *n* **Reaction** starting with the formation of a Schiff base from a **sugar** and an **amino acid**. Can form large highly colored **melanoidin molecules** or smaller highly **flavored** molecules.

main group *n* Any of the eight columns of the **periodic table** comprising the first two and the last six columns.

maize *n* *Zea mays*, a grain called "corn" in the United States.

malt 1. *n* Sprouted and dried **grain**. 2. *v* Prepare **malt** (1) from grain.

malt extract *n* Solid or syrup obtained by concentrating **wort** prepared from **grain malt**.

malt liquor *n* **Beer** brewed to have a low **concentration** of **carbohydrate** after **fermentation** and with slightly higher **ethanol** content than standard beer.

maltodextrin *n* Unfermentable additive of **hydrolyzed starch** consisting mostly of short **polymers** of alpha(1→4) linked **glucose**.

maltose *n* **Sugar** made of two **glucose** units connected by an alpha(1→4) link.

maltotriose *n* Sugar made of three **glucose** units connected by alpha(1→4) links.

maltster *n* Person or institution that makes **malt**.

manifold *n* System of **pipes, tubes**, or other conduits in which one main conduit is connected to two or more secondary conduits.

mannose *n* A six-carbon **aldose** in a specific form.

manometer *n* Device that measures **pressure** by the height of a column of liquid.

manway *n* Large enough opening to admit a person to the interior of a vessel. *See* **confined space**.

mash 1. *v* Treat **milled malt** and other starchy materials with hot water to allow **enzymes** to **hydrolyze** the **starch** to **sugar**. 2. *n* **Slurry** of water and **milled grain**.

mash conversion vessel *n* Vessel for **mashing**. Abbrev: MCV.

masher *n* **Premasher**.

mash filter *n* Device for **wort separation** by driving the wort through **filters**.

mash in *v* Mix water with grist to begin **mashing**.

mash-in temperature *n* Temperature of **mash** (2) after **mashing in**. *See* **striking temperature**.

mash thickness 1. *n* Ratio of **mass** of water to mass of **grist** in **mashing**. This is the preferred definition in commercial **brewing**. 2. *n* Ratio of mass of grist to volume of water in mashing.

mash tun 1. *n* **Mash conversion vessel**. 2. *n* (chiefly British) **Mash conversion vessel** specifically with a **false bottom** so it can also be used for the **lauter process**.

mass fraction *n* Mass of a **component** divided by the mass of the entire sample. Can be applied to the **elements** in a **compound** or to the **components** of a **mixture**.

MBT *n* 3-Methylbut-2-ene-1-thiol. Gives **lightstruck beer** a skunky **flavor**.

MCV *n* **Mash conversion vessel**.

mead *n* **Alcoholic beverage** made from **fermented** honey.

mealy *adj* Made porous by **modification**. Said of **malt** kernels. *See* **glassy**.

mean *n* Average.

mechanism 1. *n* Sequence of elementary steps accounting for a chemical **reaction**. 2. *n* Scientific doctrine holding that life processes can be accounted for by the normal laws of physics and chemistry. *See* **vitalism**.

medium 1. *n* **Growth medium**. 2. *n* **Filtration medium**. *pl* media.

melanoidin *n* Highly colored **compounds** made up of large **polymers** produced from **sugars** and **amino acids** by the **Maillard reaction**.

membrane 1. *n* Thin film. 2. *n* Thin film surrounding a **cell** or structure within a cell consisting of a **lipid bilayer** with attached **protein molecules**.

membrane filter *n* **Filtration medium** consisting of a thin plastic film with pores of a controlled size. Used to clarify samples before measurement of light transmittance and to remove or collect **microbes** from a **fluid**.

meniscus 1. *n* Curved surface made by a liquid climbing up a surface with which it is in contact. 2. *n* Something with a crescent shape.

metal *n* **Element** with loosely held **electrons**. Typically has luster and high electrical and thermal conductivity and can often be bent or shaped without cracking. Metals often react by losing **electrons** to give positive **ions**.

metallic luster *n* Shininess in **metals**.

metalloid *n* **Element** with some properties characteristic of **metals** and some properties characteristic of **nonmetals**.

methylene blue *n* Blue dye used to assess **viability** of **yeast cells**. Live cells decolor the dye; dead cells remain blue.

microbe *n* Microscopic organism such as **bacteria** or **yeast**.

microbial filtration *n* **Filtration** to remove live **microbes**.

micropyle *n* Small opening in a seed or other structure.

microRNA *n* Type of **small regulatory RNA** with about 22 **nucleotides** arranged in a hairpin conformation. Abbrev: **miRNA**.

microscope *n* Instrument to visualize very small objects.

mill 1. *n* Device to crush or grind a material. 2. *v* Mechanically grind or crush a material.

miRNA *n* **MicroRNA**.

mitochondrion *n* **Organelle** responsible for production of **ATP** by **aerobic respiration**. *pl* mitochondria.

mixture *n* Material with two or more substances not chemically bound to one another.

modification *n* Changes that a **grain** undergoes during **malting**.

molar concentration *n* Number of **moles** of a **component** of a **solution** divided by the volume of the solution in **liters**. Also called **molarity**.

molar heat capacity *n* **Energy** needed to raise the **temperature** of one **mole** of a pure substance by $1\,°C$.

molarity *n* **Molar concentration**.

molar mass *n* Mass of one **mole** of an **element** or **compound**.

mole *n* Amount of a substance containing $6.022\,05 \times 10^{23}$ particles or formula units.

molecule *n* Group of **atoms** that are much more tightly bound to one another than to any other atoms. As a result, they travel together as one distinguishable particle.

monomer *n* One of many structurally related **units** that are or can be connected to make a large **molecule**. *See* **polymer**.

monosaccharide *n* Simple **sugar**. **Aldehyde** or **ketone** with three or more carbon **atom**s and an —OH **group** on all non-**carbonyl** carbons or the **hemiacetal** or **hemiketal** resulting from internal addition of one of the —OH groups across the C=O **bond**.

montejus *n* **Hop back**.

morphology *n* Shape of an organism.

mouth feel *n* Component of **flavor** resulting from heat, cold, texture, or irritation of sensory **cells** in the mouth.

mRNA *n* **RNA** that serves as a template for synthesis of **proteins** and other **polypeptides**. For *messenger RNA*.

must *n* Fruit juice, especially grape juice, intended for **fermentation** into **wine**.

mutation *n* Change in a **DNA** or **RNA** sequence in which one or more **nucleotides** are substituted, deleted, or added.

mycotoxin *n* Toxic substance produced by **fungus**. Potential contaminant in **grain**.

NAD *n see* **nicotinamide adenosine dinucleotide**.

NADH *n see* **nicotinamide adenosine dinucleotide**.

near infrared *n* Light whose **wavelength** is slightly longer than what can be seen with the eyes.

Neck *n* Sort tube attached to the outside of the opening of a **keg** to accommodate the **spear**.

needle valve *n* **Valve** with a small orifice and an extremely narrow and tapered plug to give a high degree of control of the flow rate.

nephelometer *n* Instrument to quantify turbidity by measuring light scattered at an angle, usually 90° from the incident light.

neuron *n* Nerve **cell**.

neurotransmitter *n* Compound that carries signals to or from **neurons**.

neutral 1. *adj* Without net positive or negative electrical charge. 2. *adj* Neither **acidic** nor **basic**.

neutron *n* Electrically neutral elementary subatomic particle. Found in **atomic nucleus**.

newton *n* **SI** unit of **force** = kg·m·s^{-2}. Symbol: N.

nicotinamide adenosine dinucleotide (NAD) *n* **Molecule** involved in **redox reactions**, such as **fermentation**, by which an **electron** is carried from one **reaction** to another, found in two forms; **NAD$^+$** is **oxidized** and **NADH** is **reduced.**

nip point *n* Location where a rotating part of a machine can trap and injure a body part.

nitrogenated *adj* Containing dissolved nitrogen (typically about 20 mg/L for nitrogenated **beer**). Also called *nitrogenized*.

nitrosamine *n* Compound containing the $R_2N-N{=}O$ group. Carcinogen that sometimes occurs in **malt**.

noble gas *n* Chemical **element** in group 8A.

node *n* Joint in a plant.

nonmetal *n* **Element** that holds its **electrons** tightly. Nonmetals often react by gaining **electrons** to give negative **ions** or by forming **covalent bonds**.

normal distribution *n* Mathematical function describing the outcomes of many processes such as fabrication, filling, and measurement in which the outcome is continuous, that is, it can take any value. Characterized by a bell-shaped curve. Also called the Gaussian distribution.

normally closed 1. *adj* Of a **valve**: allows flow when electric **current** or air **pressure** is applied to the **actuator**. Blocks flow otherwise. 2. *adj* Of a **relay**: blocks flow of electric current in the main circuit when current is applied to the control circuit. Allows current otherwise.

normally open 1. *adj* Of a **valve**: blocks flow when electric **current** or air **pressure** is applied to the **actuator**. Allows flow otherwise. 2. *adj* Of a **relay**: allows flow of electric current in the main circuit when current is applied to the control circuit. Blocks current otherwise.

NPT 1. *n* American National Pipe Thread system of standards for tapered threading on pipes and fittings. 2. *adj* Following the **NPT** (1) standards.

nuclear envelope *n* **Membrane** surrounding the **nucleus** in a **eukaryote**.

nuclear pore complex *n* **Proteins** that form channels in the **nuclear envelope**.

nucleate boiling *n* Mode of boiling in which individual bubbles form on and break away from the surface. *See* **film boiling**.

nucleation site *n* Particle or other feature that attracts dissolved **molecules** and allows them to come out of **solution**.

nucleobase *n* One of the nitrogen- and oxygen-containing cyclic compounds – adenine, cytosine, guanine, thymine, or uracil – that occur in **DNA** or **RNA**. Also called "base" or "nitrogenous base."

nucleolus *n* Structure within the **nucleus** in which **ribosomes** are made.

nucleotide *n* **Compound** with a **ribose** or deoxyribose phosphate **ester** in which one of the ribose hydroxyl groups has been replaced with a **bond** to a nitrogen atom on a platelike **molecule** called a **nucleobase**.

nucleus 1. *n* Positively charged particle at the center of an **atom**. 2. **Nucleation site**. 3. *n* Membrane-enclosed region of certain **cells** containing genetic material.

null hypothesis *n* Assumption that measurements or observations do not differ significantly from the reference population. If the null hypothesis can be shown to have a sufficiently low probability, it is rejected.

numerical aperture *n* Measure of the light-gathering power of a **microscope objective** lens.

nutrient media *n* General purpose broth or **agar** that contains sufficient nutrients to support growth of a wide range of microorganisms.

oast *n* **Kiln** for drying **hops**.

objective *n* **Microscope** lens nearest the sample.

obligate *adj* Of necessity, said of **aerobic** and **anaerobic microbes**. *See* **facultative**.

OE *n* **Original extract**.

off-flavor *n* Undesired **flavor**.

off-line *adj* Requiring samples to be carried to a laboratory away from the process location. Said of an analytical instrument or procedure.

OG *n* **Original gravity**.

olfactory *adj* Related to the sense of smell.

olfactory bulb *n* Part of the brain that processes **aroma** sensations.

oligosaccharide *n* A molecule with 3–10 **sugar** units connected by **glycosidic links**, usually without branches.

on-line *adj* Located and installed to allow process operators to make frequent measurements near the process location. Said of an analytical instrument or procedure.

opaque beer *n* African family of beer styles often made with sorghum and consumed during active **fermentation**.

optical axis *n* Imaginary line passing through the centers of the lenses of an optical device.

OPV *n* Overpressure valve. *See* **relief valve**.

organelle 1. *n* **Membrane**-bound functional structure within a **cell**. 2. *n* Any specialized functional structure within a cell.

organic 1. *adj* Having carbon–carbon or carbon–hydrogen **bonds**. 2. *adj* Prepared without artificial fertilizers, pesticides, additives, or ingredients. Said of food.

organic chemistry *n* Behavior, synthesis, and study of **organic** (1) **compound**s.

organoleptic *adj* Involving qualities that make use of the sense organs (eyes, ears, tongue, skin, and nose).

original extract *n* Mass percent solids in wort before **fermentation**.

original gravity *n* **Specific gravity** of **wort** before **fermentation**.

oscillating density meter *n* Instrument used for the measurement of **density**, particularly for **wort**, which consists of a U-shaped tube filled with sample. The tube is made to vibrate, and the resonant frequency is measured. The vibrational frequency depends on the sample mass and can be used to infer density.

osmosis *n* The transfer of **molecules** through the pores of a **membrane** that can allow some molecules to pass but blocks others. *See* **reverse osmosis**.

osmotic pressure *n* Magnitude of **pressure** that stops flow through a **semipermeable membrane**.

osmotic stress *n* Physiological problems in a **cell** caused by a change in the **concentration** of dissolved substances outside the cell **membrane**.

oxidase *n* **Enzyme** that **catalyzes oxidation reactions**. The presence or absence of cytochrome C oxidase is used to characterize **bacteria** and their ability to undergo **aerobic respiration**.

oxidation *n* **Chemical reaction** in which electrons are a **product**.

oxidation number *n* Fictive electric charge determined by allocating all **electrons** shared by two **atoms** to the more **electronegative** of the pair. If the atoms are of the same element, the shared electrons are divided evenly.

oxidized *adj* Having given up **electrons**.

oxidizer *n* Substance that accepts **electrons**, causing another substance to become **oxidized**. Also called oxidizing agent.

oxonium ion *n* **Ion** with the formula R_3O^+, where the R's can be hydrogen atoms, **alkyl groups**, or a combination of both.

oxyacid *n* Inorganic **acid** containing oxygen. Usually the acidic hydrogen atoms are in **hydroxyl groups**.

packing 1. *n* Part of a **valve** that prevents escape of **fluid** through the stem. 2. *n* Powder used as a stationary phase in **chromatography**.

pallet *n* Flat structure with an upper and lower deck for supporting and transporting boxes, **barrels**, or other freestanding containers. Often handled with a **fork truck**. *See* **skid**.

papain [pa PAY en] *n* **Protein-hydrolyzing enzyme** from papaya fruit. Used as a **fining**.

Paraflow *n* (chiefly British) Plate **heat exchanger**; a trade name.

partial pressure *n* **Pressure** that a **component** of a **gas mixture** would exert if it were the only gas in the container.

particle size distribution *n* Fraction of material whose particle diameters fall into each of a series of ranges.

pasteurization unit *n* Unit of intensity of **pasteurization** equivalent to the lethality of one minute of treatment at 60 °C. The application of the term is complicated because different microbes have different sensitivities to pasteurization, and the degree of lethality is not linear with temperature.

pasteurize *v* Preserve foods by heating (typically to about 63 °C = 145 °F) to kill **microbes**. *See* **flash pasteurization**.

pathogenic *adj* Causing disease.

PBS *n* Phosphate-buffered saline. Saline solution with fixed pH used for diluting **cell cultures**.

PCR *n* Polymerase chain reaction. Process to amplify pieces of **DNA** of a specific sequence as defined by a set of **primers** (short sequences complementary to the target). *See* **thermal cycler**.

pellicle *n* Thin film produced by **bacteria** on top of a **fermenting** liquid.

pentosan *n* Structural **polysaccharide** made up of **pentoses**. Typically found in plants.

pentose *n* **Monosaccharide** with five carbon **atoms**.

peptidase *n* **Enzyme** that breaks the chains of **protein molecules**.

peptide *n* **Polymer** of 50 or fewer **amino acids**.

period 1. *n* Row of the **periodic table**. 2. *n* Accessible energy states for **elements** in a row of the periodic table. 3. *n* Amount of time required for an oscillator or a wave to make one complete oscillation.

periodic table *n* Chart of chemical **elements** in order of **atomic number** and organized in columns by the number of **valence electrons**.

perlite *n* **Filter medium** prepared from **hydrated** volcanic glass.

permanent hardness *n* **Hardness** in water that is not removed by boiling.

permeability *n* Measure of the extent to which an object can transmit fluids.

permeation *n* Penetration of a solid barrier by a **gas** or liquid.

peroxide *n* Compound with —O—O— functionality.

peroxisome *n* **Organelle** involved in breaking down **lipids** and **reduction** of **reactive oxygen species**.

personal protective equipment *n* Items worn by an individual to provide protection against workplace hazards. Some examples common in **brewing** are eye and face protection, gloves, hearing protection, hard hats, and protective shoes. Abbrev: PPE.

Petri dish *n* Shallow flat bowl used to **culture microbes**.

PGA *n* **Propylene glycol alginate**.

pH *n* Measure of the acidity or basicity of a solution. More acidity gives a lower pH, more basicity gives a higher pH.

phase *n* Region of a sample with a particular **composition** and properties.

phase diagram *n* Chart showing points at which **phases** of pure materials or **mixtures** exist at equilibrium. When there are many such points, they form lines or surfaces called *phase boundaries*. Phase diagrams for pure substances conventionally have the **temperature** on the horizontal axis and **pressure** on the vertical axis.

pH electrode *n* **Glass electrode**.

phenol 1. *n* **Compound** C_6H_5OH. 2. *n* Family of compounds with an —OH group directly attached to an **aromatic** ring.

pH meter *n* Electrochemical device to measure **pH** in water solutions.

phosphate 1. *n* PO_4^{3-} **ion**. 2. *n* **Compound** containing the $-OPO(OH)_2$ group or one if its ions.

phospholipid *n* Biological **compound** whose **molecules** include a phosphorus-containing group to which two **hydrophobic** hydrocarbon chains are attached.

photon *n* Smallest unit of light.

pH paper *n* Paper impregnated with **pH indicators**. Compared to a color scale to estimate pH.

PIKE *n* **Isomerized hop** extract that has been neutralized with potassium hydroxide (KOH) to improve solubility.

pin *n* **Unit** of volume equal to 20.46 L = 4.5 imp. gal. Used for **casks**.

pipe *n* Rigid, hollow, cylindrical tube of standardized inside and outside diameter used to convey fluids. The nominal size of pipe is the inside diameter. *See* **tubing**.

pipe fitting *n* Threaded connector that makes a joint between parts of a **fluid** handling system. The seal is made by driving the threads together, sometimes with a flexible sealing material between the outside and inside threads.

pipe scale *n* Hard deposits adhering to the insides of conduits and vessels.

pitch 1. *v* Add yeast to wort to begin **fermentation**. 2. *n* Mixture of wood resins, sometimes including artificial materials, used to line **barrels**. 3. *v* Line with or apply **pitch** (2).

pitch rate *n* Number of **yeast** cells pitched per volume of **beer**. Also called pitching rate.

plasma membrane *n* **Cell membrane**.

plate *v* Apply a sample to a **medium** as a test for **microbes**.

Plato *See* **degree Plato**.

pleomorphic *adj* Having variability is size or shape.

plump *adj* Retained by a certain size screen, usually 2.5 mm (6/64 in. = 2.38 mm in North America) for **barley**.

points *n* Measure of **specific gravity** equal $1000 \times (SG-1)$. For example, a specific gravity of 1.044 is equivalent to 44 points. Four points is equivalent to about 1% dissolved sugar. Also called **degrees Oeschle**.

polar *adj* Having a positively charged end and a negatively charged end.

pollutant *n* Undesired **component** of a **mixture**.

polyatomic ion *n* Collection of two or more **atoms** with an electric charge, e.g. **carbonate ion**.

polymer 1. *n* Substance made of **molecules** produced from subunits connected by **covalent bonds**. 2. *n* (informal) Polymeric **macromolecule** (1). *See* **monomer**.

polypeptide *n* **Polymer** of **amino acids**. *See* **protein**.

polyphenol *n* Loosely defined group of natural products from plants having multiple **conjugated** rings and many **phenolic groups**. Polyphenols are moderately soluble in water, more so at high **pH**.

polysaccharide *n* **Carbohydrate** consisting of many **monosaccharide** units connected with **glycosidic links**.

polyvinylpolypyrrolidone *n* **PVPP**.

porosity 1. *n* Condition or state of having pores. 2. *n* Volume fraction of a material that is pores.

posttranslational modification *n* **Reactions** involving making and breaking **bonds** in **protein molecules** after they have been synthesized by a **ribosome**. These can include cleaving the protein chain, attaching phosphate or **lipid**, forming disulfide bonds, etc.

potential *n* **Electrical potential**.

potential alcohol *n* Concentration of **ethanol** that could result from complete, ideal **fermentation** of a sample of **wort** or **must**.

power *n* Rate at which **energy** is transferred. Sometimes called "wattage."

precipitate 1. *v* Come out of solution, especially as a solid. 2. *n* Solid that comes out of solution.

precoat *n* **Filtration medium** applied to the filter cloth or screen before filtration begins. *See* **body feed**.

premasher *n* Device to mix **grist** with **strike water** before they are added to the **mash tun**.

prerun tank *n* Vessel to hold **wort** from the **wort separation** vessel before it is transferred to the **kettle**.

pressure *n* Force divided by area.

pressure relief valve *n* **Relief valve**. Abbrev: PRV.

primary cooling system *n* Device that uses energy to extract heat from a cold system and release it to a warm system. Usually used to cool a **secondary cooling system**. Also called a *refrigeration system*.

primary structure *n* Sequence of **amino acids** in a **protein**.

primer *n* Short segment of **DNA** or **RNA** to which **nucleotides** are added to make a new DNA strand.

processed malt *n* **Malt** treated to introduce **flavors** or other properties that do not derive from the **malting** process. For example, smoked malt.

process steam *n* Hot water **vapor** of a quality suitable for application of **heat** and **pressure**, but not for direct contact with food. *See* **culinary steam**.

product *n* Material formed in a **chemical reaction**.

prohibition *n* Laws against manufacture, sale, or use of **alcoholic beverages** or other substances. Specifically, national prohibition on alcohol in the United States from 1919 to 1933.

prokaryote *n* Single-**cell** organism with no **nucleus**.

propagator *n* Vessel used to encourage **yeast cells** to reproduce to provide yeast for **fermentation**.

propylene glycol alginate *n* Copolymer of mannuronic acid and guluronic acid esterified with propylene glycol. Used as a foam enhancer. Abbrev: PGA.

protease *n* **Enzyme** that **catalyzes hydrolysis** of **protein**.

proteasome *n* **Protein** complex that **hydrolyzes** proteins that have been tagged with **ubiquitin**.

protein *n* **Polymer** of 50 or more **amino acids**.

proton *n* Positively charged elementary subatomic particle. Found in **atomic nucleus**.

psi *n* **Pounds per square inch**. A unit of pressure equal to 0.06895 **bar**.

psychoactive *adj* Causes changes in brain function.

pub *n* **Bar** (2).

pullulanase *n* **Limit dextrinase**.

pump 1. *n* Device to drive a **fluid** from place to place. 2. *n* **Protein** structure extending through a **membrane** to drive specific **ions** or **molecules** from regions of lower **concentration** to regions of higher concentration.

punt *n* Indentation in the bottom of a bottle.

purification *n* Removal of undesired **components** from a **mixture**.

PVPP *n* **Polymer**, **polyvinylpolypyrrolidone**, used as a **fining**. Cross-linked, insoluble version of polyvinylpyrrolidone.

pycnometer *n* Instrument that can be filled with a precisely reproducible volume of a **fluid**. Used to determine **density**.

PYF *n* Premature **yeast flocculation**. **Sedimentation** of yeast before **fermentation** is complete.

pyruvic acid *n* $CH_3COCOOH$, an intermediate **compound** in **respiration** and **fermentation**.

python *n* Bundle of beverage lines refrigerated by two or more tubes carrying cold **fluid** to and from a **chiller**. The bundle is encased in insulation to minimize heat leakage and to prevent **condensation**.

quality 1. *n* Characteristic of freedom from **defects**. 2. *n* Characteristic of meeting end user's requirements or expectations. 3. *n* Fraction (of steam) that is **vapor**.

quality management *n* Policies and procedures to enhance **quality**.

quaternary ammonium ion *n* **Ion** with the formula NR_4^+. Used in **sanitizing** solutions.

quaternary structure *n* Resulting three-dimensional shape of several interacting **proteins**, distinct **polypeptides** that interact through non-covalent **bonds** to form one functional complex.

rack *v* Transfer **beer** from one vessel to another, especially to a **keg** or **cask**.

reactant *n* Material that is consumed by a chemical **reaction**.

reaction *n* Process in which one or more substances, the **reactant(s)**, are converted to one or more other substances, the **product(s)**.

reaction quotient *n* Number representing the extent or completeness of a chemical **reaction**.

reactive oxygen species *n* **Molecules**, **ions**, or **free radicals** containing oxygen in a form that reacts more easily than oxygen molecules.

real ale *n* **Ale** that meets the production and serving standards of the **Campaign for Real Ale**.

real attenuation *n* **Original extract** minus **real extract**, sometimes expressed as a fraction or percent of the **original extract**.

real degree of fermentation *n* **Degree of fermentation**. *See* **apparent degree of fermentation**. Abbrev: RDF.

real extract *n* Mass percent solids in decarbonated beer after **fermentation**. Also called **final extract**. Abbrev: RE. *See* **apparent extract**.

receptor *n* **Protein** structure that selectively binds certain **molecules** or **ions** and causes a response inside a **cell**.

redox *adj* Involving an **oxidation** and a **reduction**.

reducing end *n* End of a **polysaccharide molecule** that has a free —OH group (not bound by a glycosidic link) on the **anomeric carbon**.

reducing sugar *n* **Sugar** with a free —OH group (not bound by a glycosidic link) on the **anomeric carbon**.

reduction *n* Chemical **reaction** in which **electrons** are a **reactant**.

reference electrode *n* **Electrode** kept in electrical contact with a sample to complete the circuit for electrochemical measurements like **pH**.

reflux *v* Allow some or all of the **condensed vapor** in a **still** to return to the boiler.

reflux ratio *n* Fraction of liquid **condensed** in a **still** that is returned to the boiler.

refraction *n* Bending of a beam of light resulting from passing through materials with different **refractive indices**.

refractive index *n* Ratio of the speed of light in vacuum to that in the sample.

refractometer *n* Instrument to measure **refractive index**. Can be used to determine **sugar** content in **wort**, fruit juice, and honey.

regulator *n* **Valve** that controls the delivery **pressure** of a fluid despite variations in the inlet pressure.

Reinheitsgebot *n* Former Bavarian law restricting **beer** ingredients to water, **barley**, and **hops**.

relay *n* Electrical device that switches a main circuit on or off in response to the application or withdrawal of **current** in a control circuit. The relay allows a low **power** in the control circuit to control a much higher power in the main circuit.

relief valve *n* **Valve** that is set to open when the **pressure** exceeds a set value. Also called a *safety valve* or an *overpressure valve*. Abbrev: PRV or OPV.

repel *v* Exert a **force** driving something away.

replication *n* Process through which **DNA** makes an identical copy of itself during **cell** division.

resistance *n* Opposition to the flow of electrical **current**. Ratio of **electrical potential** to **current**.

resistance temperature detector *n* Temperature sensor consisting of a **metallic** conductor, often platinum, whose **resistance** varies in a predictable way with temperature. Abbrev: RTD.

resonance *n* Delocalization of **electrons** represented as the average of alternative forms of a **molecule** differing only by the placement of the **valence** electrons.

respiration *n* **Oxidation** of food by an electron acceptor. Releases **energy**. In **aerobic** respiration the acceptor is oxygen and the **products** are carbon dioxide and water.

rest *n* Time interval during which **temperature** is held constant to allow some chemical or biological process to occur.

reverse osmosis *n* Method of **purification** by forcing a liquid through a **semipermeable membrane** from the impure side to the pure side.

R-group 1. *n* Any group of **atoms** attached to a **molecule**, especially one attached through a carbon atom or hydrogen atom. 2. *n* Variable side chain on an **amino acid**.

ribes *n* Catty **flavor** in **beer**.

ribose *n* Specific five-carbon **aldose**. Building block of **RNA**. A dehydroxylated form is in **DNA**.

ribosome *n* **Protein–RNA** complex that synthesizes proteins from **amino acids** carried by **tRNA** according to a template provided by **mRNA**.

RNA *n* Ribonucleic acid. **Nucleotide polymer** often involved in biological **protein** synthesis.

roasted malt *n* Class of **malt** heat treated after curing to produce color and **flavor**.

rotameter *n* Instrument to measure the flow rate of a **fluid**. Comprises a tube with a tapered interior with the larger bore at the top. A ball rises in tube until the force produced by the flow balances the mass of the ball. The position of the ball is a measure of the flow rate.

rotation *n* Motion of a **molecule** involving turning about an axis.

rouse *v* Stir the contents of a **fermenter** to resuspend the **yeast**.

rRNA *n* Ribosomal ribonucleic acid, a component of the **ribosome**, responsible for **translation** and **protein** synthesis.

RTD *n* **Resistance temperature detector**.

ruh beer *n* **Green beer**.

saccharify *v* Catalytically **hydrolyze starch** to **sugar**.

Saccharomyces cerevisiae *n* Top **fermenting yeast** used to make **ale** and bread.

Saccharomyces pastorianus *n* Cold tolerant bottom **fermenting yeast** used to make **lager beer**. Formerly called *Saccharomyces carlsbergensis*.

sake *n* Unhopped rice **beer** from Japan. Also called rice **wine**. *See koji*.

sales gravity *n* Hypothetical **initial gravity** that unfermented wort used for **high-gravity** brewing would have had if dilution had occurred before rather than after **fermentation**.

saloon *n* **Bar** (2), usually with tables and chairs.

sanitary fitting *n* Temporary plumbing connection without any crevices and in which all surfaces in contact with the **fluid** are food compatible. The parts to be connected usually have flanges that are connected by a gasket. The flanges are tightened to the gasket by various mechanisms such as clamps or screws.

sanitize *v* Treat to reduce number of **microbes** to an acceptable level.

sanitizer *n* Substance used to **sanitize** surfaces.

saturated 1. *adj* Of a mixture: having the maximum amount of dissolved material at a particular **pressure** and **temperature**. 2. *adj* Of an organic compound: having no double or triple carbon–carbon **bonds**. 3. *adj* Of steam: having pressure equal to the **vapor pressure** of water at the steam temperature.

screw conveyor *n* Device that uses a spiral band to move material along a tube or channel.

scutellum *n* Separator between the **endosperm** and **embryo** compartment in a seed.

seam 1. *n* Closure on a can. 2. *v* Seal the lid on a can.

seamer *n* Device to **seam** a can.

secondary cooling system *n* System of **heat exchangers** and plumbing to carry heat from devices, like **fermenters**, and deliver it to a **primary cooling system**.

secondary fermentation *n* Any additional **fermentation** or **aging** period that occurs in beer after terminal gravity is reached.

secondary structure *n* Shape within a **protein** that comprises helices or sheets.

sediment 1. *v* Fall to the bottom of a container. 2. *n* Material that has fallen to the bottom of a container.

seed coat *n* Waxy covering for a seed beneath the **hull**. Also called **testa**.

selective media *n* Microbiological **medium** that encourages the growth of some microorganisms while preventing the growth of others through the addition of chemicals.

selectivity *n* Ability to cause a particular **reaction** to a greater extent than other reactions that are possible with the same reactants. Said of a **catalyst**. *See* **specificity**.

semimetal *n* **Metalloid**.

semipermeable membrane *n* A **membrane** through which certain **molecules** or **ions** can pass, but not others.

sequestrant 1. *n* **Compound** that binds to **ions** in solution, lowering their effective **concentration**. Used in **detergent** formulations to minimize the effect of **hardness**, and in foods to prevent the formation of **reactive oxygen species**. 2. *n* Any compound that prevents the formation of **precipitates** from **hardness ions**.

shape *n* Spatial arrangement of **atoms** in a **molecule** not including unshared **electron** pairs.

shared pair *n* **Bonding pair**.

shear *n* Force in a direction that tends to make the layers of an object slide past one another.

shell *n* **Energy level**.

shive *n* Stopper for a **cask** to close the opening in the side.

SI 1. *adj* According to the International System of Units based on the kilogram, meter, second, ampere, kelvin, mole, and candela. 2. *n* International System of Units (for Système International d'Unités).

sieve *n* Device with a screen or plate with openings of a specific size. Used to remove undesired particles or to measure **particle size distribution**.

sight glass *n* Transparent part of a liquid handling system allowing the appearance or quantity of the contents to be seen and evaluated.

signal sequence *n* Short portion at the end of a **protein molecule** indicating the destination of the molecule.

silica gel *n* Porous silicon dioxide with bound water. Used as a **fining**.

silo *n* Structure for storing unpackaged material, especially **grain** in the context of **brewing**.

Sinner circle *n* Four factors required for effective **cleaning**: chemical, mechanical, temperature, and time. Less of any of these factors must be made up by more of another.

siphon 1. *n* Tube to transfer a liquid over a barrier from a higher to a lower level. 2. *v* Transfer liquid with a **siphon** (1).

six-row barley *n* Variety of **barley** in which all three flowers in each group are fertile.

skeletal structure *n* Representation of a **molecule** in which carbon **atoms** are shown as the ends of line segments and the hydrogen atoms attached to carbon atoms are omitted.

skid *n* **Pallet** without a lower deck.

slack *adj* High in moisture. Said of **malt**.

slant *n* Solidified **growth medium** with a slanting surface in a test tube with a cap. Used to prepare and store pure **cultures** of **microbes**.

slurry *n* **Mixture** with solid particles suspended in a liquid.

small pack *n* Can or bottle sold for consumer use at home.

small regulatory RNA *n* Short noncoding RNA **molecules** that control **gene expression**. Also called small RNA. Abbrev: sRNA.

snift *v* Release counterpressure after filling a bottle or can.

soda lime glass *n* **Glass** prepared from sand with sodium carbonate and calcium carbonate used as a **flux**.

soft water *n* Water with low **concentrations** of calcium **ions** and magnesium ions.

soluble protein *n* Percent **protein** in **malt** that can be **extracted** by **mashing** using a standard procedure.

solute *n* Minor **component** of a **solution**.

solution *n* **Mixture** with the same composition throughout.

solvent *n* Component that makes up the majority of a **solution**.

sparge *v* Wash **wort** from a bed of **grain** with hot water.

sparge arm *n* Rotating pipe in a wort separation vessel to spray water for **sparging**.

spear *n* Tube that goes from the opening to the bottom of a **keg**.

specific gravity *n* **Mass** of the volume of a material that has the same volume of one mass unit of water. Equivalently, the ratio of the **density** of a sample to the density of water. Used as a measure of the **concentration** of dissolved **carbohydrates** in **wort** and **beer**.

specific heat capacity *n* Amount of **energy** to increase the **temperature** of one gram of a substance or solution by $1\,°C$. *See* **heat capacity, molar heat capacity**.

specificity *n* Property of an **enzyme** (or other **catalyst**) to cause a **reaction** in a certain **compound** more than in other potential **reactants**. *See* **selectivity**.

spigot *n* **Valve** for liquid.

spile *n* Peg driven into the **shive** of a **cask** to control the internal **pressure**.

spirit *n* **Alcoholic beverage** prepared by **distillation** to concentrate the **ethanol**.

spoilage *n* Flavor deterioration that takes place over time. Also called **staling**. *See* **aging**.

spond *v* Seal a vessel before **fermentation** is complete to allow pressure to build up.

spore *n* **Cell** that can develop into a new organism without merging with another cell.

spray ball *n* Device to apply a spray of **cleaning solution** to the interior surfaces of a vessel. Part of a **CIP** system.

spray jet *n* Device to deliver **cleaning solution** at high impact to the interior surfaces of a vessel. Part of a **CIP** system.

SRM *n* **Standard Reference Method**.

sRNA *n* **Small regulatory RNA**.

stacking force *n* Force of **attraction** between flat, ring-shaped **molecules**.

stage *n* Part of a **microscope** that positions the sample.

stain *n* Pigment applied to a sample to enhance contrast in a **microscope**.

stainless steel *n* Alloy of **steel** with a substantial fraction of chromium and often other **elements** such as nickel or manganese.

staling *n* **Spoilage**.

standard deviation *n* Measure of the scatter in a series of measured quantities.

Standard Reference Method *n* Standard method for **beer** color of the **American Society of Brewing Chemists (ASBC)**. Abbrev: SRM. Officially *ASBC Methods of Analysis, Beer 10*.

staphilo– *adj* Prefix: occurring in clusters. Said of **bacteria**.

starch *n* Long chain of **sugar molecules** in a form that is readily **hydrolyzed**. Usually the links are mainly alpha$(1\rightarrow4)$.

starter *n* **Culture** of **yeast** grown in a small batch of **wort** used to initiate **fermentation**.

stationary phase *n* Third stage of a **microbial** growth curve in which cell death is equal to cell growth.

stave *n* One of the narrow wooden boards that make up the wall of a **barrel** (1).

steam trap *n* **Valve** that automatically discharges condensed water from a steam line.

steel *n* Alloy of iron with 0.002–2.1% carbon by weight.

Steel's masher *n* Type of **premasher**.

steely *adj* Not made porous by **modification**. Said of **malt** kernels. Also called **glassy**. *See* **mealy**.

steep *v* Soak seeds in water to initiate sprouting.

stereoisomer *n* One of a set of **molecules** with the same **atoms** bonded in the same order but differing in the way they are directed in space.

sterile *adj* Free of **microbes** and **spores**.

sterile filtration *n* Process to remove **microbes** from a **fluid** by passing it through a **membrane filter**.

sterilize *v* Kill or remove all **microbes** and **spores** in a material or on an object.

steroid *n* Member of a family of **compounds** with a **molecular** plate consisting of four fused rings. Steroids are considered to be **lipids**.

sterol *n* **Steroid** with an —OH group bound to one of the rings.

stew *v* Heat (**grain** kernel) under high humidity during **malt** processing.

still 1. *n* Device to vaporize a liquid and **condense** and collect the **vapor**. If the condensed vapor is allowed to return to the original container, it is called a reflux still. 2. *adj* Not **carbonated**.

stillage *n* Platform or cart to keep goods, such as **casks** of **beer**, above the floor.

strain 1. *n* Deformation of an object in response to an applied force. 2. *n* Particular variety of a biological species.

strain gauge *n* Device whose electrical properties vary with extension, compression, or bending. Used in measuring instruments.

stratification *n* Formation of layers, especially in a **fermenter**, as a result of **drauflassen** or topping up a fermentation.

S/T ratio *n* Ratio of **soluble protein** to **total protein** in **malt**. Also called **Kolbach index**.

streak *v* Apply a **microbial** sample to the surface of a solid or gel **growth medium** with a series of dilutions to produce isolated **colonies**.

Strecker degradation *n* **Oxidation** of an **amino acid** to carbon dioxide and an **aldehyde**.

strepto– *adj* Prefix: occurring in long chains. Said of **bacteria**.

strike water *n* Water added to **grist** to begin **mashing**.

striking temperature *n* **Temperature** of water added to **grist** to begin **mashing**. *See* **mash-in temperature**.

strobile *n* Cone-like seed-bearing fruit of certain plants, like the **hop**.

structural formula *n* Representation of a **molecule** showing the **atoms** and the **covalent bonds** connecting them.

Student's *T*-test *n* Statistical method to determine, without knowledge of the **standard deviation** of the population, whether differences in **means** are significant.

stupor *n* Lowered level of consciousness and response.

substituent *n* **Atom** or **functional group** that replaces a hydrogen atom in an **organic** compound. The substituent is of lower priority than the principal **functional group** of the **molecule**.

substrate *n* **Reactant** in an **enzyme-catalyzed reaction**.

sugar *n* **Monosaccharide** or **disaccharide**.

supercritical fluid *n* Substance above its critical **pressure** and its critical **temperature**. Abbrev: SCF. see **Critical point**.

superheated *adj* Having a **pressure** lower than the **vapor pressure** of the substance.

super sack *n* Large container made of woven material designed for the transportation and storage of dry, freely flowing material, like **grain**. Capacity typically about 0.5–1 metric tonne (1000–2000 lb). Usually equipped with straps for handling by **fork lift** and often with a discharge spout at the bottom for delivery of the contents. Officially called a flexible intermediate bulk container (**FIBC**).

surface *n* Boundary where **phases** meet.

surface *n* **Interface** between a solid or liquid and a gas or vacuum.

surface energy *n* **Energy** to overcome **cohesive force**s and pull a material into parts, making **surface**.

surface tension *n* Energy needed to make a unit area of **surface**. *See* **interfacial tension**.

surfactant *n* Substance that lowers **surface tension**.

swab 1. *n* Tool to take a sample for **microbe** testing. 2. *v* Take a sample with a swab (1).

swing panel *n* Device in which **pipes** to different vessels and equipment are arranged, each with a **valve**. The pipes can be connected in pairs by U-shaped pieces of pipe that are connected to the valves with **sanitary fittings**. Also called *distribution panel, flow panel, transfer panel*, and *swing bend panel*.

tail piece *n* **Fitting** that can be inserted into a piece of **tubing** to make a connection. Also called *hose barb* or *hose nipple*.

tannin *n* Member of a family of **polyphenols** that can form stable complexes with **proteins**.

tap 1. *n* **Spigot** or fitting to take **beer** from a **cask** or **keg**. 2. *v* To install a **tap** (1).

tare *n* Weight of a container; must be subtracted from the total weight to get the weight of the contents.

taste *n* Sense that responds to **dissolved** materials.

taste bud *n* Organ containing **cells** responsible for the sense of **taste**.

tautomerism *n* Shift of a hydrogen **atom** resulting in the exchange of a **double bond** with an adjacent single bond.

tavern *n* Shop for purchase and consumption of **alcoholic beverages** and usually food on the premises. Some also provide lodging.

temperance movement *n* Social movement urging reduced consumption of **alcoholic beverages**. Followers advocated diverse programs ranging from voluntary moderation of consumption to enforced **prohibition**.

temperature *n* The property of an object such that **heat** flows spontaneously from a higher temperature to a lower temperature. The temperature is the same in bodies at thermal **equilibrium**. An intensive property, not to be confused with **heat**.

temperature controller *n* Device that operates a switch or **valve** controlling the application or removal of **heat** to control the **temperature** of a material.

temporary hardness *n* Water **hardness** that is removed by boiling, indicating that the negative **ion** is **carbonate**.

tension *n* Force that tends to increase the length or volume of an object; pull. *See* **compression**.

termination *n* Final step of **transcription** or **translation**. New chain and template dissociate from assembly mechanism.

terpene *n* Any member of a family of naturally occurring **hydrocarbons** with one or more **double bonds**.

terpenoid *n* Any member of a family **of terpenes** and modified terpenes containing oxygen, phosphorus, sulfur, etc.

tertiary structure *n* Entire three-dimensional shape of a protein that results from covalent and non-covalent interactions within the polypeptide.

testa *n* **Seed coat** inner layer.

tetrad *n* Grouping of four **bacterial cells**.

tetrahedral *adj* Three-dimensional **geometry** with four **electron groups** arranged at roughly 109.5° angles about a central **atom**.

thermal conductivity *n* Measure of the ability of a material to transmit **heat**.

thermal cycler *n* Automatic device to repeatedly raise and lower the **temperature** of a **reaction mixture** undergoing **PCR**.

thermistor *n* Electronic component whose **resistance** changes substantially with **temperature**. Used as the sensing element in electronic **thermometers**.

thermocouple *n* Junction of dissimilar **metals** that produces an **electrical potential** that can be used to measure **temperature**.

thermometer *n* Instrument to measure **temperature**.

thiol *n* **Compound** whose principal **functional group** is the —SH group.

thousand corn weight *n* **1000 kernel weight**.

thresh *v* Remove seeds of **grain** from the plant.

threshold, flavor *n* Lowest **concentration** of a **flavor compound** that can be detected by the average taster.

threshold potential *n* **Electrical potential** in a **cell** that causes **channels** to admit positive **ions** generating an electrical signal.

threshold sequestrant *n* **Compound** that modifies a **surface** to prevent **precipitation** of **hardness ions**.

thru *adj* Passing through a certain size screen, usually 5/64 in. for **barley** seeds.

titrate *v* Analyze by carefully mixing the **analyte** with a **reactant** of known composition, stopping when the amount of reactant is chemically equivalent to the amount of analyte. *See* **back-titrate**.

toast *n* Benediction recited over an **alcoholic beverage**.

Tollens' reagent *n* Basic solution of diammine silver ion, $Ag(NH_3)_2^+$, in water used to test for **reducing sugars**.

ton 1. *n* Unit of mass equal to 907.2 kilograms (2000 lb). 2. *n* Measure of refrigeration power equal to 3.52 kilowatts.

tonne *n* Unit of **mass** equal to 1000 kilograms. Also called a *metric ton* or a *megagram*.

top fermenting *adj* Rising to the top of the **fermenter** during **fermentation**. Said of **yeast**. Also called *top cropping*.

topoisomerase *n* **Enzyme** responsible for unwinding the double-helix of **DNA**.

total protein *n* Percent **protein** in a solid sample of **grain**, **malt**, or **adjunct**. *See* **soluble protein**. *See* **S/T ratio**.

tough *adj* Resistant to cracking and shattering.

trans-2-nonenal *n* Compound $CH_3(CH_2)_5CH=CHCHO$ that gives a cardboard **flavor** to stale **beer**.

transcription *n* Synthesis of a strand of **mRNA** from **nucleotides** according to genetic code on a strand of **DNA**.

transcription factor *n* **Protein** that binds to a **DNA** site to enhance or inhibit **gene expression**.

transducer *n* Device whose electrical properties change in response to a condition to be measured. An example is a **thermistor**, whose **resistance** is used to measure **temperature**.

transduction *n* Process in which a stimulus to a **cell** causes a signal that can be transmitted to another cell.

transition state *n* Configuration in the progress of a chemical **reaction** at which the potential **energy** is highest.

translation 1. *n* Synthesis of a **polypeptide** from **amino acids** according to genetic code on a strand of **mRNA**. 2. *n* Motion of **molecules** in which the whole **molecule** moves from place to place.

transmittance *n* Fraction or percentage of light entering a sample that emerges from the other side.

trehalose *n* **Disaccharide** consisting of two **glucose** molecules connected by an alpha(1→1)alpha **glycosidic link**.

triacylglycerol *n* Ester of **glycerol** with three **fatty acid** molecules.

trier *n* Device to gather a sample of seeds from a bin, bag, silo, or other container. Also called a *grain thief*.

triglyceride *n* **Triacylglycerol**.

trigonal planar *adj* **Geometry** with three **electron groups** all in the same plane arranged at roughly 120° angles about a central atom.

triple bond *n* **Covalent bond** formed by sharing three pairs of **electrons**.

triple point *n* **Temperature** and **pressure** at which solid, liquid, and **vapor** are all present. Equivalent to the freezing point of a pure substance under its own **vapor pressure**. Not to be mistaken for **critical point**.

trisaccharide *n* **Carbohydrate** with three simple **sugar** units.

tRNA *n* **RNA** molecule that carries an **amino acid** corresponding to its **anticodon**. For *transfer RNA*.

trub [troob] *n* Sediment from **wort** or **beer** that **precipitates** at various stages of **brewing**. It could include **coagulated proteins**, bits of **hops**, inactive **yeast**, and fatty solids. *See* **cold break**, **hot break**.

tubing *n* Hollow, often flexible cylindrical tube used to convey **fluid**. Tubing dimensions are based on the outside diameter. *See* **pipe**.

tun *n* **Brewhouse** vessel.

turbidity *n* Tendency for a material to scatter light. Cloudiness or **haze**.

turbulent *adj* Characterized by flow in which the velocity and direction of flow at any point varies in a chaotic manner, leading to good mixing. *See* **laminar**.

two-row barley *n* Variety of **barley** in which only the central flower in each group of three is fertile.

ubiquitin *n* Small regulatory **protein** with which cells tag proteins. The ubiquitin-tagged proteins may be destroyed by **hydrolysis**, transported to another part of the **cell**, or subjected to other processes depending on the location and number of tags.

ullage 1. *n* Difference between the total volume capacity of a container and the volume to which it has actually been filled. *See* **headspace**. 2. *n* Volume of product left in a container after as much has been dispensed as is practical.

ultraviolet *adj* Having a **wavelength** in the range of 400–10 nm. Said of light.

underback *n* **Grant**.

underlet *v* Pump water into a vessel by way of openings in the bottom.

underlying geometry *n* Spatial arrangement of **electron groups**, including the unshared **electron** pairs, in a **molecule**.

unfermentable dextrin *n* Short branched **starch polymers** that typically consist of at least one 1→6 branches and thus are unfermentable by traditional brewer's **yeast**. Contributes to the body of a **beer**.

unit 1. *n* Standard quantity of measure, like a *kilogram* (unit of mass). 2. *n* One of a series of connected or related items, like a *glucose unit* in starch. 3. *n* Device with a specific function, like a *refrigeration unit*.

unsaturated 1. *adj* Of a mixture: having less than the maximum amount of dissolved material at a particular **pressure** and **temperature**. Also called *undersaturated*. 2. *adj* Of an organic compound: having one or more double or triple carbon–carbon **bonds**.

unshared pair *n* Pair of **electrons** localized on a single **atom**. Also called a *lone pair*.

utilization 1. *n* Action of making use of something. 2. *n* Rate or efficiency with which something is converted to a usedful form. Often said of the ability of **yeast** to metabolize specific **sugars** or of the efficiency of **isomerization** of **hop alpha acids** during the brewing process.

vacuole *n* Fluid-filled, **membrane**-enclosed structure within the **cell** responsible for storage and digestion of waste products.

valence *adj* Relating to the highest occupied **energy level** in an **atom**.

valence electron *n* One of the **electrons** in the highest occupied energy level of an **atom**.

valency *n* Normal number of **covalent bonds** that an **atom** forms in neutral **molecules**.

valve *n* Mechanical device to provide a variable obstruction in the path of a **fluid** or of flowing particles to regulate flow.

vapor *n* Gas at a low enough **temperature** that it can be liquefied.

vapor pressure *n* **Pressure** at which a **vapor** is in equilibrium with its liquid at a specified **temperature**.

VDK *n* **Vicinal diketone**.

vegan 1. *adj* Free of materials derived from animal sources. 2. *n* Person who consumes or uses only **vegan** (1) materials.

venturi device *n* A pipe with a narrow restriction where a **gas** is introduced into a flowing liquid. High flow rate in the narrow section lowers the **pressure**, drawing the gas in and facilitating mixing. Often used to introduce oxygen to **beer wort** before **fermentation**.

vesicle *n* Large liquid-filled **organelle**.

viability *n* Fraction of a sample of **cells** or seeds that are alive.

viable *adj* Alive.

vibration *n* Motion of **molecules** in which **bonds** bend or stretch.

vicinal diketone *n* Collective term for **diacetyl** and 2,3-pentanedione.

viscosity *n* Resistance to flow in a **fluid**.

vitalism *n* Scientific doctrine that holds that processes in living organisms cannot be explained solely by the laws of chemistry and physics. Today, vitalism is rejected by mainstream science. *See* **mechanism** (2).

vitality *n* Metabolic state characteristic of the ability to thrive under stress.

volatile [VOLL it'l *or chiefly British* VOLE it tile] *adj* Easily evaporated or boiled.

Volstead Act *n* Former US law adopted in 1919 providing for the enforcement of national **prohibition** of **alcohol**.

voltage *n* **Electrical potential**.

voltage-gated channel *n* **Channel** in a **cell membrane** that opens in response to the **electrical potential** being less negative than the **threshold potential** allowing particular positive **ions** to cross the membrane.

vorlauf *n* Recirculate **wort** from the **lauter** or **mash tun** through the bed of **grain** to clarify the wort.

water sensitive *adj* Having the property of **germinating** in a moderate amount of water, but not in excess water. Said of **grain** seeds.

water softener *n* **Ion exchange** device that replaces undesired ions with sodium and chloride ions.

wavelength *n* Nearest distance between equivalent points on an electromagnetic wave.

weigh 1. *v* Determine the mass of something. 2. *v* To have a certain mass.

weight 1. *n* Force upon an object exerted by gravity. 2. *n* (informal) Mass.

well plate *n* Flat piece of plastic or ceramic with evenly-spaced holes part-way through to hold samples for analysis. Also called a microplate.

wet hops *n* **Fresh hops**.

whirlpool *n* Tank in which **beer wort** is made to flow in a circle, causing suspended solids to settle in the center.

widget *n* Hollow insert in packaged **nitrogenated** beer to raise **foam**.

wine 1. *n* **Alcoholic beverage** prepared from **fermented** fruit juice, especially grape juice. 2. *n* Loosely, any strong fermented beverage that has not been **distilled**.

worm 1. *n* **Screw conveyor**. 2. *n* Tubular **heat exchanger** used to condense vapors in a **spirit still**.

wort [wert] *n* **Sugar solution extracted** from **grain** by **mashing**.

wort chiller *n* Device to remove **heat** from **wort** after boiling.

wort separation *n* Process to separate **wort** from spent **grain** after **mashing**.

wort strength *n* Fraction or **concentration** of solids in **wort**.

xylose *n* Specific five-carbon **aldose**.

yeast *n* Single-cell **fungus**, certain species of which are used to make **beer**, **wine**, and bread.

yeast assimilable nitrogen *n* Concentration of nitrogen that can be taken up and used by **yeast**. This includes **amino acids** other than proline, as well as ammonia and inorganic ammonium compounds. Abbrev: YAN.

yeast brink *n* **Brink**.

yeast propagation vessel *n* **Propagator**.

zentner *n* Unit of mass equal to 50 kilograms when referring to **hops**. Abbrev: Znt.

zwickel [zvik'l] *n* **Spigot** in a **fermenter** for sampling the **beer**.

zymurgy *n* Science of **fermentation**.

INDEX

Mastering Brewing Science: Quality and Production, First Edition.
Matthew Farber and Roger Barth.
© 2019 John Wiley & Sons, Inc. Published 2019 by John Wiley & Sons, Inc.